.

New Frontiers in Photoenergy

New Frontiers in Photoenergy

Editor: Miles Carter

NY RESEARCH
P R E S S

New York

Published by NY Research Press
118-35 Queens Blvd., Suite 400,
Forest Hills, NY 11375, USA
www.nyresearchpress.com

New Frontiers in Photoenergy
Edited by Miles Carter

International Standard Book Number: 978-1-63238-767-7 (Hardback)

Cataloging-in-Publication Data

   New frontiers in photoenergy / edited by Miles Carter.
      p. cm.
   Includes bibliographical references and index.
   ISBN 978-1-63238-767-7
   1. Photons. 2. Light. 3. Wave packets. I. Carter, Miles.
QC793.5.P42 N49 2020
539.721 7--dc23

Contents

Permissions

List of Contributors

Index

Preface

Every book is a source of knowledge and this one is no exception. The idea that led to the conceptualization of this book was the fact that the world is advancing rapidly; which makes it crucial to document the progress in every field. I am aware that a lot of data is already available, yet, there is a lot more to learn. Hence, I accepted the responsibility of editing this book and contributing my knowledge to the community.

Photoenergy is the renewable energy which is derived from sunlight using a variety of technologies such as photovoltaics, solar architecture and solar heating. Photovoltaics refer to the conversion of light energy into electricity by using semiconductors which show photovoltaic effect. The solar cells in the photovoltaic system produce DC electricity from sunlight which can be used to charge batteries or power equipment. Solar architecture incorporates principles from architecture in order to harness clean and renewable solar power. Solar heaters make use of solar thermal collectors for collecting heat using sunlight. A few examples of solar heaters are solar parabolic troughs, solar water heaters and solar air heaters. This book provides comprehensive insights into the field of photoenergy. It presents researches and studies performed by experts across the globe. It is appropriate for students seeking detailed information in this area as well as for experts.

While editing this book, I had multiple visions for it. Then I finally narrowed down to make every chapter a sole standing text explaining a particular topic, so that they can be used independently. However, the umbrella subject sinews them into a common theme. This makes the book a unique platform of knowledge.

I would like to give the major credit of this book to the experts from every corner of the world, who took the time to share their expertise with us. Also, I owe the completion of this book to the never-ending support of my family, who supported me throughout the project.

Editor

Parametric Study on the Thermal Performance and Optimal Design Elements of Solar Air Heater Enhanced with Jet Impingement on a Corrugated Absorber Plate

Alsanossi M. Aboghrara[ID],[1,2] **M. A. Alghoul**[ID],[3,4] **B. T. H. T. Baharudin**[ID],[1] **A. M. Elbreki,**[5] **A. A. Ammar,**[5] **K. Sopian**[ID],[5] **and A. A. Hairuddin**[1]

[1]*Mechanical Department, Faculty of Engineering, UPM, 43400 Serdang, Selangor, Malaysia*
[2]*Physics Department, Faculty of Science Traghen, University of Sebha, Sebha, Libya*
[3]*Center of Research Excellence in Renewable Energy (CoRERE), King Fahd University of Petroleum and Minerals (KFUPM), Dhahran 31261, Saudi Arabia*
[4]*Energy and Building Research Center, Kuwait Institute for Scientific Research, Safat, 13109 Kuwait City, Kuwait*
[5]*Solar Energy Research Institute, Universiti Kebangsaan Malaysia, 43600 Bangi, Selangor, Malaysia*

Correspondence should be addressed to Alsanossi M. Aboghrara; alsanossi_15@yahoo.com, M. A. Alghoul; dr.alghoul@gmail.com, and B. T. H. T. Baharudin; hangtuah@upm.edu.my

Academic Editor: Alberto Álvarez-Gallegos

Previous works revealed that cross-corrugated absorber plate design and jet impingement on a flat absorber plate resulted in a significant increase in the performance of a solar air heater (SAH). Involving these two designs into one continuous design to improve the SAH performance remains absent in the literature. This study aimed to evaluate the achieved enhancement on performance parameters of a SAH with jet impingement on a corrugated absorber plate. An energy balance model was developed to compare the performance parameters of the proposed SAH with the other two SAHs. At a clear sky day and a mass flow rate of 0.04 kg/s, the hourly results revealed that the max fluid outlet temperatures for the proposed SAH, jet-to-flat plate SAH, and cross-corrugated plate SAH are 321, 317, and 313 K, respectively; the max absorber plate temperatures are 323.5, 326.5, and 328 K, respectively; the maximum temperature differences between the absorber plate and fluid outlet are ~3, 9, and 15 K, respectively; the max efficiencies are 65.7, 64.8, and 60%, respectively. Statistical *t*-test results confirmed significant differences between the mean efficiency of the proposed SAH and SAH with jet-to-flat plate. Hence, the proposed design is considered superior in improving the performance parameters of SAH compared to other designs.

1. Introduction

A solar air collector is a unit that captures the solar radiation by an absorbing medium and transforms it into thermal energy to heat the inlet ambient air. Solar air collector is an important configuration of solar thermal systems and is being widely utilized in many commercial, agriculture, industrial, and process applications [1]. A lot of equipment or appliances need to possess a capacity for high heat transfer performance in order to guarantee their quality and capability [2, 3]. Flat plate solar air heaters are commonly used in the case of low/moderate temperature. On the other hand, a solar air heater (SAH) reports low efficiencies due to the low convective heat transfer coefficient on the smooth absorber's surface [4–6], the air-limiting energy extraction [7] and flow rates [8, 9]. Therefore, it is necessary to develop techniques that can improve both heat and mass transfer of SAH.

Research and development (R&D) on the improvement of heat and mass transfer are extensively reported in extant literature. One reported method of improving convective heat transfer is increasing the surface area of the device by roughening the surface, which increases the turbulence within the channel [10, 11].

Reviews pertaining to heat transfer and thermal efficiency in SAH in the context of artificial roughness were discussed in [12]. Prasad and Mullick [4] confirmed that the heat transfer in SAH was improved by placing wires under the absorber plate. This increases the absorber plate's efficiency factor. The best performance of a thermohydraulic installed on top of an artificially roughened SAH was reported by [13, 14] relating the installation of rough transverse wire on top of the absorber, which tolerated the varied height of roughness [15], while [8] reported the heat transfer coefficient and friction factor correlations of a rib-roughened SAH duct in the context of transitional flow. Bhagoria et al. [16] combined the topside heat transfer upon an artificially roughened SAH to a fully developed turbulent flow. Karwa and Chitoshiya [17] compared the performance of multiple geometry of roughness elements in SAHs and proved that smaller diameter protrusion wires were superior in the context of flow. Kumar et al. [18] developed correlations for friction factor in multi V-shaped roughness of SAH. The analysis done by [19] resulted in a 12.5–20% enhancement in thermal efficiency using 60° V-down discrete rib roughness.

In the case of a corrugated absorber, Meyer et al. [20] studied the convective heat transfer in a V-trough linear SAH; while the natural convection in a channel formed by a V-shaped surface and a flat plate was studied and compared numerically and experimentally by Zhao and Li [21]. Stasiek [22] experimented on the heat transfer and fluid flow across corrugated-undulated heat exchanger surfaces, while Piao [23] experimentally investigated the natural, forced, and mixed convective heat transfers in a cross-corrugated channel of SAH. Noorshahi et al. [24] conducted a numerical study on the natural convection in a corrugated enclosure with mixed boundary conditions, while Gao [25] numerically simulated the natural convection inside the channel formed by a flat cover and a wavelike absorbing plate. El-Sebaii et al. [26] conducted both theoretical and experimental studies on a flat plate absorber and V-corrugated plate absorbers of double pass solar air heaters. Their respective thermal performances were compared, and the results are shown in Figure 1, where it can be seen that the double pass V-corrugated plate SAH was 11–14% more efficient than the flat plate SAH.

Metwally et al. [27] analysed a corrugated duct SAH and other five conventional designs, as per Figure 2. The enhancement factor of the convective heat transfer coefficient within the corrugated duct was 4-5. It was also confirmed that the heat removal factor of the corrugated duct collector had improved by an average value of 59%,

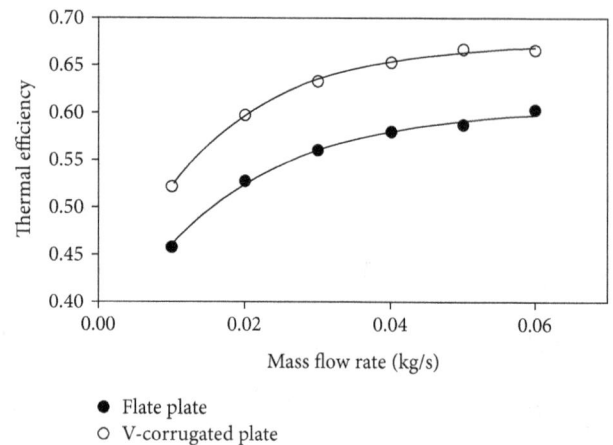

FIGURE 1: Effect of mass flow rate of flat and corrugated plates [26].

while its efficiency had been enhanced by 15–43% over the other conventional collectors at a flow rate range of 0.01–0.1 kg/sm^2 and solar insolation of 950 W/m^2. They concluded that the corrugated duct SAH can be regarded as an advanced design that is priced similar to conventional designs.

Gao et al. [2] compared the performance of three types of SAHs; two configurations were cross-corrugated, while the third was a flat plate. The thermal performance of these configurations were 58.9, 60.3, and 48.6%, respectively. It can be seen that both cross-corrugated configurations exceeded that of the flat plate collector. This can be attributed to the improved turbulence and heat-transfer rates within the air flow channel on the corrugated plate, which was ~3.25 more than those of the flat plate heater. It should also be pointed out that the difference in thermal performance between the two configurations of cross-corrugated SAH was at best quite marginal.

Liu et al. [28] analysed the effect of V-groove and cross-corrugated absorbers upon the thermal performance at multiple air mass flow rates per unit area of the collector, within 0.0025–0.5 kg/m^2·s, as per Figure 3. It was confirmed that the cross-corrugated collector exceeded that of the V-groove collector by ~7% in terms of efficiency. This enhancement could be due to the increased turbulence within the air flow channel. The supposedly more efficient heat removal via the cross-corrugated collector as opposed to that of the V-groove collector is detailed in Figures 3(a)–3(d).

Releasing a carefully directed fluid against a heat transfer surface could result in the efficient transfer or significant amounts of thermal energy and mass between the surface and the fluids, as per Figure 4. Therefore, jet impingement concept attracted many researchers, especially in the context of processes and thermal control applications such as turbine blades [29], microelectromechanical component [30], and solar heat absorbers [31]. The suitability of jet impingement for these aforementioned applications is mostly due to the high rates of heat transfer that takes place within the impingement region. Jet impingement results in heat transfer

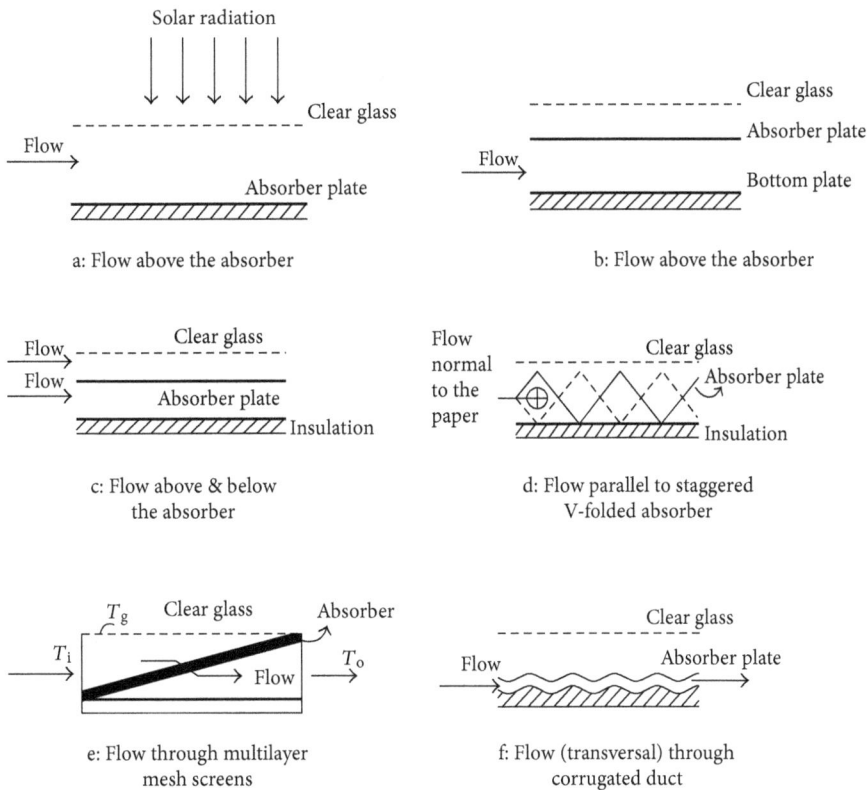

(a)

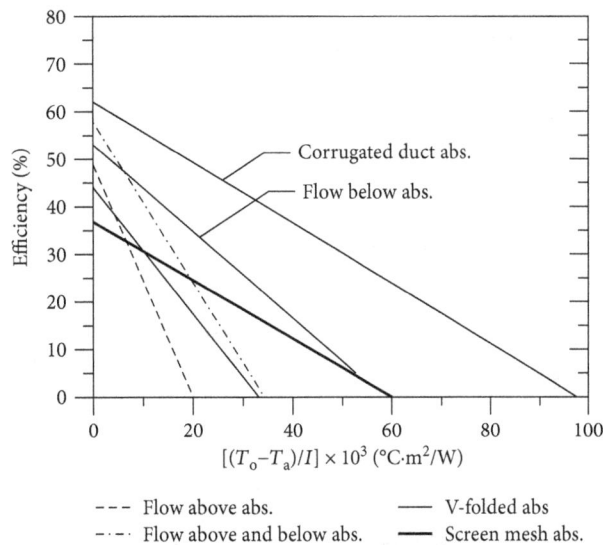

(b)

FIGURE 2: (a) Different solar air collector configurations [27]; (b) thermal efficiency and reduced temperature relationship for the tested collectors [27].

coefficients that are thrice to those of conventional convection cooling [32].

Sung and Mudawar [33] studied the cooling performance of microchannel flows equipped with jet impingement. Two configurations were employed, which are circular and single-slot jets, as per Figure 5. Both configurations confirmed that increased flow rates result in improved heat transfer coefficient. However, it should also be pointed out

that multiple circular jets are far better at cooling compared to its single-slot jet pattern counterpart.

Belusko et al. [34] determined the thermal efficiency of the unglazed SAH installed and not installed with jet impingement under normal conditions. They reported that jet impingement resulted in significant improvement to the thermal efficiency of an unglazed SAH, as per Figure 6.

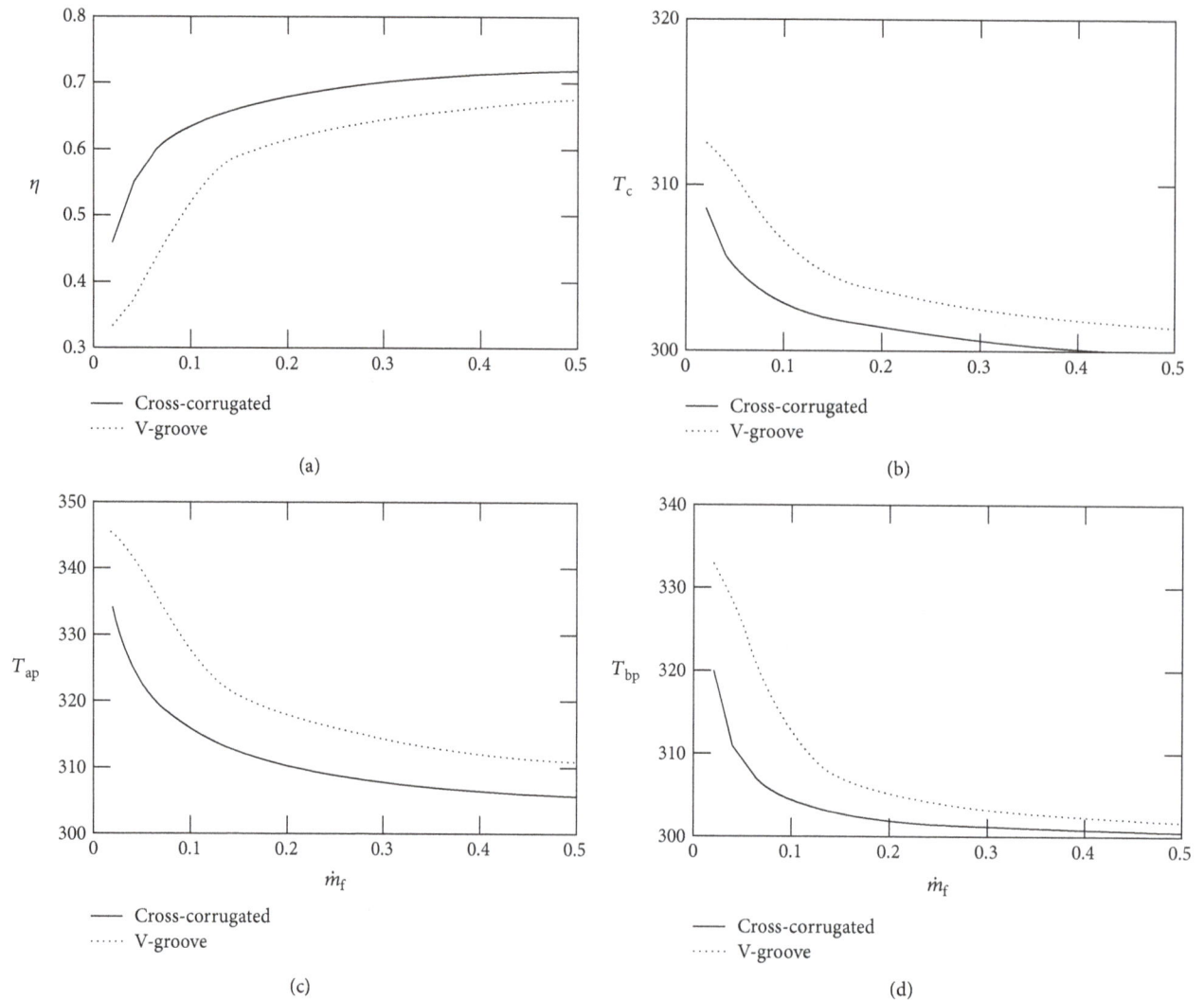

FIGURE 3: (a–d) Efficient heat removal by cross-corrugated compare to V-groove collector, Liu et al. [28].

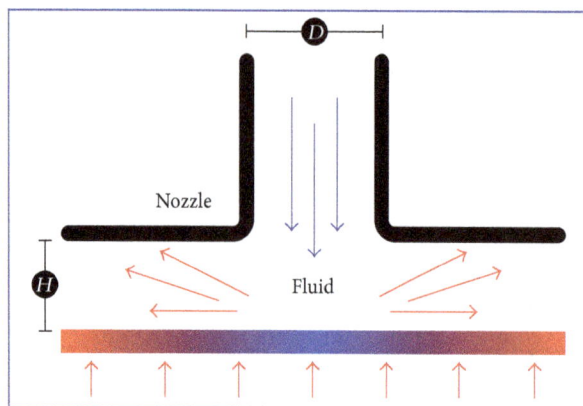

FIGURE 4: Effect of directed fluid released against the heat transferring surface.

Klein and Hetsroni [35] experimented with a wall region of the jet to improve its heat transfer using actuation against an impinging jet through a piezoelectric actuator. The setup allowed actuation at multiple amplitudes/frequencies against a steady flow of the impinging liquid jet. The chip resulted in the creation of a vortex beside the plate, and this, combined with the deflected flow, formed other vortices at both the top and bottom surfaces, which allows the deflected jet to press/divert the vortex downwards to enhance heat transfer, as per Figure 7. They reported enhancements to the heat transfer coefficients of up to 34%.

Chauhan and Thakur [7] compared the thermohydraulic performance of the impinging jet solar air heater in the form of effective efficiency with that of a conventional solar air heater. They confirmed that the impinging jet solar air heater was superior to a conventional solar air heater within a specific range of Reynolds number, resulting in an enhancement of ~34.54–57.89%, as shown in Figure 8.

As per the previous discussion, the corrugated absorber plate, or jet impingement, as individual approaches, significantly affected the optimization of the thermal performance of the SAH. The combination of both approaches is an avenue that has yet to be explored in literature. We proposed a modified SAH, involving the addition of a jet impingement

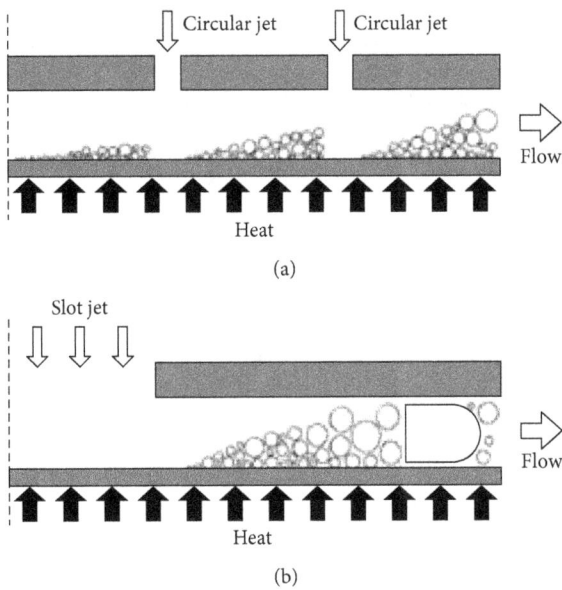

FIGURE 5: Two different jets. Multiple circular jets and single-slot jet [33].

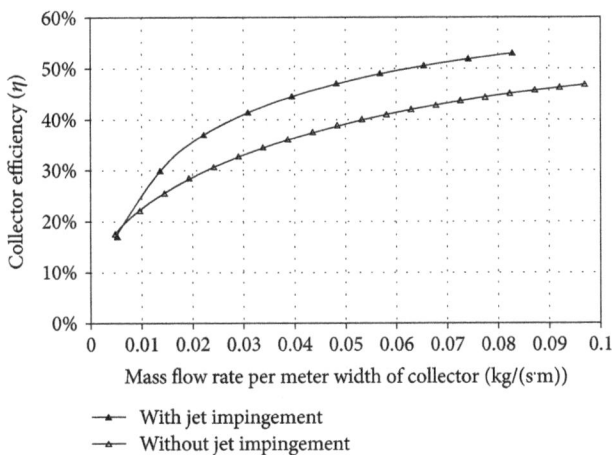

FIGURE 6: Efficiency at typical conditions of the collector with and without jet impingement [34].

on a corrugated absorber plate. The mathematical intricacies pertaining to this design will be analysed using energy balance analysis.

2. Materials and Method

2.1. Research Design. Figure 9 describes the steps involved in this study.

2.2. Materials. This study compares the proposed SAH design to two previous SAH designs to pinpoint the potential enhancement that can be realized in terms of temperature outlet, useful energy, and efficiency. The previous SAH designs with the cross-corrugated absorber plate [36] and the SAH design with jet impingement on a flat absorber plate [37] are illustrated in Figures 10(a) and 10(b), respectively.

While the schematic diagram of the proposed SAH design with jet impingement on the corrugated absorber plate is illustrated in Figure 11. The collector includes a frame of rectangular cross-section with an inlet and outlet for the passage of air, sheet of glass cover at the top of the frame, and an absorber plate mounted in the frame below the cover. The absorber is a corrugated plate coated with black paint on both sides and a jet plate spacing a distance below the absorber plate. Jets of air are directed through the holes of the jet plate and impinge on the lower surface of the corrugated absorber plate to increase the efficiency of the heat transfer, while the channels facilitate the corrugated flow of the spent jets for discharging via the outlet.

2.3. Mathematical Model Assumptions. This study represents a mathematical approach using the energy balance equations to compare and evaluate the performance of the three implemented designs of the solar air heater. The values will be numerically determined using the iteration approach. Figure 12 details the block diagram that is representative of the many steps taken for the calculation of the aforementioned values.

In order to model these behaviours, a number of assumptions need to be made to form the basis of the physical settings in this work. These assumptions include the following:

(1) Thermal performance of the SAH is at a steady-state condition.

(2) The sky is regarded as a black body for long-wavelength radiation at an equivalent sky temperature.

(3) Front and back losses occur at ambient temperatures.

(4) The collector (dust, dirt, and shading) and thermal inertia of the collector components are regarded as negligible.

(5) The operating temperatures of the collector components and mean air temperatures in the air channels are all considered to be uniform.

(6) The temperature of the air varies only in the flow direction.

(7) Thermal losses caused by the wind and insulation are negligible.

(8) Jet plate holes are circular in shape.

(9) Temperature decrease caused by the glass cover, absorber plate, and bottom plate are negligible.

2.4. Research Activities. The research activities in this study are as follows:

(1) Critical evaluation of the R&D on the available designs of SAHs was performed to propose a conceptual design that could improve the turbulence and heat transfer rates to enhance the temperature outlet, useful energy, and the efficiency of the SAH.

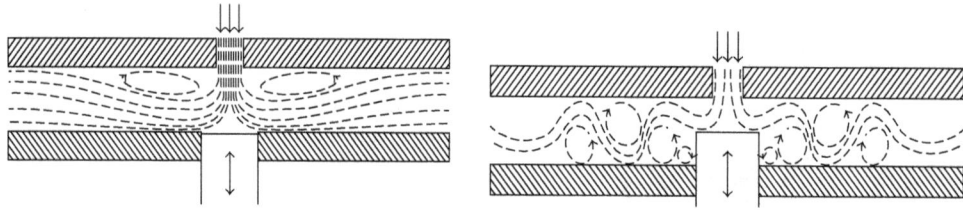

FIGURE 7: Characteristics of flow for a jet impinging on a flat surface and an actuating slab [35].

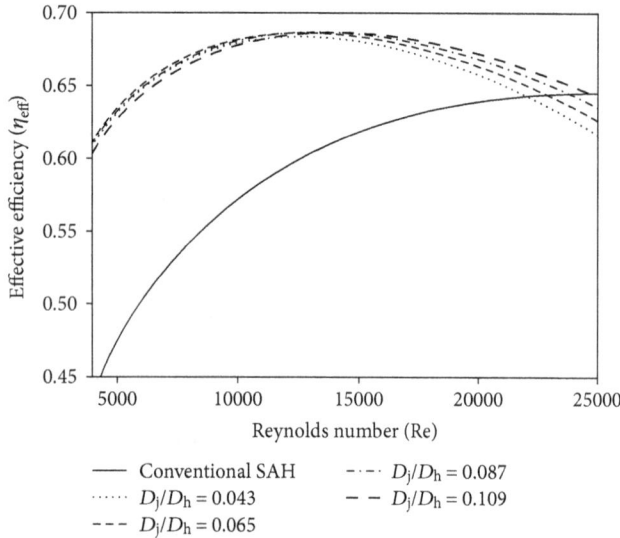

FIGURE 8: Variation of effective efficiency with Reynolds number at different values of jet diameter ratio [7].

(2) Development of a mathematical model for the proposed design of the SAH was performed.

(3) Validation of the aforementioned model (procedures and equations) versus previous designs of the SAH in [36, 37] under their reported assumptions and working conditions, as per Tables 1 and 2, respectively was performed.

(4) The thermal performance parameters of the proposed design of SAH will be compared to those of [36, 37] under the assumptions and working conditions outlined in Table 3.

2.5. Statistical Test Parameters. To measure the goodness of agreement between the values of the mathematical model and those reported in literature, the following statistical test parameter was used:

(1) EFF, which was reported to be the most accurate measure of closeness between predicted and referenced values [38]

$$EEF = 1 - \frac{\text{mean square error (MSE)}}{\text{variance of reference data}}. \quad (1)$$

EFF values that are 0-1 signify a close match between the predicted and reported values, while a negative EFF value means the exact opposite. Values that are as close as possible to unity is generally preferred, as these values imply a near perfect model.

A statistical paired *t*-test is used to pinpoint the differences in the mean values of the efficiency of the SAH pre- and postmodifications:

(2) Paired difference *t*-test

This test will determine whether the efficiency of the proposed SAH (jet impingement on a corrugated absorber plate) has significantly improved compared to the premodified SAH (jet impingement on a flat absorber plate). There are two probable results of this test, which is reliant upon the *p* value; a *p* value that is lower than the reference probability implies a statistical significance and the lack of a null hypothesis, while a *p* value that exceeds the reference probability implies an insignificant result. In the case of this work, the reference probability was assumed to be 0.05. The paired *t*-statistic was mathematically determined and subsequently converted to obtain the *p* value using the *t*-table or other viable utility programs. The test statistic for the paired difference *t*-test can be calculated using [39]

$$t = \frac{\bar{X} - \mu_0}{s/\sqrt{n}}, \quad (2)$$

where $\bar{X}$ is the average difference, s is the standard deviation of the difference, and n is the sample size. In paired testing, the null hypothesis is assumed when μ_0 is 0, which implies the lack of differences between the groups.

3. Mathematical Model of the Proposed Design

This section details the study of two adopted mathematical models. The first utilized the energy balance equations to compare the proposed SAH (jet impingement on a corrugated plate) with reference SAH (jet impingement on a flat plate), while the second utilized the Hottel method to apply parametric analysis and determine the optimum parameters of the proposed SAH.

3.1. Mathematical Model to Examine the Potential of the Proposed Design of SAH. The thermal network for the

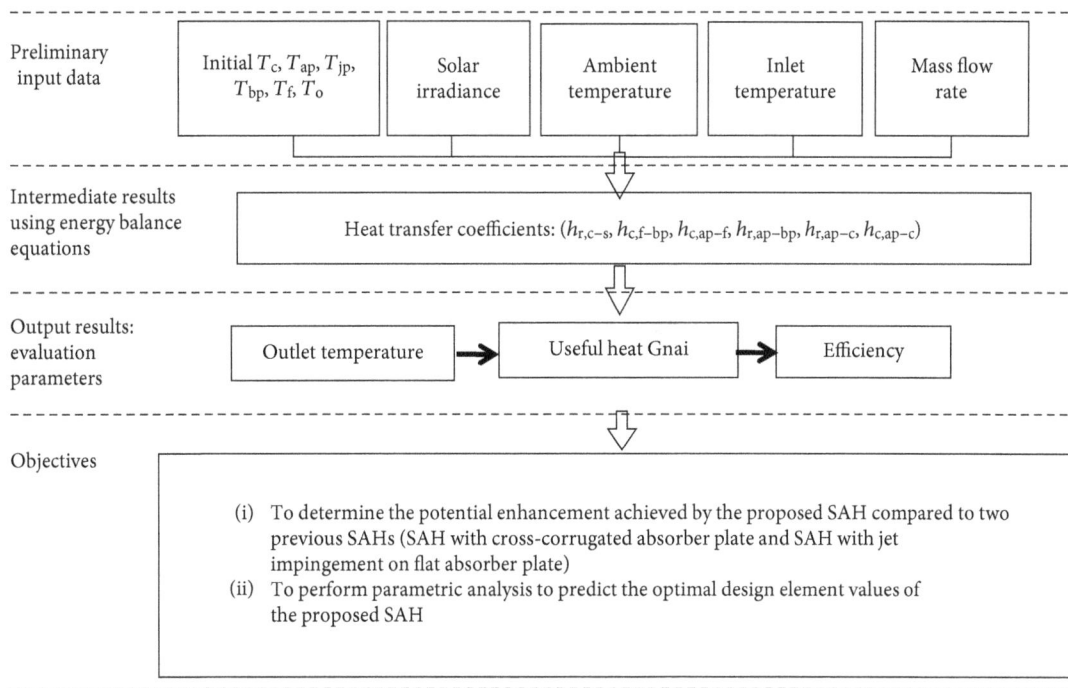

FIGURE 9: Block diagram of the research design.

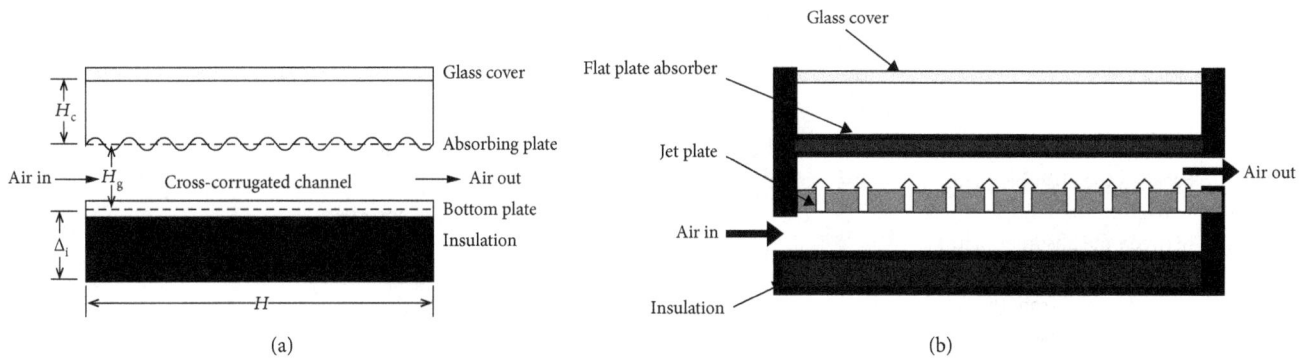

FIGURE 10: Schematic diagram of previous SAH designs (a) with a cross-corrugated absorber plate, Lin et al. [36] (b) with jet impingement on a flat absorber plate, Choudhury and Garg [37].

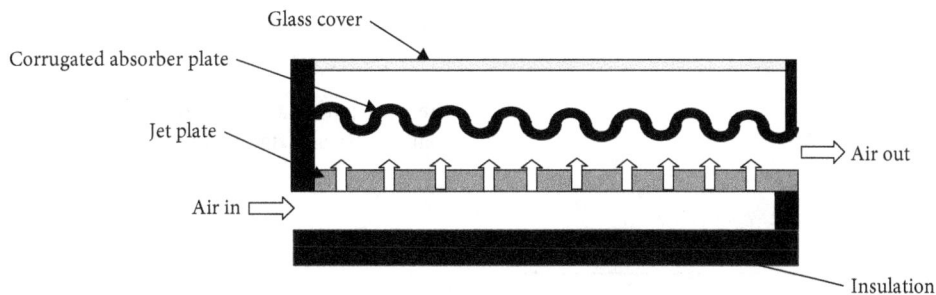

FIGURE 11: Schematic diagram of the proposed design of SAH with jet impingement on a corrugated absorber plate.

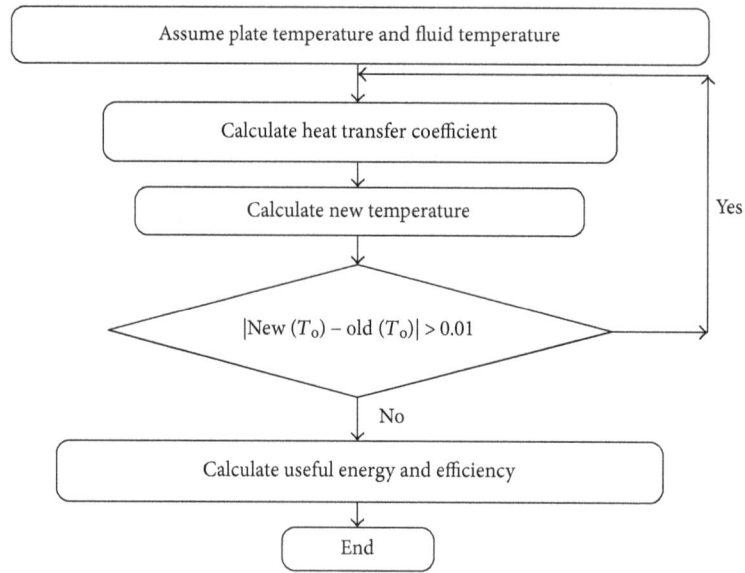

FIGURE 12: Block diagram of the calculation steps.

TABLE 1: The assumptions and working conditions of the SAH with a cross-corrugated absorber plate in [36].

Working conditions and assumptions	Value
Solar radiation (I)	800 (W/m^2)
Ambient temperature (T_A)	285 (K)
Inlet temperature (T_i)	295 (K)
Air mass flow rate per unit area of the heater (m_f)	0.1 (kg/m^2·s)
Width of the heater (W)	1 (m)
Length of the heater (L)	2 (m)
Mean gap between the absorber plate and jet plate (H_g)	0.04 (m)
Mean gap between the absorber plate and cover (H_c)	0.04 (m)
Convection heat transfer from the glass cover due to wind (h_w)	11.4 (W/m^2 K)
Thickness insulation (b)	0.05 (m)
Absorptivity of solar radiation of the absorbing plate (α_{ap})	0.95
Transmissivity of solar radiation of the glass cover (τc)	0.84
Solar radiation absorptivity of the glass cover (α_c)	0.06
Emissivity of thermal radiation of the absorbing plate (ε_{ap})	0.94
Emissivity of thermal radiation of the bottom plate (ε_{bp})	0.94
Thermal radiation emissivity of the glass cover (ε_c)	0.9
Angle of inclination of the heater (θ)	30°

TABLE 2: The assumptions and working conditions of jet impingement on a flat absorber plate in [37].

Working conditions and assumptions	Value
Solar radiation (I)	900 (W/m^2)
Ambient temperature (T_A)	300 (K)
Convection heat transfer from the glass cover due to wind (h_w)	11.4 (W/m^2 K)
Width of the heater (W)	1 (m)
Length of heater (L)	2 (m)
Mean gap between the absorber plate and jet plate (H_g)	0.05 (m)
Mean gap between the absorber plate and cover (H_c)	0.05 (m)
Absorptivity of solar radiation of the absorbing plate (α_{ap})	0.95
Transmissivity of solar radiation of the glass cover (τ_c)	0.90
Spacing between absorber and back plate (z)	0.1 (m)
Diameter of the hole or nozzle	0.01 (m)
Center to center spacing between holes/nozzles	0.06 (m)

the incident solar radiation and the optical loss and can be mathematically determined as in [2].

$$S \simeq 0.96\,\tau_c \alpha_{ap} I. \tag{3}$$

components of SAH with jet impingement on a corrugated plate is illustrated in Figure 13.

The solar radiation that is captured by the absorbing plate per unit area S (W/m^2) is equivalent to the difference between

In steady state conditions, the energy balance equations for the cover, absorber, jet plate, back plate, air in the passage between the back plate and jet plate, and air in the passage between the absorber and back plates in the case of the jet plate air heater are as follows:

TABLE 3: Working conditions and assumptions used in evaluating the potential of the proposed design.

Parameter	Value
Solar radiation (I)	Clear sky day (W/m^2)
Ambient temperature (T_A)	Associated ambient temperature of the selected clear sky day, 303–317.5 K
Air mass flow rate per unit area of the solar air heater (m_f)	0.02–0.08 (kg/s)
Width of the heater (W)	0.3 (m)
Length of the heater (L)	1.4 (m)
Mean gap between the absorber plate and jet plate (H_g)	0.04 (m)
Mean gap between the absorber plate and cover (H_c)	0.04 (m)
Convection heat transfer from the glass cover due to wind (h_w)	11.4 (W/m^2 K)
Absorptivity of solar radiation of the absorbing plate (α_{ap})	0.95
Transmissivity of solar radiation of the glass cover (τc)	0.84
Solar radiation absorptivity of the glass cover (α_c)	0.06
Emissivity of thermal radiation of the absorbing plate (ε_{ap})	0.94
Emissivity of thermal radiation of the bottom plate (ε_{bp})	0.94
Thermal radiation emissivity of the glass cover (ε_c)	0.90

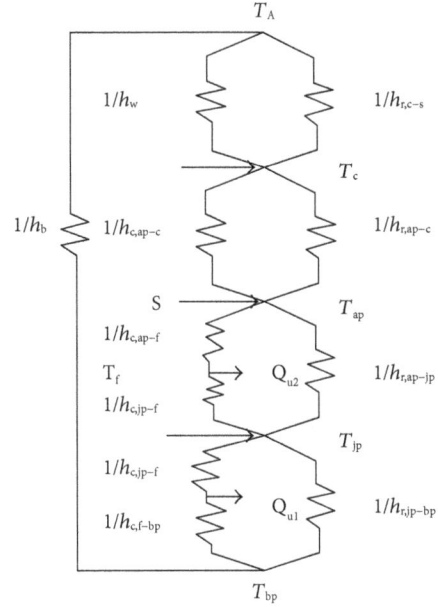

FIGURE 13: Thermal network for the single-cover solar air heater.

respectively, $h_{r,ap-j}$, represents radiation heat transfer coefficient between the absorbing and jet plates, and $h_{c,ap-a2}$ represents convection heat transfer coefficient of the fluid on the absorbing plate.

(iii) For the fluid, energy balance equations are obtained from [37].

(a) The fluid between the bottom plate and jet plate is as follows:

$$C_p \dot{m}_1 (T_{o1} - T_A) = h_{c,j-a1}(T_j - T_a) + h_{c,b-a1}(T_b - T_a). \quad (6)$$

(b) The fluid between the absorber and jet plates is as follows:

$$C_p \dot{m}_1 (T_o - T_{o1}) = h_{c,j-a2}(T_j - T_{o1})A_h + h_{c,ap-a2}(T_{ap} - T_{o1})(A_a). \quad (7)$$

(iv) On the jet plate, the energy balance equation is as follows:

$$h_{r,ap-j}(T_{ap} - T_j)\left(\frac{A_a}{A_h}\right) = h_{c,j-a2}(T_j - T_{o1}) + h_{c,j-a1}(T_j - T_a) + h_{r,j-bp}(T_j - T_{bp}). \quad (8)$$

In the case of the bottom plate, heat gains via fluids through convection are represented by $h_{c,bp-a1}$ and the jet plate via thermal radiation is represented by $h_{r,j-bp}$, both of which are offsetted by the thermal loss to the surroundings via conduction, represented by h_b (W/m^2 K),

(i) On the glass cover

The energy balance equation for the glass cover is as follows:

$$\alpha_c I + \left(h_{c,ap-c} + h_{r,ap-c}\right)(T_{ap} - T_c) = (h_w + h_{r,c-s}) \cdot (T_c - T_a), \quad (4)$$

where the absorptivity of solar radiation by the glass cover is α_c, while the energy gain is $\alpha_c I$.

(ii) On the absorbing plate

The energy balance for the absorbing plate is described as follows:

$$S\frac{A_h}{A_a} = \left(h_{c,ap-c} + h_{r,ap-c}\right)(T_{ap} - T_c) + h_{r,ap-j}(T_{ap} - T_j) + h_{c,ap-a2}(T_{ap} - T_{o1}), \quad (5)$$

where S represents the absorbed solar radiation by the absorber metal and $h_{c,ap-c}$ and $h_{r,ap-c}$ represent convection and radiation heat transfer coefficients between the absorber plate and glass cover,

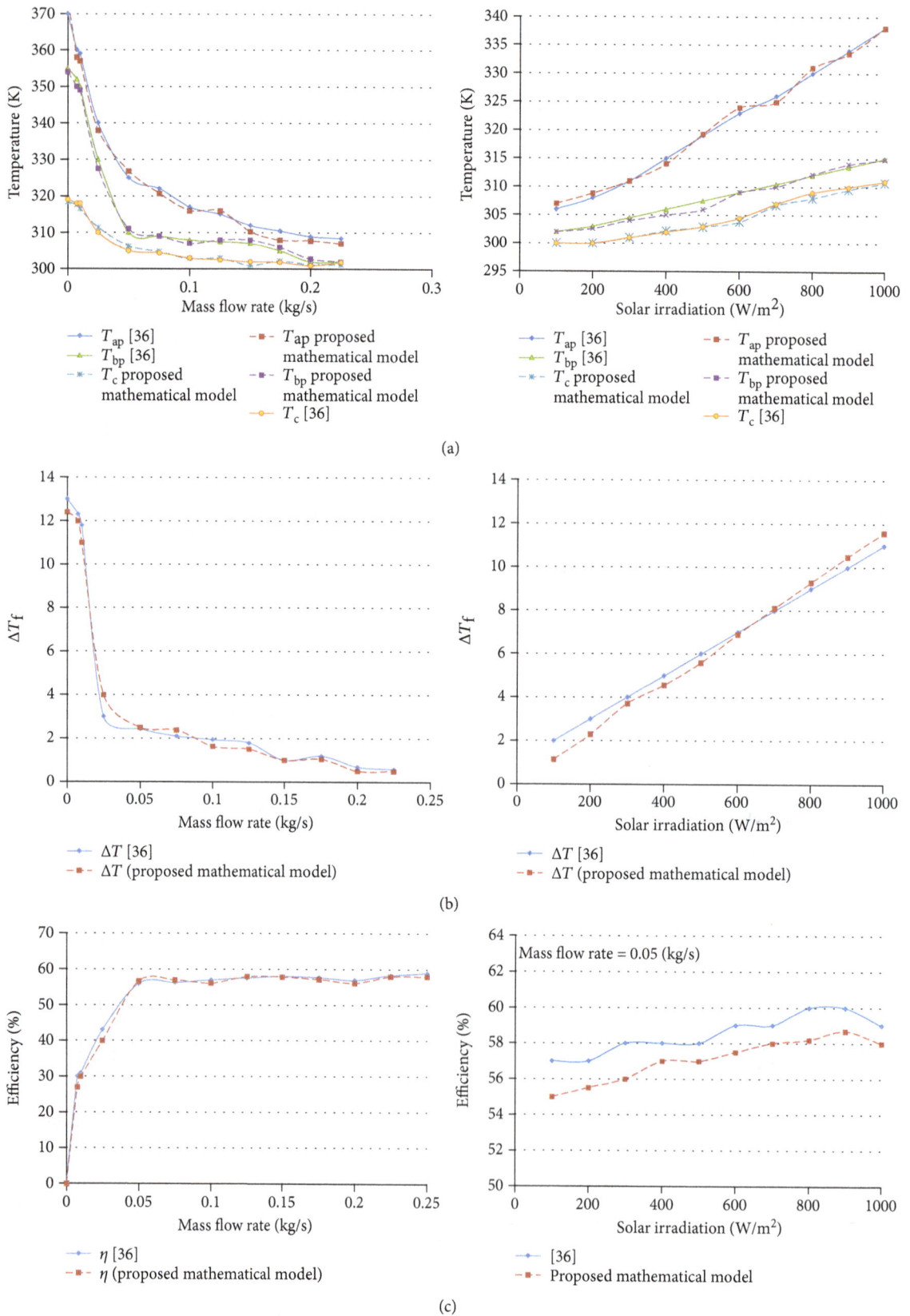

Figure 14: Comparison of solar air heater performance parameters obtained from the proposed mathematical model and that in [36]. (a) Predicted temperature of absorber plate, bottom plate, and glass cover plate of SAH versus mass flow rate and solar irradiation. (b) Predicted fluid temperature difference versus mass flow rate and solar irradiation. (c) Predicted SAH efficiency versus mass flow rate and solar irradiation.

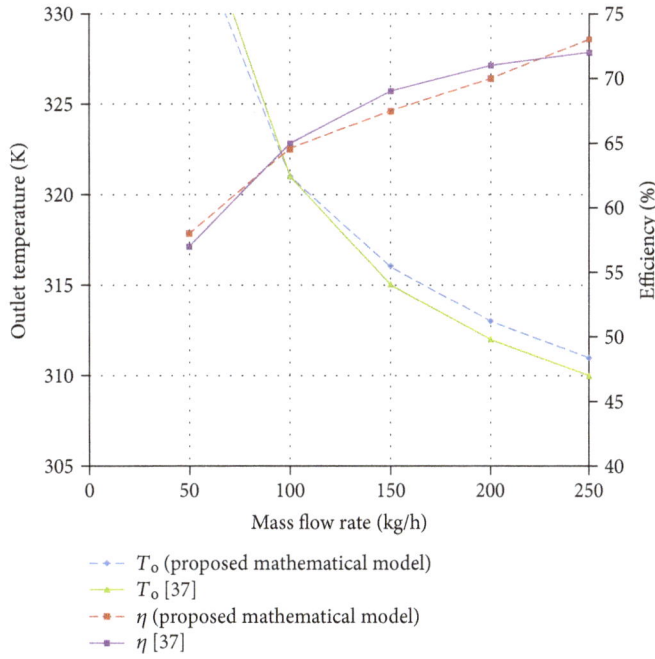

FIGURE 15: Predicted outlet temperature and efficiency of the SAH obtained from the proposed mathematical model and from reference work [37].

TABLE 4: Accuracy of the proposed mathematical model in predicting the values of different SAH evaluation parameters for the two reference works.

Reference	Parameters of SAH	Statistical EFF test result "evaluation of parameters versus mass flow rate"	Statistical EFF test result "evaluation of parameters versus solar radiation"
Lin et al. [36]	Glass cover temperature values (T_c)	0.98	0.9642
	Absorber plate temperature values (T_{ap})	0.99	0.9546
	Bottom plate temperature (T_{bp})	0.97	0.9554
	Fluid temperature difference ($\Delta T = T_o - T_i$)	0.97	0.9569
	Collector efficiency	0.27	0.2105
Choudhury and Garg [37]	Outlet temperature values (T_o)	0.98	—
	Collector efficiency	0.97	—

which is the conduction heat transfer coefficient across the insulation, where

$$h_{r,j-bp}\left(T_j - T_{bp}\right) = h_{c,bp-a1}\left(T_{bp} - T_a\right) + h_b\left(T_{bp} - T_A\right), \quad (9)$$

where T_a is $T_a = (T_A + T_{O1})/2$.

It is found from (4) that

$$T_c = \frac{\alpha_c I + \left(h_{c,ap-c} + h_{r,ap-c}\right)\left(T_{ap}\right) + \left(h_w + h_{r,c-s}\right)\left(T_a\right)}{\left(h_{c,ap-c} + h_{r,ap-c} + h_w + h_{r,c-s}\right)}, \quad (10)$$

from (5) that

$$T_{ap} = \frac{S(A_h/A_a) + \left(h_{c,ap-c} + h_{r,ap-c}\right)\left(T_c\right) + \left(h_{r,ap-j}\right)\left(T_j\right) + \left(h_{c,ap-a2}\right)\left(T_{o1}\right)}{\left(h_{c,ap-c} + h_{r,ap-c} + h_{r,ap-j} + h_{c,ap-a2}\right)}, \quad (11)$$

from (6) that

$$T_{o1} = \frac{\dot{m}_1(T_A) + h_{c,j-a1}\left(T_j - T_a\right) + h_{c,b-a1}\left(T_b - T_a\right)}{C_p \dot{m}_1}, \quad (12)$$

from (7) that

$$T_o = \frac{\dot{m}_1(T_{o1}) + h_{c,j-a2}\left(T_j - T_{o1}\right)(A_h) + h_{c,ap-a2}\left(T_{ap} - T_{o1}\right)(A_a)}{C_p \dot{m}_1}, \quad (13)$$

from (8) that

$$T_j = \frac{\left(h_{r,ap-j} T_{ap}\right) * (A_a/A_h) + \left(h_{c,j-a2} T_{o1}\right) + \left(h_{c,j-a1} T_a\right) + \left(h_{r,j-bp} T_{bp}\right)}{h_{c,j-a2} + h_{c,j-a1} + h_{r,j-bp} + \left((A_a/A_h) * h_{r,ap-j}\right)}, \quad (14)$$

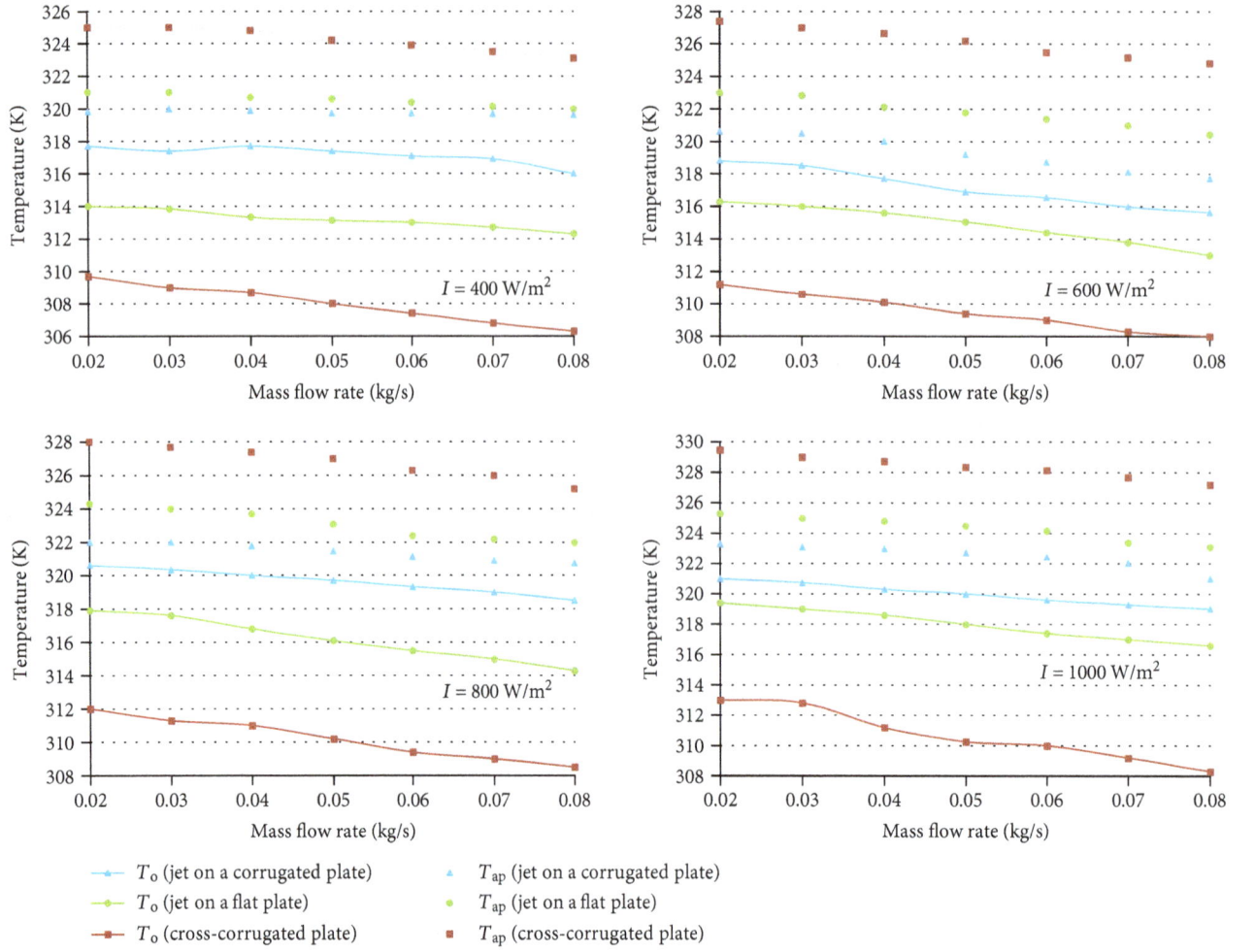

FIGURE 16: Absorber plate and outlet temperature of different collector designs versus mass flow rate for different solar radiation levels.

and from (9) that

$$T_{bp} = \frac{\left(h_b T_A\right) + \left(h_{r,j-bp} T_j\right) + \left(h_{c,a1-bp} T_a\right)}{\left(h_b\right) + \left(h_{r,j-bp}\right) + \left(h_{c,a1-bp}\right)}. \quad (15)$$

The efficiency of solar heat gain of the heater is obtained as in [37].

$$\eta = \frac{C_p (\dot{m}_1)(T_o - T_i)}{I * A_h}. \quad (16)$$

h_w is recommended by [40] as

$$h_w = 5.7 + 3.8 V_w, \quad (17)$$

where V_w (m/s) is the wind velocity of the ambient air and it is usually assumed that $V_w = 1.5$ m/s, $h_w = 11.4$ W/m^2 K. $h_{r,c-s}$ is given as in [41]

$$h_{r,c-s} = (T_c + T_s)\left(T_c^2 + T_s^2\right)\frac{(T_c - T_s)}{(T_c - T_a)}, \quad (18)$$

where $\sigma = 5.67 \times 10^{-8}$ W/m^2 K^4 is the Stefan–Boltzmann constant and the sky temperature T_s (K) is estimated by the formulation given by [42]

$$T_s = 0.0552 T_a^{1.5}. \quad (19)$$

The radiation heat transfer is given as in [37].

$$\begin{aligned} h_{r,ap-c} &= \frac{\sigma\left(T_{ap}^2 + T_c^2\right)\left(T_{ap} + T_c\right)}{\left(1/\epsilon_{ap}\right) + \left(1/\epsilon_c\right) - 1}, \\[2mm] h_{r,ap-j} &= \frac{\sigma\left(T_{ap}^2 + T_j^2\right)\left(T_{ap} + T_j\right)}{\left(1/\epsilon_{ap}\right) + \left(1/\epsilon_j\right) - 1}, \\[2mm] h_{r,j-bp} &= \frac{\sigma\left(T_j^2 + T_{bp}^2\right)\left(T_j + T_{bp}\right)}{\left(1/\epsilon_j\right) + \left(1/\epsilon_{bp}\right) - 1}, \\[2mm] h_{c,ap-a2} &= \left(h_{r,ap-j}\right) * \frac{\left(T_{ap} - T_{o1}\right)}{\left(T_{ap} - T_a\right)}. \end{aligned} \quad (20)$$

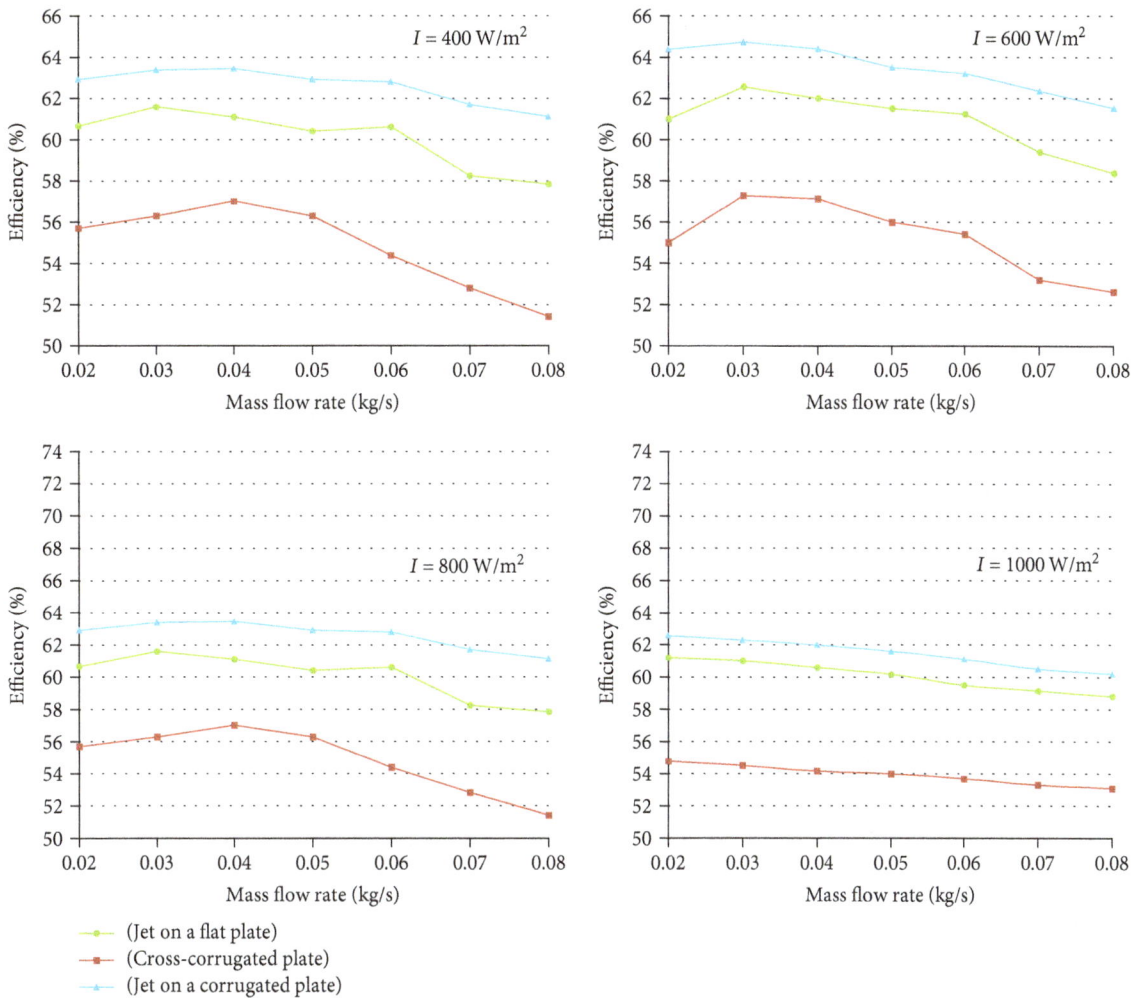

FIGURE 17: SAH Efficiency under the purview of three collector designs versus mass flow rate for different solar radiation levels.

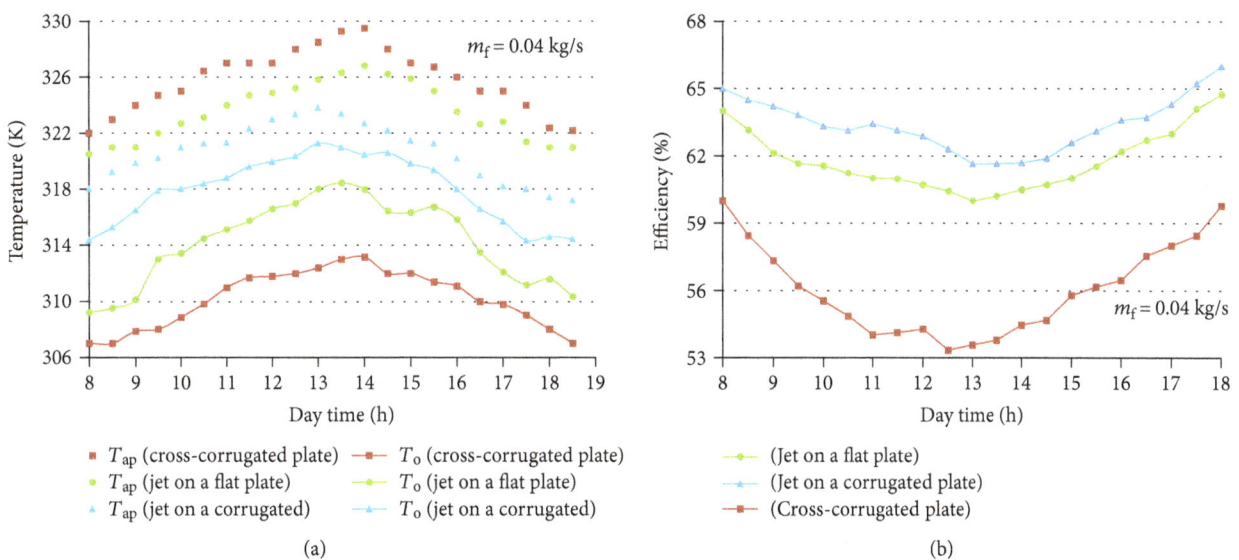

(a)

(b)

FIGURE 18: Hourly performance parameters under the purview of three designs of solar air heater. (a) Absorber plate temperature and air outlet temperature. (b) Collector efficiency.

(a)

(b)

(c)

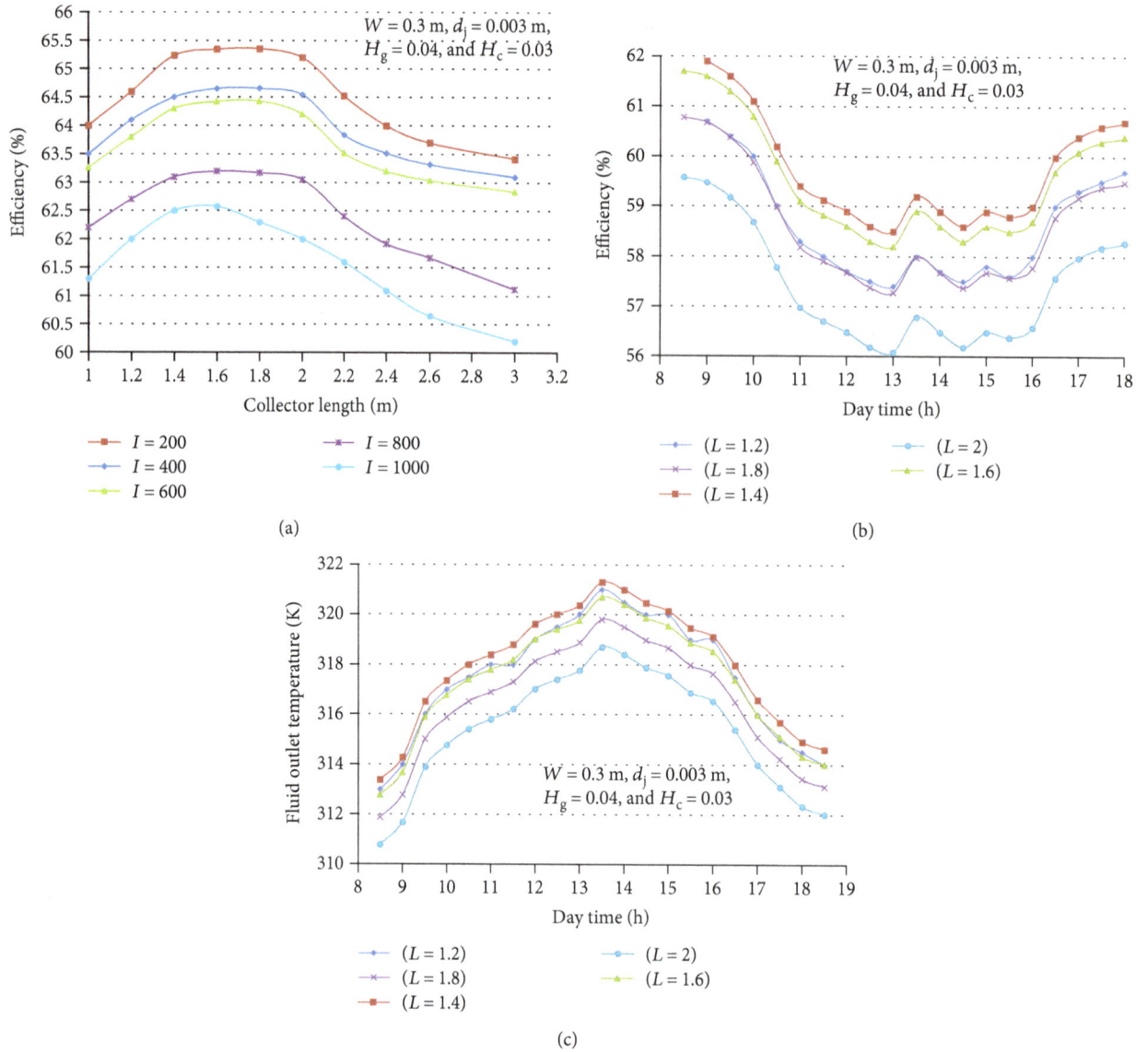

FIGURE 19: Prediction of the optimum length (L) of the proposed solar air heater. (a) SAH efficiency versus collector length at different solar radiation levels. (b) Hourly SAH efficiency over daytime at different collector lengths. (c) Hourly air outlet temperature at different collector lengths.

The conduction heat transfer coefficient across the insulation is estimated by

$$h_{\text{b}} = \frac{k_{\text{i}}}{b}, \tag{21}$$

where k_{i} (W/m K) is the thermal conductivity of the insulation and b (m) is the mean thickness of the insulation.

The convection heat transfer coefficient between the glass cover and the absorbing plate is calculated by

$$h_{\text{c,ap--c}} = \text{Nu}_{\text{ap--c}} \frac{K}{H_{\text{c}}}. \tag{22}$$

And $\text{Nu}_{\text{ap--c}}$ is the Nusselt number for natural convection in the channel formed by the cover and the absorbing plate and is given as in [37]

$$\text{Nu}_{\text{ap--c}} = 0.1673(\text{Ra} * \cos \theta)^{0.2917}, \tag{23}$$

where θ (°) is the angle of inclination of the heater and Ra is the Rayleigh number which is defined as

$$\text{Ra} = \frac{\rho^2 c_p g \beta \left(T_{\text{ap}} - T_{\text{c}}\right) H_{\text{c}}^3}{K_\mu}, \tag{24}$$

in which ρ (kg/m^3), β (1/K), and μ (kg/m s) are the density, thermal expansion coefficient, and dynamic viscosity of air and g (m/s^2) is the acceleration due to gravity.

$$\text{Nu}_{\text{ap--f}} = 0.0743\text{Re}^{0.76}, \tag{25}$$

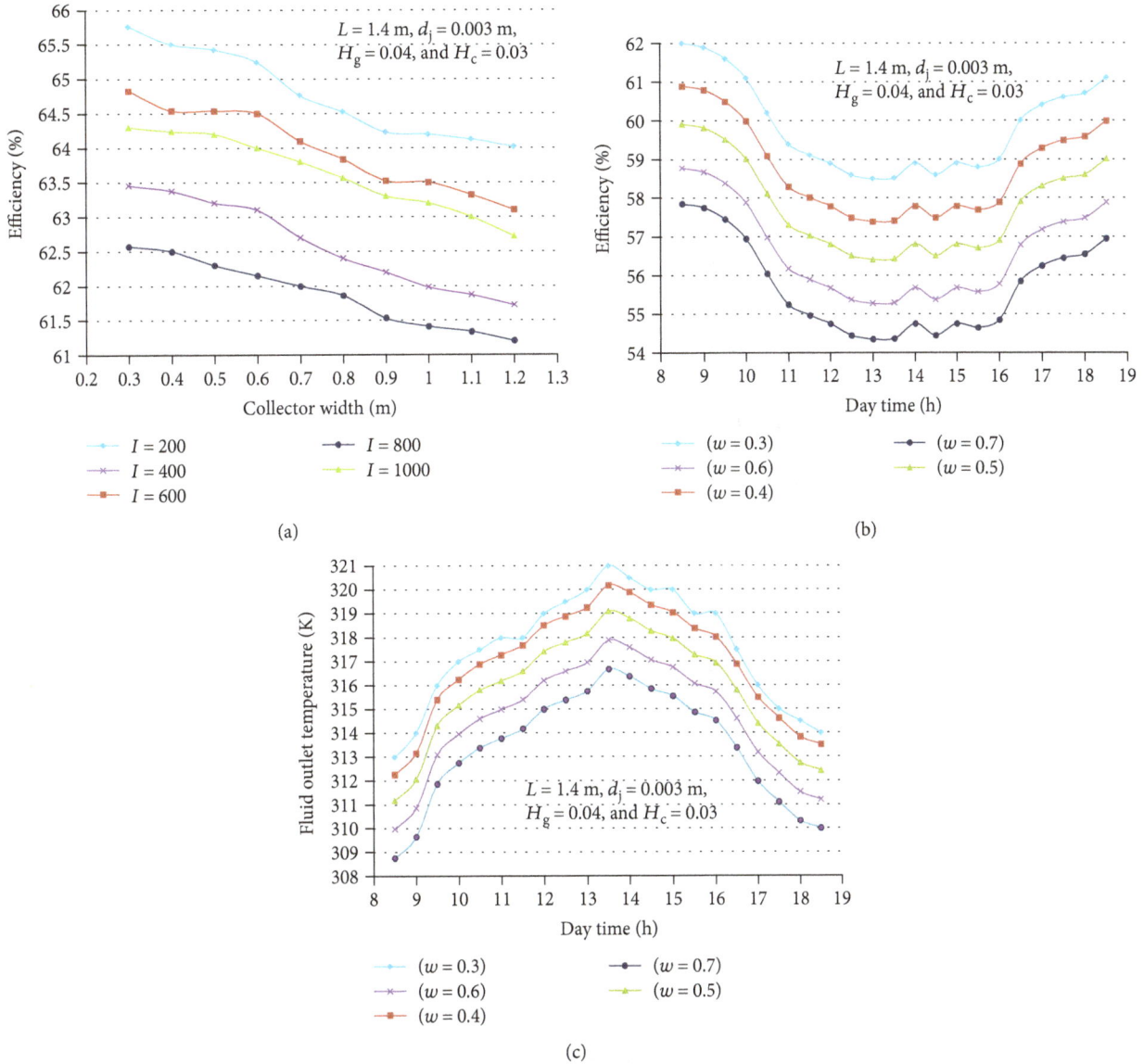

FIGURE 20: Prediction of the optimum width (W) of the proposed solar air heater. (a) SAH efficiency versus collector width at different solar radiation levels. (b) Hourly SAH efficiency over daytime at different collector widths. (c) Hourly air outlet temperature at different collector widths.

where Re is the Reynolds number, defined as follows:

$$\text{Re} = \frac{\rho \overline{V_f} D_h}{\mu}, \tag{26}$$

where $\overline{V_f}$ (m/s) is the mean velocity of fluid in the channel.

4. Validation of the Proposed Mathematical Model

The equations and procedures were used to predict the results of [36, 37] under their reported working conditions/assumptions, as per Tables 1 and 2, respectively, for the purpose of validating the proposed mathematical model.

Figures 14(a)–14(c) show the comparison between the temperatures of the absorber plates, the bottom plate, and the glass cover plate of the SAH and the fluid temperature gradient and SAH efficiency predicted by the proposed mathematical model and that in [36], while Figure 15 shows the comparison between outlet temperature and the efficiency of the SAH predicted by the proposed mathematical model and that in [37]. The values predicted by the proposed mathematical model are represented in the graph with discrete lines, while the results obtained from the aforementioned references are represented in the graph by continuous lines.

Figures 14 and 15 display excellent agreements between the values predicted by the proposed model and those reported in literature by the two aforementioned references. A statistical test was applied using the EFF test to further affirm the agreement between these values, and the results of the statistical test is tabulated in Table 4, falling between 27 and 99% for the many evaluation parameters.

(a)

(b)

(c)

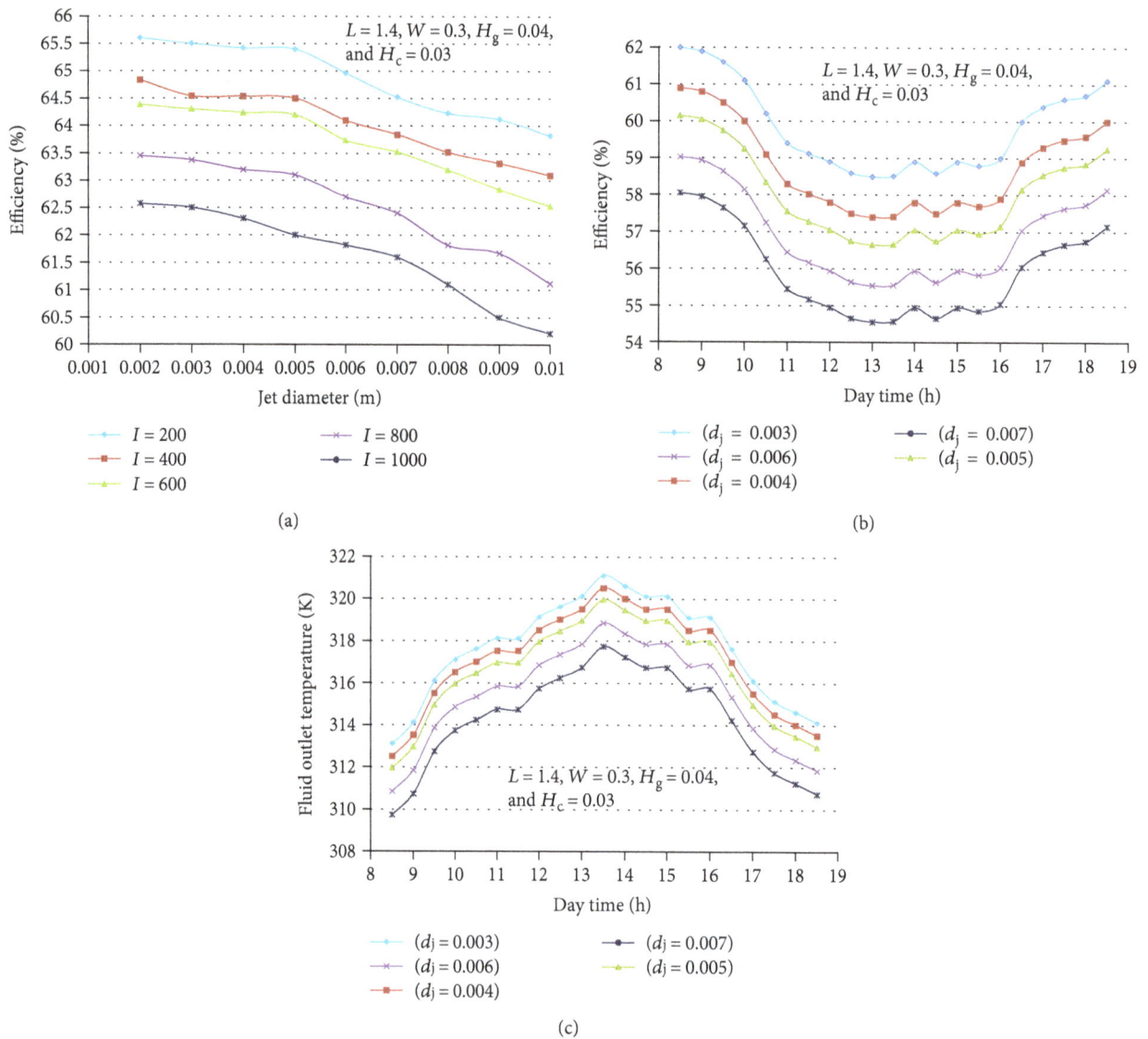

FIGURE 21: Prediction of the optimum jet diameter (d_j) of the proposed solar air heater. (a) SAH efficiency versus jet diameter at different solar radiation levels. (b) Hourly SAH efficiency over daytime at different jet diameters. (c) Hourly air outlet temperature at different jet diameters.

Based on the plots in Figures 14 and 15 and the values tabulated in Table 4, it can be confirmed that the proposed mathematical model is suitable for the prediction of the thermal performance parameters of the SAH.

5. Potential of the Proposed SAH as a Superior Design

This section details the comparison between the proposed SAH with jet impingement on the corrugated absorber plate to the jet-to-flat plate SAH and the cross-corrugated absorber plate SAH. The potential of the proposed SAH with respect to these aforementioned designs will be measured using the following evaluation parameters: temperature difference between fluid outlet and absorber plate, collector efficiency, and statistical paired t-test, and the working conditions/

assumptions that are used as input data in the three SAH designs are tabulated in Table 3.

5.1. Effect of Different SAH Designs on the Temperature of the Fluid Outlet and Absorber Plate.
Figure 16 shows the temperature of the fluid outlet and absorber plate at different mass flow rates for multiple solar radiation for a total of 3 SAH designs. The absorber plate's temperature is represented by discrete lines, while the fluid outlet's temperature is represented by continuous lines. It can be seen that the mass flow rate is inversely proportional to the temperature of the absorber plate and the fluid outlet, while it is also evident that the temperature of the absorber plate and fluid outlet is directly proportional to the solar radiation in the context of any mass flowrate. The highest temperature is achieved at solar radiation of 1000 W/m^2 and at mass flow rate falling between 0.02 and 0.04 kg/s. Figure 16 shows that the

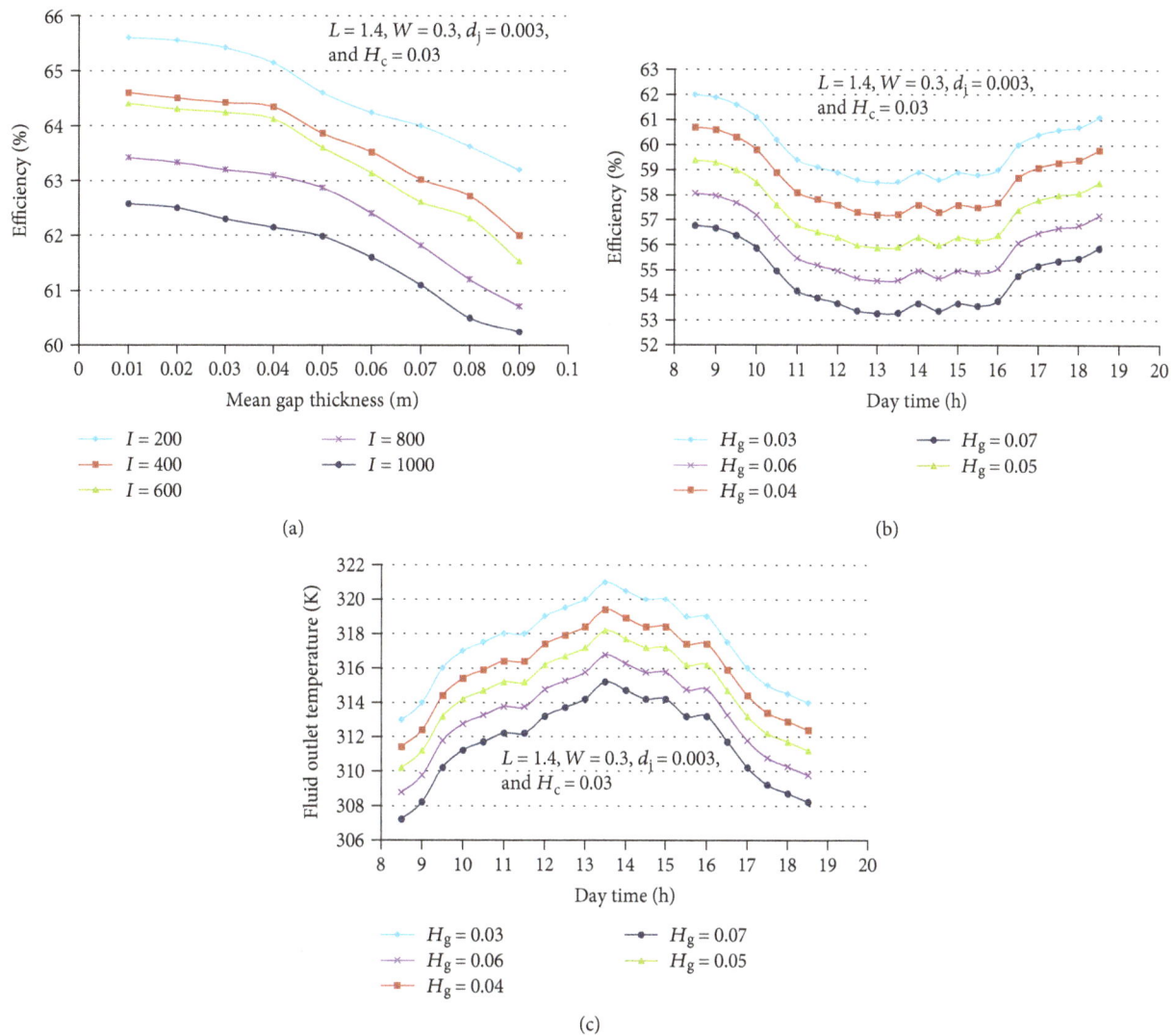

FIGURE 22: Prediction of the optimum mean gap thickness between the absorber plate and the jet plate (H_g) of the proposed solar air heater. (a) SAH efficiency versus H_g at different solar radiation levels. (b) Hourly SAH efficiency over daytime at different H_g. (c) Hourly air outlet temperature at different H_g.

proposed SAH reported the highest fluid outlet temperature, followed by SAH with the jet-to-flat plate absorber, then SAH with the cross-corrugated plate. Simultaneously, the proposed SAH reported the lowest absorber plate temperature, followed by the SAH with the jet-to-flat plate absorber, then SAH with the cross-corrugated plate. It should be pointed out that the fluid outlet and absorber plate's temperatures of the proposed SAH converged significantly at temperature differences of less than 3 K, which means that the proposed SAH design allowed the fluid outlet to gain highest heat from the corrugated absorber plate. Simultaneously, it was also shown that the fluid outlet and absorber plate temperatures for other designs remain beyond the desired convergence. Hence, it can be posited that the proposed design is superior for heat extraction when compared to the other designs.

5.2. Effect of Different Collector Designs on SAH Efficiency.
Figure 17 shows the SAH efficiency at multiple mass flow rates for different solar radiation levels under the purview

of three collector designs. It is evident that the mass flow rate or solar radiation is inversely proportional to SAH efficiency. The highest efficiency was reported to be within mass flow-rate range 0.02–0.04 kg/s. It should be pointed out that as per Figure 17, the proposed SAH showed the highest efficiency, followed by the jet-to-flat plate absorber, then SAH with the cross-corrugated plate. As per Figures 16 and 17, we can confirm that the jet impingement on the corrugated absorber is the best SAH design compared to the others being studied.

5.3. Hourly Thermal Performance of Different SAH Designs under Clear Sky.
The thermal performance of the designs is plotted hourly during daytime on a clear sky day. Figure 18(a) shows the hourly fluid outlet temperature and absorber plate temperature of different SAH designs at a mass flow rate $m_f = 0.04$ kg/s. The absorber plate temperature is represented by discrete lines, while fluid outlet temperature is represented by continuous lines. The results

(a)

(b)

(c)

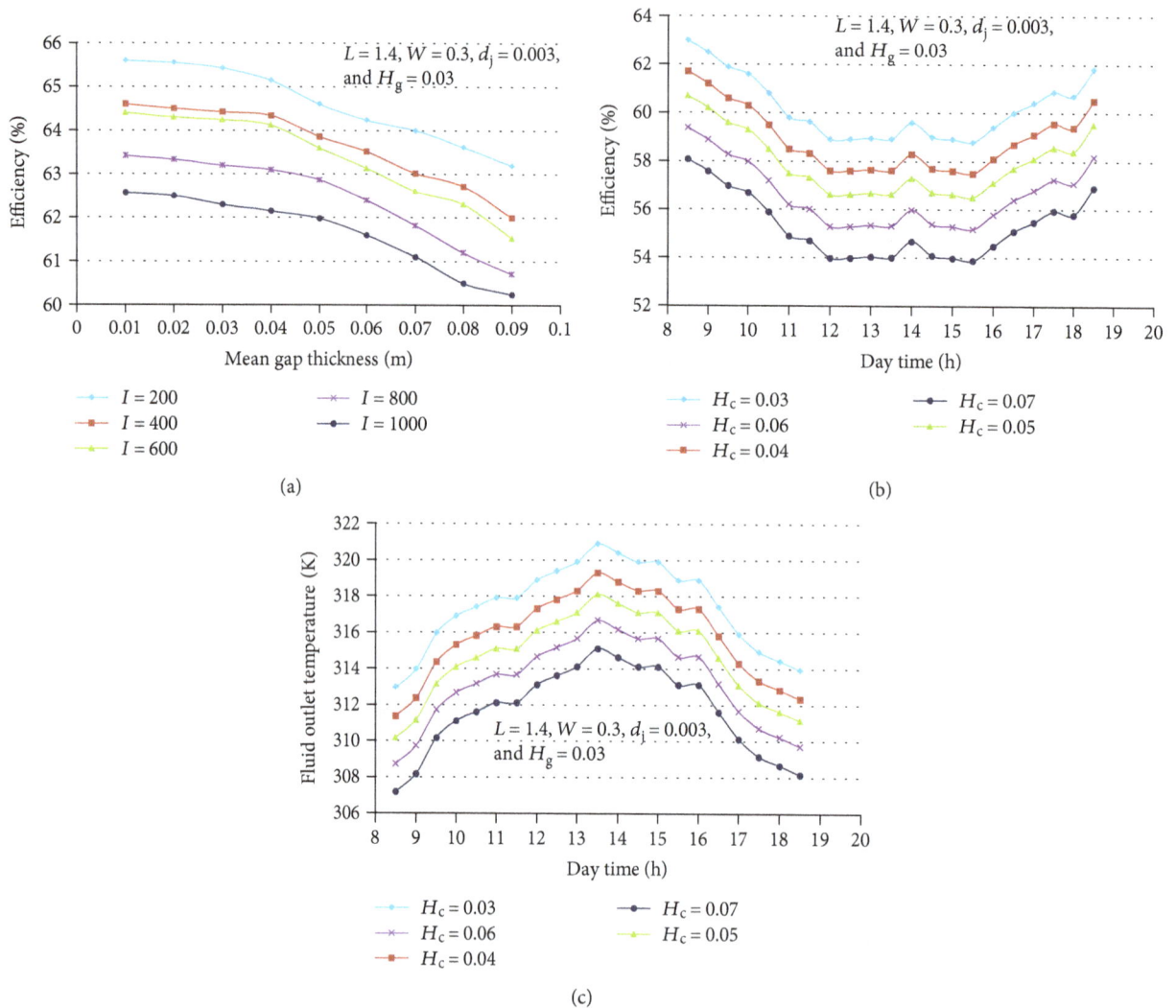

FIGURE 23: Prediction of the optimum mean gap thickness between the absorber plate and the glass cover (H_c) of the proposed solar air heater. (a) SAH efficiency versus H_c at different solar radiation levels. (b) Hourly SAH efficiency over daytime at different H_c. (c) Hourly air outlet temperature at different H_c.

showed that the min–max fluid outlet temperatures during the day for the proposed SAH, jet-to-flat plate SAH, and cross-corrugated plate SAH are 314–321 K, 309.5–317 K, and 307–313 K, respectively. It can also be seen that the min–max absorber plate temperatures during a clear sky day for the proposed SAH, jet-to-flat plate SAH, and cross-corrugated plate SAH are 317–323.5 K, 319–326.5 K, and 322–328 K, respectively. It was seen that the maximum temperature difference between the absorber plate and fluid outlet for the proposed SAH, jet-to-flat plate SAH, and the cross-corrugated plate SAH is ~3 K, 9 K, and 15 K, respectively.

Figure 18(b) shows the efficiency of multiple SAH designs at a mass flowrate $m_f = 0.04$ kg/s. It can be seen that the min–max efficiencies for the proposed SAH, jet-to-flat plate SAH, and the cross-corrugated plate SAH are 61.5–65.7, 60–64.8, and 53.6–60%, respectively. As per Figures 18(a) and 18(b), it is evident that the thermal performance of the proposed SAH during the day was significantly enhanced compared to the cross-corrugated plate SAH and the jet on a flat plate SAH. This enhancement of

the proposed SAH vis-à-vis the jet-to-flat plate SAH needs to be accurately ascertained via statistical test. The paired t-test was used to determine whether the efficiency enhancement of the proposed SAH can be regarded as significant vis-à-vis the jet-to-flat plate SAH. It was found that the statistical paired t-test proved that the efficiency enhancement is significant.

6. Parametric Performance Optimization of the Proposed SAH

In this round of analyses, the optimal values of the design parameters of SAH are going to be determined. The investigated ranges of parameters are 1–3 m for the collectors' length, 0.3–1.2 m for the collectors' width, 0.002–0.01 m for the jet diameter, and 0.01–0.09 m for the mean gap thickness between the absorbing plate and the jet plate.

6.1. Optimum Collector Length of the Proposed SAH. The influence of collector length (L) upon the efficiency of the

SAH at multiple solar radiation values are detailed in Figure 19(a). Generally, at any solar radiation value, the collector length is directly proportional to the SAH efficiency, up till the point where the collector length reaches 1.4–2 m, where at this point, the efficiency is inversely proportional to the collectors' length. In order to pinpoint the exact optimal collector length, the hourly SAH efficiency and hourly fluid outlet temperature were predicted during a clear sky day and plotted against multiple collectors' length, as per Figures 19(b) and 19(c). The results showed that the optimal collector length showing maximum efficiency and maximum fluid outlet temperature concurrently is 1.4 m.

6.2. Optimum Collector Width of SAH. The effect of the collectors' width (W) on the efficiency of the SAH at multiple solar radiation values is detailed in Figure 20(a). Generally, at each solar radiation value, it can be seen that the collectors' width is inversely proportional to the efficiency of the SAH. The hourly SAH efficiency and hourly fluid outlet temperature predicted during a clear sky day were plotted at multiple collector widths, as per Figures 20(b) and 20(c). The optimum collector width indicating maximum efficiency and maximum fluid outlet temperature concurrently is $W = 0.3$ m.

6.3. Optimum Jet Diameter (d_j). The effect of jet diameter (d_j) on the efficiency of the SAH at multiple solar radiation values is detailed in Figure 21(a). At multiple solar radiation values, the range of jet diameter between 0.002 and 0.005 m showed almost maximum efficiency; thereafter, the SAH efficiency started gradually decreasing as jet diameter increases. In order to pinpoint the exact optimal jet diameter, the hourly SAH efficiency and hourly fluid outlet temperature were predicted and plotted against multiple jet diameter values, as per Figures 21(b) and 21(c). The results showed that the optimum jet diameter showing maximum SAH efficiency and maximum fluid outlet temperature concurrently is 0.003 m.

6.4. Optimum Mean Gap Thickness between the Absorbing Plate and the Jet Plate (H_g). The influence of H_g upon SAH efficiency at multiple solar radiation values is illustrated in Figure 22(a). The range of mean gap thickness that realizes maximum efficiency at multiple solar radiation falls within 0.01–0.04 m; thereafter, it starts demonstrating an inverse relationship with the mean gap thickness. In order to precisely determine the optimal mean gap thickness, the hourly SAH efficiency and hourly fluid outlet temperature are plotted at multiple mean gap thicknesses and are shown in Figures 22(b) and 22(c). It was seen that the optimum mean gap thickness showing maximum SAH efficiency and maximum fluid outlet temperature concurrently is 0.3 m.

6.5. Optimum Mean Gap Thickness between the Absorbing Plate and the Glass Cover (H_c). The influence of H_c upon the efficiency of the SAH at different solar radiation values is presented in Figure 23(a). The range of the mean gap thickness that realizes maximum efficiency at multiple solar radiation falls within 0.01–0.04 m; thereafter, it becomes inversely related to the mean gap thickness. In order to precisely determine the optimal mean gap thickness, the

hourly SAH efficiency and hourly fluid outlet temperature seen during a clear sky day are plotted against multiple mean gap thicknesses, as per Figures 23(b) and 23(c). It was shown that the optimal mean gap thickness showing maximum SAH efficiency and maximum fluid outlet temperature concurrently is 0.3 m.

7. Conclusions

This study involves the development of an energy balance model of the proposed design of SAH and comparing it to two previously reported SAH designs. The improvement to the thermal performance of the proposed SAH was evaluated in the context of fluid outlet temperature, temperature difference between the fluid outlet and absorber plate, and thermal efficiency. The proposed SAH reported the highest fluid outlet temperature and efficiency, followed by the SAH with the jet on the flat plate absorber and the SAH with the cross-corrugated plate, respectively. The fluid outlet and absorber plate temperatures of the proposed SAH converged significantly when close to a temperature difference of less than 3 K. Moreover, the statistical paired t-test results reported significant differences in the mean values of the SAH efficiency pre and postmodifications. It can be concluded that the proposed design significantly improved the thermal performance of SAH.

Abbreviations

L: Length of heater (m)
W: Width of heater (m)
H_g: Mean gap thickness between the absorbing plate and the jet plate (m)
A_a: Absorber area (m^2)
A_h: Heater area (m^2)
d_j: Jet diameter (m)
I: Solar insulation rate incident on the glass cover (W/m^2)
V_w: Wind velocity of the ambient air (m/s)
τ_c: Transmissivity of the glass cover
α_{ap}:   Absorptivity of the absorbing plate
η: Efficiency of the air heater
S: Solar radiation absorbed by the absorbing plate per unit area (W/m^2)
m_f: Air mass flow rate per unit area of heater (kg/s)
q_u: Useful energy gain (W/m^2)
c_p: Specific heat of air (kJ/kg K)
T_a: Temperature of air flowing through channel above bottom plate (K)
T_f: Temperature of air flowing through channel above jet plate (K)
T_{o1}:   Outlet air temperature through jet, above jet plate (K)
T_o: Outlet air temperature through heater (K)
T_{ap}:   Mean temperature of absorbing plate (K)
T_c: Mean temperature of the glass cover (K)
T_A: Ambient temperature (K)
h_w: Convection heat transfer coefficient from glass cover due to the wind (W/m^2 K)

$h_{r,c-s}$: Radiation heat transfer coefficient between the cover and the sky (W/m^2 K)

$h_{c,f-bp}$: Convection heat transfer coefficient between fluid and bottom plate (W/m2 K)

$h_{c,ap-f}$: Convection heat transfer coefficient between absorbing plate and the fluid (W/m2 K)

$h_{r,ap-bp}$: Radiation heat transfer coefficient between absorbing plate and the bottom plate (W/m2 K)

$h_{r,ap-c}$: Radiation heat transfer coefficient between glass cover and absorbing plate (W/m^2 K)

$h_{c,ap-c}$: Convection heat transfer coefficient between glass cover and absorbing plate (W/m^2 K)

k_i: Thermal conductivity of the insulation (W/m K)

b: Mean thickness of the insulation (m).

Conflicts of Interest

The authors declare that they have no conflicts of interest.

References

[1] N. K. Bansal, "Solar air heater applications in India," *Renewable Energy*, vol. 16, no. 1-4, pp. 618–623, 1999.

[2] W. Gao, W. Lin, T. Liu, and C. Xia, "Analytical and experimental studies on the thermal performance of cross-corrugated and flat-plate solar air heaters," *Applied Energy*, vol. 84, no. 4, pp. 425–441, 2007.

[3] M. A. Wazed, Y. Nukman, and M. T. Islam, "Design and fabrication of a cost effective solar air heater for Bangladesh," *Applied Energy*, vol. 87, no. 10, pp. 3030–3036, 2010.

[4] K. Prasad and S. C. Mullick, "Heat transfer characteristics of a solar air heater used for drying purposes," *Applied Energy*, vol. 13, no. 2, pp. 83–93, 1983.

[5] J. C. Han, "Heat transfer and friction in channels with two opposite rib-roughened walls," *Journal of Heat Transfer*, vol. 106, no. 4, pp. 774–781, 1984.

[6] S. Singh, S. Chander, and J. S. Saini, "Heat transfer and friction factor of discrete V-down rib roughened solar air heater ducts," *Journal of Renewable and Sustainable Energy*, vol. 3, no. 1, article 013108, 2011.

[7] R. Chauhan and N. S. Thakur, "Heat transfer and friction factor correlations for impinging jet solar air heater," *Experimental Thermal and Fluid Science*, vol. 44, pp. 760–767, 2013.

[8] S. K. Verma and B. N. Prasad, "Investigation for the optimal thermohydraulic performance of artificially roughened solar air heaters," *Renewable Energy*, vol. 20, no. 1, pp. 19–36, 2000.

[9] A. M. Elbreki, M. A. Alghoul, A. N. Al-Shamani et al., "The role of climatic-design-operational parameters on combined PV/T collector performance: a critical review," *Renewable and Sustainable Energy Reviews*, vol. 57, pp. 602–647, 2016.

[10] L. Goldstein and E. M. Sparrow, "Experiments on the transfer characteristics of a corrugated fin and tube heat exchanger configuration," *Journal of Heat Transfer*, vol. 98, no. 1, pp. 26–34, 1976.

[11] L. Goldstein and E. M. Sparrow, "Heat/mass transfer characteristics for flow in a corrugated wall channel," *Journal of Heat Transfer*, vol. 99, no. 2, pp. 187–195, 1977.

[12] M. Sethi and M. Sharma, "Effective efficiency prediction for discrete type of ribs used in solar air heaters," *International Journal of Energy & Environment*, vol. 1, no. 2, pp. 333–342, 2010.

[13] D. Gupta, S. C. Solanki, and J. S. Saini, "Heat and fluid flow in rectangular solar air heater ducts having transverse rib roughness on absorber plates," *Solar Energy*, vol. 51, no. 1, pp. 31–37, 1993.

[14] R. P. Saini and J. S. Saini, "Heat transfer and friction factor correlations for artificially roughened ducts with expanded metal mesh as roughness element," *International Journal of Heat and Mass Transfer*, vol. 40, no. 4, pp. 973–986, 1997.

[15] R. Karwa, S. C. Solanki, and J. S. Saini, "Heat transfer coefficient and friction factor correlations for the transitional flow regime in rib-roughened rectangular ducts," *International Journal of Heat and Mass Transfer*, vol. 42, no. 9, pp. 1597–1615, 1999.

[16] J. L. Bhagoria, J. S. Saini, and S. C. Solanki, "Heat transfer coefficient and friction factor correlations for rectangular solar air heater duct having transverse wedge shaped rib roughness on the absorber plate," *Renewable Energy*, vol. 25, no. 3, pp. 341–369, 2002.

[17] R. Karwa and G. Chitoshiya, "Performance study of solar air heater having v-down discrete ribs on absorber plate," *Energy*, vol. 55, pp. 939–955, 2013.

[18] A. Kumar, R. P. Saini, and J. S. Saini, "Development of correlations for Nusselt number and friction factor for solar air heater with roughened duct having multi v-shaped with gap rib as artificial roughness," *Renewable Energy*, vol. 58, pp. 151–163, 2013.

[19] S. Saurav and V. N. Bartaria, "Heat transfer and thermal efficiency of solar air heater having artificial roughness: a review," *International Journal of Renewable Energy Research*, vol. 3, no. 3, pp. 498–508, 2013.

[20] B. A. Meyer, J. W. Mitchell, and M. M. El-Wakil, "Convective heat transfer in Vee-trough linear concentrators," *Solar Energy*, vol. 28, no. 1, pp. 33–40, 1982.

[21] X. W. Zhao and Z. N. Li, "Numerical and experimental study on free convection in air layers with one surface V-corrugated," in *In Proceedings of the Annual Meeting of the Chinese Society of Solar Energy*, vol. 1991, pp. 182–192, 1991.

[22] J. A. Stasiek, "Experimental studies of heat transfer and fluid flow across corrugated-undulated heat exchanger surfaces," *International Journal of Heat and Mass Transfer*, vol. 41, no. 6-7, pp. 899–914, 1998.

[23] Y. Piao, *Natural, Forced and Mixed Convection in a Vertical Cross-Corrugated Channel*, University of British Columbia, 1992, Doctoral dissertation.

[24] S. Noorshahi, C. A. Hall III, and E. K. Glakpe, "Natural convection in a corrugated enclosure with mixed boundary conditions," *Journal of Solar Energy Engineering*, vol. 118, no. 1, pp. 50–57, 1996.

[25] W. Gao, *Analysis and Performance of a Solar Air Heater with Cross Corrugated Absorber and Back-Plate [MSc Thesis]*, Yunnan Normal University, Kunming, 1996.

[26] A. A. El-Sebaii, S. Aboul-Enein, M. R. I. Ramadan, S. M. Shalaby, and B. M. Moharram, "Investigation of thermal performance of-double pass-flat and v-corrugated plate solar air heaters," *Energy*, vol. 36, no. 2, pp. 1076–1086, 2011.

[27] M. N. Metwally, H. Z. Abou-Ziyan, and A. M. El-Leathy, "Performance of advanced corrugated-duct solar air collector compared with five conventional designs," *Renewable Energy*, vol. 10, no. 4, pp. 519–537, 1997.

[28] T. Liu, W. Lin, W. Gao, and C. Xia, "A comparative study of the thermal performances of cross-corrugated and v-groove

solar air collectors," *International Journal of Green Energy*, vol. 4, no. 4, pp. 427–451, 2007.

[29] J. C. Han, "Recent studies in turbine blade cooling," *International Journal of Rotating Machinery*, vol. 10, no. 6, pp. 443–457, 2004.

[30] M. K. Sung and I. Mudawar, "Experimental and numerical investigation of single-phase heat transfer using a hybrid jet-impingement/micro-channel cooling scheme," *International Journal of Heat and Mass Transfer*, vol. 49, no. 3-4, pp. 682–694, 2006.

[31] M. Rǎȗger, R. Buck, and H. Mǎžller-Steinhagen, "Numerical and experimental investigation of a multiple air jet cooling system for application in a solar thermal receiver," *Journal of Heat Transfer*, vol. 127, no. 8, pp. 863–876, 2005.

[32] N. Zuckerman and N. Lior, "Jet impingement heat transfer: physics, correlations, and numerical modeling," *Advances in Heat Transfer*, vol. 39, pp. 565–631, 2006.

[33] M. K. Sung and I. Mudawar, "Single-phase and two-phase hybrid cooling schemes for high-heat-flux thermal management of defense electronics," *Journal of Electronic Packaging*, vol. 131, no. 2, article 021013, 2009.

[34] M. Belusko, W. Saman, and F. Bruno, "Performance of jet impingement in unglazed air collectors," *Solar Energy*, vol. 82, no. 5, pp. 389–398, 2008.

[35] D. Klein and G. Hetsroni, "Enhancement of heat transfer coefficients by actuation against an impinging jet," *International Journal of Heat and Mass Transfer*, vol. 55, no. 15-16, pp. 4183–4194, 2012.

[36] W. Lin, W. Gao, and T. Liu, "A parametric study on the thermal performance of cross-corrugated solar air collectors," *Applied Thermal Engineering*, vol. 26, no. 10, pp. 1043–1053, 2006.

[37] C. Choudhury and H. P. Garg, "Evaluation of a jet plate solar air heater," *Solar Energy*, vol. 46, no. 4, pp. 199–209, 1991.

[38] D. G. Mayer and D. G. Butler, "Statistical validation," *Ecological Modelling*, vol. 68, no. 1-2, pp. 21–32, 1993.

[39] J. L. Hintze, "NCSS Statistical Software NCSS.com 208-1 © NCSS, LLC," 1981, Chapter 208 Paired T-Test.

[40] W. H. McAdams, "Heat transmission (No. 660.28427 M32)," 1954.

[41] X. Q. Zhai, Y. J. Dai, and R. Z. Wang, "Comparison of heating and natural ventilation in a solar house induced by two roof solar collectors," *Applied Thermal Engineering*, vol. 25, no. 5-6, pp. 741–757, 2005.

[42] W. C. Swinbank, "Long-wave radiation from clear skies," *Quarterly Journal of the Royal Meteorological Society*, vol. 89, no. 381, pp. 339–348, 1963.

Hardware Implementation of a Fuzzy Logic Controller for a Hybrid Wind-Solar System in an Isolated Site

Aymen Jemaa [ID],[1] Ons Zarrad,[1] Mohamed Ali Hajjaji [ID],[2,3] and Mohamed Nejib Mansouri[1]

[1]Unit of Industrial Systems Study and Renewable Energy (ESIER), National Engineering School, University of Monastir, Monastir, Tunisia
[2]Laboratory of Electronic and Microelectronic, University of Monastir, Monastir, Tunisia
[3]Higher Institute of Applied Sciences and Technology of Kairouan, University of Kairouan, Kairouan, Tunisia

Correspondence should be addressed to Aymen Jemaa; jemaa_aymen@yahoo.fr

Academic Editor: Joaquín Vaquero

In this paper, two main contributions are presented to manage the power flow between a wind turbine and a solar power system. The first one is to use the fuzzy logic controller as an objective to find the maximum power point tracking, applied to a hybrid wind-solar system, at fixed atmospheric conditions. The second one is to respond to real-time control system constraints and to improve the generating system performance. For this, a hardware implementation of the proposed algorithm is performed using the Xilinx System Generator. The experimental simulation results show that the suggested system presents high accuracy and acceptable execution time performances. The proposed model and its control strategy offer a proper tool for optimizing the hybrid power system performance which we can use in smart house applications.

1. Introduction

Wind and solar energies present the most famous renewable energy sources which attract attention due to the decreasing fossil fuel reserves and environmental property impact. The use of wind energy systems may not be technically viable at all sites because of low wind speeds and being more unpredictable than solar energy [1, 2]. The hybridization of these renewable energy sources can ensure the continuity of energy production and can be an economic solution for countries to develop and decrease the consumption of fossil (fuel and nuclear) energy sources. Research and development efforts in solar, wind, and other renewable energy technologies are necessary to continue improving their performance. Among the most important factors to consider, we can cite the maximum power point tracking (MPPT), which presents an essential factor of system performances.

In general, MPPT methods can be classified into two principal categories. The first one uses classic algorithms such as hill-climbing (HC), perturbation and observation (P&O), and incremental conductance (IncCond). For the second type, it is based on intelligent methods such as fuzzy logic, artificial neural network (ANN), and neurofuzzy, which is a combination between the two previous methods.

The idea to use the fuzzy logic controller is to control, respectively, the proportional-integral controller for wind turbines and the duty cycle (d) for solar energy to regularize successively the optimal rotor speed and the pulse-width modulation in the boost converter. This algorithm does not require a specific detailed mathematical model or linearization, about an operating point, and it is independent from system parameter variations.

For a subsystem wind turbine, the pitch angle of the turbine is synchronized according to the measured wind speed values in a fuzzy logic controller. For the solar subsystem, the impedance of the photovoltaic cell's output is equal to the values of the impedance measured on the load impedance in the fuzzy logic controller. Both controllers are applied to boost the performance. The control of the two subsystems presents one of the main objectives of this manuscript.

To get the optimum performance and effective power of the wind-solar turbine at fixed atmospheric conditions, a second objective is studied, which is the implementation of both controllers. This implementation is designed on Xilinx, which has the Xilinx System Generator (XSG) tools to facilitate the design. The implementation on a field-programmable gate array (FPGA) circuit with XSG allows us to find the solution of complexity, parallelism, and calculation time.

The XSG handles much of the routing and placement timing. The FPGA design flow eliminates the complex and time-consuming floor planning, placement and routing, and timing analysis. The FPGA runs up to 500 MHz with a superior performance. The unprecedented logic density increases, and a host of other features like embedded processors, DSP blocks, clocking, and high-speed serial raise the performance.

This paper is planned on six sections: The first section is an introduction which contains a generality of a wind-solar system and the purpose of this work. Section 2 is the related work and the model description of the hybrid system components. Section 3 describes the XSG tools, the conception, and the hardware architecture of the different blocks of a hybrid system with an XSG design. Section 4 contains the results and discussions of the simulation. The results of a fuzzy logic controller implementation on FPGA are given in Section 5 before the conclusion which is the final part of this paper.

2. Related Work and Proposed Hybrid-System Model

2.1. Related Work. Tracking the various MPPT algorithms permits fixing and extracting the maximum power, which has been dealt with in the literature.

In [3], a photovoltaic- (PV-) wind system was controlled and compared with two approaches. The first one was a hybrid system with HC and P&O algorithms. The second one used a fuzzy logic controller. The based power generation of the system was 3.1 kW.

In [4], two connection models of a solar-wind system were studied with fuzzy logic controllers. The first model had just one hybrid system connected to the main grid. The second one had the same system with a 75 kW load connected to the grid.

In [5], a standalone PV-wind energy conversion system was sized and optimized. P&O and fuzzy logic algorithm controllers were used to get the optimum mechanical speed of turbine and duty cycles of a DC-DC converter.

2.2. Proposed Solar-Wind System. In this paper, a hybrid power system is designed with two renewable energy sources, solar and wind, which are connected with a conventional energy source. A wind system with a solar photovoltaic one is the best hybrid combination of all renewable energy systems and is suitable for most applications [6]. Furthermore, the system is dedicated to isolates sites that are not

connected to the grid, so the only energy source is the hybrid power system.

Figure 1 describes the schematic diagram of the wind-solar hybrid energy conversion system studied in this paper. Its principal blocks are a PV generator, a wind turbine, a permanent magnet synchronous generator (PMSG), a rectifier, boost converters, a continuous DC-DC bus, and fuzzy logic controllers.

2.3. Wind Turbine Model. Depending on the aerodynamic characteristics, the wind turbine mechanical power is given by

$$P = \frac{1}{2}\pi\rho C_p(\lambda, \beta)R^2 V_v^3, \quad (1)$$

where ρ is the air density, C_p is the power coefficient, λ is the tip-speed ratio, β is the pitch angle, R is the turbine radius, and V_v is the wind speed.

The form of the used power coefficient C_p is given in

$$C_p(\lambda) = -0.2121\lambda^3 + 0.0856\lambda^2 + 0.2539\lambda. \quad (2)$$

It depends only on one variable, which is the tip-speed ratio λ. The pitch angle β is usually the angle between the turbine blades and its longitudinal axis. In this work, β is set to zero. The tip-speed ratio λ is considered as the linear speed form of the rotor to the wind speed. The power value extracted from the wind turbine system will be at its utmost when the power coefficient C_p is at its maximum at a defined value of the tip-speed ratio λ.

Accordingly, for each wind speed, there is an optimum rotor speed value where the maximum power is extracted from the wind. Therefore, we can say that C_p is maximal at a particular λ_{opt}. The expression of the tip-speed ratio λ is presented in

$$\lambda = \frac{\Omega R}{V_v}, \quad (3)$$

where Ω is the turbine's angular speed.

The variable-speed wind form studied in this work is expressed in (4) and illustrated in Figure 2.

$$V_v(t) = 7.5 + 0.2 \sin(0.1047t) + 2\sin(0.2665t) \\ + \sin(1.2930t) + 0.2\sin(3.6645t). \quad (4)$$

The analyzed wind turbine and PMSG have their electric specifications given, respectively, in Tables 1 and 2.

Figure 3 presents the plot of the $C_p(\lambda)$ characteristic where a maximum point wind speed with the coordinate (C_{pmax}, λ) is equal to (0.15, 0.78).

2.4. Permanent Magnet Synchronous Generator (PMSG). To define the PMSG model, we use the following simplifying assumptions [7]:

(i) The stator is connected on a star and is neutral in the air to eliminate the homopolar component of currents.

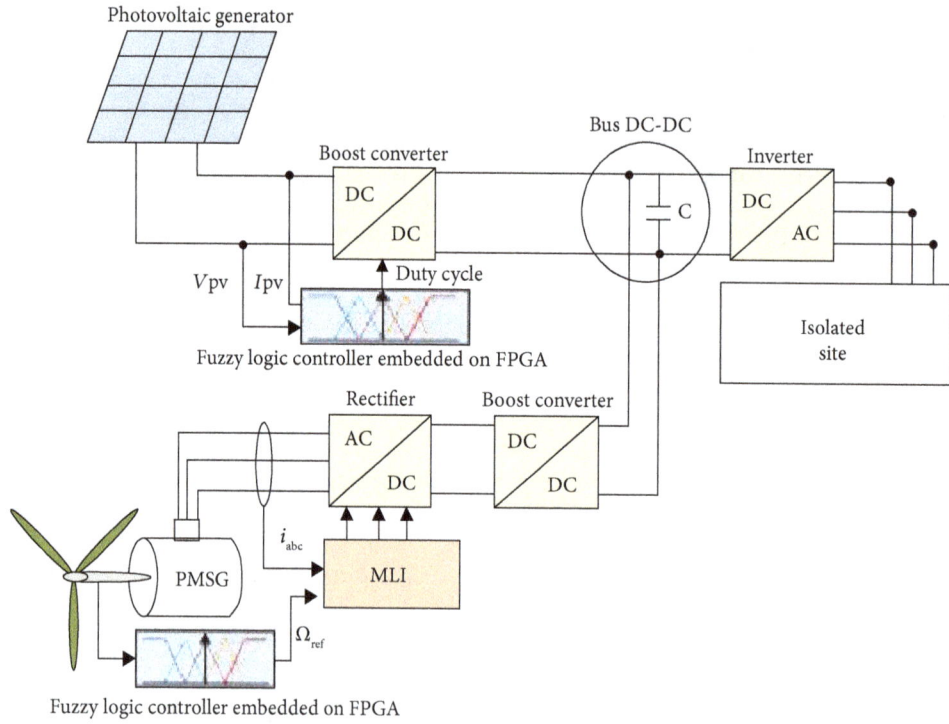

FIGURE 1: Hybrid wind-solar energy conversion system for an isolated site.

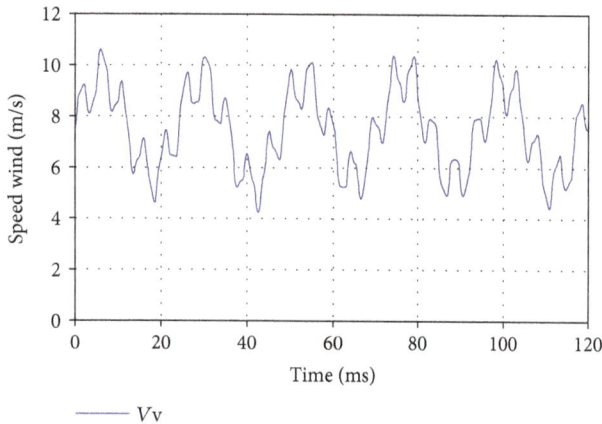

FIGURE 2: Allure of variable-speed wind.

TABLE 1: Wind turbine specifications.

Air density "ρ"	1.2 kg/m^3
Turbine radius "R"	0.5 m
Turbine height "H"	2 m
Inertia constant "J"	16 kg/m^2
Friction factor "f"	0.01 kg·m/rad
Maximum coefficient of power "C_{pmax}"	0.15
Optimal tip-speed ration "λ_{opt}"	0.78

TABLE 2: PMSG specifications.

Number of rotor pole pairs	2
Stator phase resistance	0.137 Ω
Stator phase inductance	0.0027 H
Inertia constant	0.1 kg·m^2
Friction factor	0.06 kg·m/rad

(ii) The saturation of the magnetic circuit is neglected, which leads to expressing the magnetic fluxes as linear functions of the phase currents.

(iii) The distribution of the electromotive force (f.e.m) in the air gap is sinusoidal, and the harmonics of space are then neglected.

(iv) The variation in resistance as a function of temperature is neglected.

(v) Hysteresis and current losses are neglected.

The generator is modeled by the following voltage equations given by (5) in the rotor reference frame (d, q) axes.

$$V_d = R_s i_d + L_d \frac{di_d}{d_t} - p\Omega L_q i_q,$$

$$V_q = R_s i_q + L_q \frac{di_q}{d_t} + p\Omega L_d i_d + p\Omega \phi_f, \tag{5}$$

where i_d, i_q, v_d, and v_q are the currents and voltages, respectively; L_d and L_q are the equivalent stator inductances in the (d, q) axes, respectively; R_s is the stator resistance; $\omega = p\Omega$

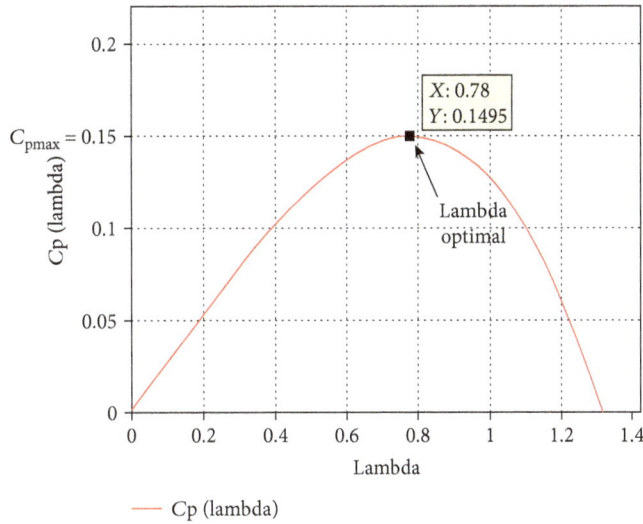

FIGURE 3: Characteristic curve of $C_p(\lambda)$ in MATLAB/SIMULINK environment.

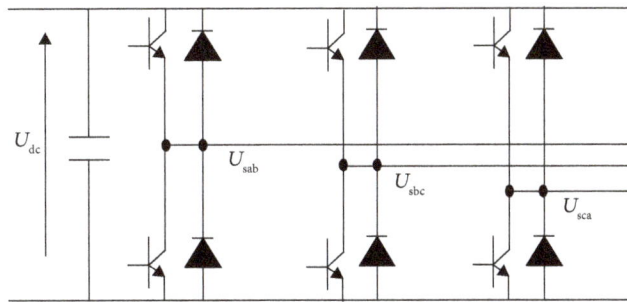

FIGURE 4: PWM rectifier circuit diagram.

is the electric frequency related to the mechanical speed, and ϕ_f is the permanent magnetic flux produced by the rotor magnets.

The electrical torque (T_e) applied to the PMSG rotor can be expressed by

$$T_e = p i_q \left(\left(L_d - L_q \right) i_d + \phi_f \right),\tag{6}$$

where p is the pair pole number of the machine. The magnetic flux is a constant that depends on the material used for the realization of the magnets.

The mechanical expression is represented by

$$J \frac{d\Omega}{dt} + f\Omega = T_e - T_r \tag{7}$$

where J is the value of the total inertia of the rotor, f is the viscous friction coefficient, and T_r is the electromagnetic torque.

2.5. *Pulse-Width Modulation.* Pulse-width modulation (PWM) is a static converter used to rectify an alternating signal and then transform it into a continuous signal. In order to reduce the simulation time and simplify the modeling, the rectifier is modeled by ideal switches where there is a

zero resistance in the on state, an infinite resistance in the off state, and an instant response to control signals.

PWM is composed of six components; each one comprises two switching cells consisting of a transistor and a diode, as depicted in Figure 4. Accordingly, the current passes in both directions.

The main function of the PWM switches is to provide the connection between the AC voltage produced by the wind subsystem and the DC bus.

The switches' states are complementary, as defined by

$$\begin{aligned} S = 0 \quad &\text{if } i_j = +1, \\ S = 1 \quad &\text{if } i_j = -1, \\ &j = a, b, c. \end{aligned} \tag{8}$$

In general, the voltage equation of the outputs can be written in the form expressed by

$$U_n = U_{dc} \left(S_n - \frac{1}{3} \sum_{n=a}^{c} S_n \right), \tag{9}$$

where S_n is equal to 0 or 1 depending on the state of the switches, and n is a, b, or c.

The equations of voltages for the balanced three-phase system without a neutral are given by

$$\begin{bmatrix} e_a \\ e_b \\ e_c \end{bmatrix} = R \begin{bmatrix} i_a \\ i_b \\ i_c \end{bmatrix} + L \frac{d}{dt} \begin{bmatrix} i_a \\ i_b \\ i_c \end{bmatrix} + \begin{bmatrix} U_a \\ U_b \\ U_c \end{bmatrix}, \tag{10}$$

where e_n and i_n are, respectively, the voltages and currents of the inputs a, b, and c.

2.6. Solar Model. The solar cells can be divided into two types, bulk and thin film [2, 8–11]. Solar cells are made of different materials and have different efficiency values. Based on cost and generation efficiency, we choose monocrystalline silicon for our work among the different materials provided in Table 3.

The characteristic *I-V* for a PV module is given by

$$\begin{aligned} I = n_p I_{PV} \\ - n_p I_0 \left(\frac{T_c}{T_{ref}}\right)^3 e^{\left(qE_g/ak\right)\left((1/T_{ref})-(1/T_c)\right)} \left[e^{\left(q(VI+IR_S)/akt_c n_S\right)-1}\right] \\ - \frac{V + IR_S}{R_p}, \end{aligned} \tag{11}$$

where I_0 is the reverse saturation current of the diode, E_g is the value of the energy band of the diode material, a is the ideality factor of the diode, k is the Boltzmann's constant, q is the electron charge, and T_c is the cells' temperature in Kelvin.

The analyzed PV generator has the electric specifications given in Table 4.

2.7. Boost DC-DC. Generally, the objective of the DC-DC boost converter is to help the power renewable energy voltage [13], which is in our work the solar and wind energies, to converge the reference value provided by the MPPT algorithm. Perturbing the duty ratio of a PV generator signal, which is fed into the converter, can minimize the proportion between the input voltage and the output desired voltage. The general topology of the boost DC-DC converter is shown in Figure 5.

The equation system of the boost DC-DC converter is done by

$$\begin{aligned} i_L &= i - c_1 \frac{dv_i}{d_t}, \\ i_0 &= (1-d)i_L - c_2 \frac{dv_0}{dt}, \tag{12} \\ v_i &= (1-d)v_0 + L \frac{di_L}{dt}. \end{aligned}$$

We use two DC-DC boost converters in the output of the solar and wind subsystems to rebuild optimized voltages. The currents of the DC-DC boost converters are transmitted to the DC-DC bus.

TABLE 3: Comparison of solar cell efficiency [10].

Solar cell	Monomer efficiency (%)	Module efficiency (%)
Monocrystalline silicon	22	10–15
Polycrystalline silicon	18	9–12
Boron-phosphorus compound	30	17
Thin-film amorphous silicon	13	10
Thin-film Cu-In	19	12
Thin-film Cd-Te compound	16	9

TABLE 4: Photovoltaic generator specifications [12].

Rated power	60 W
Current at maximum point	3.25 A
Voltage at maximum point	16.8 V
Short circuit current	3.56 A
Short circuit voltage	21.6 V
Number of cells in parallel	1
Number of cells in series	36

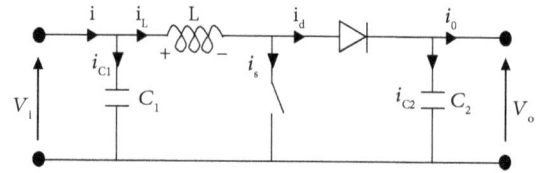

FIGURE 5: Circuit diagram of boost converter.

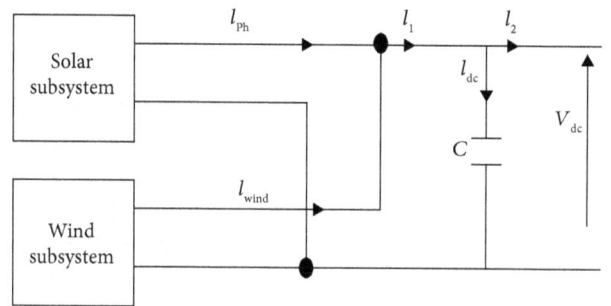

FIGURE 6: Electrical diagram of the continuous bus.

2.8. DC-DC Bus. The coupling of the two subsystems (solar and wind) is provided via a continuous bus. As illustrated in Figure 6, the DC bus is represented by the capacitor C connected to both subsystems. The role of the C capacitor integrated in the bus ensures the regulation of the ripple.

In our system, the photovoltaic energy is connected to the load via an MPPT-controlled DC-DC converter. The wind energy is connected to the load by the controlled PWM rectifier.

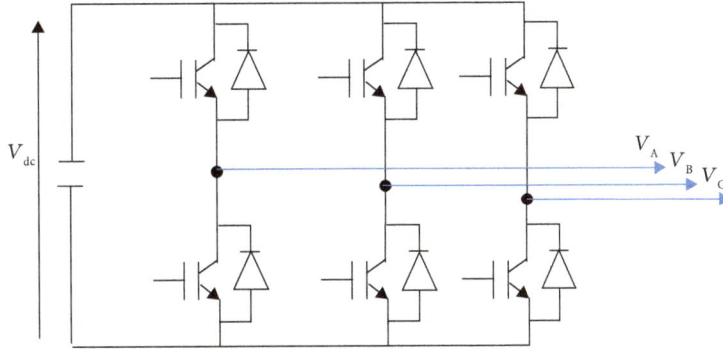

FIGURE 7: Structure of a three-phase voltage PWM inverter.

On the basis of Figure 6 and the equation of meshes, we can establish the following relations.

$$\frac{dV_{dc}}{dt} = \frac{1}{c}I_{dc},$$

$$I_1 = I_{Ph} + I_{wind} = I_{dc} + I_2,$$

$$I_{dc} = I_1 - I_2,$$

$$V_{dc} = \frac{1}{c}\int_{t2}^{t_1} I_{dc}dt + V_{dc0}.$$

(13)

2.9. Inverter. The inverter is the last output of our system. The role of this block is to transform the single-phase to a three-phase signal.

The inverters are static converters of electrical energy from DC to AC. The function of an inverter is the inverse of a rectifier; that is, for a DC voltage given at the input, the inverter performs the voltage cutting by means of semiconductors (transistors or thyristors) in order to obtain an AC voltage that can be adjusted in frequency and effective values.

The AC waveform of the output voltage is determined by the ripple system. The most used inverters are in PWM. The type of inverter chosen for our work is presented in Figure 7.

By applying the mesh equations, we obtain the compound voltages between phases given by

$$V_{AB} = (F_1 - F_2)V_{dc},$$

$$V_{BC} = (F_2 - F_3)V_{dc},$$

$$V_{CA} = (F_3 - F_1)V_{dc},$$

(14)

where F_1, F_2, and F_3 respectively represent the state of the switches K1, K2, and K3, and V_{dc} presents the output of a continuous bus.

Assuming that the input load is balanced, the output constitutes then a balanced system given in

$$V_A + V_B + V_C = 0.$$

(15)

The input current I_{dc} is expressed as a function of current outputs I_A, I_B, and I_C, as expressed in

$$I_{dc} = F_1I_A + F_2I_B + F_3I_C.$$

(16)

2.10. MPPT Fuzzy Logic Controllers. The implementation of classic algorithms is simple and is independent from turbine characteristics [14], but there still exist issues like the selection of the step size. The use of a big step size can define the MPPT fast, but it can result in severe oscillations around the MPPT. Reducing the perturbation step size slows down the MPPT process mostly where the wind speed varies fast despite the fact that the oscillations around the MPPT can be minimized [15, 16]. To solve this conflicting situation, we use in this paper a fuzzy logic control algorithm that can realize a variable step-size control.

The fuzzy logic control can use a large step size when the operating point is far away from the maximum power point, whereas the step can be minimized when the algorithm converges to the maximum power point [17, 18]. Thus, we can say that the fuzzy logic control can dynamically change its step depending on the energy input conditions.

Generally, The MPPT controller is a functional element of the PV and wind systems, which enables searching the operating point of the PV generator and the wind turbine under variable load and atmospheric conditions.

For the solar subsystem, the MPPT is based on the circuit maximum power transfer requirements: the objective is to check the point where the PV cell's output impedance is equal to the load impedance. The duty cycle is controlled by the MPPT controller for the pulse-width modulation block. This controls the power converter (DC-DC) to deliver a maximum power to the DC load bus.

The structure of the fuzzy controller used in the solar subsystem is depicted in Figure 8. In a MATLAB/SIMULINK (V.R2012b) environment, there is a fuzzy toolbox that allows the user to manage this structure and formulate fuzzy rules. Using this tool, we can configure the used command.

A controller based on a fuzzy logic algorithm is composed of three stages: fuzzification, rule base, and defuzzification. During fuzzification, the numerical input variables P_{pv} and V_{pv} are converted into linguistic variables based on a membership function. There is a block for calculating the error (E) and the change of the error (dE), expressed, respectively, in (17) and (18), at sampling instant k, as follows:

$$E(k) = \frac{dP}{dV} = \frac{P(k) - P(k-1)}{V(k) - V(k-1)},$$

(17)

$$dE(k) = E(k) - E(k-1),$$

(18)

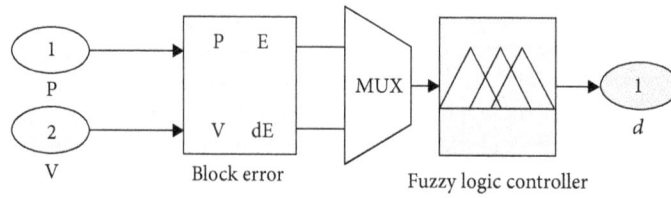

FIGURE 8: SIMULINK model of fuzzy logic controller of solar subsystem.

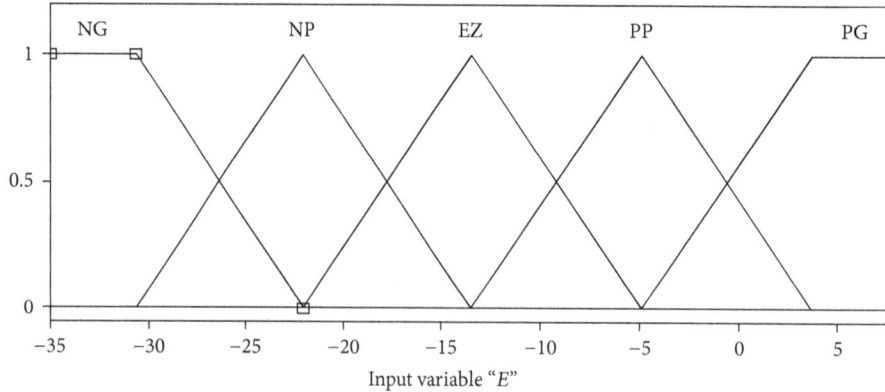

FIGURE 9: Member functions of solar input E.

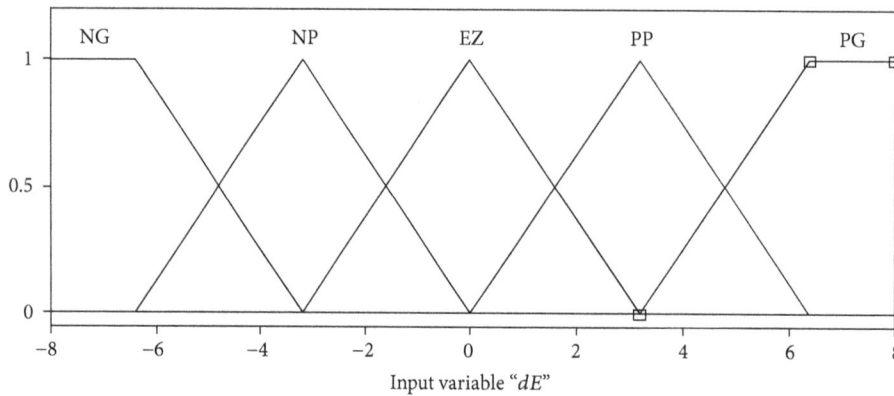

FIGURE 10: Member function of solar input dE.

where $P(k)$ and $V(k)$ are the power and the terminal voltage delivered, respectively, by a PV module.

The value of the error $E(k)$ determines the exact MPPT controller output according to the sign. The MPPT controller can decide in this way what the variation in the duty cycle (decrease or increase the speed of convergence) will be, which must be imposed on the DC-DC boost converter to approach the maximum power point. The value result of $E(k)$ and $dE(k)$ are converted to the linguistic variables, then the output of the fuzzy logic controller, which is d, can be looked up in a rule-base table.

The member function of the input variables (E and dE) of the solar fuzzy logic controller with MATLAB fuzzy tools is five member functions for each one. They are parametrized for E and dE, as shown in Figures 9 and 10, respectively.

TABLE 5: Fuzzy logic controller inference rules.

E			dE		
	NB	NS	ZE	PB	PS
NB	PB	PB	PB	PB	PB
NS	PB	PS	PS	PS	ZE
ZE	PS	PS	ZE	NS	NS
PB	NS	NS	NS	NS	NS
PS	NB	NS	NS	NS	ZE

The linguistic variables assigned to the duty factor d for the different combinations of $E(k)$ and $dE(k)$ are established according to our knowledge given in Table 5.

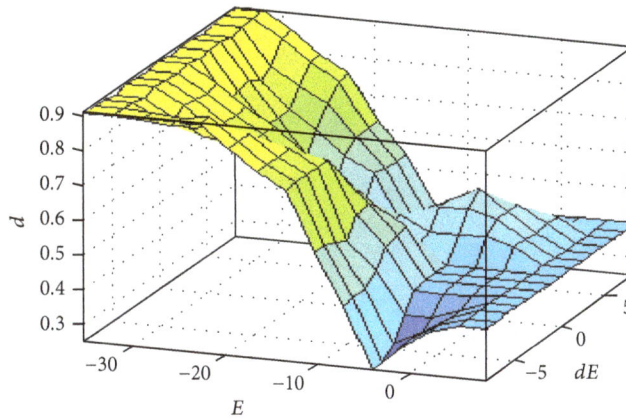

FIGURE 11: 3D surface of output duty cycle (d) values.

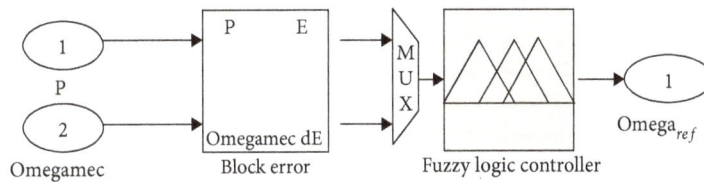

FIGURE 12: SIMULINK model of a fuzzy logic controller of a wind subsystem.

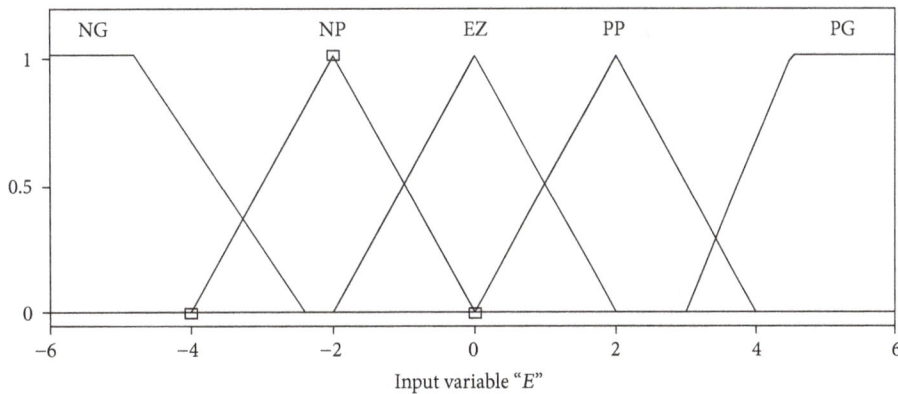

FIGURE 13: Member function of wind input E.

In this table, the linguistic variables used for the fuzzy logic controller are PB (positive big), PS (positive small), ZE (zero), NS (negative small), and NB (negative big).

The Mamdani type and inference rules with logical Min–Max operators are chosen for the fuzzy logic controller.

By applying the rules, the fuzzy tools generate the result output of each pair (E, dE), as described in Figure 11.

In the second case of the wind subsystem, to control the MPPT, we use the same block error inputs where $P(k)$ is the power output delivered by the wind turbine module and $\Omega(k)$ is the wind speed of the module. According to the sign of $E(k)$, the MPPT controller can decide what the variation in the wind speed (Ω_{ref}) will be, which must be imposed.

We choose the same five linguistic variables for the MPPT controller [19]. For the inference rules of the MPPT

controller, we choose the Mamdani type inference rules with logical Min–Max operators.

On defuzzification, the fuzzy logic controller output is converted from a linguistic variable to a numerical one while still using the membership function [20].

In Figure 12, the structure of the MPPT fuzzy logic controller designed to command the wind subsystem is shown.

Each member function of the input variables (E and dE) of the wind fuzzy logic controller with MATLAB fuzzy tools is parametrized for E and dE, as illustrated in Figures 13 and 14, respectively.

By applying the rules, the fuzzy tools generate the result output of each pair (E, dE), as described in Figure 15.

All the other blocks (the PV generator, the wind turbine, the PMSG, the rectifier, the boost converters, the DC-DC

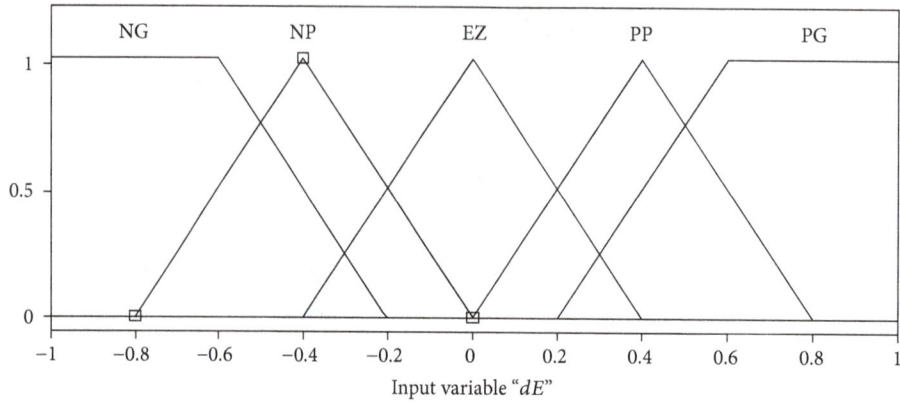

FIGURE 14: Member function of wind input dE.

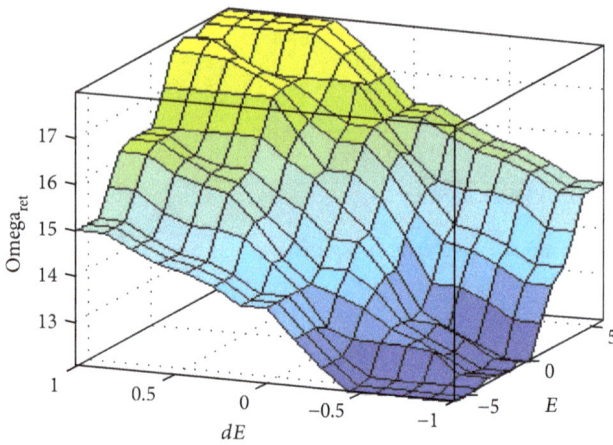

FIGURE 15: 3D surface of output wind speed reference values.

bus, etc.) are designed according to the mathematical equations of each component, as described above.

3. XSG Conception

3.1. Introduction. The classic FPGA implementation methodology consists of two main steps [21]. In the first one, the command is modeled and simulated using the MATLAB/ SIMULINK software tools in our work. The second step is dedicated to the hardware architecture design and the HDL description which is performed manually. The designers that are not familiar with the HDL-coding process can consume a lot of time in this step. In this paper, we use the XSG tool for the development of a digital signal processor (DSP). It allows a high-level implementation of DSP algorithms on an FPGA circuit. It consists of a graphical interface library on MATLAB/SIMULINK used for modeling and simulating a dynamic system process [22]. We can find the library of logic cores, specific for FPGA implementation, provided to be configured and used according to the designer's requirements. Next, the implementation of system generator blocks, the Xilinx Integrated Software Environment (ISE) is used to create "netlist" files. The latter serves to convert the logic design into a physical file and generate the bitstream that will

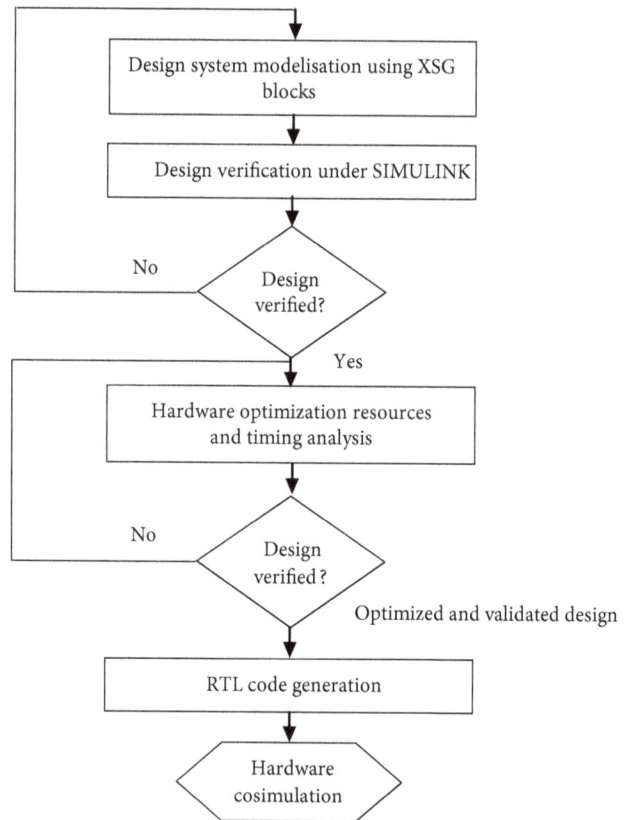

FIGURE 16: Flow of system generator tool.

be transferred to the target device through a standard JTAG (Joint Test Action Group) connection. The simulation in MATLAB XSG allows the test of the FPGA application with the same conditions as with the physical device and with the same data width of variables and operators. The simulation result can be bit-to-bit and cycle accurate, which means that each single bit is accurately processed and each operation will take the exact number of cycles to process.

Figure 16 presents the processing flow of the system generator tool from the hardware architecture design to the generation and cosimulation process. In this paper, the

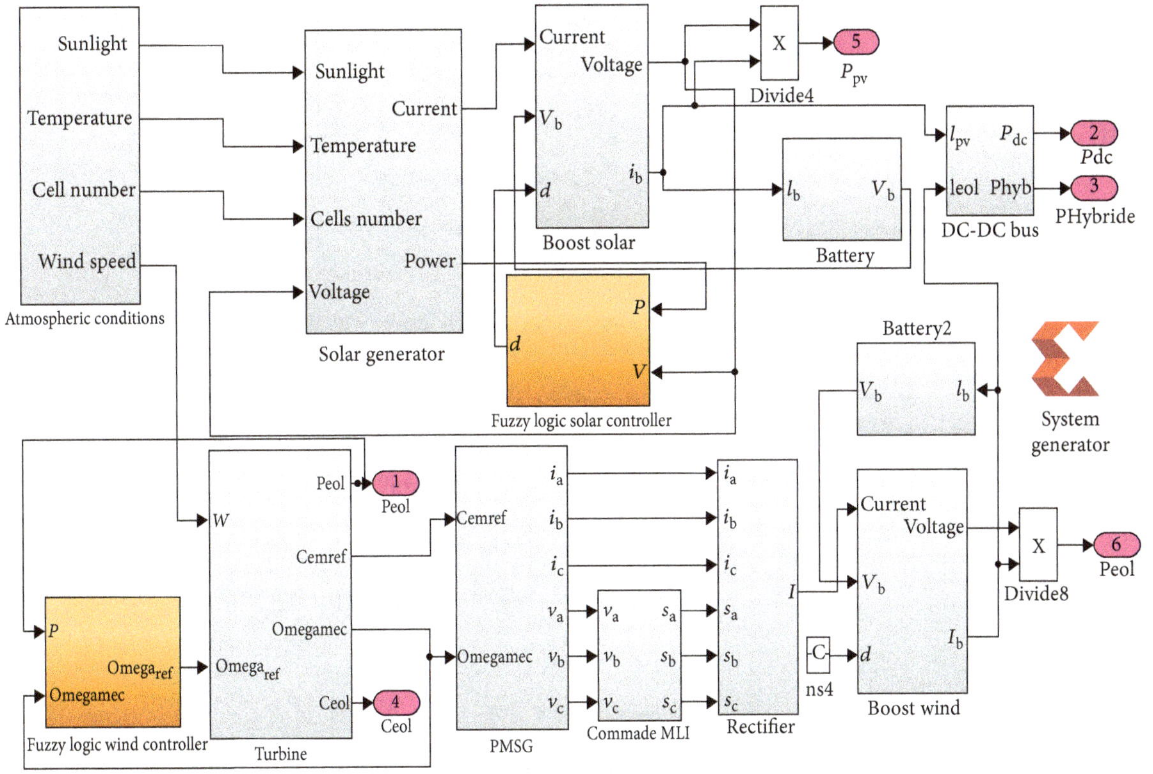

FIGURE 17: Wind-solar system designed model using XSG block.

architecture is designed using XSG components and the blocks will be simulated under SIMULINK to compare the results obtained in both cases, the architecture designed with a MATLAB code algorithm and the one designed with a hardware module.

To simulate our wind-solar system, we must choose support processing. We can use, for example, FPGA or DSP. Both approaches are markedly different, so we have studied it to contribute the right choice for our system in this work.

Firstly, translating the block diagram to the FPGA may well be simpler than converting it to a C code for the DSP. Furthermore, the FPGA does not have a fixed hardware structure as in the DSP; it is defined by the user concept. Finally, the FPGA, contrary to the DSP, allows the processes to be done simultaneously, which means that parallel processing is done according to the HDL code. Seeing that our system requires parallel processing and a flexible structure, we choose the FPGA support.

By using the XSG tool block, maximum precision is necessary on configuring all blocks, between the gateways, to run with a full output type. Word-length optimization is a key parameter that permits reducing the utilization of FPGA hardware resources while maintaining a satisfactory level of accuracy. For that reason, successive simulation iteratively reduces the output word length, which is made until achieving the minimum word length while ensuring the same maximum precision. The fundamental scalar signal type in MATLAB/SIMULINK is a double-precision floating-point number, and only the processing block is made based on system generator blocks that operate on Boolean and fixed-point values. A step of adaptation and interfacing

between the input and output blocks is necessary. Fortunately, XSG offers a simple interfacing using the predefined "gateway-in" and "gateway-out" blocks provided by the Xilinx blockset library.

3.2. General Structure of Wind-Solar System with XSG. To translate our system for the simulation with XSG, we need to change the specific blocks with an equivalent existing Xilinx library. The blockset in Figure 17 describes the global wind-solar system model designed with fuzzy logic controllers in a MATLAB/SIMULINK environment with XSG blocks.

Adding the system generator and translating the two blocks, the fuzzy logic controllers of the solar and wind subsystems are necessary to pass the XSG simulation. We use the Xilinx library to change the MATLAB/SIMULINK component with adequate XSGs and to add gateway inputs and gateway outputs. The fuzzy toolbox does not exist in the XSG library, so we build it. It is composed of Min–Max inference rules, linguistic variables (PB, PS, ZE, NS, and NB) and defuzzification. We use a MUX from the Xilinx library to convert the output to numeric values with the gravity center method. The fuzzy built controller is integrated on wind and solar subsystems in the same way because it depends on E and dE. A description of the function of each subsystem (wind and solar fuzzy logic controllers) will be given in the following sections.

3.3. Hardware Architecture of Fuzzy Logic Controller. To build the solar fuzzy logic controller, we use the Xilinx library. As shown in Figure 18, we have two gateway inputs

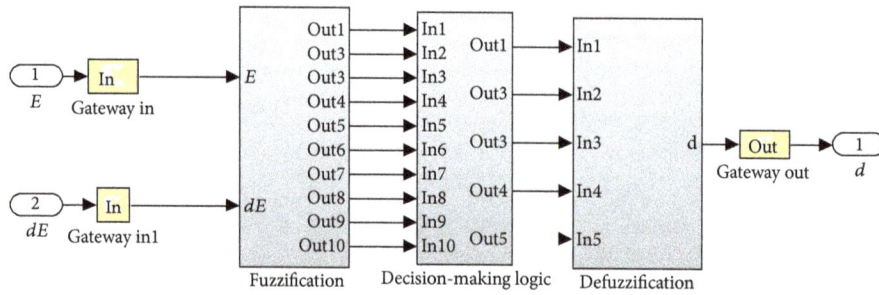

FIGURE 18: Architecture of solar fuzzy logic controller with XSG.

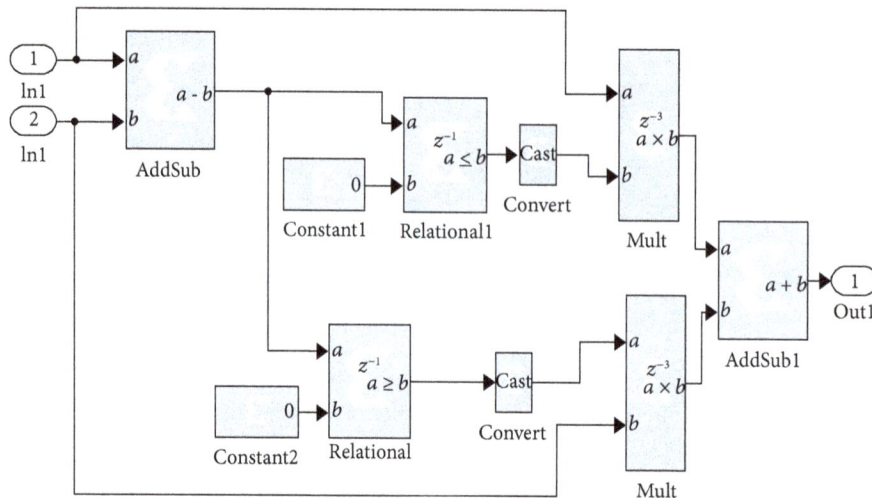

FIGURE 19: Min block built with Xilinx library.

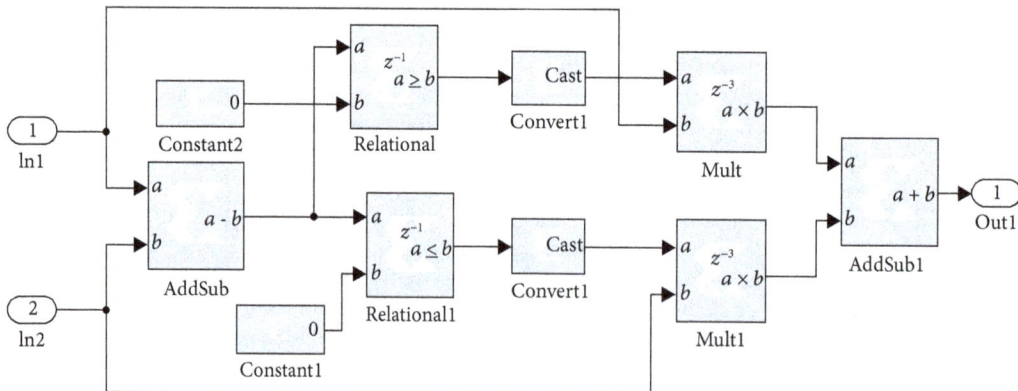

FIGURE 20: Max block built with Xilinx library.

(E, dE), a gateway output (d), and three subsystems which are fuzzification, decision-making logic, and defuzzification, where they function with the same steps used by the MATLAB/SIMULINK tool. The same procedure is applied to control the bitch angle for the wind fuzzy logic controller.

In the subsystem fuzzification, we model the five linguistic variables; then in the decision-making logic subsystem, we apply the Min–Max inference to the linguistic variables. For each gateway input, we have 25 rules, hence 50 rules for both

gateway inputs. We reuse the Min–Max inference until we have only one gateway output.

Figures 19 and 20 respectively present the Min and Max blocks built with the Xilinx library.

For linguistic variables, we cite PS as an example, which is described in Figure 21.

During the defuzzification, a MUX is used to convert the gateway output to a numeric value. Figure 22 shows the defuzzification block built with XSG components.

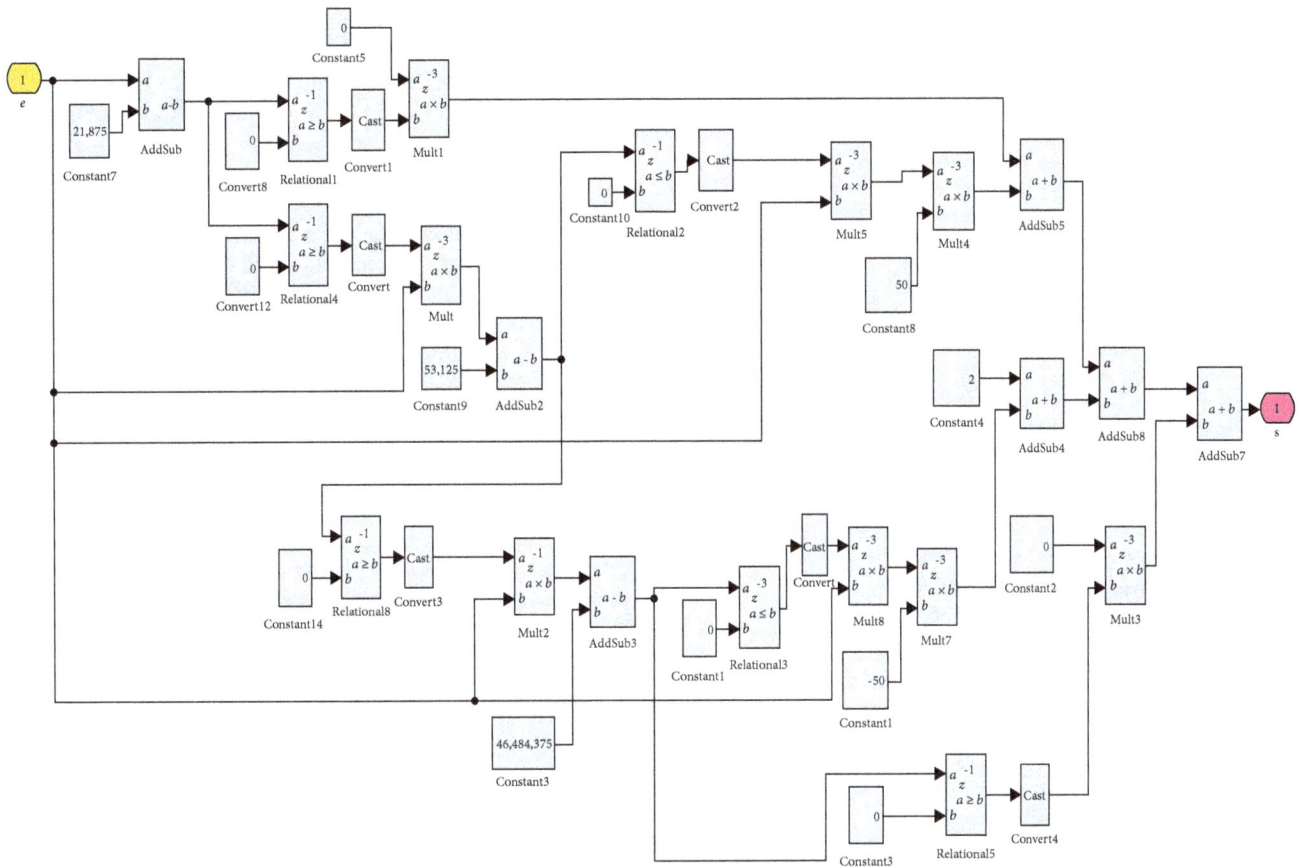

FIGURE 21: Linguistic variable PS built with Xilinx library.

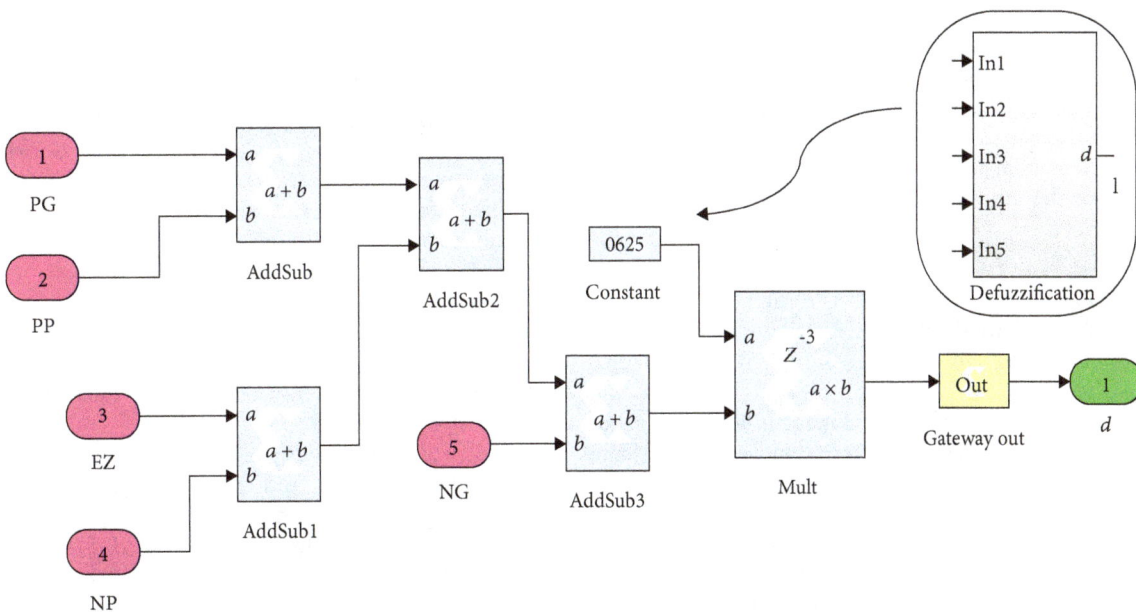

FIGURE 22: Defuzzification block built with Xilinx library.

4. Simulation Results

4.1. Simulation of Controller Output. Figures 23 and 24 respectively present the simulation results of the duty cycle output of the solar fuzzy logic controller and the wind speed (omega$_{ref}$) output of the wind fuzzy logic controller. As can be seen, the outputs are numerical values applied to adjust the performance of subsystems.

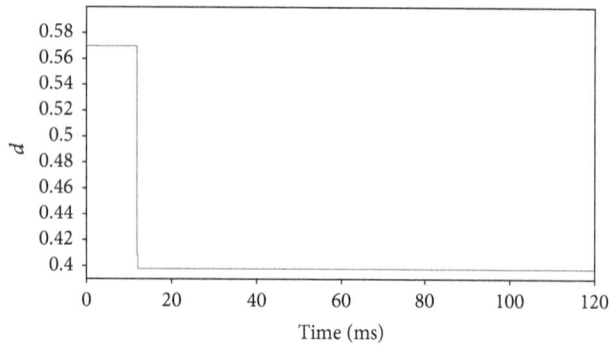

FIGURE 23: Fuzzy logic controller output of solar subsystem.

The duty cycle (d) and the wind speed (omega$_{ref}$) stabilization values are, respectively, 0.41 and 15 m/s.

4.2. Simulation of Power Output. To analyze the results, we compare the system with XSG and the system designed with MATLAB/SIMULINK block simulations on fixed atmospheric conditions.

The values of the different atmospheric conditions used for the simulation of the designed wind-solar system are given on Table 6.

Figure 25 compares between power versus time evolution obtained with MPPT controllers in the two cases: with simple block MATLAB/SIMULINK (PHyb) and with XSG. As is noted, the maximum power value is obtained at the same time, but the stabilization is quicker for the system with XSG. Thus, we can say that the system model with XSG uses its advantage of parallel processing to reach the stabilization form.

The curve of the hybrid power output with XSG using fuzzy logic controllers shows a good accuracy of the proposed algorithm since the stabilization at time is 12 ms.

In the system with the XSG library, the forms are the same as in the system with simple blocks. Consequently, we can conclude that the synchronization between the XSG blocks is perfect.

4.3. Simulation of Inverter Output. The intersection of the signals (carrier and modulator) gives the switching times. Figure 26 illustrates the switch output signal of the inverter switches.

The output signal is 1 if the modulator is larger than the carrier and 0 in the opposite case. The output signal therefore changes the state (0 or 1) at each intersection of both signals. Figure 27 describes the two signals to be compared.

The three voltages of the resulting phases of the inverter are phase shifted by $2\Pi/3$. The frequencies of the carrier and modulating signals are successively 50 Hz and 5 Hz. The purpose of this choice is to extract a maximum intersection. Figure 28 schematically depicts the appearance of a phase voltage at the output of the inverter.

These results of the inverter output coincide with the system modeling by using the MATLAB/SIMULINK library. In this step, we can say that the cosimulation with XSG is satisfied.

The picture of the experimental setup is shown in Figure 29.

To ensure greater efficiency of our model, with fuzzy logic controllers, we compare it with other work results.

In Table 7, a comparison with some references in terms of power delivered by the hybrid system and the inverter signal output is provided.

As represented in Table 7, a comparison with some references uses different topologies in terms of tracking efficiency and response time. After ensuring that the system progress is validated and evaluated, we can move now to implement the commands on FPGA circuits.

5. Implementation on FPGA

In the first time, we start with a standard 36-bit fixed-point format in various blocks of the XSG wind-solar system so as to ensure the same performance as the floating-point format of MATLAB simulation [24]. In a second step, by using the test-error approach, we try to reduce the width of the data while keeping the performance and the synchronization between blocks. The utilized general fixed-point format is 24 bits [25, 26] which guarantees the same performance as MATLAB.

Once completed and verified, the XSG architecture can be automatically mapped to hardware implementation on FPGA. We choose Virtex-6-XC6VLX315T and we use the "generate" button from the settings window of the system generator tool. After that, the VHDL is automatically generated [27]. The characteristics of the FPGA Virtex-6-XC6VLX315T that we use in the prototype are shown on Table 8.

We use an HDL code generated into the XILINX ISE design software to prepare the transfer of the command controller to the FPGA support.

Synthesis is the step where the code HDL transforms to electronic components and the register transfer level will be generated.

The different resources of device utilization provided by the Xilinx ISE are shown in Table 9.

6. Conclusion

In this paper, we have put forward a hardware implementation of fuzzy logic controllers for a solar-wind system targeted for isolated sites.

First, we have designed the hybrid system with MATLAB/SIMULINK and XSG blocks. In general, the maximum power point has achieved forms of the wind-solar system with simple blocks: MATLAB/SIMULINK and XSG blocks are practically the same. The system with XSG has stabilized more quickly because of the parallel advantage of the XSG architecture. The maximum power point has achieved time in two analyzed cases, which has been about 0.4 ms.

In the second step, the performance of the proposed hardware implementation in terms of processing latency has been evaluated relatively to the first achieved maximum time and stabilization form. The XSG tool has been used

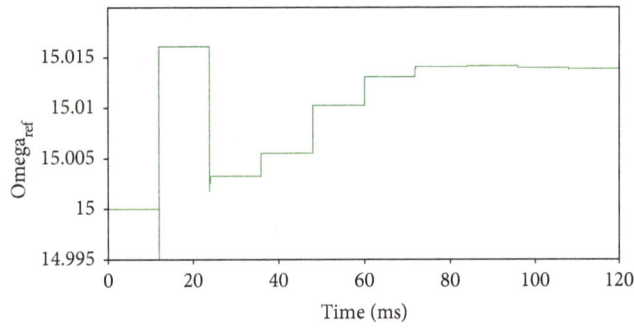

FIGURE 24: Fuzzy logic controller output of wind subsystem.

TABLE 6: Atmospheric wind and solar values.

Atmospheric condition	Values
Wind speed	10 m/s
Sunlight	1000 W/m^2
Temperature	300 K

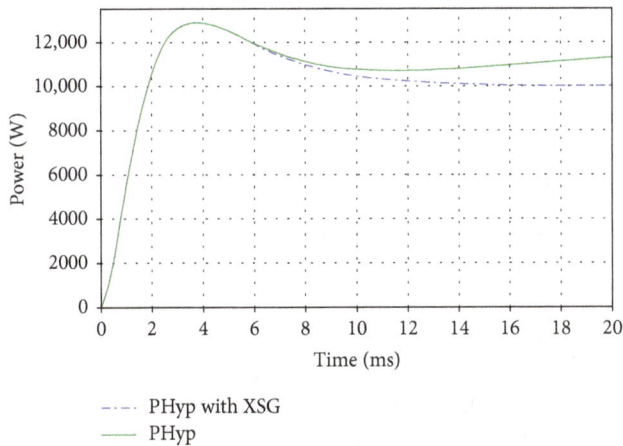

FIGURE 25: Power of wind-solar system with XSG and MATLAB/SIMULINK on fixed atmospheric conditions.

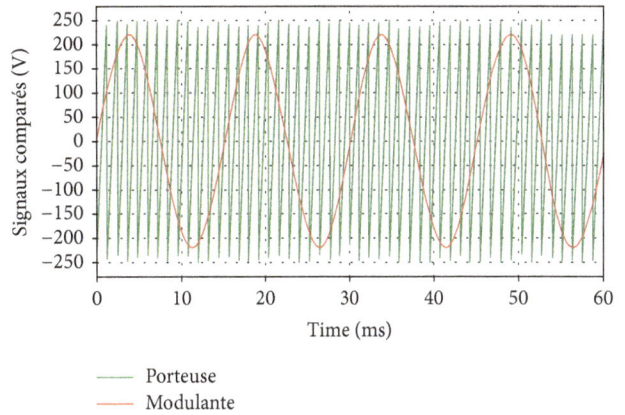

FIGURE 26: Switching signal of inverter switches.

for the system development. The use of this tool demonstrates that it has benefits in terms of conception time, since the developed design has been utilized firstly for the software validation and for the hardware system generation.

FIGURE 27: Aspect of modulated and carrier signals.

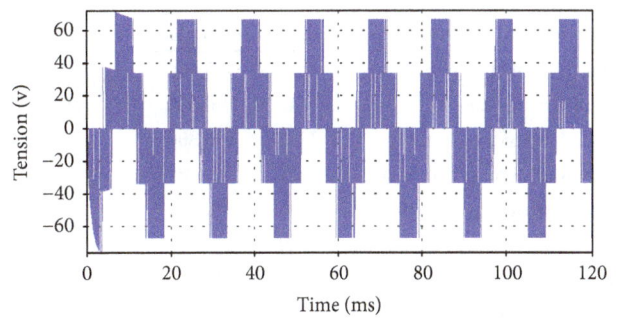

FIGURE 28: Voltage of one phase at inverter output.

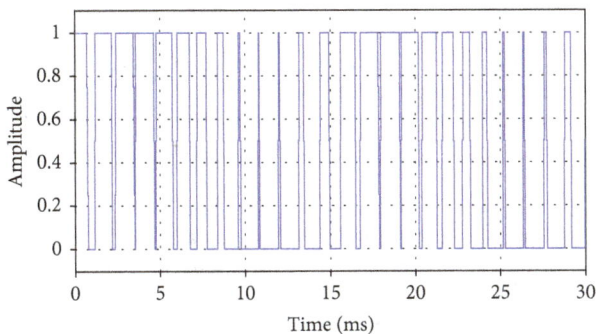

FIGURE 29: Picture of the experimental setup.

TABLE 7: Performance comparison between proposed work and other works.

References	Algorithms	Tracking efficiency η_{MPPT} (%)	Response time (s)
Proposed algorithm	Fuzzy logic	99.61	0.40
Izadbakhsh et al. [23]	ANFIS for solar fuzzy logic for wind subsystems	99.34	0.80
Suresh and Dhandapani [3]	Fuzzy logic	99.40	0.58
	HC and P&O	98.32	1.7
Rezvani et al. [4]	Fuzzy logic without load	99.25	0.80
	Fuzzy logic with load		1.1
Belmili et al. [5]	Fuzzy logic	99.22	0.80
	P&O	96.98	2.95

TABLE 8: Characteristics of the FPGA Virtex-6-XC6VLX315T.

Logic cells	314.880	
Configurable logic blocks (CLBs)	Slices	49.200
	Max distributed RAM (Kb)	5.090
DSP48E1 slices	1.344	
Interface blocks for PCI express	2	
Total I/O banks	18	
Max user I/O	720	
Total number of configuration bits	104,465,888	
Price	$667.29	

TABLE 9: FPGA circuit device utilization.

Slice logic utilization	Used	Available	Utilization (%)
Number of slice registers	21,286	393,600	5
Number of slice LUTs	37,009	196,800	18
Number used as logic	27,480	196,800	13
Number used as memory	5851	81,440	7
Number of occupied slices	11,300	49,200	22
Number with unused flip flop	22,350	39,794	56
Number with unused LUT	2785	39,794	6
Number of bonded IOBs	257	600	42
Number of DSP48E1s	1056	1344	78

Moreover, the timing response and tracking efficiency of the complete design on the XSG architecture indicate that a high performance in terms of execution time can be reached. Indeed, the suggested architecture largely respects the timing and tracking constraints of the used hybrid system. The performance of the proposed solar-wind system supports increases with respect to material specifications. The experimental simulation results in terms of achieved tracking time, efficiency, and algorithm utilization are compared to those of other existing systems.

At the end of the paper, we have implemented and optimized the designed system on an FPGA circuit, Virtex-6-XC6VLX315T, by using the Xilinx ISE.

Conflicts of Interest

The authors declare no conflict of interest.

Authors' Contributions

All authors helped in conceiving the experiments. Aymen Jemaa designed and performed the experiments. At the same time, Aymen Jemaa and Mohamed Ali Hajjaji wrote the main part of the paper. Ons Zarrad and Mohamed Nejib Mansouri contributed in interpreting the results and revising and writing of the paper.

Acknowledgments

This work was partially supported by ESIER.

References

[1] P. Nema, R. K. Nema, and S. Rangnekar, "A current and future state of art development of hybrid energy system using wind and PV-solar: a review," *Renewable and Sustainable Energy Reviews*, vol. 13, no. 8, pp. 2096–2103, 2009.

[2] J.-H. Chen, H.-T. Yau, and J.-H. Lu, "Implementation of FPGA-based charge control for a self-sufficient solar tracking power supply system," *Applied Sciences*, vol. 6, no. 2, 2016.

[3] P. Suresh and K. Dhandapani, "Modeling of an efficient hybrid wind and PV system for DC bus voltage regulation," *IJCTA*, vol. 9, pp. 2517–2525, 2016.

[4] A. Rezvani, M. Izadbakhsh, and M. Gandomkar, "Enhancement of hybrid dynamic performance using ANFIS for fast varying solar radiation and fuzzy logic controller in high speeds wind," *Journal of Electrical Systems*, vol. 11, pp. 11–26, 2015.

[5] H. Belmili, S. Boulouma, B. Boualem, and A. M. faycal, "Optimized control and sizing of standalone PV-wind energy conversion system," *Energy Procedia*, vol. 107, pp. 76–84, 2017.

[6] S. K. A. Shezan, N. H. Khan, M. T. Anowar et al., "Fuzzy logic implementation with MATLAB for solar-wind-battery-diesel hybrid energy system," *Imperial Journal of Interdisciplinary Research*, vol. 2, pp. 574–583, 2016.

[7] M. Rosyadi, S. M. Muyeen, R. takahashi, and J. Tamura, "A design fuzzy logic controller for a permanent magnet wind generator to enhance the dynamic stability of wind farms," *Applied Sciences*, vol. 2, no. 4, pp. 780–800, 2012.

[8] Z. Ons, J. Aymen, A. Craciunescu, and M. Popescu, "Comparison of hill-climbing and artificial neural network maximum power point tracking techniques for photovoltaic modules," in *2015 Second International Conference on Mathematics and Computers in Sciences and in Industry (MCSI)*, pp. 19–23, Sliema, Malta, 2015.

[9] K. K. Tse, B. M. T. Ho, H. S. H. Chung, and S. Y. R. Hui, "A Comparative study of maximum-power-point trackers for photovoltaic panels using switching-frequency modulation scheme," *IEEE Transactions on Industrial Electronics*, vol. 51, no. 2, pp. 410–418, 2004.

[10] C.-H. Huang, H.-Y. Pan, and K.-C. Lin, "Development of intelligent fuzzy controller for a two-axis solar tracking system," *Applied Sciences*, vol. 6, no. 5, 2016.

[11] C. Marisarla and K. R. Kumar, "A hybrid wind and solar energy system with battery energy storage for an isolated system," *International Journal of Engineering and Innovative Technology*, vol. 3, pp. 99–104, 2013.

[12] J. Aymen, Z. Ons, A. Crăciunescu, and M. Popescu, "Comparison of fuzzy and neuro-fuzzy controllers for maximum power point tracking of photovoltaic modules," *Renewable Energy and Power Quality Journal*, vol. 1, pp. 796–800, 2016.

[13] S. Hassan, H. Li, T. Kamal, U. Arifoğlu, S. Mumtaz, and L. Khan, "Neuro-fuzzy wavelet based adaptive MPPT algorithm for photovoltaic systems," *Energies*, vol. 10, no. 3, 2017.

[14] X.-P. Li, W.-L. Fu, Q.-J. Shi, J.-B. Xu, and Q.-Y. Jiang, "A fuzz logical MPPT control strategy for PMSG wind generation systems," *Journal of Electronic Science and Technology*, vol. 11, pp. 72–77, 2013.

[15] E. Cam, G. Gorel, and H. Mamur, "Use of the genetic algorithm-based fuzzy logic controller for load-frequency control in a two area interconnected power system," *Applied Sciences*, vol. 7, no. 3, 2017.

[16] M. Lotfy, T. Senjyu, M. Farahat, A. Abdel-Gawad, and H. Matayoshi, "A polar fuzzy control scheme for hybrid power system using vehicle-to-grid technique," *Energies*, vol. 10, no. 8, 2017.

[17] A. Aksjonov, K. Augsburg, and V. Vodovozov, "Design and simulation of the robust ABS and ESP fuzzy logic controller on the complex braking maneuvers," *Applied Sciences*, vol. 6, no. 12, 2016.

[18] S. Cheng, W.-b. Sun, and W.-L. Liu, "Multi-objective configuration optimization of a hybrid energy storage system," *Applied Sciences*, vol. 7, no. 2, 2017.

[19] W. Na, P. Chen, and J. Kim, "An improvement of a fuzzy logic-controlled maximum power point tracking algorithm for photovoltic applications," *Applied Sciences*, vol. 7, no. 4, 2017.

[20] P. Suresh and B. C. Sujatha, "Maximum power point tracking of a hybrid solar-wind power generation system for a smart grid using fuzzy logic control," *International Journal of Scientific & Engineering Research*, vol. 7, no. 5, pp. 165–170, 2016.

[21] R. Hmida, A. Ben Abdelali, and A. Mtibaa, "Hardware implementation and validation of a traffic road sign detection and identification system," *Journal of Real-Time Image Processing*, vol. 15, no. 1, pp. 13–30, 2018.

[22] R. Hmida, A. Ben Abdelali, F. Comby, L. Lapierre, A. Mtibaa, and R. Zapata, "Hardware implementation and validation of 3D underwater shape reconstruction algorithm using a stereo-catadioptric system," *Applied Sciences*, vol. 6, no. 9, 2016.

[23] M. Izadbakhsh, A. Rezvani, and M. Gandomkar, "Improvement of microgrid dynamic performance under fault circumstances using ANFIS for fast varying solar radiation and fuzzy logic controller for wind system," *Electrical Engineering*, vol. 63, no. 4, pp. 551–578, 2014.

[24] M. Bahoura, "FPGA implementation of blue whale calls classifier using high-level programming tool," *Electronics*, vol. 5, no. 1, 2016.

[25] R. Nagaraj and B. K. Panigrahi, "Simulation and hardware implementation of FPGA based controller for hybrid power system," *International Journal of Electrical Energy*, vol. 3, no. 2, pp. 86–93, 2015.

[26] P. Balamurugan, A. D. Praveenraj, I. Kaleeswarn, R. M. Karthik, and A. Tamilarasi, "FPGA implementation of MPPT hybrid algorithm for photovoltaic system," in *International Conference on Engineering Innovations and Solutions (ICEIS-2016)*, pp. 107–113, Tamilnadu, India, 2016.

[27] A. P. Ruiz, M. Cirstea, W. Koczara, and R. Teodorescu, "A novel integrated renewable energy system modelling approach, allowing fast FPGA controller prototyping," in *2008 11th International Conference on Optimization of Electrical and Electronic Equipment*, pp. 395–400, Brasov, Romania, 2008.

3

Development and Application of a Fuzzy Control System for a Lead-Acid Battery Bank Connected to a DC Microgrid

Juan José Martínez,[1] José Alfredo Padilla-Medina ⓘ,[2] Sergio Cano-Andrade,[3] Agustín Sancen,[4] Juan Prado,[2] and Alejandro I. Barranco ⓘ[2]

[1]Mechatronics Engineering Department, Technological Institute of Celaya, Av. Tecnológico y G. Cubas, s/n, 38010 Celaya, GTO, Mexico

[2]Electronics Engineering Department, Technological Institute of Celaya, Av. Tecnológico y G. Cubas, s/n, 38010 Celaya, GTO, Mexico

[3]Department of Mechanical Engineering, Universidad de Guanajuato, 36885 Salamanca, GTO, Mexico

[4]Department of Engineering Sciences, Technological Institute of Celaya, Av. Tecnológico y G. Cubas, s/n, 38010 Celaya, GTO, Mexico

Correspondence should be addressed to José Alfredo Padilla-Medina; alfredo.padilla@itcelaya.edu.mx

Academic Editor: Joaquín Vaquero

This study presents the development and application of a fuzzy control system (FCS) for the control of the charge and discharge process for a bank of batteries connected to a DC microgrid (DC-MG). The DC-MG runs on a maximum power of 1 kW with a 190 V DC bus using two photovoltaic systems of 0.6 kW each, a 1 kW bidirectional DC-AC converter to interconnect the DC-MG with the grid, a bank of 115 Ah to 120 V lead-acid batteries, and a general management system used to define the operating status of the FCS. This FCS uses a multiplexed fuzzy controller, normalizing the controller's inputs and outputs in each operating status. The design of the fuzzy controller is based on a Mamdani inference system with AND-type fuzzy rules. The input and output variables have two trapezoidal membership functions and three triangular membership functions. LabVIEW and the NI myRIO-1900 embedded design device were used to implement the FCS. Results show the stability of the DC bus of the microgrid when the bank of batteries is in the charging and discharging process, with the bus stabilized in a range of 190 V ± 5%, thus demonstrating short response times to perturbations considering the microgrid's response dynamics.

1. Introduction

Electrical energy plays an important role in our daily lives, due to the fact that it can be universally applied for conversion into other forms of energy, such as heat, lighting, and mechanical energy. However, its storage and, frequently, the process of transforming it into other forms of energy are complex. One example of systems that convert energy for storage is the electrochemical sources of energy known as batteries, which use chemical compounds as storage media, employing chemical reactions, and produce or store electrical energy [1]. The most frequently used chemical compounds are lead-acid, nickel-cadmium, and ion-lithium, as well as molecular hydrogen and methanol in the case of

fuel cells. Their main applications are in mobile electronic devices, electric automobiles, and direct current microgrids (DC-MGs). These applications require a system for controlling the charge and discharge of the battery, which ensures their efficiency, security, and reliability.

Dai et al. and Passino et al. [2, 3] propose fuzzy control systems (FCSs), as they are useful for modeling the nonlinear behavior of battery charge and discharge as a result of temperature and current changes and the ageing process. Alternatively, control systems for battery charge and discharge using artificial neural networks to model the multivariable behavior of the battery bank have been reported [4–7]. Although both control techniques are efficient and provide acceptable results, FCSs are more widely used due to the ease of their implementation; however, on some occasions, they require a greater amount of memory, which depends on the number of fuzzy rules and membership functions [8].

With regard to the use of FCS, [9, 10] used an FCS for the rapid charging of nickel-cadmium batteries, with their results demonstrating that it is possible to completely charge the battery bank in a time period of 10 minutes with a 6 A current. Meanwhile, [11–13] propose the use of FCSs in DSPs to charge Li-ion batteries, obtaining more efficient and secure control systems with charging times ranging from 15 minutes [12] to 4.5 hours [13].

FCSs have been successfully used for charging and discharging batteries in electric vehicles. These vehicles use a fuel cell that functions as an electrical generator, with the energy generated delivered to the battery by means of a power converter, which, in turn, delivers the energy to the motor of the vehicle by means of another power converter. Li et al. [14] used a fuzzy controller that operates based on the charge level and voltage of the battery, producing the working status of the battery, which was compared with the state of charge (SoC) of the battery, in order to avoid overestimating the SoC of the battery and generating excessive discharge. The references [15, 16] used an FCS to determine the point of operation of the power converter in an electrical vehicle. One of the most recently reported studies on the management of energy in electric vehicles used an adaptive FCS to determine the energy that is transferred to the battery [17]. The control system was implemented using the LabVIEW software with the NI CompactRIO module and operates based on the SoC of the battery and the power required.

In the area of the design of energy management systems used in DC-MGs, FCSs have also been used for the control of energy in batteries. Chen et al. [18] designed an FCS that monitors the SoC of the battery and the level of power in the microgrid in order to define the current that is either injected into or extracted from the battery bank.

Sikkabut et al. [19] presented a small-scale experimental prototype of a DC-MG with a 60 V bus that uses a photovoltaic array of 800 W, a battery bank of between 11.6 Ah and 24 V, and a 100 F supercapacitor bank. This prototype uses a management system implemented in the card dSPACE DS1104 platform, which includes an FCS to stabilize the DC bus. This study presented the experimental results, obtaining 5% variations in the DC bus due to the load variations when connected to the bus.

The majority of studies presented the use of the SoC of the battery in order to manage the energy, with the SoC obtained by subtracting the charge flow out of the battery from the initial existing charge, obtaining a result which shows nonlinear behavior and which requires compensation for the effects of the temperature changes [20].

This paper presents the design and implementation of a control system based on fuzzy logic to control the charge and discharge states of a bank of lead-acid battery in a DC-MG. A Mamdani-type fuzzy control with two input variables and an output variable with five membership functions each was used in this study to control the three states of operating statuses of the battery bank. In order to achieve this, a general control system was designed to multiplex the states of control

using the same fuzzy controller for each state, thus enabling the stabilization of the DC bus in adequate time periods with a decrease in the computational burden required for the implementation of the charge and discharge controller for the batteries. Furthermore, the SoC of the battery bank is not used as an input variable of the fuzzy controller as proposed in [14] and defined by the OCV-ampere-hour counting method; instead, the battery bank voltage and current are used as input variables, which are considered to generate the set of inference rules of the controller. Also, at each controller operating state, the technical characteristics for charging and discharging of the batteries provided by the manufacturer are considered, in order to extend the useful life of the battery bank. The controller was implemented using the LabVIEW virtual instrumentation software and the NI myRIO-1900 embedded design device.

The paper is organized as follows: Section 2 presents the description of the DC-MG system, Section 3 presents the different management systems used in the analysis, Section 4 presents the results obtained from the study, and finally, Section 5 concludes the paper.

2. Description of the DC-MG

The DC-MG was designed to use a maximum power level of 1 kW, with a DC bus of 190 V. Figure 1 presents the general diagram for the proposed DC-MG, which defines the energy flow between the different elements that comprise the system. The principal energy source for the DC-MG is two photovoltaic panels (PPs), each of which supplies a DC-DC converter with a capacity of 0.6 kW. Two photovoltaic panel simulators (PPSs) (Agilent—model E4360A) were used to simulate the behavior of the PPs. The second element that forms part of the proposed DC-MG is the bidirectional DC-AC converter, which enables the DC-MG to be interconnected to the grid allowing it to supply energy to the grid or take energy from it. The DC-MG has an energy storage system (ESS) composed of ten CALE-SOLAR batteries connected in a series, with each battery designed for deep-cycle applications comprising 12 V and a 115 Ah capacity. This bank is connected to the DC bus by means of a bidirectional DC-DC converter. Moreover, it has a bank of 0.230 F supercapacitors connected in parallel to the DC bus, with the objective of maintaining the voltage level of the DC bus during the connection or disconnection intervals of the DC-AC converter, the ESS, the PPs, and the loads.

The FCS proposed in this study is applied to the bidirectional DC-DC converter used to interconnect the bank of batteries with the DC bus of the DC-MG, the topology of which is shown in Figure 2. This converter comprises a buck converter and a boost converter connected in parallel. The converter's switches enable the change from one mode of operation to another, as the mode of operation depends directly on the switch used. This topology comprises an inductor (L), a capacitor (C), and two power transistor switchers (Q1 and Q2).

In order to evaluate the functionality of the DC-MG under different scenarios, loads with both linear and nonlinear characteristics were used. For a linear load, a 600 Ω

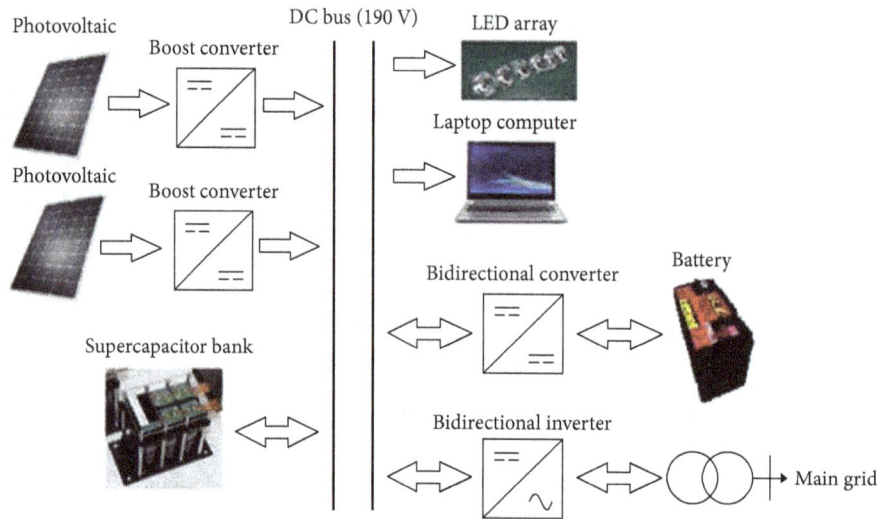

FIGURE 1: General scheme of the proposed DC microgrid.

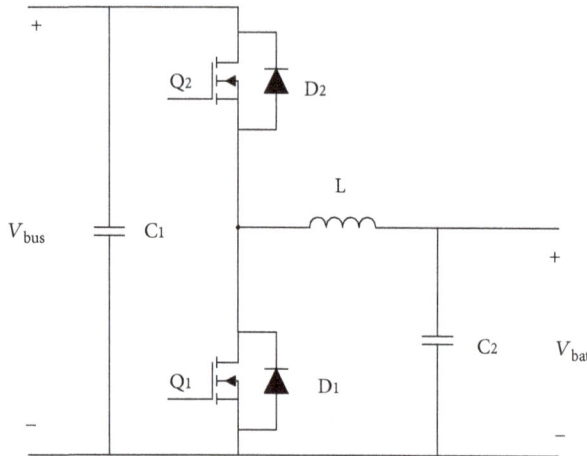

FIGURE 2: Topology of the bidirectional DC-DC converter.

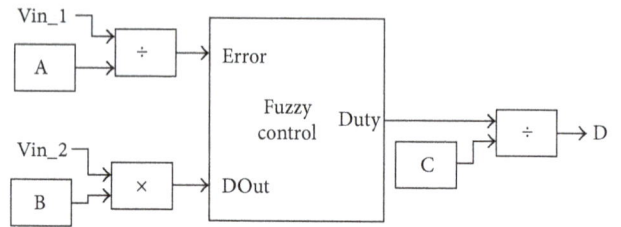

FIGURE 3: Schematic representation of the fuzzy control system.

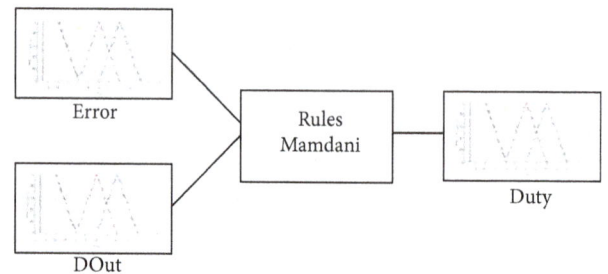

FIGURE 4: Schematic representation of the structure of the fuzzy controller.

resistive load bank was used, while two types of lamps and a laptop were used for nonlinear loads. The first group of lamps consists of seven commercial LED-type lamps of 23 W each, while the second group comprises four louver-type lamps with three 28 W T5 fluorescent lamps, each with a commercial electronic ballast, as well as a laptop that requires a maximum power level of 150 W. Nonlinear loads used at this supply voltage are able to operate correctly at 190 V DC, with the manufacturer's specifications indicating that their supply voltage is within the universal input voltage range of 100–264 V AC.

3. Management System

The management system proposed for the DC-MG uses controllers that operate in each of the elements of the DC-MG (PPs, ESS, and the interconnection system with the grid). The general management system (GMS) is described below, as well as the design of the control system for the charge and discharge of the battery bank.

3.1. General Management System. The principal objective of the general management system (GMS) of the DC-MG is to make maximum use of the energy generated by the PPs, supplying the energy necessary for the loads connected to the DC bus and injecting the remaining energy into the battery bank or the grid. This system uses the five modes of operation described below:

(i) Mode I: the system startup mode, in which the PPs function in voltage control (VC) mode, while the ESS and the grid are disabled.

(ii) Mode II: the PPs work with the maximum power point tracking (MPPT) algorithm [20], while the

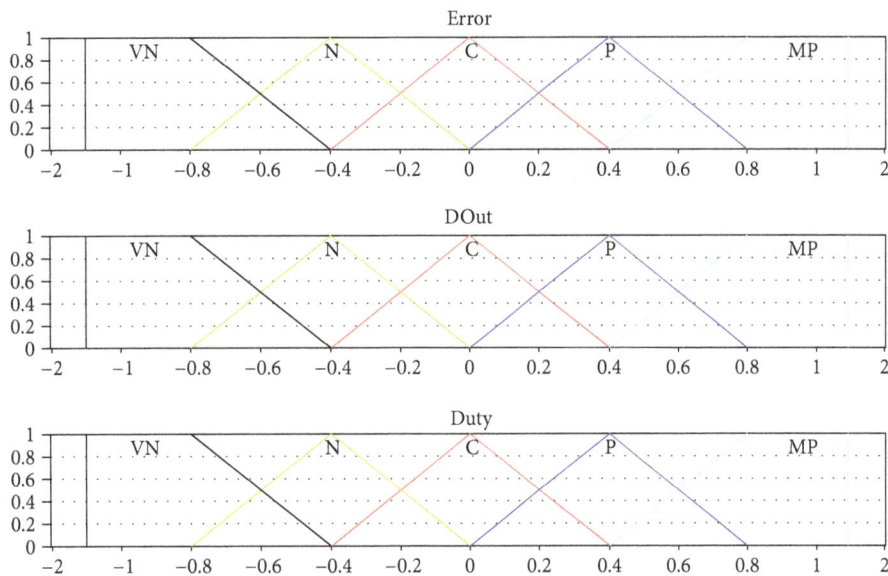

FIGURE 5: Membership functions for the input and output variables.

converter for the ESS functions as a boost converter, supplying energy to the loads connected to the bus, while the grid is disabled.

(iii) Mode III: the PPs work with the MPPT algorithm, while the bidirectional DC-AC converter is used as a rectifier, taking energy from the grid and injecting it into the loads connected to the bus, while the ESS is disabled. In this operating mode, it is possible that energy is required from the grid in order to inject it into the ESS.

(iv) Mode IV: the PPs work with the MPPT algorithm, and the bidirectional DC-DC converter works as a buck, injecting energy into the ESS, while the grid is disabled.

(v) Mode V: the PPs work with the MPPT algorithm, while the bidirectional DC-AC converter works as an inverter for injecting energy into the grid, while the ESS is disabled.

The GMS always starts up in operation mode I and changes the operation mode depending on the voltage level of the bus, the power levels of the PPs, the battery bank, and the grid.

3.2. Control System for the Battery Bank.
For the control of the energy that is injected into or extracted from the bank of batteries, an FCS was designed and implemented based on the current and the voltage of the battery bank. This control system is divided into three states of operation that could work in modes II, III, and IV of the GMS.

3.2.1. State I (Battery Charge).
Energy is extracted from the PPs and injected into the battery bank, while the FCS operates based on the voltage level of the DC bus. If the voltage level increases, the energy injected into the battery bank

increases, while, if the voltage level decreases, the energy injected into the battery bank decreases. Thus, the balance of the voltage level of the DC bus is maintained. Figure 3 shows the diagram for the general control used to control the bidirectional converter (buck topology). For this state of operation, the input variable Vin_1 corresponds to the difference between Vbus and the SP (Vin_1 = Vbus − SP), where the SP is equal to 190 V; the second input variable Vin_2 corresponds to the voltage variations occurring in the DC bus. The Duty output variable defines the variations that will occur in the duty cycle, variations which are integrated by means of the integral control action. The values for the variables A, B, and C were defined heuristically on testing the fuzzy controller online with the buck converter, in an effort to maintain the stability of the bus and eliminate the effects of the perturbations on the bus in the least time possible. In this state, $A = 15$, $B = 1$, and $C = 1000$.

3.2.2. State II (Battery Charge).
In this state of operation, further to using the energy generated by the PPs, the energy from the grid is used to charge the battery bank. To control the energy injected into the battery bank, the FCS works on the current (Ibat) injected into the bank. This current is defined by the power required by the bidirectional DC-AC converter before injecting energy into the battery bank, with the intention of not exceeding the maximum capacity of the converter designed to work at 1 kW. The control scheme is represented in Figure 3. For the state Vin_1 = SP − Ibat, Vin_2 corresponds to the variations in the current injected into the battery bank, while the Duty output variable generates the variations that are integrated in order to obtain the duty cycle of the converter operating in its buck topology. For this state, $A = 10$, $B = 2$, and $C = 500$.

3.2.3. State III (Battery Discharge).
In those circumstances where the energy generated by the PPs would not be sufficient to supply the energy required for the loads, the battery

bank is used to supply the missing energy. The bidirectional power converter operates through its boost topology during the discharge process. For this state of operation, the controller shown in Figure 3 is also used. For this state, the variable Vin_1 = SP − Vbus, where Vin_2 corresponds to the voltage variations for the bus and the Duty output is integrated for the control of the converter's duty cycle. For this state, $A = 10$, $B = 0.5$, and $C = 800$.

The states of operation previously described are defined based on the following criteria for the operation of the battery bank: (1) the charge and discharge processes of the battery bank must always be concluded, (2) the continuous current charge process is completed once the battery bank attains a maximum voltage value of 135 V, (3) the continuous charge voltage process is completed when the current injected into the battery bank is less than 0.5 A, and (4) the discharge process is completed when the battery voltage is 100 V.

It should be noted that the bidirectional power converter used in the battery bank has a maximum capacity of 1 kW, for which reason the maximum current that can be extracted or injected into the bank is 10 A.

3.3. Multiplexed System.

3.3. Multiplexed System. The input and output variables for the FCS (Vin_1, Vin_2, and *D*) are multiplexed, and their selection is determined for each state of operation described in Section 3.2. The input variables are obtained by measuring both the voltage level of the DC bus and the current that is injected into the battery bank in order to obtain an Error signal and define Vin_1. The input variable Vin_2 is obtained by calculating the variations in the voltage of the DC bus and the current injected into the battery bank. The output signal is used as a signal for the power transistor switch that controls either the converter when it operates in its buck topology (injecting energy into the battery bank) or the power transistor switch that operates when the converter functions in its boost topology (extracting energy from the battery bank). This enables the use of an FCS to maintain the stability of the three states of operation of the ESS.

3.4. Fuzzy Controller. The structure of the fuzzy controller is shown in Figure 4. The fuzzy sets for the input and output variables are very negative (VN), negative (N), zero (Z), positive (P), and very positive (VP).

The triangular and trapezoidal membership functions are the most popular functions due to their ease of representation and the fact that they require less computer processing capacity, providing satisfactory results. For the proposed fuzzy controller, the input and output variables use two trapezoidal membership functions and three triangular membership functions (see Figure 5). The group of fuzzy rules was designed under the working principle of a fuzzy PD + I controller [18], which eliminates the error that exists between the variable measured and the desired value for the point of reference (set point), considering the variations in the variable measured. For example, if the "Error" input is very negative and the "DOut" input is very positive, then the "Duty" output is negative. Table 1 illustrates the 21 rules used to design the fuzzy controller.

TABLE 1: AND-type fuzzy rules.

		Error				
		VN	N	Z	P	VP
DOut	VN	VN	N	VP	VP	
	N	VN	N	P	P	
	Z	N	N	Z	P	VP
	P	N	VN	N	P	
	VP	N	VN	VN	P	

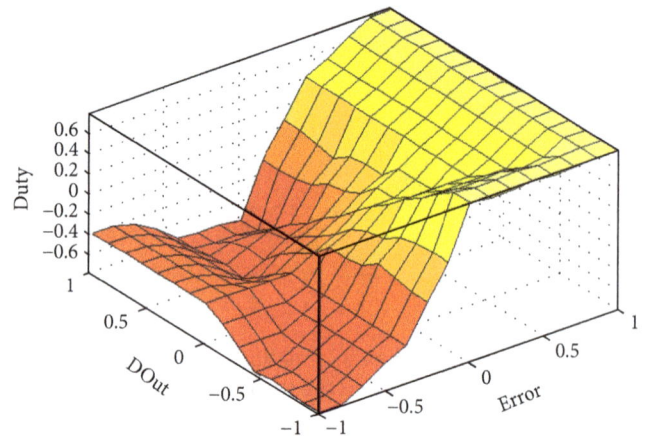

FIGURE 6: Relationship between the input and output variables of the FCS.

The Mamdani-type fuzzy inference system is used for the defuzzification process, in which the centroid method is used to obtain the variation of the duty cycle (Duty) in the form of a numerical value based on a fuzzy output. The graph that relates the input variables to the output variable is shown in Figure 6.

4. Results

The experimental results obtained for the proposed FCS are presented below, using the LabVIEW virtual instrumentation software and the NI myRIO-1900 embedded design device.

4.1. State of Operation I. The GMS directs energy to the battery bank when it is discharged or in the process of charging. Figure 7(a) shows the behavior of the voltage in the DC bus and the current that is injected into battery bank. In this case, the PPs generate 568 W, of which 116 W is supplied to the load connected to the bus and 452 W is injected into the battery bank. The time required for the transient response is 3 seconds. When the battery bank is totally charged, it is necessary that the PPs operate in VC mode and that the bidirectional converter is turned off. The behavior of the DC bus and the current that is injected into the battery bank during this change is illustrated in Figure 7(b). Undertaking these changes of state generates perturbations in the bus of up to 8 V, with a change established after 5 seconds.

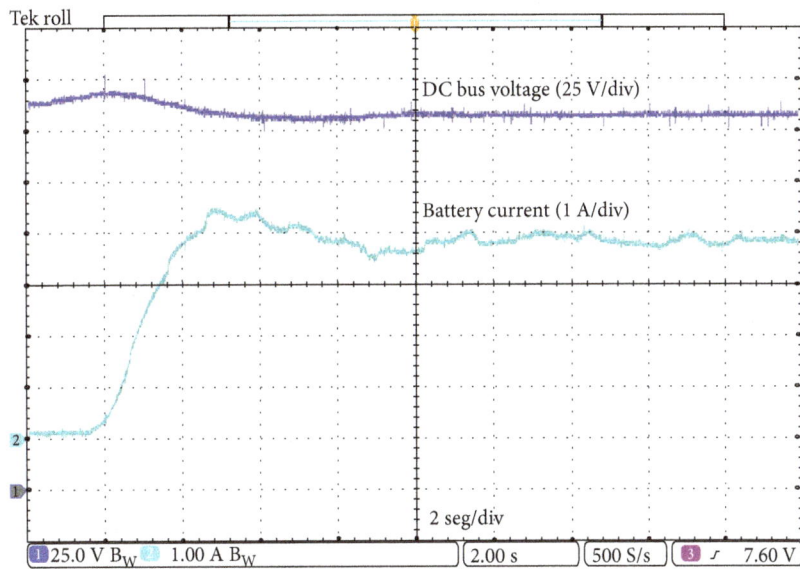

(a)

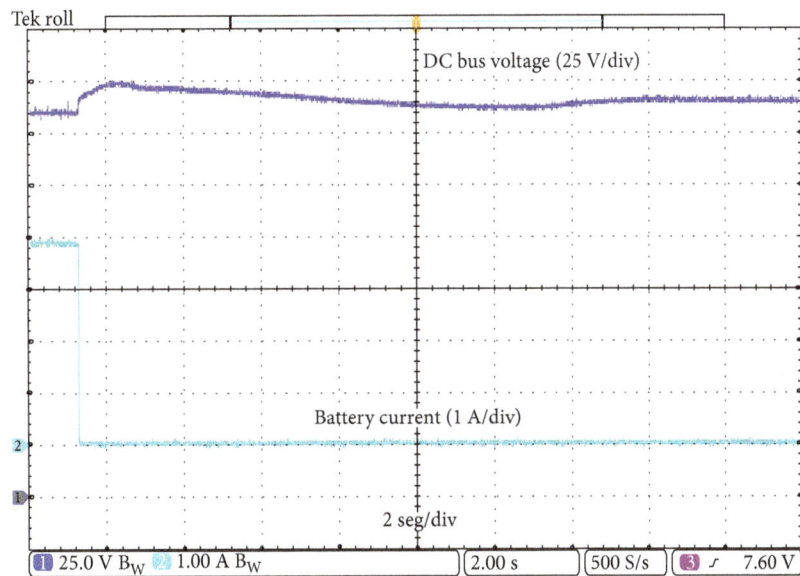

(b)

FIGURE 7: (a) Battery bank charging in process and (b) battery bank charging turned off.

The charge process for the battery bank starting with the continuous current charge is presented below, considering a current level of 8.8 A, where the voltage of the bank increases over time until reaching 135 V. When the battery bank attains this voltage level, the control changes to constant-voltage mode, generating a decrease in the current that is injected into the batteries. It is considered that the battery bank is totally charged when the current injected into the bank is 500 mA. For these experiments, it was decided to inject a constant current of 8.8 A, due to the PPs' generation capacity and the fact that a 1 kW converter interconnects the DC-MG with the grid. Figure 8(a) illustrates the behavior of the voltage of the battery bank while the charge is conducted in constant-current mode and constant-voltage

mode. The charge in constant-current mode has a duration of 8 hours, while the charge in constant-voltage mode was 8.5 hours. Figure 8(b) illustrates the behavior of the current in both modes.

4.2. State of Operation II. In this state of operation, energy is extracted from the grid and is injected into the battery bank. Figure 9(a) illustrates the behavior of the DC bus and the current that is injected into the battery bank at the moment of activating the bidirectional converter for the battery bank in buck topology. When the current injected into the battery bank increases, there is a 2 V drop in the DC bus voltage. The time in which the FCS takes in injecting the 2 A into the battery bank is 20 seconds, a time

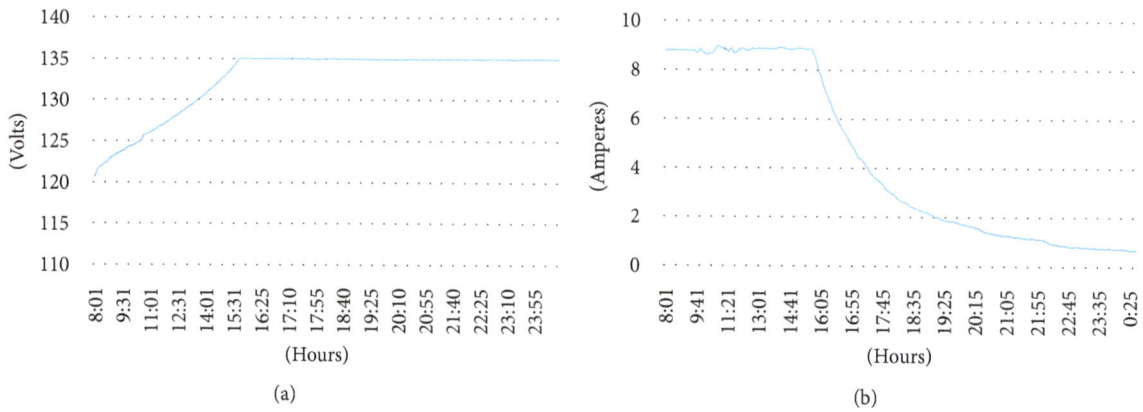

FIGURE 8: (a) Behavior of the battery bank voltage in charge mode and (b) behavior of the battery bank current in charge mode.

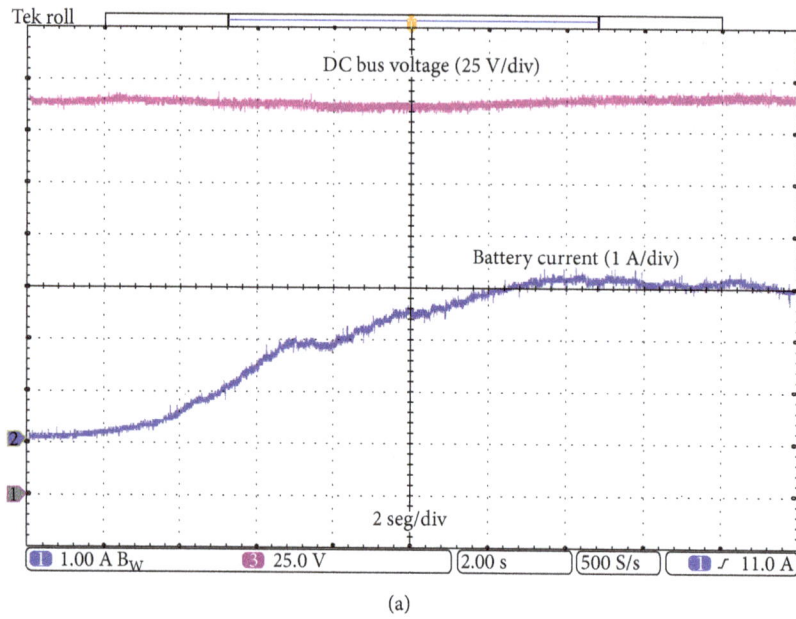

(a)

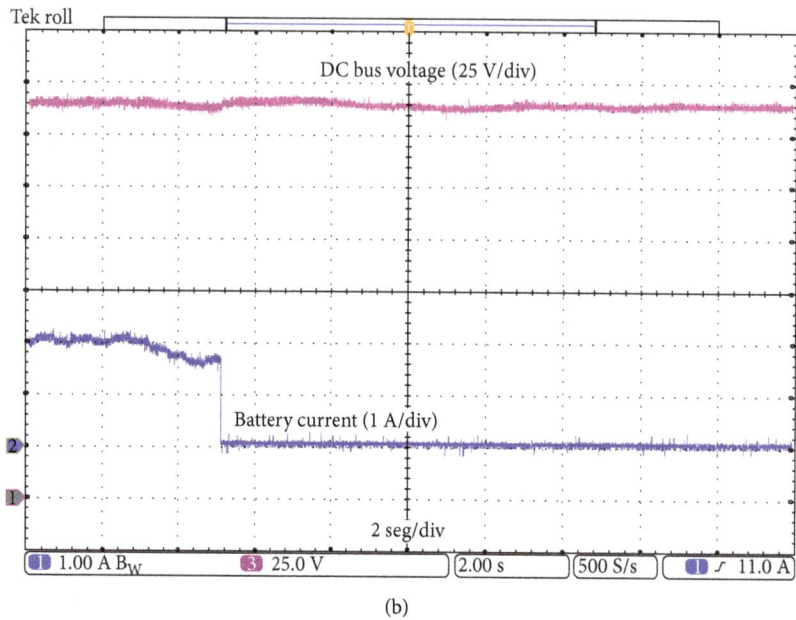

(b)

FIGURE 9: (a) Activation of the bidirectional converter in buck topology and (b) bidirectional converter turned-off.

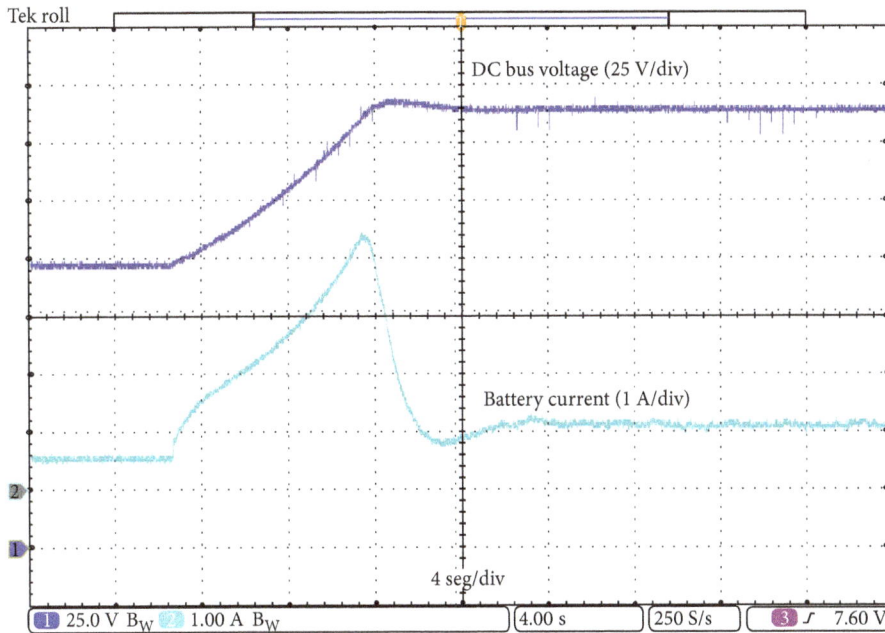

FIGURE 10: Behavior on starting-up the microgrid with the battery bank.

period which does not destabilize the DC bus. When it is injecting energy into the battery bank and is totally charged, the current that is injected into the bank is cancelled. Figure 9(b) shows the behavior of the current injected into the battery bank and the DC bus at the moment of changing the mode of operation.

4.3. State of Operation III. This state of operation uses the energy generated by the PPs and the energy from the battery bank for supplying the energy required for the loads connected to the DC bus. When the PPs do not generate energy, the battery bank supplies all the energy required by the loads. Figure 10 shows the behavior of the DC bus during the start-up of the converter in order to obtain a voltage level of 190 V. The start-up time for the battery bank is 9.6 seconds. Furthermore, Figure 10 illustrates the behavior of the current that is extracted from the battery bank. This image shows an excess current caused by the demands of the bank of supercapacitors. When the response passes to the stable state, the current injected by the converter is 0.647 A (the output current for the bidirectional converter). The output for the bidirectional converter is 190 V and 0.647 A, while the converter input (battery bank) has a voltage of 123 V and delivers a current of 1 A. The load connected to the bus is a resistive load of 123 W.

Once the FCS establishes the DC, tests are undertaken to ascertain the response of the system. Figure 11(a) shows the behavior of both the voltage of the bus and the current in the battery bank on connecting a 125 W LED lamp, while Figure 11(b) illustrates the behavior of both the voltage of the bus and the current in the battery bank on disconnecting the lamps. These images show that the drop and overshoot in the bus on connecting and disconnecting the load are lower

than 4 V, which represents 2.1% of the desired value for the bus, and that, furthermore, the settling time for the system is 2 seconds.

Further to the tests undertaken with LED lamps, the behavior was tested using a laptop with a 150 W load. Figures 12(a) and 12(b) show the behavior of the connection and disconnection of the laptop. These images demonstrate that the dynamic energy consumption of the laptop generates neither overshoot nor drops in the voltage of the bus. However, the current signal presents variations due to the laptop's energy consumption. The settling time is 2 seconds.

The behavior of the system was also tested using fluorescent lamps with an energy requirement of 500 W. Figures 13(a) and 13(b) illustrate the behavior on connecting and disconnecting these lamps, an effect which generates a drop and an overshoot of 5 V, representing 2.6% of the desired voltage for the DC bus. The settling time that generates the connection and disconnection of the load is 3 seconds.

In the same state, if the energy generated by the PPs is lower than the energy required by the load connected to the bus, the voltage level of the bus will decrease. If the battery bank has energy, it will be used to supply the energy missing for the loads. Figure 14(a) presents the behavior of the DC bus at the moment that the bidirectional converter operates in voltage boost mode, taking energy from the battery bank and injecting it into the DC bus. During the change of mode of operation in the DC-MG, the bus suffers a 10 V drop in voltage (5.2% of the desired voltage). The same figure shows the behavior of the current provided by the battery bank. In order to undertake this change of mode, a fluorescent lamp requiring 375 W is connected, of which the PPs supply 240 W, while the remaining 135 W is supplied by the battery

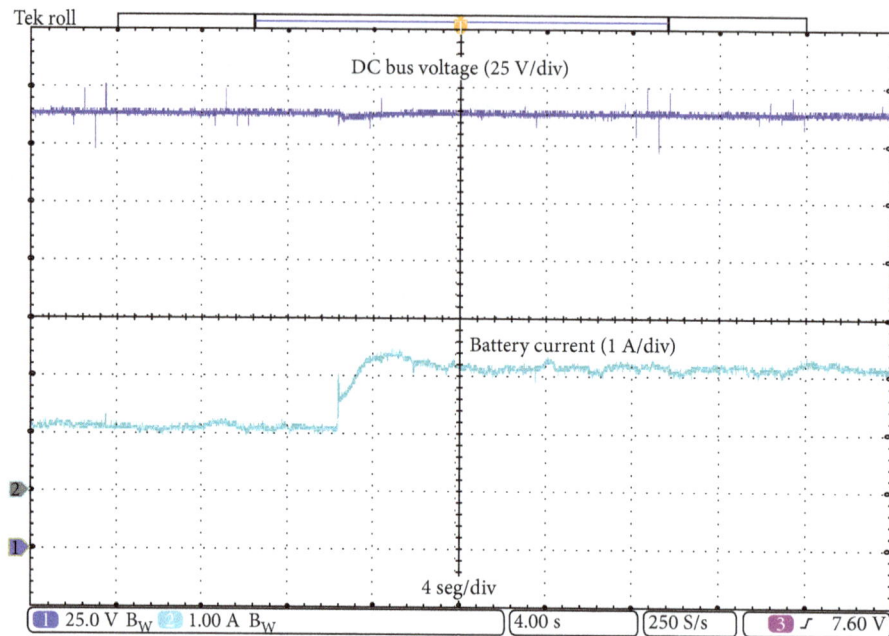

(a)

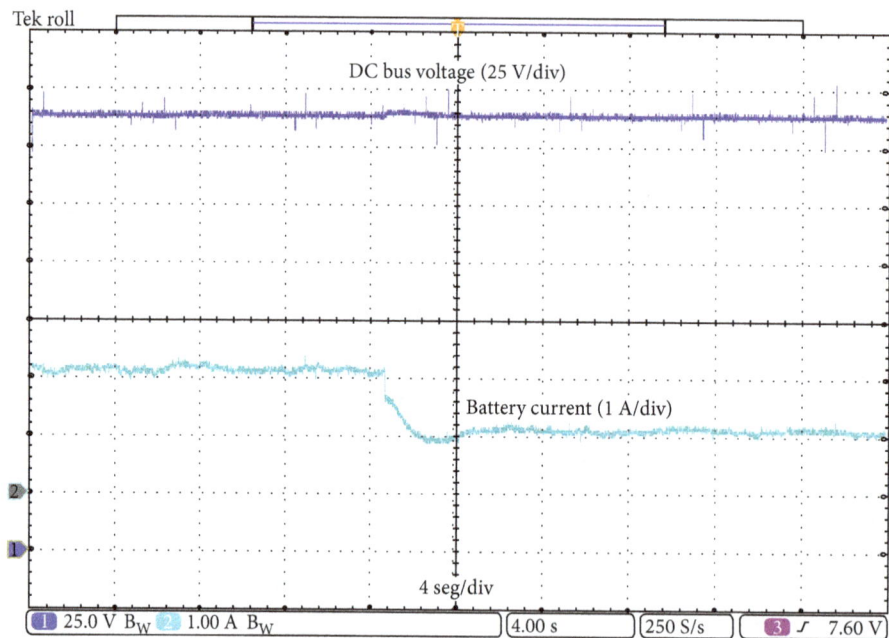

(b)

FIGURE 11: (a) Connection of 125 W LED lamps and (b) disconnection of 125 W LED lamps.

bank. Figure 14(b) presents the behavior of the DC bus and battery bank current on the disconnection of the fluorescent lamps. This disconnection generates an increase in the DC bus, in turn causing the deactivation of the bidirectional converter. These changes generate a 7.5 V overshoot in the bus that represents a 3.9% increase in the desired value for the bus. The settling time is 4 seconds.

The discharge process for the battery bank is shown below, with the PPs and the grid disabled for this discharge, leaving the battery bank as the sole energy supply to the DC bus. The load that was connected requires 1 kW of power, while the DC bus uses a voltage level of 190 V. Figure 15(a) shows the behavior of the voltage of the battery bank, with an initial voltage of 121.2 V, which reduces over time, generating an increase in the current supplied by the bank due to the fact that it requires constant power for the load connected to the bus. Figure 15(b) presents the behavior of the current of the battery bank on discharge. The battery bank was discharged until the voltage fell to the level of 100 V, in order to avoid damaging the battery bank. Under

(a)

(b)

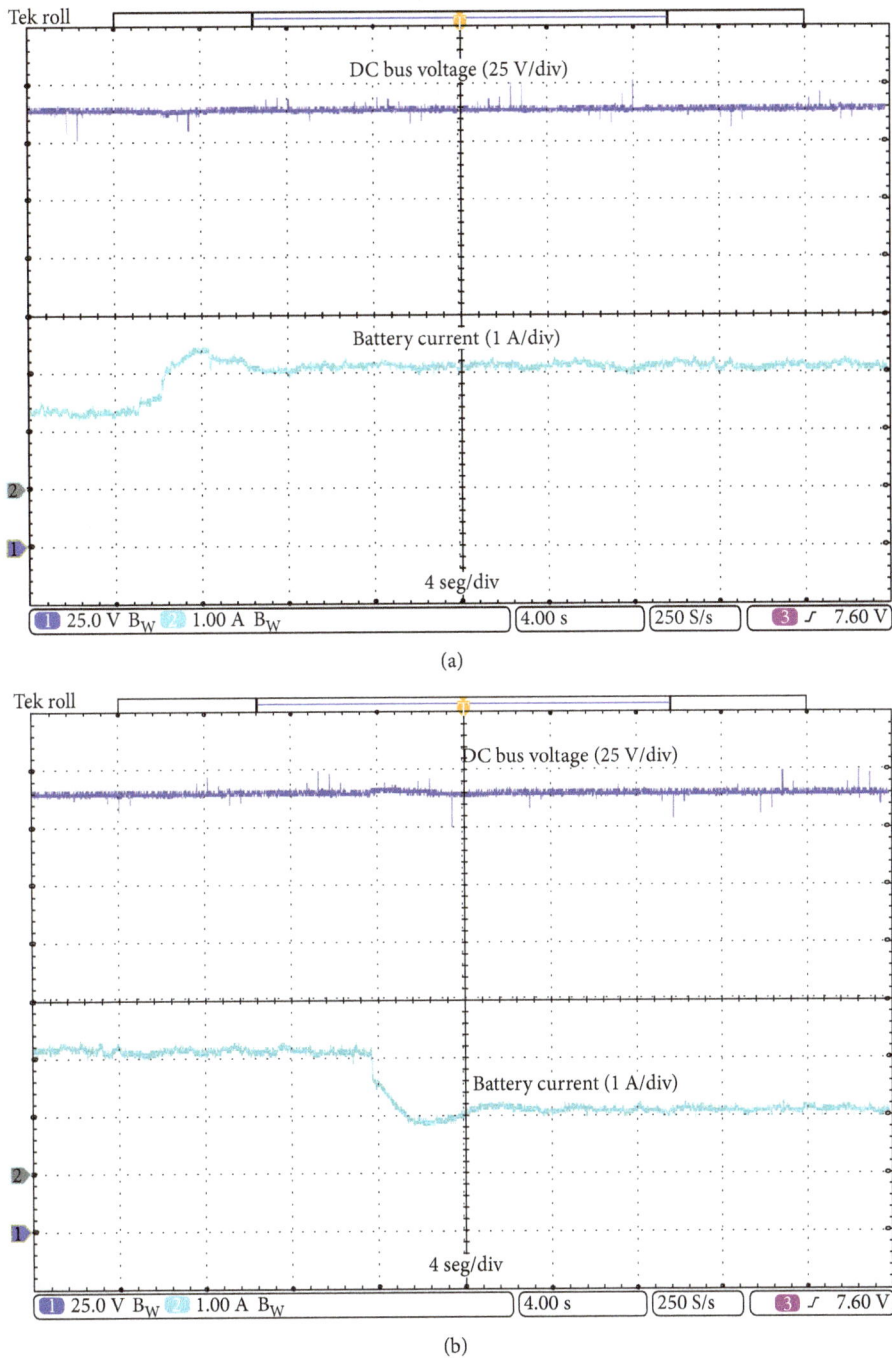

FIGURE 12: (a) Connection of the DC bus to the laptop and (b) disconnection of the DC bus from the laptop.

these conditions, the bank was able to maintain the voltage level of the bus while it supplied energy at a load of 1 kW for approximately 10 hours.

4.4. Comparison of the PD + I Fuzzy Controller with a Classical PI Controller. The comparison between the PD + I fuzzy controller and the classical PI controller is developed by measuring the DC bus voltage and the battery bank current during the state of operation III (see Section 4.3), in which the control of the converter used in the PPs changes

from the voltage control algorithm to the MPPT algorithm and the voltage controller for the DC-DC bidirectional converter of the battery bank is activated.

An advantage of the fuzzy control system is the smooth current signal that is obtained from the battery bank, which directly affects in a positive manner the behavior of the DC bus. Table 2 shows that the fuzzy controller system provides a faster response and a higher overshoot than the classical PI controller. With respect to the DC bus voltage, Table 3 shows that the fuzzy controller provides a faster response, a lower

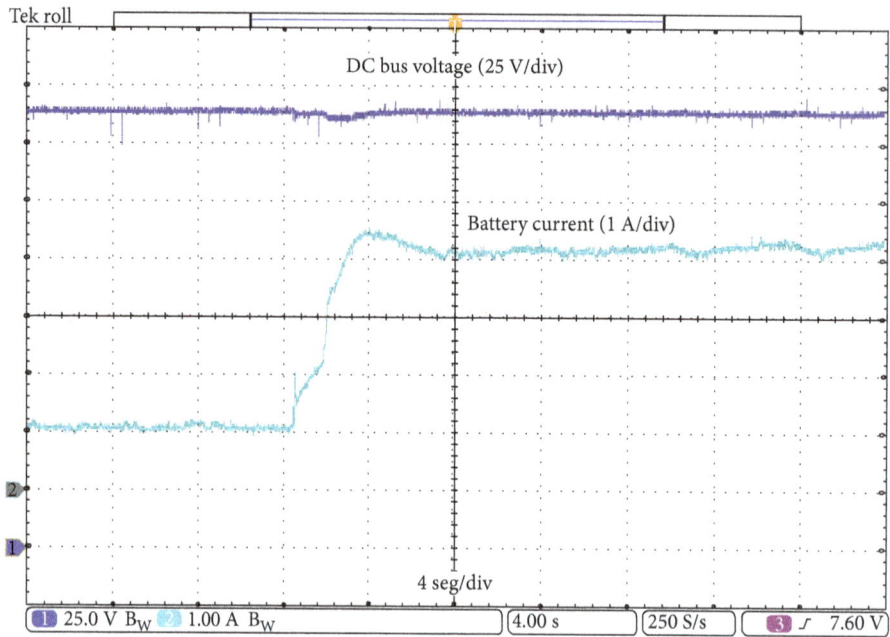

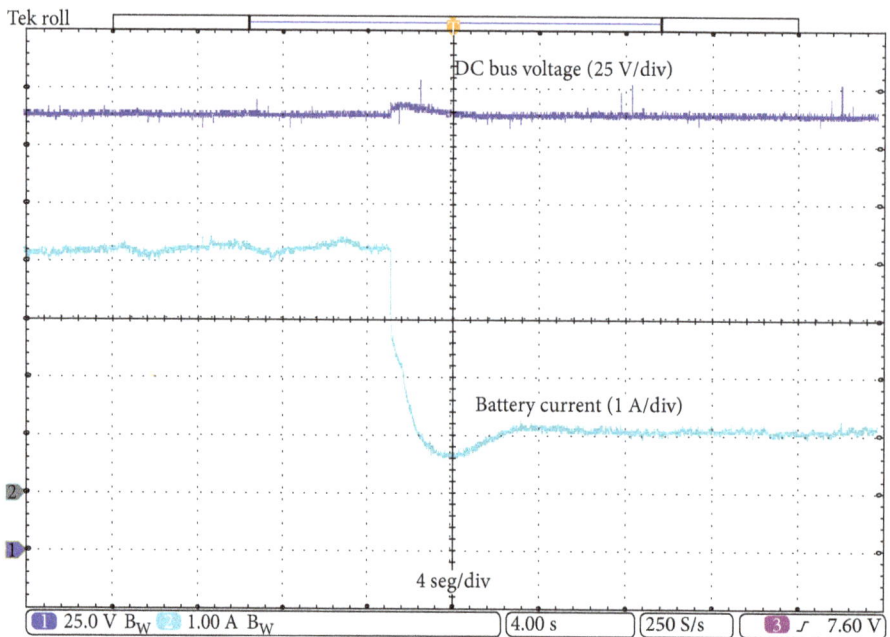

FIGURE 13: (a) Connection of 500 W fluorescent lamps and (b) disconnection of 500 W fluorescent lamps.

error at a steady-state operation, and a lower overshoot than the classical PI controller.

5. Discussion and Conclusions

This study presents the experimental tests on the behavior of the DC bus of a DC-MG considering the charge and discharge of the battery bank. The experiments considered real loads connected to the DC bus, such as fluorescent lamps, LED lamps, and computers, with the objective of ascertaining the behavior of the DC-MG using an FCS applied to the storage system.

Recent studies have designed classic control systems for the control of the charge and discharge of battery banks. Yu [21] designed an autonomous experimental system for a DC-MG with a 5 kW capacity, with experimental results showing the behavior of the charge of the battery bank in constant-current and constant-voltage modes. Analysis of the experimental results presented reveals that the current presents a high frequency loop with an approximate amplitude of 1.5 A in constant-current mode and 1 A in constant-voltage mode.

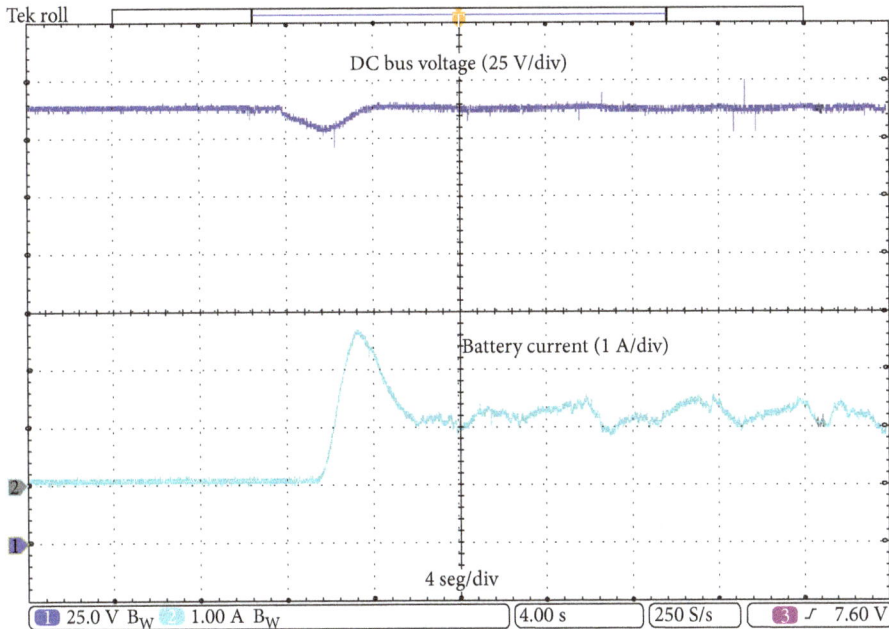

(a)

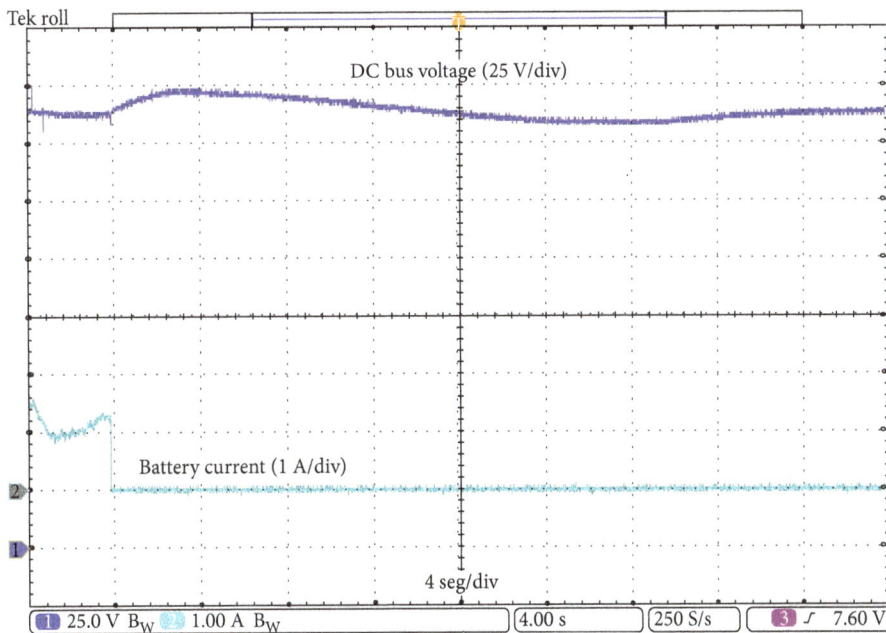

(b)

FIGURE 14: (a) Change of control mode for photovoltaic panels and the activation of the battery bank and (b) change of operational mode in the photovoltaic panels and battery bank.

The results presented in this study are centered on the DC-MG's different modes of operation, which operate in different elements. Furthermore, the high frequency loop for the proposed DC-MG's current and voltage curves is lower than that presented in [21].

In order to define whether the battery bank is operating in either charge or discharge mode, the power difference between the load connected to the DC-MG bus and energy-generating sources was considered. Chen et al. [18] used an FCS which took into account this difference and the SoC of the batteries in order to control their SoC and modify the current that is injected or extracted from the battery bank. The proposed FCS output generates a variation in the current that is injected into or extracted from the battery bank depending on the mode of operation. The experimental results show a high frequency loop in the current for the batteries of approximately 1.5 A, the same as that found in [21]; moreover, the behavior presents slight oscillations in the

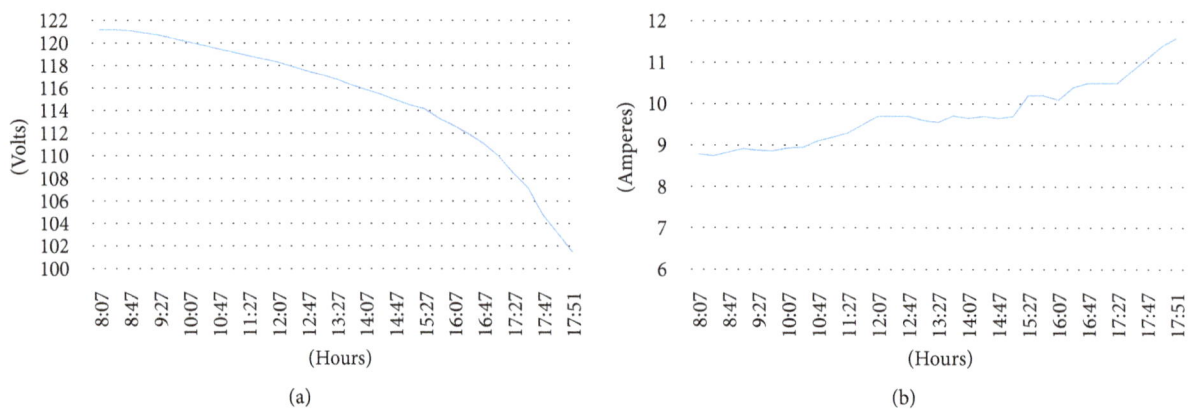

FIGURE 15: (a) Behavior of the voltage of the battery bank in discharge mode and (b) behavior of the current of the battery bank in discharge mode.

TABLE 2: Response parameters of the current extracted from the battery bank.

Controller type	Rise time	Settling time	Overshoot	Peak time
PI	2.4 seconds	4.8 seconds	71%	2.6 seconds
PD + I fuzzy	1.4 seconds	4 seconds	78%	1.6 seconds

TABLE 3: Response parameters of the DC bus voltage.

Controller type	Settling time	Overshoot	Peak time	Error at steady state
PI	8 seconds	5.52%	2.4 seconds	1.31%
PD + I fuzzy	4 seconds	5.2%	1.6 seconds	0.65%

stable state; however, this type of oscillation softens with the use of the FCS proposed in this study.

The experimental tests of the prototype presented show that the behavior of the system when used with the fuzzy controller remains stable with the connection and disconnection of sources and loads in the DC-MG, maintaining a value of 190 V ± 5% for the bus. The stabilization times recorded during the perturbations generated are short, considering the response dynamic of the microgrid. These times depend on the capacitance of the DC bus and the FCS, with the bus connected to a bank of supercapacitors, which eliminates the peaks that generate the connections or disconnections of sources or loads. The speed of the response of the battery bank current depends on the stabilization time of the DC bus, for which reason the FCS enables a soft response in both the DC bus and the battery bank current.

6. Future Studies

After proving, experimentally, the correct functioning of the fuzzy controller for the control of the charge and discharge of the battery bank, the embedding of the fuzzy control system in an FPGA is proposed, with the objective of creating an economical control system.

Abbreviations

FCS: Fuzzy control system
DC-MG: Direct current microgrid

DSP: Digital signal processor
DC: Direct current
AC: Alternating current
PP: Photovoltaic panel
PPS: Photovoltaic panel simulator
ESS: Energy storage system
VC: Voltage control
SP: Set point
GMS: General management system
SoC: State of charge
MPPT: Maximum power point tracking.

Conflicts of Interest

The authors declare that they have no conflicts of interest.

References

[1] H. A. Kiehne, *Battery Technology Handbook*, vol. 118, CRC Press, Boca Raton, 2003.

[2] H. Dai, P. Guo, X. Wei, Z. Sun, and J. Wang, "ANFIS (adaptive neuro-fuzzy inference system) based online SOC (state of charge) correction considering cell divergence for the EV (electric vehicle) traction batteries," *Energy*, vol. 80, pp. 350–360, 2015.

[3] K. M. Passino, S. Yurkovich, and M. Reinfrank, *Fuzzy Control*, vol. 20, Addison-wesley, Reading, MA, 1998.

[4] G. Capizzi, F. Bonanno, and C. Napoli, "Recurrent neural network-based control strategy for battery energy storage in

generation systems with intermittent renewable energy sources," in *2011 International Conference on Clean Electrical Power (ICCEP)*, pp. 336–340, Ischia, Italy, 2011.

[5] A. Eddahech, O. Briat, and J. M. Vinassa, "Neural networks based model and voltage control for lithium polymer batteries," in *8th IEEE Symposium on Diagnostics for Electric Machines, Power Electronics & Drives*, pp. 645–650, Bologna, Italy, 2011.

[6] M. Ashari and D. K. Setiawan, "Inverter control for phase balancing of diesel generator — battery hybrid power system using diagonal recurrent neural network," in *2011 21st Australasian UniversitiesPower Engineering Conference (AUPEC)*, pp. 1–5, Brisbane, QLD, Australia, 2011.

[7] L. Ciabattoni, G. Cimini, M. Grisostomi, G. Ippoliti, S. Longhi, and E. Mainardi, "Supervisory control of PV-battery systems by online tuned neural networks," in *2013 IEEE International Conference on Mechatronics (ICM)*, pp. 99–104, Vicenza, Italy, 2013.

[8] F. Chekired, A. Mellit, S. A. Kalogirou, and C. Larbes, "Intelligent maximum power point trackers for photovoltaic applications using FPGA chip: a comparative study," *Solar Energy*, vol. 101, pp. 83–99, 2014.

[9] Z. Ullah, B. Burford, and S. Dillip, "Fast intelligent battery charging: neural-fuzzy approach," *IEEE Aerospace and Electronic Systems Magazine*, vol. 11, no. 6, pp. 26–34, 1996.

[10] H. Surmann, "Genetic optimization of a fuzzy system for charging batteries," *IEEE Transactions on Industrial Electronics*, vol. 43, no. 5, pp. 541–548, 1996.

[11] C. E. Lyn, N. A. Rahim, and S. Mekhilef, "DSP-based fuzzy logic controller for a battery charger," in *TENCON'02. Proceedings. 2002 IEEE Region 10 Conference on Computers, Communications, Control and Power Engineering*, vol. 3, pp. 1512–1515, Beijing, China, 2002.

[12] M. W. Cheng, S. M. Wang, Y. S. Lee, and S. H. Hsiao, "Fuzzy controlled fast charging system for lithium-ion batteries," in *2009 International Conference on Power Electronics and Drive Systems PEDS*, pp. 1498–1503, Taipei, Taiwan, 2009.

[13] G. C. Hsieh, L. R. Chen, and K. S. Huang, "Fuzzy-controlled Li-ion battery charge system with active state-of-charge controller," *IEEE Transactions on Industrial Electronics*, vol. 48, no. 3, pp. 585–593, 2001.

[14] S. G. Li, S. M. Sharkh, F. C. Walsh, and C. N. Zhang, "Energy and battery management of a plug-in series hybrid electric vehicle using fuzzy logic," *IEEE Transactions on Vehicular Technology*, vol. 60, no. 8, pp. 3571–3585, 2011.

[15] M. Kim, Y. J. Sohn, W. Y. Lee, and C. S. Kim, "Fuzzy control based engine sizing optimization for a fuel cell/battery hybrid mini-bus," *Journal of Power Sources*, vol. 178, no. 2, pp. 706–710, 2008.

[16] C. Y. Li and G. P. Liu, "Optimal fuzzy power control and management of fuel cell/battery hybrid vehicles," *Journal of Power Sources*, vol. 192, no. 2, pp. 525–533, 2009.

[17] H. Yin, W. Zhou, M. Li, C. Ma, and C. Zhao, "An adaptive fuzzy logic-based energy management strategy on battery/ultracapacitor hybrid electric vehicles," *IEEE Transactions on Transportation Electrification*, vol. 2, no. 3, pp. 300–311, 2016.

[18] Y. K. Chen, Y. C. Wu, C. C. Song, and Y. S. Chen, "Design and implementation of energy management system with fuzzy control for DC microgrid systems," *IEEE Transactions on Power Electronics*, vol. 28, no. 4, pp. 1563–1570, 2013.

[19] S. Sikkabut, P. Mungporn, C. Ekkaravarodome et al., "Control of high-energy high-power densities storage devices by Li-ion battery and supercapacitor for fuel cell/photovoltaic hybrid power plant for autonomous system applications," *IEEE Transactions on Industry Applications*, vol. 52, no. 5, pp. 4395–4407, 2016.

[20] S. Malkhandi, "Fuzzy logic-based learning system and estimation of state-of-charge of lead-acid battery," *Engineering Applications of Artificial Intelligence*, vol. 19, no. 5, pp. 479–485, 2006.

[21] B. Yu, "Design and experimental results of battery charging system for microgrid system," *International Journal of Photoenergy*, vol. 2016, Article ID 7134904, 6 pages, 2016.

4

Renewable Generation (Wind/Solar) and Load Modeling through Modified Fuzzy Prediction Interval

Syed Furqan Rafique ⓘ,[1,2,3] Zhang Jianhua,[2] Rizwan Rafique,[4] Jing Guo,[2] and Irfan Jamil ⓘ[5]

[1]Department of Electrical Engineering, National University of Science and Technology, Islamabad, Pakistan
[2]Department of Electrical Engineering, North China Electric Power University, Beijing, China
[3]Goldwind Technology, Beijing, China
[4]Department of Electrical Engineering, Norwegian University of Science and Technology, Trondheim, Norway
[5]College of Energy & Electrical Engineering, Hohai University, Nanjing, China

Correspondence should be addressed to Syed Furqan Rafique; 08beefrafique@seecs.edu.pk

Academic Editor: Philippe Poggi

The accuracy of energy management system for renewable microgrid, either grid-connected or isolated, is heavily dependent on the forecasting precision such as wind, solar, and load. In this paper, an improved fuzzy prediction horizon forecasting method is developed to address the issue of intermittence and uncertainty problem related to renewable generation and load forecast. In the first phase, a Takagi-Sugeno type fuzzy system is trained with many evolutionary optimization algorithms and established coverage grade indicator to check the accuracy of interval forecast. Secondly, a wind, solar, and load forecaster is developed for renewable microgrid test bed which is located in Beijing, China. One day and one step ahead results for the proposed forecaster are expressed with lowest RMSE and training time. In order to check the efficiency of the proposed method, a comparison is carried out with the existing models. The fuzzy interval-based model for the microgrid test bed will help to formulate the energy management problem with more accuracy and robustness.

1. Introduction

Intermittent, disperse, and dilute nature of renewable sources like wind and solar give new challenges for the integration of these sources in the microgrid planning and control [1, 2]. The task is even harder in islanded operation of the microgrid. One of the proposed solutions is to use the energy storage system for smoothing the uncertain behavior of generation for energy management system and to minimize the operational cost of the system. Nevertheless, the aforementioned problem is still not accurately solved and reliable unless or until a proper generation prediction method is not employed to address these issues [3].

In islanded operation of a microgrid, the task for smoothing the fluctuations produced by renewable sources is more challenging because limited generation options are available in the system [4]. Therefore, in standalone systems, this creates a critical demand supply and power quality problem unlike big geographical power systems.

To address the uncertainty and intermittence problem, many researches have been carried out under stochastic and robust forecast system [5–9], but the reliability of the system and actual implementation as compared to simulation results are the big concern in all over the world which will lead in doing more efforts in this field.

The forecasting of renewable resources such as wind and PV is strongly coupled with accuracy of the weather prediction model because usually the wind speed and irradiance information is utilized to forecast wind and PV power, respectively, using empirical formula. Factors such as location, surrounding terrain, and climatology showed a strong relation with accuracy of weather parameter, namely, temperature, vapor pressure, precipitate, cloud coverage, solar radiation, wind speed/direction, and humidity. A short-term wind power forecast is shown in [1]; the authors used sparse vector autoregressive sVAR model for the wind farm in Australia. A logit normal transformation is combined with the spatiotemporal weather data and compared with ARIMA and VAR methods in order to show the authority of the proposed method. Wind speed is predicted in [10], authors mentioned, to develop a weather warning system for more extreme wind speed predictions which leads to high fluctuation in wind power. However, the paper did not address neither implementation nor simulation details. The ARMA-based-enhanced boosting technique is mentioned for day-ahead wind power prediction in a wind farm at Jiangsu province, China, in [11]. The method showed improvement in 15.5% MAE with respect to the ARMA model and 3.21% MAE for the persistence model. Nonetheless, the validation accuracy check is not performed against more accurate techniques such as wavelet-ARIMA and hybrid Kalman filter in statistical domain. An interval-based fuzzy inference wind forecast is performed in [12], using empirical and nonparametric approach. Proposed method is applied on a Danish wind farm to predict different prediction intervals on adaptive resampling-based coverage rates for confidence interval. Authors used beta probability distribution function to calculate wind power forecast error in [13]; they argued that the fat-tailed forecast error pdf is more accurately modelled by beta distribution. However, the model is restricted to the particular dataset only.

Wavelet decomposition is utilized for solar radiation into low- and high-frequency band then SVM is applied to classify the pattern by Xiyun et al. in [14]. Historical solar radiation and atmospheric pressure are successfully used to get 3.78% RMSE and 12.83 MAE. Another technique [15] is mentioned in literature for 6 h ahead the backpropagation neural network-based solar power predictor; authors performed correlation analysis on several weather parameters and finally solar radiation and air temperature are used to develop predictor for solar power forecast, and the proposed method is deployed as a software package at Ljubljana, Slovenia, for testing and validation. Autoregressive exogenous ARX model is successfully integrated with numerical weather prediction which has 35% more accuracy than persistence models in [16]; AR works as a short-term predictor and the model can predict 4 h to several days of solar thermal power. Another ANN-NWP framework is mentioned in [17] with a confidence interval up to 95%.

Machine learning techniques are very useful in solving forecasting issues particularly neural network application in load prediction [2, 18]. Neural network has a high capability to capture nonlinear effects in the dataset, but the selection of proper neurons, hidden layers, computational time for training, and the possibility to fall in local minima is the main challenge to improve in NN type models. Felice et al. in [2] discussed a neural network with regularized negative correlation learning RNCL methodology, as the neural network has high sensitivity for its initial condition; therefore, this method is combined with the output of different networks which gives better performance over error reduction. Intrusive and nonintrusive load monitoring and prediction approaches are mentioned in [18]. Fuzzy regression model is proposed by Hong and Wang [19] for STLF; the authors used historical, weekday, and temperature information for the 3-year dataset and then formulated three fuzzy linear regression models. Mean absolute percentage error (MAPE)

was in the range of 2.6–4.58% for hourly peak load. However, the authors did not show enough data to support the complete effectiveness of the proposed method and gathering large dataset is a difficult task in practical systems. The Takagi-Sugeno method is used to generate scenarios in the grid-connected microgrid by adding error probability distribution values in point forecast [20]; the method reduced the forecasting error in RG and load, plus authors also developed a robust EMS with two-step relaxation and Benders algorithm in order to maximize the exchange cost between microgrid and utility. Finally, the system is tested with the Monte Carlo simulation for the feasibility study. An efficient and model independent support vector machine-based method is discussed in [21], but the selection of kernel function, computation time, and parameter selection is the shortcoming for SVM-based approaches. A variety of statistical prediction methods is applied for building load demand, such as ARIMA and multiple exponential smoothing functions which are also applied using dry bulb temperature data in [22, 23] for long-term forecast, but LTF is mostly utilized for investment and planning purposes. Hybrid approaches for STLF also applied in various researches, such as a cooling load of a building, are predicted using ARIMA followed by SVM in [24]; it is used to reduce the error by 50% as compared to the individual algorithm. A mean square error is also minimized in [25] using general NN + wavelet transform + genetic algorithm- (GA-) based fuzzy inference system (GNN-W-GAF) for real-time data, where the NN is used to get initial prediction using wavelet information, and then GA is used to adjust the weights of the fuzzy inference system in the second stage of prediction. Hybrid approach using PSO-NN is applied in [26] by clustering data into three inputs, namely, weekday, weekend, and holidays. However, none of the above paper represents the uncertainty associated with modeling of the forecast problem and STLF issues where the density forecast is irrelevant and sequential time stamps are highly correlated.

Stochastic and uncertain behavior of wind and PV are not mentioned in most of the literatures as discussed earlier; hence, this study will show the details for modelling uncertainty by using fuzzy interval prediction-based approach. A similar method is used in the study [27], where the covariance of the error vector is used to develop the fuzzy regression model without mentioning the training method which will deeply affect the accuracy of the forecast, whereas this paper proposed evolutionary-based training for linear regressors in fuzzy interval in order to get better accuracy than existing least square and backpropagation-based method of [27], plus performance indicators such as coverage grade and interval band for lower and upper intervals are also introduced in order to check the quality of the forecast.

The rest of the paper is arranged in the following order. Section 2 provides the description of a grid-connected microgrid and the importance of prediction in energy management efficiency enhancement. Section 3 provides the detail on the mathematical model of the fuzzy interval and fuzzy PSO-based prediction interval algorithm. Section 4 provides the results and application description for wind, PV, and load forecasts which are tested on the Goldwind Microgrid test

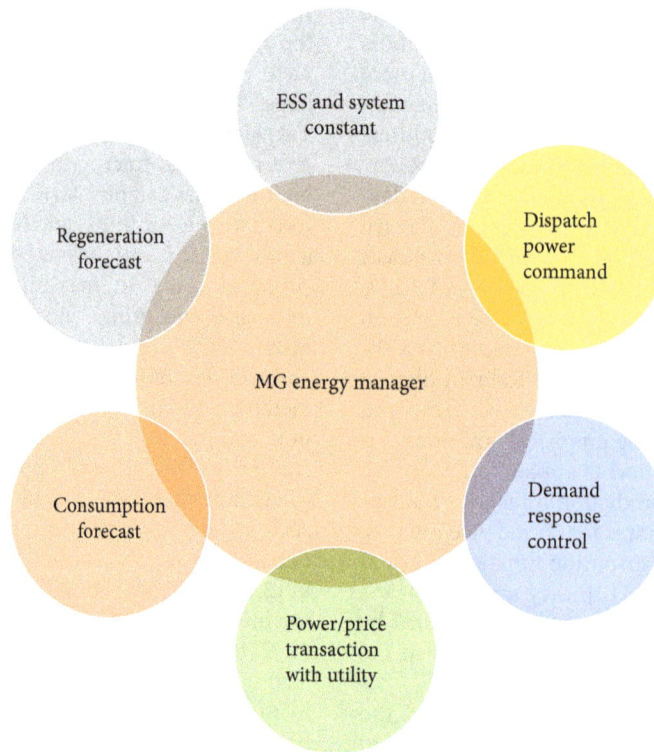

FIGURE 1: Microgrid energy manager.

bed, Beijing. Finally, Section 5 gives conclusion and future recommendation for this work.

2. Importance of Prediction in Energy Management

Energy management system (EMS) is the backbone of any microgrid infrastructure; it is responsible for a reliable and economical operation. It has three types of control methods, one is centralized in which all the data gathered at a one central processing unit, then it dispatches the control commands under a system constraint back to the primary controllers for execution. Second method is decentralized, where all units collect local information and communicate with each other for proper decision-making. Last method is hybrid control, where local controllers collect local information and also execute central controller commands as an upper-level controller and provide a perfect coordination among all the controllers. Also in Figure 1, a smart energy manager (SEM) features are highlighted for the grid-connected and isolated microgrid. The responsibilities of EMS in a microgrid system include solving the optimization problem for economic dispatch, gathering technical and physical constraints of the network, generating units and loads, then selecting the best possible dispatching signals for satisfying objectives which are set by the operator. Centralized secondary control is more suitable for an isolated microgrid where the infrastructure is fixed plus the demand and supply problem is crucial. Whereas a distributed scheme is more suitable for a plug-in play type of functionalities just like in a grid-connected

microgrid with multiple owners. The following are the operations of EMS:

(i) Generation prediction for all uncontrollable DERs, such as wind turbine and PV panels, is usually done by the EMS/secondary controller in next day/hour-based timestamps.

(ii) Load forecast and demand-side management policies for all consumers are usually developed at the secondary controller in next day/hour-based timestamps.

(iii) State of charge (SOC) for battery and other battery management constraints usually handled by the battery management module (BMM).

(iv) Connection status for grid and price forecast for utility in next hour/day format is performed at the secondary controller.

(v) System diagnostic, security, state estimation, black start, self-healing, and other active/reactive power adjustment checks are usually done by the secondary controller.

Considering uncertainties in the renewable generation source, spinning reserve becomes a necessary part of the microgrid as an energy buffer to mitigate the fluctuation plus smoothing the overall voltage and power profile. This spinning reserve also creates economic and management-related hurdles which lower the motivation to invest in microgrid-related systems. Hence, an accurate prediction of these uncertain units is necessary in a sliding horizon fashion in order to get the forecast from several minutes to several days. Similarly, load uncertainty is purely unavoidable, due to the fact that usually in a traditional power system the customer has no participation in the decision-making process, which makes it very difficult to predict and control. Forecasting techniques can be divided into different time horizons, such as very short-term (1 min–1 hr), short-term (1 hr–1 week), medium-term (1 month–1 year), and long-term (1 year and above), different spatial resolutions (0.01 km to 100 km), and techniques, for example, statistical, machine learning, and physical/numerical. Numerical weather prediction (NWP) is concerned with the weather forecast and related with the analysis of a number of variables in multidimensional calculation space; it followed the relation of thermodynamics, fluid dynamics, and chemical reaction of the air particles. For lower temporal and spatial resolution, NWP prediction is very accurate. Statistical forecasting methods rely on consistent historical data and are very computationally efficient and fast but for nonlinearity machine learning techniques they are appropriate. Machine learning usually falls into an overfitting problem for training dataset, but up to this date it is very reliable, as well as it is improving by the passage of time. Hence, an EMS for the microgrid system must be accounted for the uncertainties in the forecast in order to take better decisions.

3. Fuzzy Takagi-Sugeno Modelling

Fuzzy prediction interval (FPI) modelling is utilized in a variety of studies in the past, and it is used to forecast power output for nondispatchable sources. FPI is useful for approximation of nonlinear dynamic system, and it facilitates to develop robust EMS formulation which is the ultimate scope of this study. The number of rules is minimized in this study which makes the partition of output variable space that is projected onto the input variable space in order to get the optimal solution for fuzzy sets and rules. Fuzzy clustering approach is used to make that partition and get the premise parameter. Fuzzy c-mean clustering is applied to develop initial clusters, minimizing the intracluster variance, and assigning initial weights that will be adjusted in later stages in order to avoid local minima. The Takagi-Sugeno-based fuzzy model is helpful to achieve consequence parameters based on the least square method.

A linear membership-based fuzzy rule space is developed by Takagi-Sugeno in [28, 29], which is capable of estimating a variety of nonlinear systems. The idea is to partition the space of input andoutput variables into separate linear fuzzy subspaces and then combine into normalized membership function β; this method is called premise parameter selection by Takagi-Sugeno in [28]. Let us suppose that a linear relationship is established between the input and output variable.

$$
\begin{aligned}
z &= \sum_{i=1}^{n} \beta_i \left(a_0^i + a_1^i x_1 + a_2^i x_2 + \cdots + a_k^i x_k \right), \\
&= \sum_{i=1}^{n} \left(a_0^i \beta_i + a_1^i x_1 \beta_i + a_2^i x_2 \beta_i + \cdots + a_k^i x_k \beta_i \right),
\end{aligned}
\tag{1}
$$

where β_i is the normalized membership function of ith rule R_i ($i = 1 \cdots n$), input/output dataset $x_{1j}, x_{2j}, \ldots, x_{kj}$ maps to z_j at interval k, j is the measurement vector from 1 to m, the total output vector is $Z = [z_1, z_2, \ldots, z_m]^T$, and the coefficients of linear functions $a_0^i, a_1^i, \ldots, a_k^i$ are solved by the least square method to find the optimal minima of these weights and n denotes the total number of rules. Extension of this rule (1) for multivariable can be written as $\hat{\mathbf{z}} = [x, y, w]$; here, $\hat{\mathbf{z}}$ is the predicted vector which depends on x, y are inputs, and w is a control vector.

$$
\begin{aligned}
\hat{\mathbf{z}} = \sum_{i=1}^{n} \beta_i &\left(a_0^i + a_1^i x_1 + a_2^i x_2 + \cdots + a_k^i x_k + b_0^i \right. \\
&\left. + b_1^i y_1 + b_2^i y_2 + \cdots + b_k^i y_k \right).
\end{aligned}
\tag{2}
$$

More explicitly for the given problem of forecasting in this paper, the T and S framework (2) can be extended to matrix notation t timestamps.

$$
\begin{aligned}
\hat{\mathbf{z}}_t &= f^{\mathrm{TS}}(x_{t-1}, y_{t-1}, w_{t-1}), \\
&= \beta_i x_{t-1} [1 \; y_{t-1} \; w_{t-1}] A, \\
&= \Upsilon^T A,
\end{aligned}
\tag{3}
$$

where the $t-1$ subscript shows the historical data vector, $A = [a_0, \ldots, a_k, b_0, \ldots, b_k]^T$, where $a_0 = [a_0^1, \ldots, a_0^n]^T$ which are the weight vectors which are used for error reduction in the proposed method later in this paper and Υ is the fuzzy regression matrix. The set of predictions can be shown as error vector e for forecast:

$$
\hat{\mathbf{z}}_t = \Upsilon^T A + e_j.
\tag{4}
$$

Fuzzy partitioning method is used in this paper to minimize the number of rules based on fuzzy c-mean clustering approach.

(1) *Interval identification*: In order to approximate function families for various sets of intervals because the deterministic solution is not reliable in renewable/load predictions, hence, the author calculated forecasting intervals with certain interval bandwidth σ and fuzzy covariance model of error.

$$
\hat{\mathbf{z}}_t^U = f^{\mathrm{TS}}(x_{t-1}, y_{t-1}, w_{t-1}) + \sigma^U \mathrm{Cov}_e^{\mathrm{TS}}(x_{t-1}, y_{t-1}, w_{t-1}),
\tag{5}
$$

$$
\hat{\mathbf{z}}_t^L = f^{\mathrm{TS}}(x_{t-1}, y_{t-1}, w_{t-1}) + \sigma^L \mathrm{Cov}_e^{\mathrm{TS}}(x_{t-1}, y_{t-1}, w_{t-1}),
\tag{6}
$$

where σ is the interval width and can be adjusted for the given dataset with certain coverage grade CG and $\mathrm{Cov}^{\mathrm{TS}}$ is the covariance of the target and predicted data model as $\mathrm{Cov}^e = (y_j - \hat{y}_j)$ same as T and S mentioned earlier.

Hence, the fuzzy regression model Υ parameters are identified by clustering and with (5) and (6), the coefficient A and covariance matrix Cov^e is trained using evolutionary search algorithms such as the least square combined back propagation (FR), particle swarm optimization (PSO), firefly (FF) optimization, cultural algorithm (CA), and genetic algorithm (GA)

(2) *Interval band*: Interval band is introduced in term of σ, where the lower the value of σ indicates the smaller bandwidth of the interval and lower the probability of missing real measurement value from the predicted values.

$$\gamma = \frac{\text{RMSE}_{\text{max}} - \overbrace{\left(\sum_{p=1}^{q} \text{RMSE}_{\text{error}_p}\right)}^{\alpha}}{\text{RMSE}_{\text{max}}}. \tag{7}$$

In the above relation, α represents the point error RMSE of the historical prediction q. RMSE_{max} is the maximum training root mean square value. Interval band is directly related with γ such as $\sigma = 1 - \gamma$.

(3) *Coverage grade*: In order to classify the forecast system based on performance, the authors in this paper introduced the coverage grade CG system. This performance evaluation method with interval band gives the insight into the accuracy of predictions. The levels of CG is divided into 3 levels, namely, A, B, and C.

$$\text{CG} = \frac{\sum_{j=1}^{m} \kappa}{m}. \tag{8}$$

Coverage grade is calculated to adjust the σ value and is the interval; κ is the binary parameter which shows that whether the measurement data lies inside of the interval or not. Lower values of RMSE in point forecast gives the bandwidth for respective grades such as A grade (90% <CG ≤ 100%) coverage, B grade (70%<CG ≤ 90%) coverage, and C grade (low < CG ≤ 60%) coverage. Based on defined parameters, the proposed method will constantly improve the performance of the forecast using interval band and coverage grade tuning with (7) and (8). For example in terms of CG_q, a forecaster with A_{30} label indicates the operator that the algorithm is running on best performance for past 30 historical points.

3.1. Particle Swarm Optimization for FPI. Swarm-based optimization techniques are very much dependent on the initial conditions, but the ability of avoiding local minima is much higher provided that the parameters are carefully selected and initialized. Particle swarm is a well-known global optimization technique which mimics the bird flocking approach like other swarm optimization techniques. All birds are looking for food (min/max objective) in different directions, and a bird close to the food will be followed (distance and velocity are set on every step) by the swarm and finally, it converges on the solution in [30].

$$\begin{aligned} \text{Vel}_{\text{particle}}^{t+1} = w_{\text{int}} \text{Vel}_{\text{particle}}^{t} + c_{\text{pl}} r_{\text{dist}}^{1} \left(p_{\text{partlocal}}^{t} - x_{\text{particle}}^{t} \right) \\ + c_{\text{gl}} r_{\text{dist}}^{2} \left(p_{\text{partglobal}}^{t} - x_{\text{particle}}^{t} \right), \end{aligned} \tag{9}$$

where Vel, p, and x are the velocity, best position, and position, respectively, for each particle at the t interval. Furthermore, w_{int}, c_{pl}, and c_{gl} are inertial weight, local learning coefficient, and global learning coefficient, respectively. r_{dist}^{1} and r_{dist}^{2} are the random distribution of the particle. The complete PSO-based training algorithm for fuzzy model tuning is mentioned in Algorithm 1.

4. Case Study

4.1. Microgrid Test Bed. This paper used the data from the Goldwind Microgrid test bed which is located (3945′23.6′N, 11631′56.3″E) in the suburb of Beijing, China. The microgrid system consists of a wind turbine which is rated at 2.5 MW; two photovoltaic system, one is 250 kW and the other one is 200 kW capacity; two battery systems connected with bidirectional converters; four fuel generators include diesel and natural gas fuel type; and the maximum rated load of 3 MW for the office building on the site. Simplicity of this forecast scheme is the reduce input variable set which is easily available and faster to predict, and it can modify for the next step predictions, for example, wind velocity and historical wind power are used for future wind power prediction; similarly, historical load data is used for future load forecast. The impact of other weather parameters is embedded in the historical power output, but in case of load, it is more related to the time of the day rather than the weather data. All the NWP [weather research and forecasting model (WRF)] information for the next day such as wind velocity and solar irradiance are acquired from the FTP server of Goldwind Technology, whereas the other historical information is collected from the SCADA system. PSO algorithm is used to modify the weight matrix of the fuzzy set of functions which are separated into 5 clusters and then interval bands are obtained with different coverage grades in order to check the forecast performance. The results are presented in 10 min sampling time with one step to one day (144 steps) ahead predictions. One day ahead prediction is performed every 10 min interval based on the input data for real and predicted values.

Flowchart is shown in Figure 2 for the PSO-based fuzzy prediction interval scheme; two step training is applied on every time stamp. In the first step, historical data is acquired and fuzzy set of membership functions is initialized in clusters using fuzzy c-mean clustering method. After that, PSO is applied in order to get optimal-trained fuzzy model with lowest RMSE value, and it sends the power reserve signal to the EMS module in case of a higher RMSE than the set points. In the second step, the model acquires next day ahead weather parameters in order to get the initial forecast; later on, the coverage grade is calculated and checked against the set point of CG value (in percentage), if the value is above the set point then the interval band adjusts and again calculates the next step ahead the forecast else, the model will train and update again accordingly. Finally, an Intel i5, 2.53 GHz quad-core processor with 4 GB RAM laptop is used to compute the results of this study.

4.2. Wind Prediction. Wind power prediction is done by FPI, June 11, 2016, to July 26, 2016, data are used to train and July

Get Initial Parameters (fuzzy model weights, cost function, variables count, variables range, max iteration, population size, inertial weight and damping ratio, personal and global learning coefficient)
for 1 **to** population size **do**
    Initialize the position [30], velocity, personal and global best of each particle based on the cost function.
end for
repeat
    for 1 **to** population size **do**
        Update velocity of a particle using (9)
        Update position, velocity and cost of each particle
        Apply minor limits for the velocity of each particle
        if particle cost ≤ particle best cost **then**
            update personal best
            if particle best cost ≤ best solution cost **then**
                Update global best
            end if
        end if
    end for
    Best cost of iteration ← best solution cost
    Update inertial weight
until Maximum iteration & minimum error achieve

ALGORITHM 1

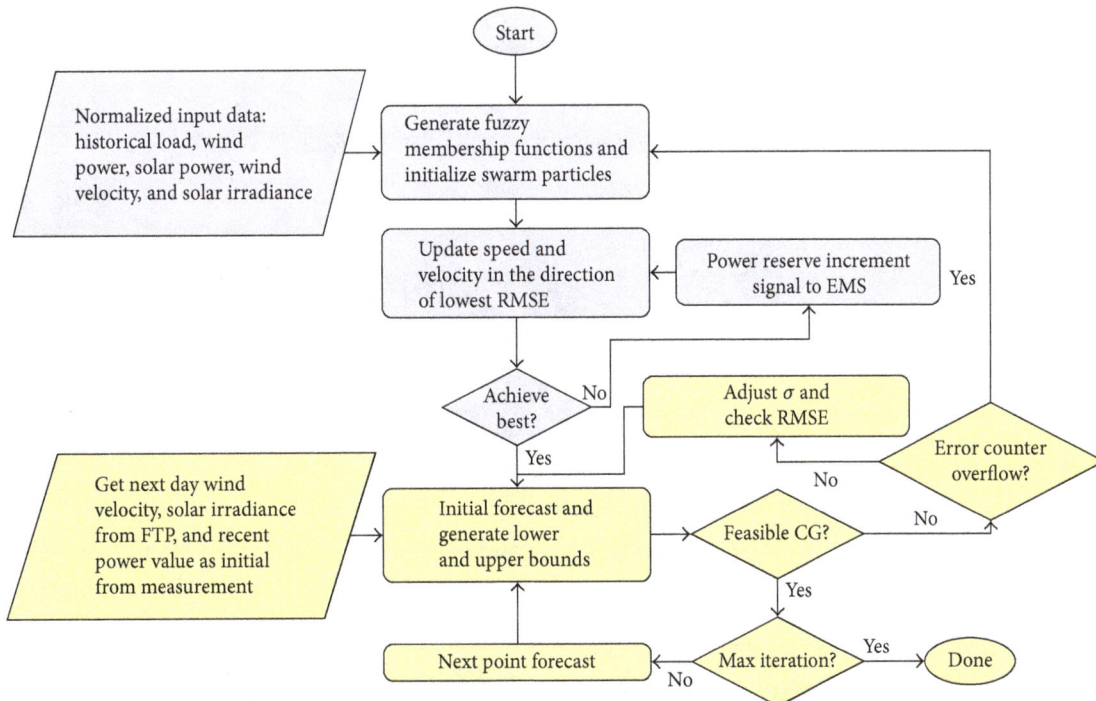

FIGURE 2: Flowchart of the proposed method.

27 to Aug 12, 2016, data are used for validation purposes. All the data are gathered from the SCADA of the Goldwind Microgrid test bed, Beijing. The rated capacity of this single PMSG type wind turbine is 2.5 MW. Wind velocity and historical power are the training inputs with 10 min of sampling time. The linear regressor for input variable (wind speed) and output variable (wind power) are v_r and p_r, respectively, in discrete time. As in interval prediction, the output of exogenous variable is related to the last historical values $(t-1)$ then, in accordance with the notation $z_r = [v_{r-1}^3, p_{r-1}]$ where r is the discrete variable. Hence, the ith rule for the membership function in terms of linear regressors for the wind

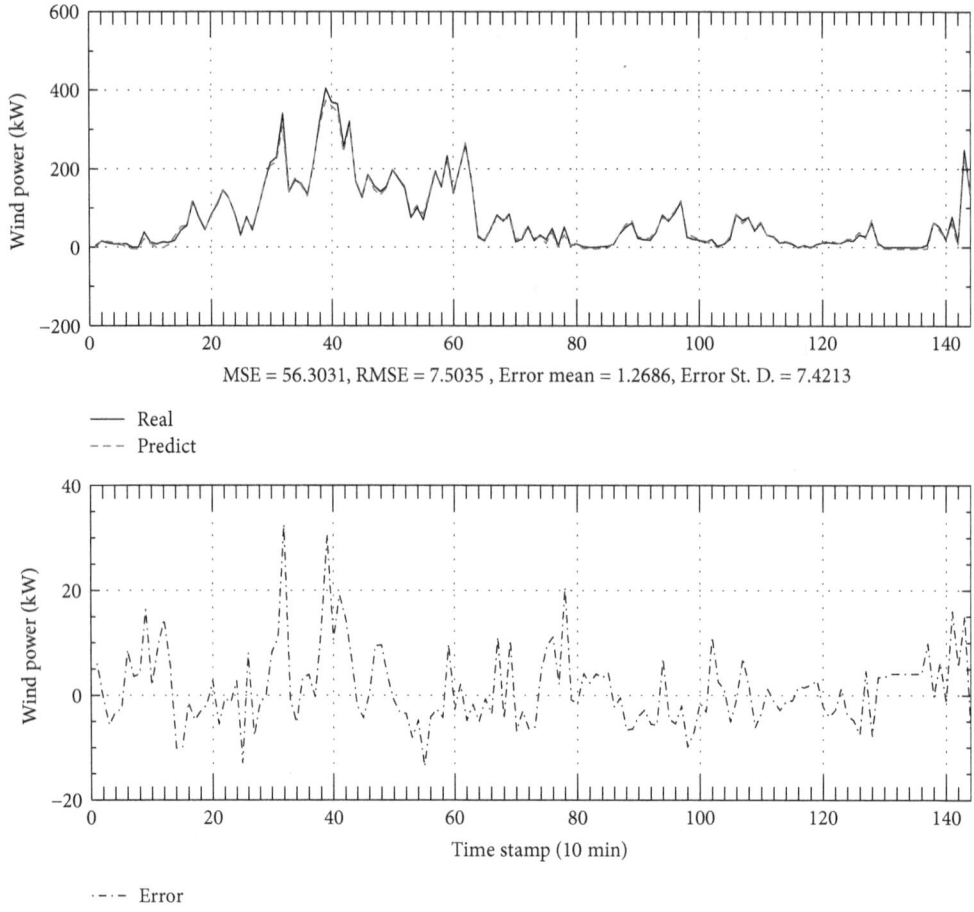

$$\text{MSE} = 56.3031, \text{RMSE} = 7.5035, \text{Error mean} = 1.2686, \text{Error St. D.} = 7.4213$$

—— Real
--- Predict

--- Error

FIGURE 3: One day ahead RMSE value of wind forecast.

energy prediction is defined by clustering approach which is mentioned in Section 3.

$$\hat{z}_r = P_w(r) = \sum_{i=1}^{n} \beta_i \left(a_0^i + a_1^i v_{r-1}^3 + a_2^i v_{r-2}^3 + \cdots + a_k^i v_{r-k}^3 \right.$$
$$\left. + b_0^i + b_1^i p_{r-1} + b_2^i p_{r-2} + \cdots + b_k^i p_{r-k} \right). \tag{10}$$

Here, $A = [a_0, \ldots, a_{r-144}, b_0, \ldots, b_{r-144}]^T$ is same as mentioned earlier and considers $P_w^U(r)$ and $P_w^L(r)$ as the lower and upper bounds related to $P_w(r)$ such that $P_w^L(r) \leq P_w(r) \leq P_w^U(r)$ for all r in discrete space. Furthermore, $A^U = [a_0^U, \ldots, a_{r-144}^U, b_0^U, \ldots, b_{r-144}^U]^T$ is for $P_w^U(r)$ and $A^L = [a_0^L, \ldots, a_{r-144}^L, b_0^L, \ldots, b_{r-144}^L]^T$ is for $P_w^L(r)$. These parameters, namely, A^U and A^L can be adjusted based on training with PSO. The prediction intervals now can be written as

$$P_w^U(r) = f^{TS}\left(v_{r-1}^3, p_{r-1}\right) + \sigma^U \text{Cov}_{\text{wind}}^{TS}\left(v_{r-1}^3, p_{r-1}, \text{Cov}_{r-1}^e\right), \tag{11}$$

$$P_w^L(r) = f^{TS}\left(v_{r-1}^3, p_{r-1}\right) + \sigma^L \text{Cov}_{\text{wind}}^{TS}\left(v_{r-1}^3, p_{r-1}, \text{Cov}_{r-1}^e\right), \tag{12}$$

$$\text{Cov}_{\text{wind}}^{TS} = \sum_{i=1}^{n} \beta_i(z_{r-1}) \text{Cov}_{r-1,i}^e, \tag{13}$$

where covariance of past error Cov_{r-1}^e is integrated with the fuzzy interval of input and error as $\text{Cov}_{\text{wind}}^{TS}$. Next, the lower and upper bounds for the fuzzy interval-based wind power P_w are calculated based on the coverage grade, and interval band is adjusted accordingly.

Figure 3 shows the one day ahead forecast of wind power and error value for 144 steps. This can be easily judged from the graph that the error increases more at the sharp changes in measured wind power. The value of mean square error (MSE), root mean square error (RMSE), error mean, and standard deviation gives insight about the forecast, for example, the RMSE [kW] is 7.5035 which is very satisfactory result achieved through the PSO integration in the T and S model.

In Figure 4, case 1 presents the gradual variation in one day ahead wind forecast and measured power with different coverage grades based on different values of σ. It should be noted here that the CG of level C has a narrower band with some data points whereas the grade A level CG can be accounted for almost all data points but the uncertainty is higher, whereas level B has fewer data points compared to

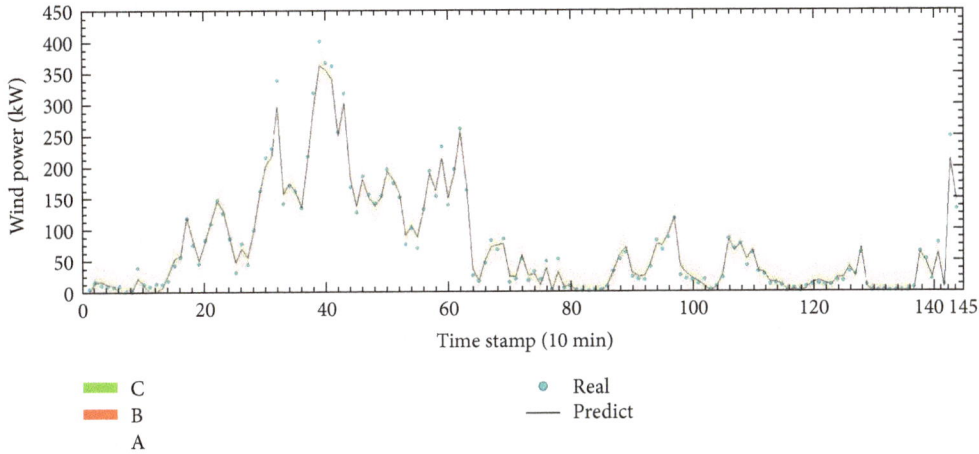

FIGURE 4: Fuzzy interval prediction for one day ahead (case 1) wind power with coverage grades.

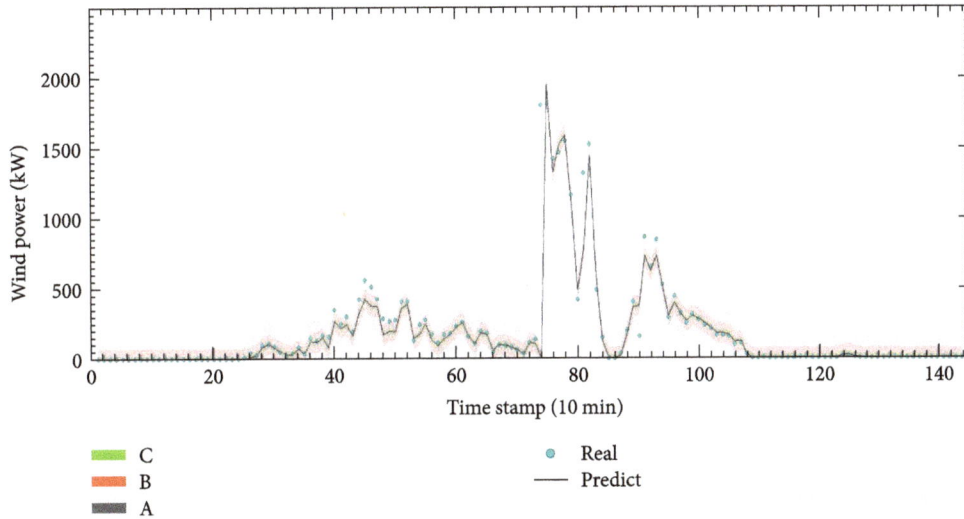

FIGURE 5: Fuzzy interval prediction for one day ahead (case 2) wind power with coverage grades.

A but the uncertainty is lower than A. Hence, for a normal day with gradual change of wind velocity forecast, an operator can go with grade C or B level predictions but in case 2, the abrupt wind velocity alerts from day ahead of the NWP forecast data where the level of uncertainty is higher, so one should choose grade A or B as shown in Figure 5. Furthermore, higher variability in real values is captured accurately with the help of grade A prediction intervals; hence, the operator or EMS can adjust the spinning reserves based on the level of uncertainty required. Power reserve capability can cope against forecasting uncertainty by leaving an adequate margin for controllable units to compensate the mismatch. Here, the mechanism will only alert EMS for the possible mismatch.

Table 1 shows the comparative results obtained by applying other techniques to check the forecast error for one step (10 min) and one day ahead. The table mentioned train time, mean absolute error (MAE), and RMSE for 4 different approaches including fuzzy regression (FR) [27], cultural algorithm (CA), firefly (FF) algorithm, and genetic

TABLE 1: Comparison of prediction interval forecast for wind power.

Model	One step ahead		One day ahead		Train time (s)
	RMSE (kW)	MAE (kW)	RMSE (kW)	MAE (kW)	
FR	5.97	9.47	8.52	8.91	11.4
CA	189.9	10.15	39.5	26.78	131.3
FF	7.83	2.41	9.42	8.94	145
GA	1.88	8.72	10.19	8.35	60.1
PSO	4.28	6.67	7.5	9.8	56.4

algorithm (GA) apart from PSO. It should be noted that longer horizon has a higher error than one step ahead except CA. Based on the results obtained, fuzzy prediction model with PSO outperforms other techniques in terms of lowest RMSE in long-horizon forecast whereas CA has the worst RMSE value and train time due to the fact of falling in local minima. GA also shows a better result in one step

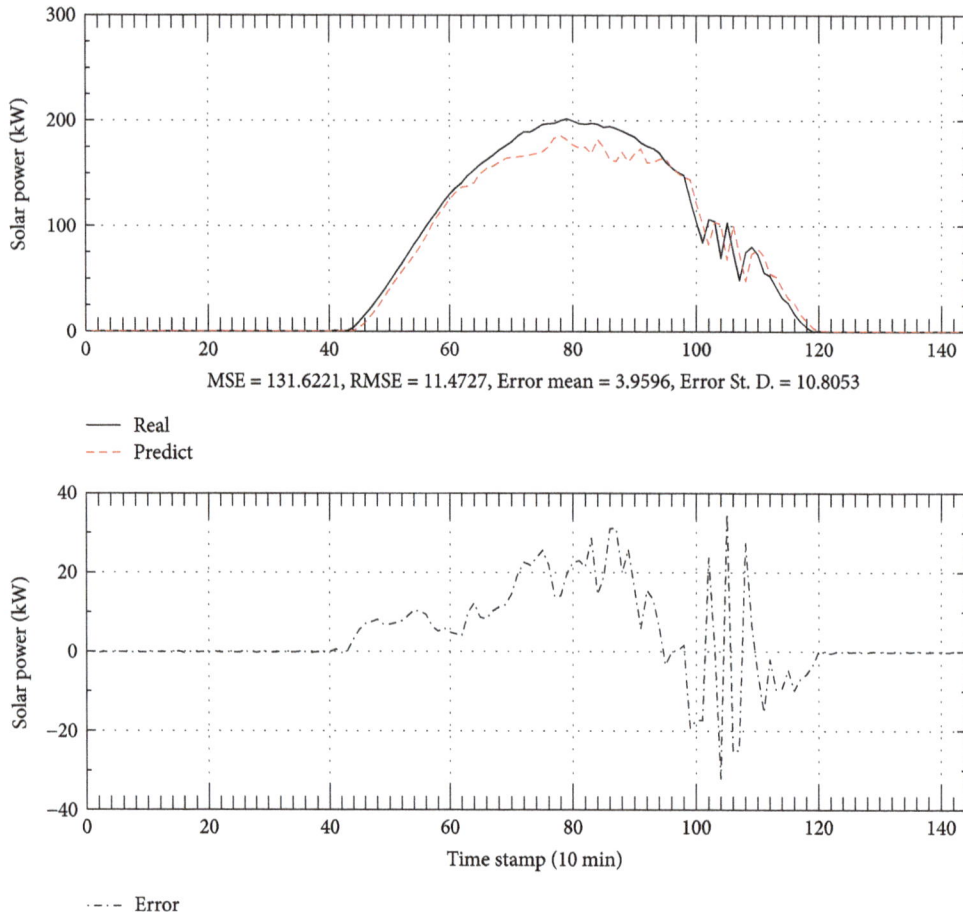

MSE = 131.6221, RMSE = 11.4727, Error mean = 3.9596, Error St. D. = 10.8053

—— Real
----- Predict

----- Error

FIGURE 6: One day ahead RMSE value of solar forecast.

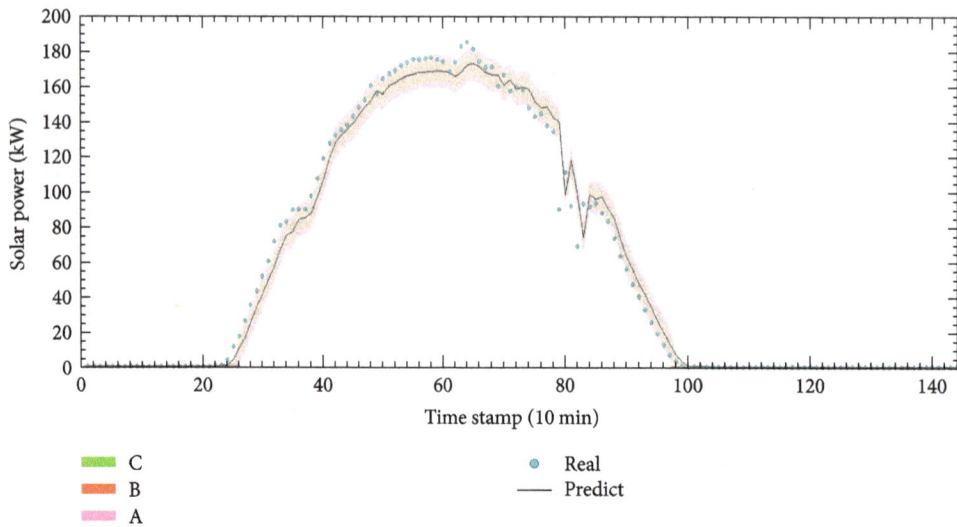

FIGURE 7: Fuzzy interval prediction for one day ahead (case 1) solar power with coverage grades.

ahead than PSO but for longer run it has higher RMSE than PSO. FR shows significant fast train time as mentioned in [27], but PSO shows superior results for one-day head forecast for local dataset.

4.3. Solar Prediction. Solar power fuzzy model is described in this section using the data acquired from June 11, 2016, to July 26, 2016, for training and July 27 to Aug 12, 2016, for validation. The rated capacity of the PV panel at the top

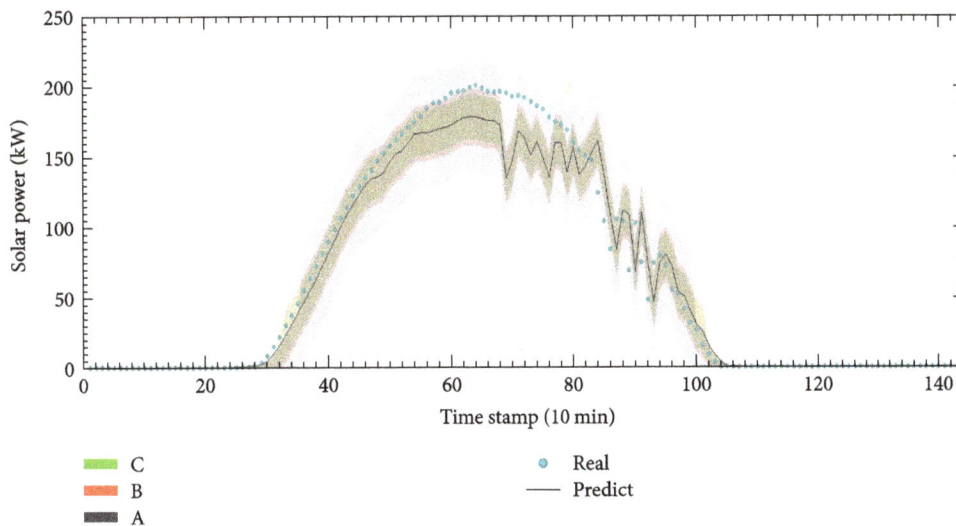

FIGURE 8: Fuzzy interval prediction for one day ahead (case 2) solar power with coverage grades.

of the Goldwind Technology office building is 450 kW, but due to efficiency issues, we can only observe power close to 200 kW at peak sunny days. Here, the input variables for the training of the proposed model are irradiance data from NWP and historical solar power output at 10 min interval (144 points in a day) $p_r = p_s$: hence, $z_r = [ir_{r-1}, p_{r-1}]$ is the relation for the identification procedure mentioned in Section 3.

$$\hat{\mathbf{z}}_r = P_s(r) = \sum_{i=1}^{n} \beta_i \left(a_0^i + a_1^i ir_{r-1} + a_2^i ir_{r-2} + \cdots + a_k^i ir_{r-k} \right. $$
$$\left. + b_0^i + b_1^i p_{r-1} + b_2^i p_{r-2} + \cdots + b_k^i p_{r-k} \right). $$
$$(14)$$

Here, $A = [a_0, \ldots, a_{r-144}, b_0, \ldots, b_{r-144}]^T$ is the weight vector which is tuned with PSO and considers $P_s^U(r)$ and $P_s^L(r)$ as the lower and upper bounds related to $P_s(r)$, such that $P_s^L(r) \leq P_s(r) \leq P_s^U(r)$ for all r in discrete space. Hence, A^U and A^L are calculated for certain coverage grade and interval band.

Figure 6 shows the predicted solar output and the error value for 144 points ahead. The day starts from midnight where zero values are recorded. Additionally, the error increases at the peak period of the day and RMSE (kW) value is recorded as 11.4727.

A normal sunny day solar power forecast result is shown in Figure 7 as case 1. The forecast shows zero values at midnight and evening intervals while the peak power output is recorded around the afternoon time. Note that the predicted value is slightly lower than that of the real one in the rise up and peak period and slightly higher than the real value at the falling period due to the inaccuracy in the NWP value for day-ahead irradiance. However, the interval with grade B (70–90%) is sufficed to capture uncertainty at peak points of the day which is a more important time for the operator or EMS to

properly allocate the extra power using spinning reserves. Another scenario in case 2 is shown in Figure 8; here, the abrupt and unwanted change in solar power due to irregular sunlight is completely missed by predicted value at the peak point of the day, whereas grade A (90–100%) level interval band captured this trend at the cost of higher uncertainty.

Table 2 shows the comparison between PSO-based fuzzy model and other fuzzy models in terms of RMSE and MAE values for solar power forecast. The results show one day ahead and 10 min ahead forecast, lowest training time obtained with FR, whereas the lowest RMSE for one day and one step ahead is achieved with PSO, which again showed better performance than other techniques.

4.4. Load Prediction. For load predictor modeling, a dataset consisting of 10 min samples is acquired from 4 May to 28 April 2016 for training and 28 April to 16 May 2016 for the validation purpose. Load data consist of historical power demand of the office building at Goldwind Technology, Beijing. The maximum load demand of the building is about 3 MW. With the help of correlation analysis, the temperature correlation with demand is almost vanished because of very short duration for spring and autumn in Beijing; hence, the air conditioning and heating loads are

TABLE 2: Comparison of prediction interval forecast for solar power.

Model	One step ahead		One day ahead		Train time (s)
	RMSE (kW)	MAE (kW)	RMSE (kW)	MAE (kW)	
FR	5.44	5.41	12.97	8.2	1.22
CA	5.65	5.63	12.57	7.73	46.69
FF	5.3	5.28	13.42	8.28	45.7
GA	5.53	5.50	12.32	7.65	23.43
PSO	4.05	4.03	11.47	7.04	20.95

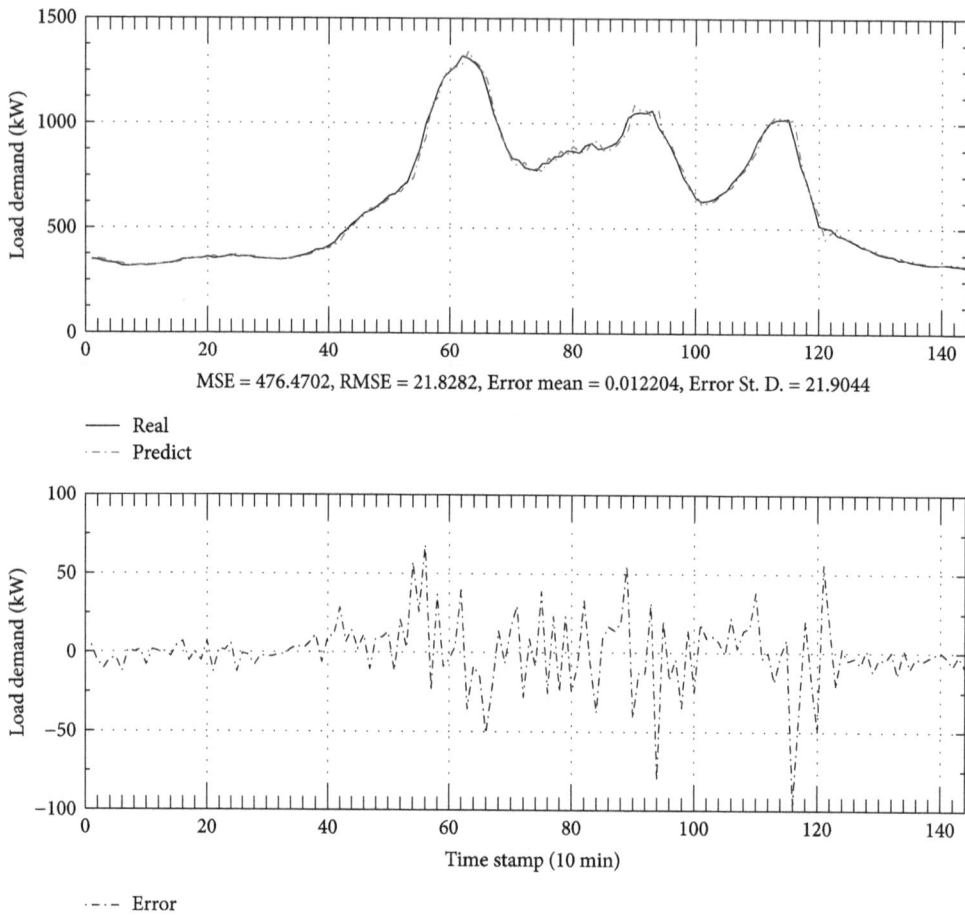

MSE = 476.4702, RMSE = 21.8282, Error mean = 0.012204, Error St. D. = 21.9044

—— Real
- - - Predict

- - - Error

FIGURE 9: One day ahead RMSE value of demand forecast.

balanced out and only start as well as end of time for work are correlated with the demand. Using the identification method, we can safely consider the historical power demand pattern $p_r = p_l$ in order to predict next day ahead power demand $z_r = [p_{r-1}]$ in order to avoid complexity.

$$\hat{\mathbf{z}}_r = P_l(r) = \sum_{i=1}^{n} \beta_i \left(a_0^i + a_1^i p_{r-1} + a_2^i p_{r-2} + \cdots + a_k^i p_{r-k} \right), \quad (15)$$

where $A = [a_0, \ldots, a_{r-144}]^T$ is trained by PSO and similarly considers $P_l^U(r)$ and $P_l^L(r)$ as the lower and upper bounds related to $P_l(r)$ such that $P_l^L(r) \le P_l(r) \le P_l^U(r)$ for all r in discrete space. Hence, A^U and A^L are calculated for certain coverage grade and interval band. Weekdays and holiday effects can be integrated with the exogenous variable of the fuzzy interval model for better accuracy, but this job can be done by tuning the model everyday with the past 30-day data point windows in order to refresh the membership parameters of the model.

Figure 9 represents the load power prediction for one day ahead (1–144) data and the RMSE value is recorded as 21.82. It is worth noting that the base load around midnight and evening is accurately captured by the fuzzy model where the small deviations are observed at the peak levels of the predicted demand power, but this issue can be tackled with the

rolling horizon type of EMS such as the model predictive controller-based energy manager, because at each time step EMS will trigger for new values and similarly the forecast module also updates its calculation which changes to the new operating points.

In Figure 10, case 1 is shown for the one day ahead load prediction of a weekday using the last day power values as the input. The day starts with the midnight flat load demand around 00–06, then the rise up time starts from 07–10 where the office usually starts, then there have two peaks around 11–13 and 14–16 corresponds to the peak hours pre- and postlunchtimes, after that, a falling edge starts followed by a constant load indicates the off work time. This scenario is accurately predicted by the fuzzy model because the constant pattern in load demand of working days plus the coverage grade C and B contain the majority of the data points with very narrow interval band which shows very low RMSE values. In another load demand scenario shown in Figure 11 as case 2, the pattern is same but the sudden changes at peak time or maximum demand time is very crucial in predicting optimal reserve allocation. As seen here, grade A is successful in capturing more than (95%) of measurement values.

Table 3 shows the comparison between PSO-based fuzzy model and other evolutionary-based fuzzy models in terms of RMSE and MAE values for load power forecast. PSO again performs better in terms of RMSE for one day ahead forecast,

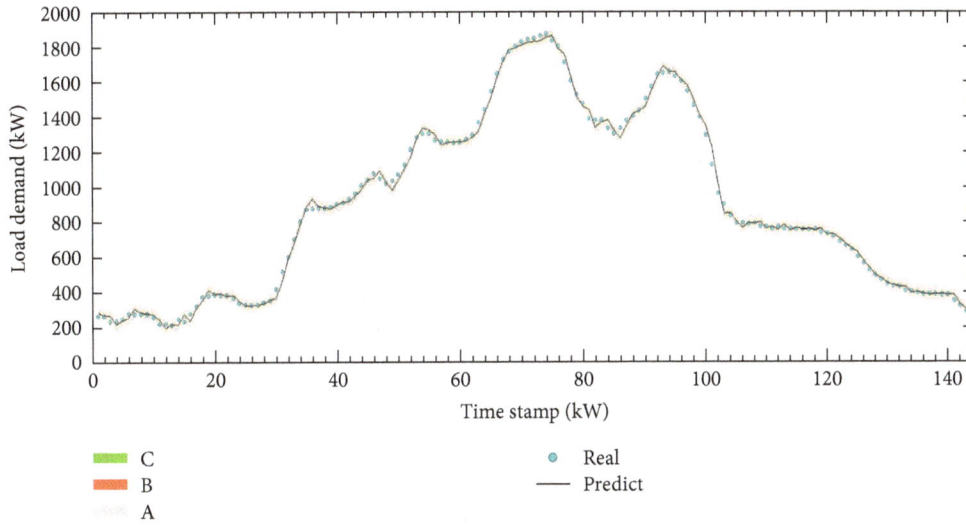

FIGURE 10: Fuzzy Interval prediction for one day ahead (case 1) load power with coverage grades.

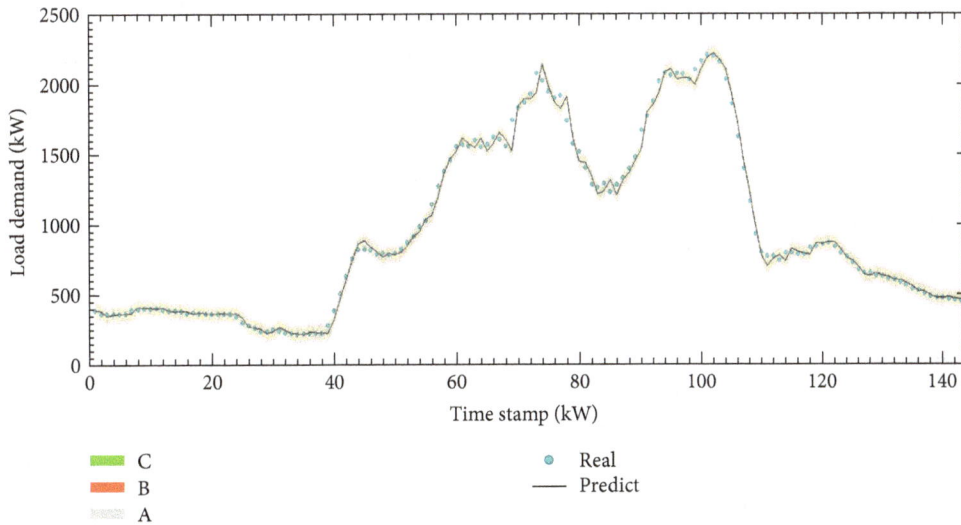

FIGURE 11: Fuzzy interval prediction for one day ahead (case 2) load power with coverage grades.

TABLE 3: Comparison of prediction interval forecast for demand power.

Model	One step ahead		One day ahead		Train time (s)
	RMSE (kW)	MAE (kW)	RMSE (kW)	MAE (kW)	
FR	20.67	18.28	23.38	14.86	20.22
CA	31.11	26.27	51.16	34.59	112.9
FF	16.25	14.10	22.81	15.19	309.8
GA	18.9	17.09	22.47	14.9	153.7
PSO	5.85	46.7	21.83	14.58	115.3

but the training time takes much longer than that of the FR-based training model. Overall, PSO performs well in demand forecast and can be improved more with the additional input of training parameters including time of the day, work, and weekend information.

5. Conclusion

Improved fuzzy interval prediction model trained with meta-heuristic algorithms is proposed in this paper, which is the premise study for energy management system of the micro-grid test bed in Beijing. The fuzzy interval prediction model is helpful in getting close to real results for the uncertainties associated with the nondispatchable renewable generation and consumer demand. Fuzzy prediction intervals are generated for wind, solar, and load for one day ahead prediction using real-time values, and the results are characterized into different coverage grades based on the accuracy of the forecast. Wind and solar showed the higher level of accuracy in the forecast due to the fact that these sources are directly related with their respective input variables whereas load forecast accuracy can be improved with more information and active interaction of the consumer in decision-making. All the results are shown with the traditional fuzzy regression and metaheuristic techniques, where the proposed method showed superior results in terms of lower RMSE and evaluation scores than the other approaches. Furthermore, the error and coverage level information achieved through the proposed scheme will improve the utilization of reserves more robustly in the microgrid. Future work will address the implementation details of energy management system based on the input of this prediction algorithm for the same microgrid test bed.

Conflicts of Interest

The authors declare that there are no conflicts of interests regarding the publication of this paper.

Acknowledgments

The authors would like to thank the Microgrid R&D staff members at Goldwind Technology for the collaboration and assistance in this research study.

References

[1] J. Dowell and P. Pinson, "Very-short-term probabilistic wind power forecasts by sparse vector autoregression," *IEEE Transactions on Smart Grid*, vol. 7, no. 2, pp. 1–770, 2016.

[2] M. De Felice and X. Yao, "Short-term load forecasting with neural network ensembles: a comparative study [application notes]," *IEEE Computational Intelligence Magazine*, vol. 6, no. 3, pp. 47–56, 2011.

[3] M. Soshinskaya, W. H. Crijns-Graus, J. M. Guerrero, and J. C. Vasquez, "Microgrids: experiences, barriers and success factors," *Renewable and Sustainable Energy Reviews*, vol. 40, pp. 659–672, 2014.

[4] S. Mei, Y. Wang, and Z. Sun, "Robust economic dispatch considering renewable generation," in *2011 IEEE PES, Innovative Smart Grid Technologies Asia (ISGT)*, pp. 1–5, Perth, WA, Australia, 2011.

[5] W. Shi, X. Xie, C.-C. Chu, and R. Gadh, "Distributed optimal energy management in microgrids," *IEEE Transactions on Smart Grid*, vol. 6, no. 3, pp. 1137–1146, 2015.

[6] F. Valencia, J. Collado, D. Sáez, and L. G. Marn, "Robust energy management system for a microgrid based on a fuzzy prediction interval model," *IEEE Transactions on Smart Grid*, vol. 7, no. 3, pp. 1486–1494, 2016.

[7] M. Hosseinzadeh and F. R. Salmasi, "Robust optimal power management system for a hybrid ac/dc micro-grid," *IEEE Transactions on Sustainable Energy*, vol. 6, no. 3, pp. 675–687, 2015.

[8] Q. Jiang, M. Xue, and G. Geng, "Energy management of microgrid in grid-connected and stand-alone modes," *IEEE Transactions on Power Systems*, vol. 28, no. 3, pp. 3380–3389, 2013.

[9] W. Wei, F. Liu, S. Mei, and Y. Hou, "Robust energy and reserve dispatch under variable renewable generation," *IEEE Transactions on Smart Grid*, vol. 6, no. 1, pp. 369–380, 2015.

[10] Ahlstrom, L. Jones, R. Zavadil, and W. Grant, "The future of wind forecasting and utility operations," *IEEE Power and Energy Magazine*, vol. 3, no. 6, pp. 57–64, 2005.

[11] Y. Jiang, C. Xingying, Y. Kun, and L. Yingchen, "Short-term wind power forecasting using hybrid method based on enhanced boosting algorithm," *Journal of Modern Power Systems and Clean Energy*, vol. 5, no. 1, pp. 126–133, 2017.

[12] P. Pinson and G. Kariniotakis, "Conditional prediction intervals of wind power generation," *IEEE Transactions on Power Systems*, vol. 25, no. 4, pp. 1845–1856, 2010.

[13] H. Bludszuweit, J. A. Domnguez-Navarro, and A. Llombart, "Statistical analysis of wind power forecast error," *IEEE Transactions on Power Systems*, vol. 23, no. 3, pp. 983–991, 2008.

[14] X. Yang, F. Jiang, and H. Liu, "Short-term solar radiation prediction based on SVM with similar data," in *2nd IET Renewable Power Generation Conference (RPG 2013)*, Beijing, China, 2013.

[15] A. Rashkovska, J. Novljan, M. Smolnikar, M. Mohorčič, and C. Fortuna, "Online short-term forecasting of photovoltaic energy production," in *2015 IEEE Power & Energy Society Innovative Smart Grid Technologies Conference (ISGT)*, pp. 1–5, Washington, DC, USA, 2015.

[16] C. Yang and L. Xie, "A novel ARX-based multi-scale spatiotemporal solar power forecast model," in *2012 North American Power Symposium (NAPS)*, pp. 1–6, Champaign, IL, USA, 2012.

[17] S. K. Chow, E. W. Lee, and D. H. Li, "Short-term prediction of photovoltaic energy generation by intelligent approach," *Energy and Buildings*, vol. 55, pp. 660–667, 2012.

[18] I. Abubakar, S. Khalid, M. Mustafa, H. Shareef, and M. Mustapha, "Application of load monitoring in appliances' energy management – a review," *Renewable and Sustainable Energy Reviews*, vol. 67, pp. 235–245, 2017.

[19] T. Hong and P. Wang, "Fuzzy interaction regression for short term load forecasting," *Fuzzy Optimization and Decision Making*, vol. 13, no. 1, pp. 91–103, 2014.

[20] Y. Xiang, J. Liu, and Y. Liu, "Robust energy management of microgrid with uncertain renewable generation and load," *IEEE Transactions on Smart Grid*, vol. 7, no. 2, pp. 1034–1043, 2016.

[21] N. I. Sapankevych and R. Sankar, "Time series prediction using support vector machines: a survey," *IEEE Computational Intelligence Magazine*, vol. 4, no. 2, pp. 24–38, 2009.

[22] J. L. Mathieu, P. N. Price, S. Kiliccote, and M. A. Piette, "Quantifying changes in building electricity use, with application to demand response," *IEEE Transactions on Smart Grid*, vol. 2, no. 3, pp. 507–518, 2011.

[23] J. W. Taylor, "Short-term load forecasting with exponentially weighted methods," *IEEE Transactions on Power Systems*, vol. 27, no. 1, pp. 458–464, 2012.

[24] L. Xuemei, D. Lixing, D. Yuyuan, and L. Lanlan, "Hybrid support vector machine and ARIMA model in building cooling prediction," in *2010 International Symposium on Computer, Communication, Control and Automation (3CA)*, vol. 1, pp. 533–536, Tainan, Taiwan, 2010.

[25] D. Chaturvedi, A. Sinha, and O. Malik, "Short term load forecast using fuzzy logic and wavelet transform integrated generalized neural network," *International Journal of Electrical Power & Energy Systems*, vol. 67, pp. 230–237, 2015.

[26] A. G. Abdullah, G. M. Suranegara, and D. L. Hakim, "Hybrid PSO-ANN application for improved accuracy of short term

load forecasting," *WSEAS Transactions on Power Systems*, vol. 9, pp. 446–451, 2014.

[27] D. Sáez, F. Ávila, D. Olivares, C. Cañizares, and L. Marn, "Fuzzy prediction interval models for forecasting renewable resources and loads in microgrids," *IEEE Transactions on Smart Grid*, vol. 6, no. 2, pp. 548–556, 2015.

[28] T. Takagi and M. Sugeno, "Fuzzy identification of systems and its applications to modeling and control," *IEEE transactions on systems, man, and cybernetics*, vol. SMC-15, no. 1, pp. 116–132, 1985.

[29] M. Sugeno and T. Yasukawa, "A fuzzy-logic-based approach to qualitative modeling," *IEEE Transactions on Fuzzy Systems*, vol. 1, no. 1, pp. 7–31, 1993.

[30] P. Li, D. Xu, Z. Zhou, W.-J. Lee, and B. Zhao, "Stochastic optimal operation of microgrid based on chaotic binary particle swarm optimization," *IEEE Transactions on Smart Grid*, vol. 7, no. 1, pp. 66–73, 2016.

5

Experimental Study on a Forced-Circulation Loop Thermosiphon Solar Water Heating System

Tao Zhang (ID)

School of Energy and Mechanical Engineering, Shanghai University of Electric Power, Shanghai 200090, China

Correspondence should be addressed to Tao Zhang; zhtyn86@163.com

Academic Editor: Alberto Álvarez-Gallegos

Wickless gravity loop thermosiphons (LTs) have been widely used in heat collection for distances up to several meters. This two-phase closed device, which is operating under reduced pressure, is useful in solar water heating (SWH) systems because it could address the freezing problem during winter. Compared to the normal type, forced-circulation wickless LTs have significant advantages in the long-distance heat transfer and installation freedom of condensation section. In this study, a pump-forced wickless LT-SWH system with a remolded flat-plate solar collector was put forward. Solar collector acted as the evaporation section of the wickless LT, while the spiral heat exchanger in the water tank acted as the condensation section. R600a was employed as the working fluid, and long-term outdoor experiments were carried out. Results show that the instantaneous and daily average photothermal efficiency of the proposed system can reach 69.54% and 58.22%, respectively. Temperature differences between the top and bottom and the middle and bottom of the evaporation section of the wickless LT were small, and it usually ranged between 1.1 and 3.9°C. Linear fittings of the collector and system overall performance of the pump-forced wickless LT-SWH system demonstrate the promising potential application of the system.

1. Introduction

Solar water heating (SWH) systems have been widely applied in both domestic and industrial levels, and SWHs have been proven to be readily available technologies that can directly substitute renewable energy for conventional water heating. Based on the type of involved circulation, the SWHs can be categorized into two groups: natural circulation and forced circulation. Different circulation types of the SWH systems are expected to be available and suitable for different applications [1, 2]. Forced-circulation SWHs are usually used in commercial and industrial heat processing [3]. The solar collector is of great importance and works as the heart of the SWH system. The most diffused solar technologies are evacuated tube collectors in China and are flat-plate solar collectors in the rest of world [4]. However, for a conventional flat-plate solar collector, there are freezing problems when the ambient temperature is lower than zero, as well as corrosion problems when the water temperature is high and the pH departs from the neutral level [5].

Loop thermosiphons (LTs) [6], vapordynamic thermosiphons (VDTs) [7, 8], and pulsating heat pipes (PHPs) [9] are alternative solutions to the aforementioned problems. Among these heat transfer devices, LTs are most appreciated when integrating with solar collector for the simple manufacturing process. LT is a two-phase closed device that provides an alternative solution to the aforementioned problems. The evaporation and condensation sections of a LT are separated; thus, it can realize effective and remote heat transfer with the assistance of capillary pumping or gravity [10, 11]. It has excellent heat transfer characteristic due to the vacuum seal and phase-change heat transfer. It also has excellent isothermal characteristics on the basis of inner two-phase heat transfer. The LT-SWHs have been intensively investigated in recent decades. Soin et al.'s [12, 13] experiments examined the photothermal performance of a LT solar evaporator charged with acetone and petroleum ether. The photothermal efficiency of the two-phase solar collector was only approximately 6%–11%, which was lower than that of a water-based SWH system. M. Esen and H. Esen [14]

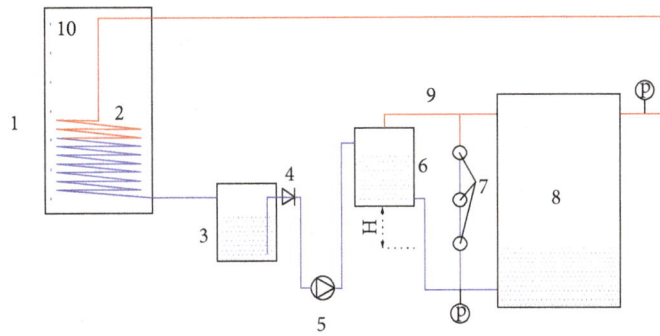

FIGURE 1: Schematic of the forced-circulation wickless LT-SWH system. (1) Water tank, (2) condensation spiral coil pipe, (3) condensation liquid storage tank, (4) check valve, (5) refrigerant pump, (6) evaporation liquid storage tank, (7) sight glass, (8) evaporation solar collector, (9) pressure balance pipe, (10) thermocouples

investigated the thermal performance of a LT solar collector by using different working fluids. Results showed that all of them had a similar maximum collection efficiency of around 50%. Chien et al. [15] developed a theoretical model validated with experimental data; the result indicated that the best instantaneous thermal efficiency was 82%, which was higher than that of conventional SWH systems. Joudi and Al-tabbakh [16] also revealed that the photothermal efficiency of a LT-SWH system was approximately 20% higher than that of a water-based collector. Arab et al. [17] employed pulsating heat pipes (PHPs) in a SWH system. An extra-long PHP was designed, and the results showed that the configuration achieves satisfactory performance. Mathioulakis and Belessiotis [18] theoretically and experimentally presented their investigations on the energy behavior of a new type of solar collector that uses a wickless LT filled with ethanol. An instantaneous efficiency of up to 60% was observed. Hussein [19, 20] presented a theoretical and experimental analysis of a LT flat-plate solar water heater. Different meteorological conditions, initial water tank temperatures, hot water withdrawal load patterns, ratios of the storage tank volume to the collector area, storage tank dimension ratios, and height of the heater storage tank and collector were examined. Ordaz-Flores et al. [21] presented the indirect performance of a LT-SWH system using acetone and methanol as the working fluids. The results showed that the average experimental photothermal efficiencies during the test day were 48% and 50%, respectively. Zhang et al. [22] used supercritical CO_2 as the working fluid; an average collecting efficiency of 58% during whole-year tests was presented. Pei et al. [23] compared a LT-PV/T collector and a normal PV/T collector by using R600a as the working fluid. The results showed that the LT-PV/T collector had higher photovoltaic conversion efficiency but lower photothermal efficiency and a smaller temperature difference among the cells of the collector than the normal PV/T collector. However, both systems showed almost the same energy efficiency. Albanese et al. [24] presented a heat pipe-assisted solar wall, and experimental and computer models were developed to evaluate the performance of the system. Some studies using LT on the BIPV/T system are also presented. Zhao et al. [25] theoretically investigated the performance of an LT-SWH for a typical apartment building in Beijing.

The relationship between the efficiency of the system and the operating parameters was established, analyzed, and discussed in detail. Wang et al. [26] examined a novel facade-based solar LT water heating system by using both theoretical and experimental methods. Various operational parameters and two types of glass cover were discussed. The experimental and simulated results were in good agreement.

The two-phase circulation flow in the LTs is driven by the temperature difference with the assistance of gravity or the capillary force. The wick situated in the evaporator is normally made by the porous structure, which needs a complex and expensive manufacturing process. It produces the capillary force to drive the liquid back to the evaporator and ensures that the working liquid is evenly distributed over the heat transfer surface in the evaporator. However, it is costly to set up wicks in the copper tubes behind a flat-plate solar collector.

Considering that the solar collectors are usually inclined and installed, a wickless gravity-assisted LT is more preferred due to its low cost and simple structure. Nevertheless, the wickless LT-SWH system cannot be used when the water tank is lower than the solar collector, which means it is hard to integrate with buildings and is not suitable for household using. Learning from the active-cycle water-based SWHs, a refrigerant pump can be introduced to a wickless LT-SWH system. Although the system requires additional power consumption, the water tank has no position limitation, the pipes have no length limitation, and the system can be easily integrated into the buildings. However, few experiments or tests about the forced-circulation wickless LT-SWH system have been conducted. The forced-circulation wickless LT-SWH is new and the system behavior has yet to be reported. Therefore, in this study, a prototype of the novel forced-circulation wickless LT-SWH system is initially proposed. It consists of a refrigerant pump, an evaporation solar collector, a condensation water tank, an evaporation liquid storage tank, and a condensation liquid storage tank. A long-term outdoor test was performed; system performance was studied.

2. System Design and Experiment Setup

The schematic of the proposed system is shown in Figure 1. It comprises a solar collector acting as the evaporation section

TABLE 1: Detailed information of the parts and the copper pipes.

Device and the copper tube	Dimension (mm)	Remarks
Water tank	Φ450*1570	150 ± 1 L
Condensation fluid storage tank	Φ58*440	5.9 L
Evaporation fluid storage tank	Φ58*440	5.9 L
Pump	/	60 W (rated power)
Solar collector (water-based)	1000*2000 *95	$\eta = 0.751 - 4.206((T_{ic} - \bar{T}_a)/G)$
Absorber plate	1960*950* 0.4	Absorption rate ≥ 0.95 Emissivity ≤ 0.05
Vapor pipe	Φ28*1000	/
Connection pipe	Φ4	/
Condensation spiral coil pipe	Φ12*18000	/
Sight glass	/	Emerson

of LT, an evaporation liquid storage tank, a condensation liquid storage tank, a water tank with a spiral coil pipe functioning as the condensation section of LT, four sight glasses, and a refrigerant pump. Information about the devices and the copper pipes used are presented in detail in Table 1. R600a was employed as the working fluid.

Refrigerant pump supplies fluid from the condensation liquid storage tank to the evaporation liquid storage tank. The condensation liquid storage tank collects and separates the condensed fluid and vapor to avoid idling of the pump. The pressure balance pipe has two functions. When the pump is working, extra pressure is exerted to the refrigerant, it disturbs the intrinsic evaporation process occurring in the solar collector, and exacerbates the oscillatory heat transfer of the loop thermosiphon. The pressure balance pipe can mitigate the fluctuation caused by the pump. Besides, pressure difference comes up between the top and the bottom of the solar evaporator; the pressure balance pipe can promote the refrigerant flow with the aid of it. When the pump is not working, the refrigerant evaporates when there are solar irradiation incidents upon the solar collector, which will raise the inner pressure of the evaporator. The refrigerant in the evaporation liquid storage tank cannot flow down when the pressure is big enough. The pressure balance pipe can avoid this and ensures the R600a liquid continuously flows down by gravity. The evaporated R600a vapor will be condensed by the water and collected in the condensation liquid storage tank.

By using the pump, the water tank can be flexibly installed, which means the system can be easily integrated with the building. The check valve prevents the reverse flow of the fluid when the pump stops. Cycle process of the system is as follows: firstly, incident solar irradiation is absorbed by the absorber plate; soon afterwards, the heat energy of the absorber plate is absorbed by the R600a fluid, which vaporizes the fluid; then, R600a vapor

flows along the vapor pipe to the spiral coil in the water tank, wherein it is to be condensed into liquid and finally flows to the condensation liquid storage tank. The evaporation liquid storage tank acts as the liquid supplement of the solar collector, whereas the pump acts as the supplement for the evaporation liquid storage tank.

The experiments were all outdoor tested in Hefei City (31.52°N, 117.17°E). The height of the water tank, condensation fluid storage tank, and evaporation storage tank were 150 mm, 500 mm, and 1050 mm above the ground, respectively. The solar collector, with an inclination of 40°, was installed facing south. With the consideration of working tilt angle of the wickless LT evaporation section, the collector inclination was a little bigger than the city latitude. Pressures of solar collector liquid inlet and vapor outlet were measured. The straight-line distances between the sight glasses' location and the bottom edge of the collector were 350 mm, 750 mm, 1150 mm, and 1550 mm. The sight glasses, as well as the pyranometer, were parallelly installed to the collector. Seven T-type thermocouples were arranged to measure the water temperature variation. A frequency modulator was employed to adjust the flow rate, and a power sensor was used to measure the pump and modulate the power consumption. The thermocouples were set at the back of the copper pipes, surface and back of the absorber plate, and surface of the glass cover. Figure 2 presents the system physical setup and thermocouple setup on the copper pipes. Ambient temperature, as well as solar irradiation, was also recorded. All the measured data were recorded through Agilent 34970A. Precisions of the devices are listed in Table 2. The R600a mass charge of the system was 4.05 kg, which means the R600a volume filling ratio of the whole volume was 50%. The experiments were carried out from 8:00 to 16:00. The flow of the pump was set at 571.2 mL/min (theoretical, the same followed) from 8:00 to 9:00, 761.6 mL/min from 9:00 to 10:00, 952 mL/min from 10:00 to 14:00, 571.2 mL/min from 14:00 to 15:00, and 380.8 mL/min from 15:00 to 16:00.

The maximum volume flow rate of the refrigerant is estimated based on the following assumptions:

(1) The maximum solar irradiation at noon is 1000 W/m².

(2) The collector efficiency is 75%, the same as the efficiency provided by the manufacturer.

(3) The R600a liquid and vapor in the solar evaporator are both under saturation state; it means that all the absorbed energy is converted to the R600a latent heat.

(4) The values of the physical property parameters of R600a are based on the temperature of 30°C.

(5) The values of the physical property parameters of R600a vapor and liquid are constant when calculating the pressure loss.

(6) The values of the flow rate at other time periods are estimated based on the photothermal performance of the gravitative loop thermosiphon.

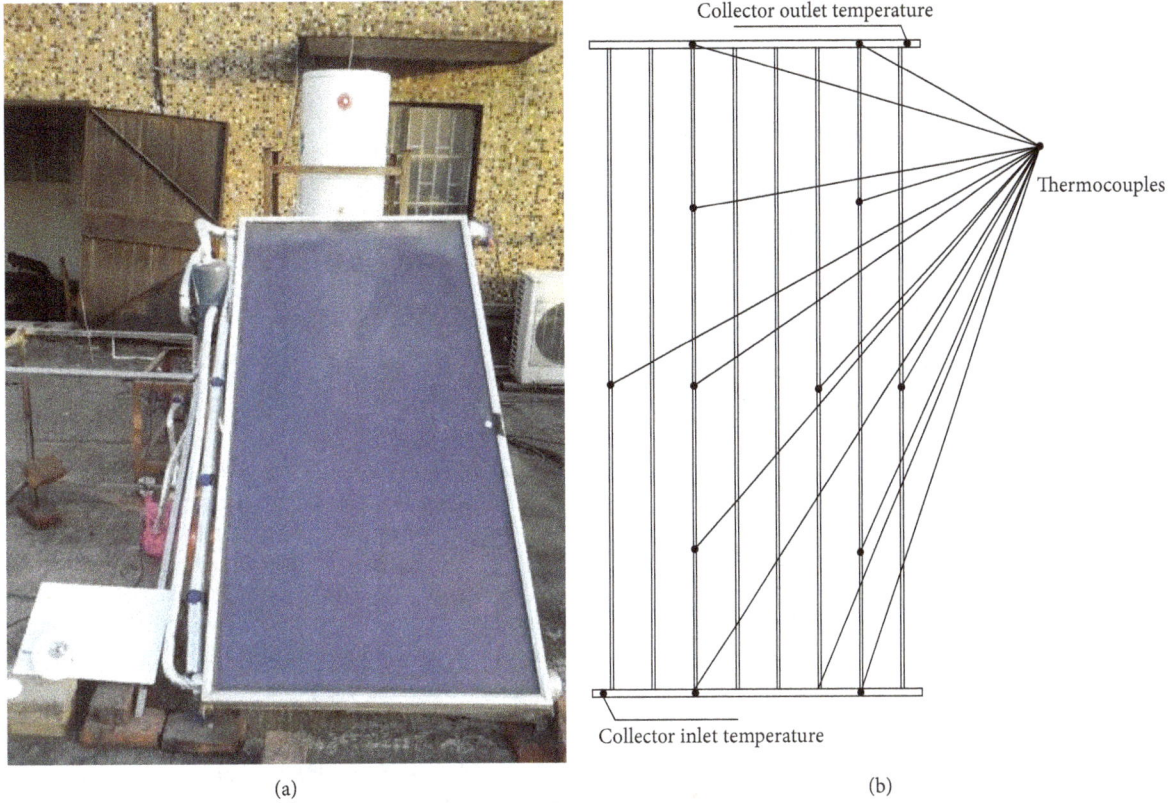

FIGURE 2: System physical setup and the detailed locations of the thermocouples' setup.

TABLE 2: Manufacturers, models, and precisions of monitoring devices.

Device	Specification	Precision
Power sensor	WBI021S91 (Weibo, China)	0.5%
Pyranometer	TQB-2 (Sunlight, China)	≤2%
Thermocouple	0.2 mm T-type (USTC, China)	±0.2°C
Data collection	34970A (Agilent, USA)	/
Pressure sensor	P3308 (Germany)	≤0.5% of FS

3. System Evaluations and Error Analysis

The daily performance of the water-based SWH system is evaluated with the daily average efficiency, which is expressed as

$$\eta_{\text{sys}} = \frac{CM(T_f - T_i) - W_p t}{HA}, \tag{1}$$

where C is the water specific heat, J/(kg·K); M is the water mass in the water tank, kg; T_f and T_i are the final and initial water temperature in the water tank, respectively, °C; W_p is the power of the pump, W; t is the working time, s; H is the total or average incident solar irradiation on the surface of glass cover during the experiment, J/m^2; and A is the aperture area of the collector, m^2. T_i and T_f in this study are the average values of the thermocouples in the water tank. System daily average photothermal efficiency can also be calculated according to (1), where H is the accumulation of solar irradiation of the whole day.

Photothermal performances of water-based SWH systems are affected by many factors, including solar irradiation, system design, environment temperature, collector inclination, wind speed, and initial water temperature. A method to evaluate the natural convection performance for a water-based SWH system is suggested by Huang and Du [27]. It is expressed as

$$\eta^* = \alpha - U \frac{T_i - \overline{T_a}}{H^*}, \tag{2}$$

where η^* is the estimated system photothermal efficiency; $\overline{T_a}$ is the average ambient temperature, °C; α is the typical photothermal efficiency when T_i equals $\overline{T_a}$; U is the system heat loss coefficient; and H^* is the daily total solar irradiation per area, MJ/(m^2·day). Equation (2) is also used to estimate the proposed system performance in this study.

The same method as that used in (2) is applied to evaluate the collector performance η^*_{col}, where T_i is the initial R600a fluid temperature that returns to the collector, °C, and H^* is the average solar irradiation during different time periods, W/m^2.

Inaccuracies in the test devices will cause some errors in the test data. The test error of the independent variables, such as the solar radiation intensity, water mass, and temperature, is determined based on the accuracy of the corresponding test devices in this study. For the dependent variables, such as the collector and system photothermal efficiencies, their test errors can be confirmed based on the test error of the independent variables. Given a dependent variable y, its function can be expressed as follows:

$$y = f(x_1, x_2 \dots x_n), \tag{3}$$

where $xi\,(i = 1, \dots, n)$ is the variable of the function.

Then, the relative error (RE) of the dependent variable can be confirmed as follows [28]:

$$\mathrm{RE} = \frac{dy}{y} = \frac{\partial f}{\partial x_1}\frac{dx_1}{y} + \frac{\partial f}{\partial x_2}\frac{dx_2}{y} + \cdots + \frac{\partial f}{\partial x_n}\frac{dx_n}{y}, \tag{4}$$

where $\partial f / \partial x_1 i$ is the error transferring coefficient of the variables. Therefore, the experimental relative mean error (RME) during the test period can be calculated as

$$\mathrm{RME} = \frac{\sum_1^N |\mathrm{RE}|}{N}. \tag{5}$$

Based on (4) and (5), the RMEs of the independent and dependent variables are calculated, and the details are presented in Table 3.

4. Experimental Results and Discussions

4.1. Details of Daily Performance of the System. To present the details of the daily performance of the system, a typical day is selected. The ambient environment, including the ambient temperature and solar irradiation fluctuation on that day, is illustrated in Figure 3. Total solar irradiance was 22.7 MJ/m^2 on that day, and the average ambient temperature was 20.5°C.

4.1.1. Instantaneous Photothermal Performance. Variations of the instantaneous photothermal efficiency (calculated based on (1) every 30 min) and average water temperature are shown in Figure 4. Photothermal efficiency firstly increases and then decreases. The highest instantaneous photothermal efficiency was registered at 10:15, with a value of $69.54 \pm 11.10\%$, which was better than the work of M. Esen and H. Esen [14] and Mathioulakis and Belessiotis [18], whereas the lowest photothermal efficiency was recorded at the end of the test, with a value of $28.77 \pm 15.77\%$.

System photothermal performance is determined by the heat energy absorbed by the solar collector. The combined action of absorptivity and heat loss of the absorber plate determines the effective heat gain. The absorptivity of the absorber plate is mainly determined by the transmittance of glass when the coating is fixed. The heat loss is mainly determined by the temperature difference between the absorber plate and ambient. Before 10:30, the temperature difference between the water and environment, positive or negative, was small. The transmittance of the glass increased with the incident angle decreased and thereby increased photothermal

TABLE 3: Experimental RMEs of all variables.

Variable	T_w	H	M^*	η_{sys}	η_{col}^*
RME	0.062%	2%	0.667%	4.013%	22.818%

$^*M$ is the water mass in the water tank.

efficiency. Between 10:30 and 14:15, although the temperature difference was relatively large and resulted in substantial heat loss, the efficiency slowly decreased with the help of a larger irradiation density and a smaller solar incident angle. After 14:15, the efficiency was sharply decreased because the temperature difference as well as the incident angle was absolutely large. An average efficiency of $58.22 \pm 12.81\%$ was obtained on this day, which was higher than the value obtained in the works of Arab et al. [17] and Ordaz-Flores et al. [21] and was similar to that of Zhang et al. [22].

The water temperature gradually increased from 11.5°C to 49.7°C during the test, but it presented a different growth gradient in different times. One can see from Figure 4 that the growth gradient was gradually increased during the first two hours, then stable and sustained in the following four hours, and gradually decreased in the last two hours. The fluctuation trend can be explained similar to that of the instantaneous photothermal efficiency. During the first two hours of the test, the solar irradiation was weak and the temperature difference between the absorber plate and the surrounding was small; the growth gradient increased with the transmittance of the glass increased;. During the middle time of the test, the stronger solar irradiation and bigger glass transmittance caused the quick rising of temperature. During the last two hours, the growth gradient decreased with the photothermal efficiency decreased.

Figure 5 qualitatively describes the liquid level of the solar evaporator. Note that it only presents the relative location of the liquid level, not the concrete values. As shown in Figure 5, the liquid level was usually low in the beginning of the test; the reason was that when the pump stopped working in the end of the last test, no fluid entered into the solar evaporator after that; however, the R600a liquid will be slowly evaporated after the last test and before this test, although the solar irradiation at these times was weak. The liquid level gradually raised when the pump began to work and then stayed nearby the second sight glass for about 3 hours; after that, the liquid level began to raise and continued to raise until the end of test, although the pump had turned down two times; the liquid level finally reached to a height higher than the fourth sight glass, that is, the evaporation liquid storage tank was full of R600a liquid at the end of the test.

As shown in Figure 5, one can also observe that the variation of the pump power was segmented by the set of flow rate. The pump flow rate was adjusted five times during the test; the pump power was presented five changes also, and it can be easily observed that the pump power was relatively stable under the same flow rate. Based on the test data, an average power of 29.64 W was obtained on that day.

According to Zhao et al. [25], the loop thermosiphon has the best filling ratio that lies between 30%~50% of the whole volume and has a negative performance when overcharged.

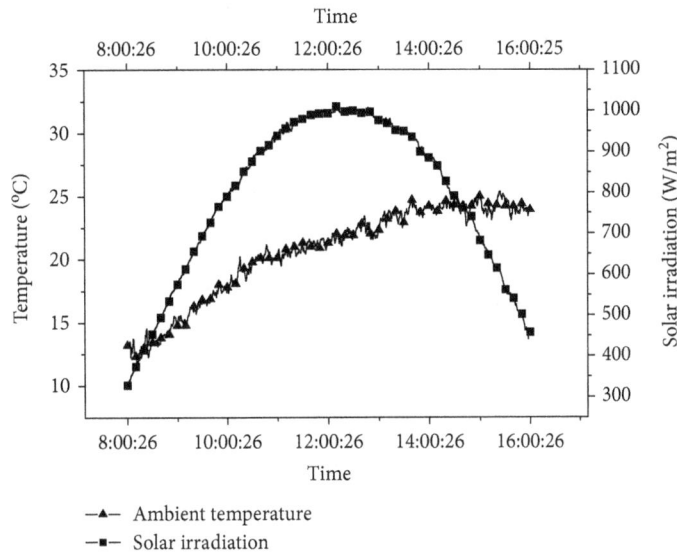

FIGURE 3: Variations in ambient temperature and solar irradiation on the typical day.

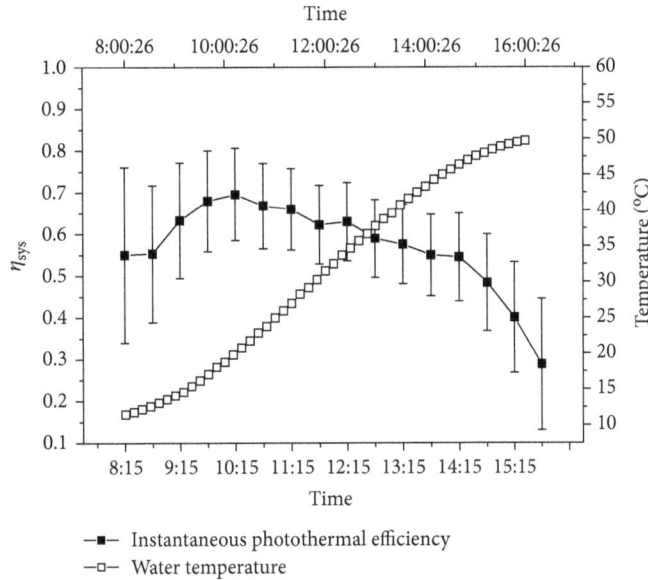

FIGURE 4: Variations in instantaneous photothermal efficiency and water temperature.

The higher the filling ratio, the smaller the photothermal efficiency achieved and the lower the efficiency fluctuation presented. Although the volume filling ratio in this study used the previous optimal result, the existence of the pump, condensation liquid storage tank, and evaporation liquid storage tank made the actual evaporation liquid level higher than Zhao et al. [25] under the same filling ratio; since the condensation liquid storage tank and evaporation liquid storage tank contribute a big amount of volume; however, those parts had limited contributions on the overall heat transfer.

At the same time, as shown in Figure 1, according to Aung and Li [30], the driving force for the circulation of working fluid in a loop thermosiphon system is mainly related to hydraulic head between the highest liquid level in the liquid return pipe and evaporation liquid level in the solar collector. From Figure 5 and the location of the evaporation liquid storage tank, one can conclude that a high liquid level also means a small hydraulic head. Therefore, the high position of the evaporation liquid storage tank will be beneficial or it can change to a slender one. Since a certain amount of volume should be provided to avoid solar collector drying out at the end of the test or at the beginning of the next test, depth-width ratio of the evaporation liquid storage tank and pump flow rate control should be further optimized. Besides, the location of the evaporation liquid storage tank and its influences on the system photothermal performance, system integrability, and convenience of installation should also be considered.

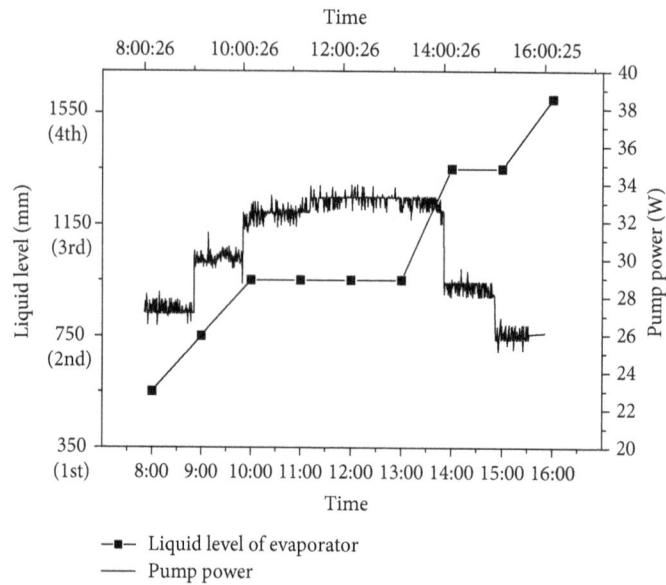

FIGURE 5: Qualitative variation of the liquid level of the evaporator and pump power.

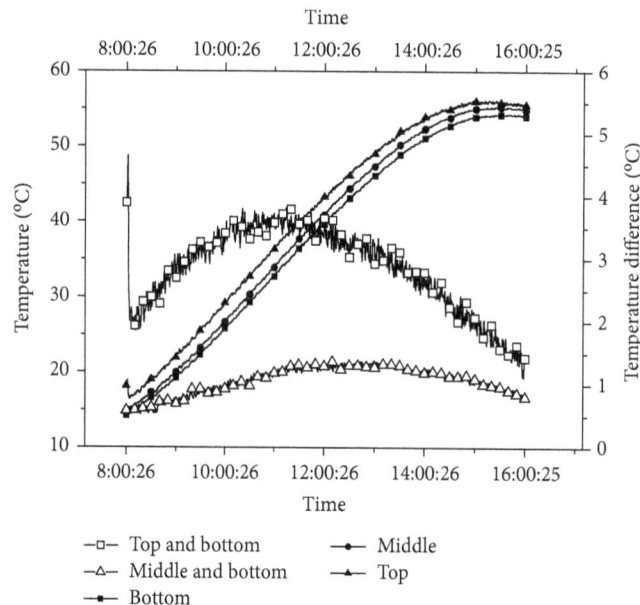

FIGURE 6: Variations in the bottom, middle, and top temperatures of the LT evaporation section and the temperature difference between the top and bottom parts of the LT.

To sum up, the instantaneous photothermal efficiency no less than the previous passive LT-SWH system was achieved. Although there was additional power consumption, the water tank has no location limited, which made the forced wickless LT-SWH easier integrated with buildings. However, system design, including the pump flow rate and evaporation liquid storage tank location, still has to be further optimized.

4.1.2. Temperature Performance of the Solar Collector. Isothermality of the evaporation section is one of the most significant characteristics of the loop thermosiphon. Thermocouples were placed to monitor it and to investigate the influence of the existence of the pump. The temperatures of the points, displayed on the right side of Figure 2, at bottom, middle, and top of the solar evaporation section (i.e., solar collector) were recorded. Average temperatures of the points at the same height are presented in Figure 6. Temperature differences between the top and bottom and middle and bottom are also plotted in Figure 6 to show the isothermality of the LT evaporation section.

From Figure 6, one can see that the collected temperatures at the bottom, middle, and top of the solar evaporator were all gradually increased with time; it was because the condensing temperature was gradually increased with time;

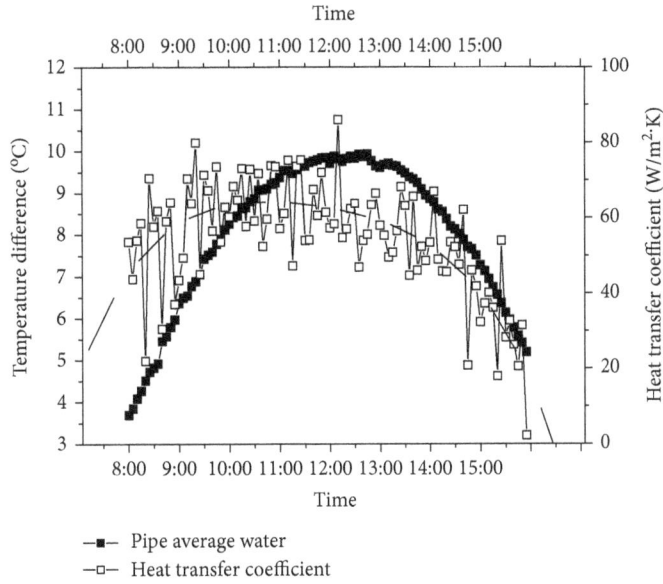

FIGURE 7: Variations in the heat transfer temperature difference and heat transfer coefficient.

TABLE 4: Experimental results for collector performance linear fitting.

Time	T_{ic}°C	$\overline{T_a}$°C	G (W/m²)	η_{col}(%)
9:45	21.9	17.214	745.698	0.679
10:15	25.2	18.724	831.796	0.695
10:45	28.8	20.2	908.420	0.668
11:15	32.3	20.5	960.682	0.659
11:45	36.0	21.2	990.689	0.622
12:15	39.4	21.7	998.936	0.630
12:45	42.8	22.3	994.909	0.590
13:15	45.8	23.3	966.365	0.575
13:45	48.8	23.9	921.590	0.549
14:15	51.1	24.1	843.782	0.545
14:45	52.8	24.1	748.462	0.484

the temperature differences between the top and bottom and middle and bottom both firstly increased and then decreased. The absorbed solar energy before the test began was responsible for the big temperature difference between the top and bottom at the beginning of the test; the solar collector that was immersed in the two-phase region was responsible for the small temperature difference between the middle and bottom. The temperature difference between the top and bottom with the maximum and minimum values was 3.9°C ± 0.4°C and 1.1°C ± 0.4°C, respectively, while the values were 1.4°C ± 0.4°C and 0.5°C ± 0.4°C between the middle and bottom.

In Figure 6, it can be observed that the temperature difference between the top and the bottom of the collector presented a similar trend compared to the instantaneous photothermal efficiency, both had a trend of increased firstly and then decreased, and both deserved a maximum value around 10:30. At the beginning of the test, the low liquid level

caused a high degree of superheat between the top and the bottom of the collector, and the value increased with the solar irradiation increased. At the end of the test, the value of superheat degree decreased with the solar irradiation decreased and liquid level increased.

However, the temperature difference between the middle and the bottom of the collector was always small, because the liquid level of R600a was around the middle or top position of the collector most of the time.

The daily average temperature differences between the top and bottom and middle and bottom were 2.8 ± 0.4°C and 1.1 ± 0.4°C, respectively. Note that the R600a vapor at the middle and top was under different dryness; heat resistance between the copper pipe and the R600a vapor at the top was higher than that of the middle. Therefore, it can be found that the collector temperature was almost evenly distributed.

4.1.3. Heat Transfer Coefficient of Pump-Forced Wickless LT. The performance of the forced wickless LT will directly influence the system performance; therefore, the heat transfer coefficient of the forced wickless LT is calculated and plotted in Figure 7. The ΔT is assumed as the temperature difference between the copper pipe and the water here; the copper pipe temperature is assumed as the average value of all the thermocouples pasted on the pipes (see Figure 2), and the water temperature is assumed as the average value of the thermocouples in the water tank.

As shown in Figure 7, the heat transfer temperature difference firstly increased and then decreased; the curve was smooth. The maximum and minimum value was 9.9 ± 0.4°C and 3.6 ± 0.4°C, respectively, and the average value was 8.0 ± 0.4°C. However, the heat transfer coefficient fluctuated obviously and had a foreseeable trend of first increased and then decreased based on the polynomial fitting.

Both the fluctuations of the heat transfer coefficient and the temperature difference between the top and the bottom

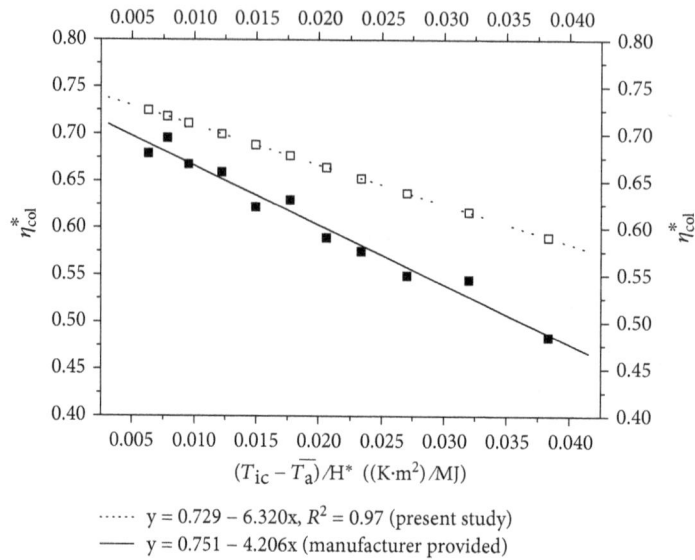

$$y = 0.729 - 6.320x, R^2 = 0.97 \text{ (present study)}$$
$$y = 0.751 - 4.206x \text{ (manufacturer provided)}$$

FIGURE 8: Graphical plot of the experimental results and the linear fitting.

TABLE 5: Solar collector performance compared with that in previous works.

System name	Collector type	$M^{\bullet}$ (kg/s)	α	U
Present work	Flat plate	/	0.729	6.320
Hussein [29]	Flat plate	0.02A	0.735	6.67
Aung and Li [30]	Flat plate	0.0125	0.457	3.711
Aung and Li [30]	Flat plate	0.0292	0.618	4.420
Aung and Li [30]	Flat plate	0.0458	0.586	3.526
Aung and Li [30]	Flat plate	0.0625	0.571	3.967

TABLE 6: Daily experimental results (arranged by date).

Date (D/M)	T_i (°C)	T_f (°C)	$\triangle T$ (°C)	$\overline{T_a}$ (°C)	H (MJ/m²·day)	η_{sys}
17/12	7.4	30.3	22.8	5.3	15.790	0.500
19/12	7.9	28.4	20.5	7.5	14.099	0.504
23/12	7.8	29.7	21.9	5.9	15.326	0.495
14/03	10.8	42.5	31.7	17.6	20.356	0.540
24/03	11.2	50.3	39.1	17.3	24.085	0.561
25/03	11.5	49.7	38.2	22.7	20.518	0.582
26/03	13.7	48.8	35.1	21.0	21.573	0.563
01/04	15.4	52.2	36.8	22.4	23.058	0.552
03/04	14.4	48.7	34.4	17.6	22.332	0.533
05/04	15.3	46.9	31.5	22.0	19.527	0.558
06/04	16.0	46.4	30.3	24.3	18.250	0.575
07/04	17.5	47.8	30.3	24.4	18.588	0.564

of the solar collector (as shown in Figure 6) are due to the oscillatory heat transfer characteristic of the LT [31]. When there is a high degree of superheat between the R600a liquid and the copper pipes, it caused intense nucleate boiling in the evaporation section/solar collector of the LT. The bubbles will grow bigger when they flow up along the copper pipe, which changed the quantity of R600a steam and then caused the oscillatory heat transfer. The oscillatory heat transfer makes the temperature fluctuate slightly, as well as the heat transfer coefficient.

4.2. Photothermal Performance of the Solar Collector. On the basis of GB/T 4271-2007 [32], test data obtained only when solar irradiation was larger than 700 W/m² were selected for analysis. The inlet temperature (T_i), commonly used in curves for water-cooled units, was replaced by the inlet R600a temperature (T_{ic}). The other parameters are given in Table 4. The linear fitting, which is calculated based on (2), is plotted in Figure 8. A curve of the collector that uses water as the working fluid is also provided for comparison.

The curve shows that the solar evaporator can provide a satisfactory photothermal performance, slightly lower than that of the water based, with absorption of 0.729. The heat loss coefficient was 6.320, which was higher than that of the normal solar collector, because there was extra heat loss caused by the long connection pipes and devices. Although the modified wickless LT solar collector had a large heat loss coefficient in this study, it can work without any freezing problem and corrosion, and most importantly, it can be easily remolded. Therefore, the improvement shows great potential for solar application and can serve as a good substitute for the normal solar collector when it is used in an area with high latitude.

Table 5 presents the comparison of collector performance between the current study and previous works with natural-convection recirculation. The results show that the collector performance of the present study was very close to that found in the study of Hussein et al. [33], with a flow rate of 0.02 A, and was better than that in the work of Nada et al. [34], under different flow rates. However, the heat loss ratio in the present study was bigger. Besides, the linear fitting results indicated that a reasonable similarity was obtained between the presented experimental results and those of

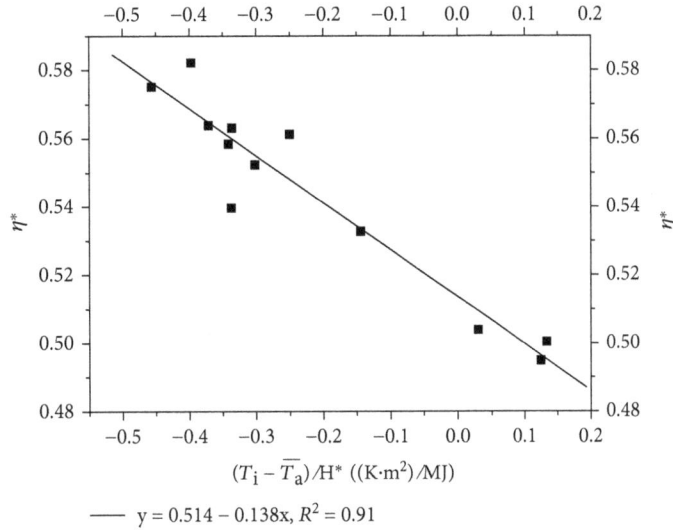

FIGURE 9: Graphical plot of the experimental results and the linear fitting.

TABLE 7: System thermal performance compared with those in previous works.

System name	A (m^2)	M (kg)	α	U
Present work	1.82	150	0.514	0.138
Chang et al. [37], B	3.80	294	0.55	0.14
Chang et al. [37], C	3.71	287	0.54	0.15
Chang et al. [37], D	1.89	120	0.45	0.14

other natural-convection recirculation two-phase collectors investigated, such as those by Hammad [35], M. Esen and H. Esen [14], and Rittidech and Wannapakne [36]. In summary, the forced wickless LT collector of the present work had a similar thermal performance compared to that of the other two-phase experiment assemblies, but its water tank had no position limitation. This feature indicates the promising prospect for solar application.

4.3. Thermal Performance of the System. To examine the overall performance and evaluate the daily performance of the system under different weather conditions, several tests were performed during winter and early spring in Hefei, China. Linear fitting based on (2) is conducted. Daily effective experimental results of the forced wickless LT-SWH system are listed in Table 6.

A wealth of information can be found in the experimental results detailed in Table 6. The system can supply water with a temperature of almost 50°C during early spring in Hefei, and it was sufficient for domestic use. An efficiency of approximately 55%, which is not less than that of a water-cooled water heating system, was obtained. An efficiency of approximately 50% was achieved even during winter. Besides, as mentioned before, no freezing problems occurred during the tests.

The plot of η^* against $(T_i - \overline{T_a})/H^*$ of the system, as well as the curve obtained under standard conditions with the

use of water, is shown in Figure 9. Based on these data, the linear regression line is expressed as

$$\eta = 0.514 - 0.138 \frac{T_i - \overline{T_a}}{H^*}, \tag{6}$$

where 0.514 was the daily average thermal efficiency when the initial water temperature equals the daily average ambient temperature. Normally, the initial water temperature is larger than the daily average ambient temperature in winter, and this condition results in a low thermal efficiency. The overall heat loss coefficient of the system was 0.138. Based on (2), the performance of the forced wickless LT-SWH under different solar irradiations, ambient temperatures, wind conditions, initial temperatures, and so on can be evaluated.

To the best of our knowledge, studies that examine the linear fitting of the system thermal performance of a two-phase SWH system are scarce. A comparison of earlier works [33] with water as the working fluid and the present study is given in Table 7. The results show that the forced wickless LT-SWH system had a similar thermal performance as the traditional SWH system. However, the proposed LT-SWH system has no corrosion and freezing problems during winter, and it has low heat loss at night due to the peculiarity of thermionic diode of LT. Above all, the proposed system demonstrates many advantages despite its low thermal efficiency and high heat loss efficiency. The system is therefore a good substitute for the normal water-based SWHs when it is used in a high-latitude area.

5. Conclusion

Compared to normal gravity LT-SWH, although there is additional power consumption, the location of the water tank of the forced-circulation wickless LT-SWH system becomes more freely. Long-term outdoor tests were conducted, and useful data were achieved in this study. The performance of the proposed wickless LT-SWH system and collector was presented. The conclusion can be drawn as the following:

(1) Based on the long-term field test data, the experimental fitting equations of the evaporation collector and the proposed system were presented. The proposed system and the solar evaporator achieved a satisfactory overall thermal performance. With the equation, the performance of the forced wickless LT-SWH system and the evaporation collector can be evaluated under different climatic conditions.

(2) The solar collector had an even temperature distribution as phase change occurred in the evaporation section of the wickless LT; the average temperature difference between the top and bottom and middle and bottom was $2.8 \pm 0.4°C$ and $1.1 \pm 0.4°C$, respectively.

Nomenclature

A: The aperture area of the collector
C: Water specific heat, J/(kg·K)
G: Instantaneous solar irradiation (W/m^2)
H: Total or average incident solar irradiation on the surface (MJ or MJ/m^2)
H^*: Daily total solar irradiation on per area, MJ/(m^2·day)
M: Water mass (kg)
$\dot{M}$: Mass flow rate (kg/s)
N: Number of independent variables
RE: Relative error
RME: Relative mean error
T: Temperature (°C)
t: Working time (s)
$\bar{T}_a$: Average ambient temperature (°C)
Wp: Pump power (W)
U: Heat loss coefficient.

Greek Symbols

η: Efficiency
η^*: System photothermal efficiency
α: Typical photothermal efficiency
$\triangle T$: Temperature difference (°C).

Subscripts

col: Collector
f: Final
i: Initial
ic: Collector initial
sys: System
w: Water.

Conflicts of Interest

The author declares no conflict of interests regarding the publication in this manuscript.

Acknowledgments

This work was sponsored by the Shanghai Sailing Program (18YF1409100) and Shanghai Local Capacity Building Program (18020501000), which are gratefully acknowledged by the author.

References

[1] The US Department of Energy, DOE/GO-10098-570, "Federal technology alert, solar water heating, a series of energy efficient guides prepared by the New Technology Demonstration Program," 2006, http://www.eren.doe.gov/femp/.

[2] S. Jaisankar, J. Ananth, S. Thulasi, S. T. Jayasuthakar, and K. N. Sheeba, "A comprehensive review on solar water heaters," *Renewable and Sustainable Energy Reviews*, vol. 15, no. 6, pp. 3045–3050, 2011.

[3] D. Y. Goswami, F. Kreith, and J. F. Kreider, *Principles of Solar Engineering*, Taylor & Francis, USA, 2000.

[4] M. A. Sabiha, R. Saidur, S. Mekhilef, and O. Mahian, "Progress and latest developments of evacuated tube solar collectors," *Renewable and Sustainable Energy Reviews*, vol. 51, pp. 1038–1054, 2015.

[5] "The U.S. Department of Energy," 2013, http://energy.gov/energysaver/articles/heat-transfer-fluids-solar-water- heating-systems.

[6] M. Mochizuki, A. Akbarzadeh, and T. Nguyen, "A review of heat pipe practical applications and innovative opportunities application for global warming, heat pipes and solid sorption transformation: fundamentals and practical applications," in *Heat Pipes and Solid Sorption Transformations*, pp. 145–212, Taylor & Francis/CRC Press, 2013.

[7] A. S. Zhuravlyov, L. L. Vasiliev, and L. L. Vasiliev Jr, "Horizontal vapordynamic thermosyphons. Fundamental and practical applications," *Heat Pipe Science and Technology, An International Journal*, vol. 4, no. 1-2, pp. 39–52, 2013.

[8] L. L. Vasiliev, L. L. Vassiliev Jr, M. I. Rabetsky et al., "Long horizontal vapordynamic thermosyphons for renewable energy sources," *Heat Transfer Engineering*, pp. 1–9, 2018.

[9] K.-H. Chien, Y.-T. Lin, Y.-R. Chen, K.-S. Yang, and C.-C. Wang, "A novel design of pulsating heat pipe with fewer turns applicable to all orientations," *International Journal of Heat and Mass Transfer*, vol. 55, no. 21-22, pp. 5722–5728, 2012.

[10] S. Launay, V. Sartre, and J. Bonjour, "Parametric analysis of loop heat pipe operation: a literature review," *International Journal of Thermal Sciences*, vol. 46, no. 7, pp. 621–636, 2007.

[11] D. Reay and P. Kew, *Heat Pipe*, Elsevier Ltd, 5th edition, 2006.

[12] R. S. Soin, K. S. Rao, D. P. Rao, and K. S. Rao, "Performance of flat plate solar collector with fluid undergoing phase chance," *Solar Energy*, vol. 23, no. 1, pp. 69–73, 1979.

[13] R. S. Soin, S. Raghuraman, and V. Murali, "Two-phase water heater model and long term performance," *Solar Energy*, vol. 38, no. 2, pp. 105–112, 1987.

[14] M. Esen and H. Esen, "Experimental investigation of a two-phase closed thermosyphon solar water heater," *Solar Energy*, vol. 79, no. 5, pp. 459–468, 2005.

[15] C. C. Chien, C. K. Kung, C. C. Chang, W. S. Lee, C. S. Jwo, and S. L. Chen, "Theoretical and experimental investigations of a two-phase thermosyphon solar water heater," *Energy*, vol. 36, no. 1, pp. 415–423, 2011.

[16] K. A. Joudi and A. A. Al-tabbakh, "Computer simulation of a two phase thermosyphon solar domestic hot water heating system," *Energy Conversion and Management*, vol. 40, no. 7, pp. 775–793, 1999.

[17] M. Arab, M. Soltanieh, and M. B. Shafii, "Experimental investigation of extra-long pulsating heat pipe application in solar water heaters," *Experimental Thermal and Fluid Science*, vol. 42, pp. 6–15, 2012.

[18] E. Mathioulakis and V. Belessiotis, "A new heat-pipe type solar domestic hot water system," *Solar Energy*, vol. 72, no. 1, pp. 13–20, 2002.

[19] H. M. S. Hussein, "Transient investigation of a two phase closed thermosyphon flat plate solar water heater," *Energy Conversion and Management*, vol. 43, no. 18, pp. 2479–2492, 2002.

[20] H. M. S. Hussein, "Optimization of a natural circulation two phase closed thermosyphon flat plate solar water heater," *Energy Conversion and Management*, vol. 44, no. 14, pp. 2341–2352, 2003.

[21] A. Ordaz-Flores, O. García-Valladares, and V. H. Gómez, "Findings to improve the performance of a two-phase flat plate solar system, using acetone and methanol as working fluids," *Solar Energy*, vol. 86, no. 4, pp. 1089–1098, 2012.

[22] X.-R. Zhang, Y. Zhang, and L. Chen, "Experimental study on solar thermal conversion based on supercritical natural convection," *Renewable Energy*, vol. 62, pp. 610–618, 2014.

[23] G. Pei, T. Zhang, Z. Yu, H. Fu, and J. Ji, "Comparative study of a novel heat pipe photovoltaic/thermal collector and a water thermosiphon photovoltaic/thermal collector," *Proceedings of the Institution of Mechanical Engineers, Part A: Journal of Power and Energy*, vol. 225, no. 3, pp. 271–278, 2011.

[24] M. V. Albanese, B. S. Robinson, E. G. Brehob, and M. Keith Sharp, "Simulated and experimental performance of a heat pipe assisted solar wall," *Solar Energy*, vol. 86, no. 5, pp. 1552–1562, 2012.

[25] X. Zhao, Z. Wang, and Q. Tang, "Theoretical investigation of the performance of a novel loop heat pipe solar water heating system for use in Beijing, China," *Applied Thermal Engineering*, vol. 30, no. 16, pp. 2526–2536, 2010.

[26] Z. Wang, Z. Duan, X. Zhao, and M. Chen, "Dynamic performance of a façade-based solar loop heat pipe water heating system," *Solar Energy*, vol. 86, no. 5, pp. 1632–1647, 2012.

[27] B. J. Huang and S. C. Du, "A performance test method of solar thermosyphon systems," *Journal of Solar Energy Engineering*, vol. 113, no. 3, p. 172, 1991.

[28] P. Gang, F. Huide, Z. Huijuan, and J. Jie, "Performance study and parametric analysis of a novel heat pipe PV/T system," *Energy*, vol. 37, no. 1, pp. 384–395, 2012.

[29] T. Zhang, G. Pei, Q. Zhu, and J. Ji, "Investigation on the optimum volume-filling ratio of a loop thermosyphon solar water-heating system," *Journal of Solar Energy Engineering*, vol. 138, no. 4, article 041006, 2016.

[30] N. Z. Aung and S. Li, "Numerical investigation on effect of riser diameter and inclination on system parameters in a two-phase closed loop thermosyphon solar water heater," *Energy Conversion and Management*, vol. 75, pp. 25–35, 2013.

[31] C. Yanze, Z. Yihui, and D. Xinwei, "An investigation of the heat-transfer fluctuation characteristics in gravity heat pipes and an exploration of methods for their restraint," *Journal of Engineering for Thermal Energy & Power*, vol. 18, no. 4, pp. 334–336, 2003.

[32] GB/T 4271, "Test methods for the thermal performance of solar collectors," *The Newest National Standard of China, Which Ruled the Test Method and the Calculation Program of the Solar Collector's Thermal Performance under Steady State and Dynamic State*, 2007.

[33] H. M. S. Hussein, M. A. Mohamad, and A. S. El-Asfouri, "Transient investigation of a thermosyphon flat-plate solar collector," *Applied Thermal Engineering*, vol. 19, no. 7, pp. 789–800, 1999.

[34] S. A. Nada, H. H. El-Ghetany, and H. M. S. Hussein, "Performance of a two-phase closed thermosyphon solar collector with a shell and tube heat exchanger," *Applied Thermal Engineering*, vol. 24, no. 13, pp. 1959–1968, 2004.

[35] M. Hammad, "Experimental study of the performance of a solar collector cooled by heat pipes," *Renewable Energy*, vol. 36, no. 3, pp. 197–203, 1995.

[36] S. Rittidech and S. Wannapakne, "Experimental study of the performance of a solar collector by closed-end oscillating heat pipe (CEOHP)," *Applied Thermal Engineering*, vol. 27, no. 11-12, pp. 1978–1985, 2007.

[37] J. M. Chang, J. S. Leu, M. C. Shen, and B. J. Huang, "A proposed modified efficiency for thermosyphon solar heating systems," *Solar Energy*, vol. 76, no. 6, pp. 693–701, 2004.

6

An Intelligent Maximum Power Point Using a Fuzzy Log Controller under Severe Weather Conditions

Khaled Bataineh [ID]

Department of Mechanical Engineering, Jordan University of Science and Technology, Irbid, Jordan

Correspondence should be addressed to Khaled Bataineh; k.bataineh@just.edu.jo

Academic Editor: Giulia Grancini

This study is aimed at providing a comparison between fuzzy systems and convectional P&O for tracking MPP of a PV system. MATLAB/Simulink is used to investigate the response of both algorithms. Several weather conditions are simulated: (i) uniform irradiation, (ii) sudden changing, and (iii) partial shading. Under partial shading on a PV panel, multipeaks appeared in P-V characteristics of the panel. Simulation results showed that a fuzzy controller effectively finds MPP for all weather condition scenarios. Furthermore, simulation results obtained from the FLC are compared with those obtained from the P&O controller. The comparison shows that the fuzzy logic controller exhibits a much better behavior.

1. Introduction

Due to global warming along with high prices of fossil fuel and its hazards on the environment, searching for other renewable green sources of energy drew the attention of the world. Solar energy is considered the main source of renewable energy. Solar energy is a permanent, nonpolluting, and low-running-cost source of energy. Photovoltaic (solar cell) systems are one of the most favorable systems, and their installation is spreading widely. Photovoltaic (PV) systems can be connected to a grid or can be used as stand-alone systems [1, 2]. The power generated from a PV panel depends on the amount of solar irradiance, cell temperature, and load [3–5].

Maximum power point tracking is essential to keep the system operating at its optimal power. Up to date, the overall PV efficiency reaches around 15%. Raising the power generated from PV systems can be achieved by tracking the maximum power point of the output power-voltage curve. This curve may contain multilocal maximum points under partially shaded conditions [6]. Employing MPPT techniques is considered the most economic way of improving the overall efficiency of the system compared with methods that rely on improving solar cell fabrication [7].

Several methods for finding MPP are developed over the last three decades [7–14]. These methods vary in terms of requiring sensors, cost, efficiencies, complexity, correct tracking when sudden shading or temperature changes, and convergence speed. Esram and Chapman presented a review of 19 methods for finding MPP [8]. Generally, the techniques are classified into three classes: offline, online, and hybrid methods. Offline methods are the short-circuit current (SCC) method, open-circuit voltage (OCV) method, look-up table method, and curve-fitting-based [9] and artificial intelligence (AI) algorithms. Online methods are the extremum seeking control (ESC) method, ripple correlation control (RCC) method, hill climbing (HC) method [10], incremental conductance (IC) method, and perturbation and observation (P&O) method [11] and modified P&O [12, 13].

As mentioned previously, when a PV system is subjected to a partially shaded condition, multiple peaks appeared in the P-V curve. Conventional methods such as P&O, hill climbing, IC, direct search algorithm, and line search algorithm with the Fibonacci sequence method could miss the global MPP [8, 14–16]. To overcome the problem with the conventional methods, several researchers have proposed several improvements. Because of the fact that the P&O method uses a fixed step size when tracking MPP, variable

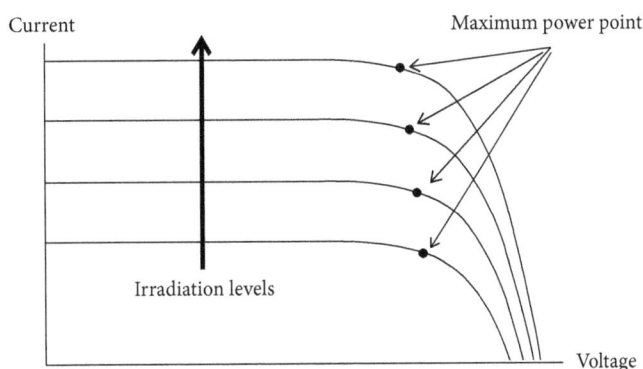

FIGURE 1: Effect of irradiance changes on *I-V* characteristics.

step size methods are developed to improve the steady-state performance and dynamic response of the PV system [17]. It has been shown that the small step size causes low oscillations during steady-state weather conditions but with a slower response. On the other hand, a larger step size leads to a faster response but with higher oscillations at steady-state conditions. Moreover, the performance of such systems reduces significantly due to the random nature of weather conditions. Tafticht et al. proposed a new method that combines nonlinear expression that is based on the open-circuit voltage with the P&O method to improve the tracking efficiency of MPP [18]. Recently, a new approach has been suggested by Heydari-doostabad et al. based on extremum seeking control (ESC) [19]. It takes the advantage of a band-pass filter (BPF) using a high-pass filter (HPF) and a low-pass filter (LPF) by only passing input power frequencies that include derivatives of PV with respect to its voltage, and so the system will operate at the global maximum point. Noguchi et al. [20] proposed a short-circuit pulse-based MPPT with fast scan on the *P-V* curve to identify the proportional parameter which is commonly used in a current-based MPPT [21]. Although the proposed method successfully found the global maximum point, momentary power loss is accompanied with additional extra cost. To avoid this extra power loss, Kazmi et al. proposed a controller that fluctuates the converter's duty cycle from zero to one to measure the open-circuit voltage and the short-circuit current and then computes the optimum voltage and current [22]. Based on the computed values, in single step, the operating point is moved to the optimal operating point. The conventional hill climbing algorithm is utilized to keep the system operating around the maximum point. Although the system finds global maximum, significant loss in power is experienced. Utilizing the particle swarm optimization algorithm under abnormal weather conditions results in a long computation time to reach the maximum operating point [23].

Intelligent systems such as neural networks (NN) and fuzzy logic controllers (FLC) have been used successfully in tracking the maximum power point of PV to decrease computation power requirement, while increasing the speed and efficiency of the tracking [2]. They are robust and relatively simple to design. However, they require complete knowledge of the operation of the PV system by the designer. Othman et al. validated the ability of FLC to find MPP

compared to the P&O algorithm [24]. Punitha et al. used a modified IC method with NN to supply V_{ref} and compared the results with those of the FLC and P&O approach to validate their proposed method [25]. Results showed the highest performance with the least response time when using ANN with IC. Subiyanto et al. presented a new method using a Hopfield neural network (HNN) to tune FLC parameters to enhance robustness and accuracy [26, 27]. Although fuzzy control has a good ability dealing with the nonlinear system, its main drawback is the generation of cumulative error due to continuous integral calculus. Most FLC-based MPPT techniques take the error ($e(t)$) and the change in error ($de(t)/dt$) as inputs. However, the requirement of differentiation not only increases the complexity of calculation but also may induce large amounts of errors from merely small amounts of measurement noise. The main objective of this study is to compare between conventional P&O and fuzzy logic controller MMPT algorithms under several extreme weather conditions. To carry out the comparison, computer-aided simulations are used to validate the results.

2. PV Modeling and Characteristics

Detailed description of PV modeling can be found in [2]. *I-V* nonlinear characteristic curves are shown in Figure 1. Partial shading occurs when radiation is not equally distributed on PV cells. The current generated by shaded cells is decreased. This leads to the reduction in the overall power generated from the PV system. In order to understand such phenomena, a PV array system with modules connected in series is considered. Under partial shading conditions, multiple peaks are presented in the *P-V* characteristic (see Figure 2).

3. MMPT Algorithm

In this section, a brief description about the P&O controller and fuzzy logic controller is given.

3.1. P&O Controller. A detailed description of the P&O controller can be found in [28–30]. The flow chart of the P&O algorithm is shown in Figure 3. The basic idea of this controller is to provoke perturbation by acting (decrease or increase) on the PWM duty cycle and observing the effect on the output PV power. The algorithm can be summarized as follows;

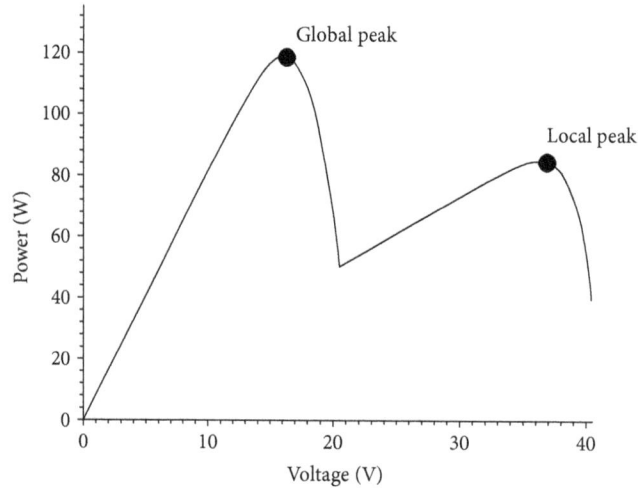

FIGURE 2: Characteristic curves of PV under partial shading.

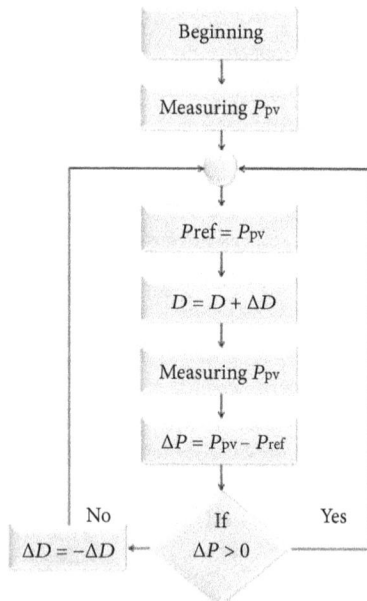

FIGURE 3: Flowchart of the P&O algorithm.

after recording the present power levels produced by the system, the algorithm performs a perturbation to the operating point by means of changing the duty cycle and measures the resulting power accordingly. If there is an increase in the power levels, another iteration is performed in the same direction. Otherwise, iteration in the reverse direction is carried out. The peak is detected when the power oscillates about a certain value; i.e., increasing and decreasing the duty cycle result in less power levels.

To generate maximum power, a DC-DC boost converter is used and placed between the source and the load. To simulate the P&O algorithm, the PV system composed of a PV panel, a boost DC-DC converter, MPPT, and resistive load is built as shown in Figure 4. In this work, a boost converter controlled by the MPPT algorithm is used to track MPP.

Voltage gain of the converter is given as [31]

$$\frac{V_o}{V_s} = \frac{1}{1 - D}. \tag{1}$$

The minimum values of inductance and capacitance of the converter necessary for stability (listed in Table 1) are given as [31]

$$L_{min} = \frac{DR(1 - D)^2}{2F_s},$$

$$C_{min} = \frac{V_o D}{F_s \Delta V_o R}. \tag{2}$$

The P&O algorithm has been implemented in a Simulink model to control the duty cycle of the switching signal of the converter. ΔV and ΔP are used to detect irradiation variations, and the algorithm determines the value of the duty cycle necessary to attain the maximum power point on the load. It changes the duty cycle value by a step size (s) which is determined by the designer.

3.2. Proposed Fuzzy Logic Controller. The proposed FL MPPT logic diagram shown in Figure 5 has two inputs (power and ΔV) and one output (D). Triangular shape membership functions have been used. The fuzzy inference is carried out using a Mamdani-type system. The defuzzification uses the center of gravity to compute the output of this FLC which is the duty cycle:

$$D = \frac{\sum_{j=1}^n \mu(D_j) - D_j}{\sum_{j=1}^n \mu(D_j)}. \tag{3}$$

The proposed fuzzy rules of the system are shown in Table 2. These two input variables and the control action for the tracking of the maximum power point are illustrated in Figure 5.

Based on the results of the Simulink model, tuning of the rules is performed to design the fuzzy logic controller. Values

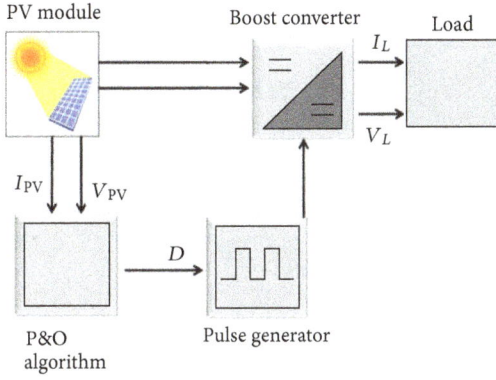

FIGURE 4: Schematic diagram of the PV system with MPPT.

TABLE 1: DC-DC booster converter design values.

Electrical characteristic	Values
Inductance	9×10^{-4} H
Output capacitor	0.001 F
Input capacitor	1×10^{-9} F
Resistance load	28.18 Ω

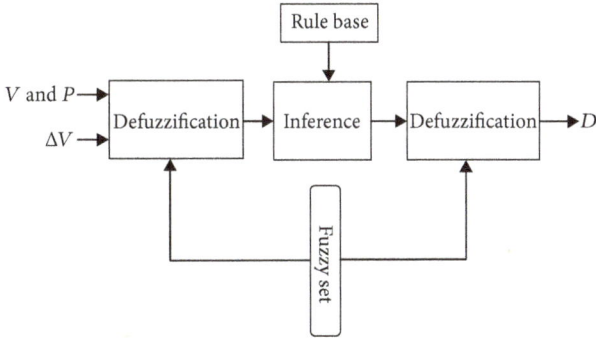

FIGURE 5: General diagram of a fuzzy controller.

of fuzzy controller inputs are compared with twenty-five rules of the system and are implicated with the membership functions. The implication has been chosen to be an "and" operator that chooses minimum values of membership functions. The implicated rules were aggregated using a maximum method. Figure 6 shows the surface view for the relation between fuzzy inputs and output. A centroid method has been selected for defuzzification by calculating the center of mass of the aggregated membership function. The crisp value represents the duty cycle of the switching signal that triggers the IGBT in the boost converter.

4. Simulation Results and Discussion

The performance of the proposed FLC is evaluated using MATLAB/Simulink. The electrical specifications of the PV cell are $I_{sc} = 4.75$ A, $V_{oc} = 0.6$ V, and $R_s = 5.1 \times 10^{-3}$ Ω. The simulated model consists of 72 cells connected in series to form a PV module. A partially shaded model is simulated by connecting two modules in series allowing each module

TABLE 2: Rules of the fuzzy controller.

$\Delta V\downarrow\backslash P\rightarrow$	Very low	Low	Medium	High	Very high
Very low	Very low	Very low	Medium	High	Very high
Low	Very low	Very low	High	High	Very high
Medium	Medium	Medium	Medium	Medium	Very high
High	Medium	Medium	High	Medium	Very high
Very high	Low	Low	Medium	High	Very high

to receive different levels of irradiation. The P&O MPPT controllers and FLC were simulated under the following tests: (i) uniform irradiation, (ii) sudden changing, and (iii) partial shading. The values of power listed in Table 3 demonstrate that a significant increase in the power output is obtained by using FLC.

Figure 7 shows that the FLC has a great ability to find MP in extremely short time when a PV panel is subjected to sudden changes of irradiation (full shading conditions). To accurately investigate the performance of the FLC, the tracking error and tracking efficiency are defined [3]:

$$\text{efficiency} = \frac{P_{\text{pv}}}{P_{\text{mpp}}} \times 100\%,$$

$$\text{error} = \frac{P_{\text{mpp}} - P_{\text{pv}}}{P_{\text{mpp}}} \times 100\%. \tag{4}$$

Power values on the load after applying FLC have been compared with the nominal values of the maximum power points for several cases of uniform irradiation and partial shading. Results listed in Table 4 show that the FLC has efficiently found the MPP for all uniform irradiation and partial shading scenarios studied.

To further prove the ability of FLC to track the MPP under partial shading, a comparison between the proposed algorithm and the P&O algorithm is conducted. Results listed in Table 5 show that the P&O algorithm is trapped at local peaks while the fuzzy controller finds the global peak. Figure 7 shows the output power for the fuzzy controller compared with the P&O algorithm at different cases. The irradiation applied on the simulated model at this case is equal on the whole panel and suddenly decreased partially at 0.2 s. The fuzzy controller found the peak at the first part with accuracy greater than that of the P&O algorithm. After partial shading being applied on 36 cells of the PV panel, the fuzzy controller tracked the global maximum power point while the P&O algorithm is trapped at the local peaks. The fuzzy controller has reached the global point with an efficiency of 99% while the P&O controller failed to reach this point. The fuzzy controller detects irradiation changing by detecting changes in voltage and power values of the PV system. This controller has the ability to change the duty cycle by a precise difference depending on membership functions. P&O changes the value of the duty cycle by a fixed step every time irradiation changes. Because of the fixed step value, it does not reach the necessary precise value of the duty cycle to get a maximum power from the system and gets trapped

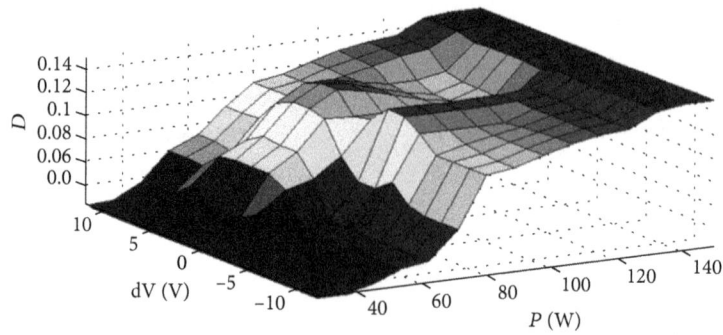

FIGURE 6: Surface view for fuzzy inputs (P, dV) vs. output (D) (adappted from Allataifeh et al. [2]).

TABLE 3: Comparison between output power values with and without MPPT.

Case no.	Irradiance level (W/m^2)	Output power (W) (without MPPT)	Output power (W) (with MPPT)	% increase
1	800	102.5	123.2	20.7%
2	600	64.9	87	34%
3	500	45	70.8	57%
4	300	21.4	38.7	80%

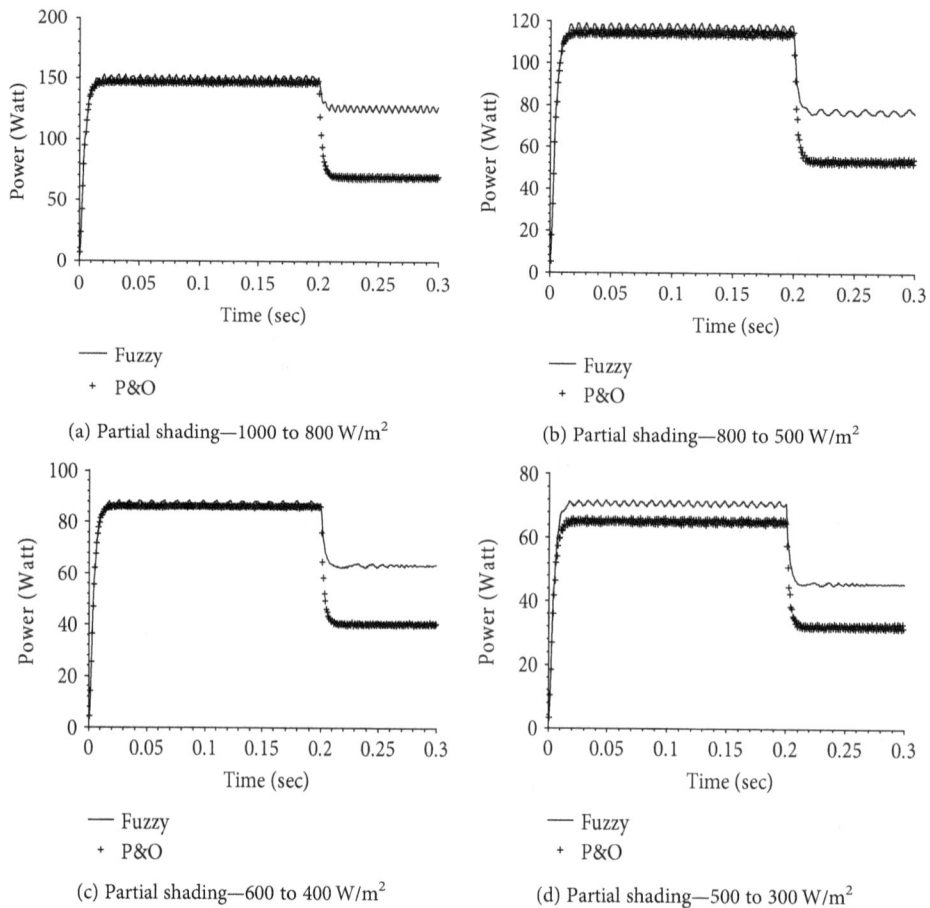

(a) Partial shading—1000 to 800 W/m^2

(b) Partial shading—800 to 500 W/m^2

(c) Partial shading—600 to 400 W/m^2

(d) Partial shading—500 to 300 W/m^2

FIGURE 7: Comparison between fuzzy and P&O partial shading.

TABLE 4: Values of model outputs.

Case no.	Condition	Ir1 (W/m^2)	Ir2 (W/m^2)	Nominal power (W)	Power after fuzzy (W)	Efficiency
1	Uniform irradiation	1000	1000	149	149	100%
2	Uniform irradiation	900	900	133	133	100%
3	Uniform irradiation	800	800	118	117.5	99.6%
4	Uniform irradiation	700	700	102.5	101.5	99%
5	Uniform irradiation	600	600	87.2	87	99.8%
6	Uniform irradiation	500	500	71.3	70.8	99.3%
7	Uniform irradiation	400	400	56.5	56	99%
8	Uniform irradiation	300	300	41.5	41	99%
9	Uniform irradiation	200	200	26.6	24.5	92%
10	Partial shading	1000	800	126.3	126	99.8%
11	Partial shading	900	600	95.7	94	98%
12	Partial shading	800	500	79.5	79	99.4%
13	Partial shading	700	500	78.2	78	99.7%
14	Partial shading	600	400	62.3	62	99.5%
15	Partial shading	900	300	61.4	60	97.7%
16	Partial shading	700	300	48	47.5	99%
17	Partial shading	500	300	46.3	46	99.4%

TABLE 5: Comparison between the fuzzy controller and P&O.

Ir1 (W/m^2)	Ir2 (W/m^2)	Local peak (W)	Global peak (W)	Output of P&O (W)	Output of fuzzy (W)
1000	800	71.24	126.3	68	126
800	500	55.7	79.5	53	79
600	400	41.5	62.3	39	61
500	300	34	46.3	31	46

at local maximum under partial shading conditions due to its way of tracking by looking for the first peak in power values.

5. Conclusions

In this study, the fuzzy logic controller (FLC) for maximum power point tracking (MPPT) of a photovoltaic system under variable insolation conditions has been developed to track the maximum power point of the PV system. A MATLAB/Simulink model consists of a PV panel and a boost converter with FLC connected to a resistive load that has been built in order to evaluate the performance of the proposed controller. The proposed system showed its ability to recover from sudden changes and maintain stability under partial shading conditions. A comparison between the performance of the proposed FLC and perturbation and observation controller has been made. The results show that the fuzzy logic controller has reached the global point with an efficiency of 96% while the P&O controller failed to reach this point.

Nomenclature

A: Ideal factor of the diode

C_{min}:   Minimum capacitance value (F)

D: Duty cycle

dV: Voltage difference (V)

FF: Fill factor

F_s: Switching frequency (Hz)

I_m: Maximum current (A)

I_{pv}:   Photovoltaic current (A)

I_s: Saturation of dark current (A)

L_{min}:   Minimum inductance value (H)

P_m: Maximum power (W)

P_{max}: Maximum power (W)

P_{pv}:   Photovoltaic power (W)

R: Resistance (Ω)

R_s: Series resistance (Ω)

R_{sh}:   Shunt resistance (Ω)

V_o: Output voltage (V)

V_{pv}:   Photovoltaic voltage (V)

V_s: Input voltage of the converter (V)

η: Efficiency (%).

Conflicts of Interest

There is no conflict of interest regarding the publication of this article.

References

[1] K. Bataineh and D. Dalalah, "Optimal configuration for design of stand-alone PV system," *Smart Grid and Renewable Energy*, vol. 3, no. 2, pp. 139–147, 2012.

[2] A. A. Allataifeh, K. Bataineh, and M. Al-Khedher, "Maximum power point tracking using fuzzy logic controller under partial conditions," *Smart Grid and Renewable Energy*, vol. 6, no. 1, pp. 1–13, 2015.

[3] C. F. Lu, C. C. Liu, and C. J. Wu, "Dynamic modelling of battery energy storage system and application to power system stability," *IEE Proceedings - Generation, Transmission and Distribution*, vol. 142, no. 4, p. 429, 1995.

[4] H. J. Möller, *Semiconductors for Solar Cell*, Artech House, Inc, Norwood, MA, 1993.

[5] R. Gottschalg, M. Rommel, D. G. Ineld, and H. Ryssel, "Comparison of different methods for the parameter determination of the solar cell's double exponential equation," in *14th European Photovoltaic Science and Engineering Conference (PVSEC)*, pp. 321–324, Barcelona, Spain, 1997.

[6] Y. Cuia, W. Yaoa, and J. Luoa, "A research and improvement on a maximum power point tracking method for PV system under partially shaded conditions," *Procedia Engineering*, vol. 29, pp. 2583–2589, 2012.

[7] Z. Salam, J. Ahmed, and B. S. Merugu, "The application of soft computing methods for MPPT of PV system: a technological and status review," *Applied Energy*, vol. 107, pp. 135–148, 2013.

[8] T. Esram and P. L. Chapman, "Comparison of photovoltaic array maximum power point tracking techniques," *IEEE Transactions on Energy Conversion*, vol. 22, no. 2, pp. 439–449, 2007.

[9] P. Bhatnagar and R. K. Nema, "Maximum power point tracking control techniques: state-of-the-art in photovoltaic applications," *Renewable and Sustainable Energy Reviews*, vol. 23, pp. 224–241, 2013.

[10] C.-H. Lin, C.-H. Huang, Y.-C. Du, and J.-L. Chen, "Maximum photovoltaic power tracking for the PV array using the fractional-order incremental conductance method," *Applied Energy*, vol. 88, no. 12, pp. 4840–4847, 2011.

[11] A. Reza Reisi, M. Hassan Moradi, and S. Jamasb, "Classification and comparison of maximum power point tracking techniques for photovoltaic system: a review," *Renewable and Sustainable Energy Reviews*, vol. 19, pp. 433–443, 2013.

[12] E. Mamarelis, G. Petrone, and G. Spagnuolo, "A two-steps algorithm improving the P&O steady state MPPT efficiency," *Applied Energy*, vol. 113, pp. 414–421, 2014.

[13] R. Ahmed, A. Namaane, and N. K. M'Sirdi, "Improvement in perturb and observe method using state flow approach," *Energy Procedia*, vol. 42, pp. 614–623, 2013.

[14] J. W. Kimball and P. T. Krein, "Discrete-time ripple correlation control for maximum power point tracking," *IEEE Transactions on Power Electronics*, vol. 23, no. 5, pp. 2353–2362, 2008.

[15] M. Miyatake, T. Inada, I. Hiratsuka, H. Zhao, H. Otsuka, and M. Nakano, "Control characteristics of a Fibonacci-search-based maximum power point tracker when a photovoltaic array is partially shaded," in *Power Electronics and Motion Control Conference, 2004. IPEMC 2004. The 4th International*, vol. 2, pp. 816–821, Xi'an, China, 2004.

[16] T. L. Nguyen and K. S. Low, "A global maximum power point tracking scheme employing DIRECT search algorithm for photovoltaic systems," *IEEE Transactions on Industrial Electronics*, vol. 57, no. 10, pp. 3456–3467, 2010.

[17] K.-H. Chao and C.-J. Li, "An intelligent maximum power point tracking method based on extension theory for PV systems," *Expert Systems with Applications*, vol. 37, no. 2, pp. 1050–1055, 2010.

[18] T. Tafticht, K. Agbossou, M. L. Doumbia, and A. Chériti, "An improved maximum power point tracking method for photovoltaic systems," *Renewable Energy*, vol. 33, no. 7, pp. 1508–1516, 2008.

[19] H. Heydari-doostabad, R. Keypour, M. R. Khalghani, and M. H. Khooban, "A new approach in MPPT for photovoltaic array based on extremum seeking control under uniform and non-uniform irradiances," *Solar Energy*, vol. 94, pp. 28–36, 2013.

[20] T. Noguchi, S. Togashi, and R. Nakamoto, "Short-current pulse-based maximum-power-point tracking method for multiple photovoltaic-and-converter module system," *IEEE Transactions on Industrial Electronics*, vol. 49, no. 1, pp. 217–223, 2002.

[21] M. A. S. Masoum, H. Dehbonei, and E. F. Fuchs, "Theoretical and experimental analyses of photovoltaic systems with voltage and current-based maximum power-point tracking," *IEEE Transactions on Energy Conversion*, vol. 17, no. 4, pp. 514–522, 2002.

[22] S. Kazmi, H. Goto, O. Ichinokura, and H. J. Guo, "An improved and very efficient MPPT controller for PV systems subjected to rapidly varying atmospheric conditions and partial shading," in *Power Engineering Conference, 2009. AUPEC 2009. Australasian Universities*, pp. 1–6, Adelaide, SA, Australia, 2009.

[23] M. Miyatake, M. Veerachary, F. Toriumi, N. Fujii, and H. Ko, "Maximum power point tracking of multiple photovoltaic arrays: a PSO approach," *IEEE Transactions on Aerospace and Electronic Systems*, vol. 47, no. 1, pp. 367–380, 2011.

[24] A. M. Othman, M. M. M. el-arini, A. Ghitas, and A. Fathy, "Realworld maximum power point tracking simulation of PV system based on fuzzy logic control," *NRIAG Journal of Astronomy and Geophysics*, vol. 1, no. 2, pp. 186–194, 2012.

[25] K. Punitha, D. Devaraj, and S. Sakthivel, "Artificial neural network based modified incremental conductance algorithm for maximum power point tracking in photovoltaic system under partial shading conditions," *Energy*, vol. 62, pp. 330–340, 2013.

[26] S. Subiyanto, A. Mohamed, and M. A. Hannan, "Intelligent maximum power point tracking for PV system using Hopfield neural network optimized fuzzy logic controller," *Energy and Buildings*, vol. 51, pp. 29–38, 2012.

[27] M. Seyedmahmoudian, S. Mekhilef, R. Rahmani, R. Yusof, and E. Renani, "Analytical modeling of partially shaded photovoltaic systems," *Energies*, vol. 6, no. 1, pp. 128–144, 2013.

[28] J. L. Santos, F. Antunes, A. Chehab, and C. Cruz, "A maximum power point tracker for PV systems using a high performance boost converter," *Solar Energy*, vol. 80, no. 7, pp. 772–778, 2006.

[29] X. Liu and L. A. C. Lopes, "An improved perturbation and observation maximum power point tracking algorithm for PV arrays," in *2004 IEEE 35th Annual Power Electronics Specialists Conference (IEEE Cat. No.04CH37551)*, pp. 2005–2010, Aachen, Germany, June 2004.

[30] G. J. Yu, Y. S. Jung, J. Y. Choi, and G. S. Kim, "A novel two-mode MPPT control algorithm based on comparative study of existing algorithms," *Solar Energy*, vol. 76, no. 4, pp. 455–463, 2004.

[31] R. Kharb, M. D. Ansari, and S. Shimi, "Design and implementation of ANFIS based MPPT scheme with open loop boost converter for solar PV module," *International Journal of Advanced Research in Electrical, Electronics and Instrumentation Engineering*, vol. 3, pp. 2320–3765, 2014.

7

Small-Sized Parabolic Trough Collector System for Solar Dehumidification Application: Design, Development, and Potential Assessment

Ghulam Qadar Chaudhary[D],[1,2] Rubeena Kousar,[1] Muzaffar Ali[D],[1] Muhammad Amar,[2] Khuram Pervez Amber,[2] Shabaz Khan Lodhi,[1] Muhammad Rameez ud Din,[1] and Allah Ditta[1,2]

[1]University of Engineering and Technology, Taxila, Punjab, Pakistan
[2]Mirpur University of Science and Technology (MUST), Mirpur, 10250 AJK, Pakistan

Correspondence should be addressed to Ghulam Qadar Chaudhary; gq.chaudhary@hotmail.com

Academic Editor: Gabriele Battista

The current study presents a numerical and real-time performance analysis of a parabolic trough collector (PTC) system designed for solar air-conditioning applications. Initially, a thermodynamic model of PTC is developed using engineering equation solver (EES) having a capacity of around 3 kW. Then, an experimental PTC system setup is established with a concentration ratio of 9.93 using evacuated tube receivers. The experimental study is conducted under the climate of Taxila, Pakistan in accordance with ASHRAE 93-1986 standard. Furthermore, PTC system is integrated with a solid desiccant dehumidifier (SDD) to study the effect of various operating parameters such as direct solar radiation and inlet fluid temperature and its impact on dehumidification share. The experimental maximum temperature gain is around 5.2°C, with the peak efficiency of 62% on a sunny day. Similarly, maximum thermal energy gain on sunny and cloudy days is 3.07 kW and 2.33 kW, respectively. Afterwards, same comprehensive EES model of PTC with some modifications is used for annual transient analysis in TRNSYS for five different climates of Pakistan. Quetta revealed peak solar insolation of 656 W/m^2 and peak thermal energy 1139 MJ with 46% efficiency. The comparison shows good agreement between simulated and experimental results with root mean square error of around 9%.

1. Introduction

Global energy consumption trends are increasing progressively over the past few decades, and fossil fuels are leading with 80% share [1]. Solar energy is one of the best eco-friendly renewable energy resources available in the world. The PTC is amongst one of the advance concentrating thermal technologies. However, PTC application is limited to medium and elevated temperature range, that is, from 150–300°C and 300–400°C, respectively [2]. In this regard, various numerical and experimental research studies have been conducted on design and development of PTC. In a

research study, the thermo-mathematical model was proposed using differential and nonlinear algebraic correlations with a concentration ratio of 9.37 [3]. The results were in good agreement with experimental data of the Sandia National Laboratory (SNL). In another study, model development of PTC with the concentration ratio of 12.7 and its simulation in solid works was carried out by using finite element method. Heat transfer phenomenon and efficiency of the system were predicted and compared with the model [4]. Similarly, a detailed numerical model based on the finite volume method was developed to analyze the heat transfer characteristics of the evacuated receiver tube [5]. By using discretization technique, the receiver was divided into small segments and energy balance was applied to each control volume. The model was tested against test results of the Sandia National Laboratory. Moreover, a complex dynamic model of PTC and its dynamic simulations for validation were presented with root mean square error of 1.2% [6]. Likewise, a novel parabolic trough solar collector model results were validated with published data from the National Renewable Energy Laboratory (NERL) and Sandia National Laboratory (SNL) [7]. It was observed that the developed model reduced the uncertainty from 1.11% to 0.64% as compared to the EES code by NREL. Similarly, in another study, a numerical model was developed to study the thermal behavior of the single and double pass receiver tube of PTC [8]. Numerical analysis revealed that double pass tube results in enhanced thermal efficiency as compared to the single pass.

The climate conditions in term of solar radiation intensity and availability strongly influence the performance of PTC systems. Hence, a research study was conducted to access the potential of PTC for industrial heat application in Cyprus using TRNSYS [9]. The system was capable to meet 50% of annual load and proved the viability of the system for high-energy consuming industries. Additionally, E-W tracking was preferred in terms of energy gain over N–S for Mediterranean climate. A similar research study was conducted to estimate the solar thermal potential and PTC performance for Algeria [10]. Heat transfer and temperature evolutions were compared for different climates. The hot desert climate was declared the best suitable for PTC.

Moreover, an experimental study was conducted, according to ASHRAE 93-1986 standard using fiberglass-reinforced PTC for hot water application [11]. A test methodology was developed to investigate the effect of different operating parameters on thermal efficiency.

Keeping in view the intermittent nature of solar energy resource, a research study was conducted to access the annual performance of mini PTC [12]. A nonevacuated tube with glass cover and black paint coating was applied in the model for transient analysis. The collector achieved maximum instantaneous efficiency of 66.78%. Similarly, a quasi-dynamic simulation model in TRNSYS was developed to analyze the performance of PTC for a direct steam generation [13]. The basic modeling approach and results comparison were discussed in the study.

However, from literature review, it is seen that the majority of the abovementioned systems are high concentration ratio PTCs which produce high temperatures, associated with high cost. Furthermore, these PTCs are not compatible with hot water applications like air-conditioning due to strong space constraints subjected to commercial and industrial buildings [14]. Moreover, geographically, Pakistan is located such that more than 95% of its area receives an average global irradiance of 5–7 kWh/m² /day with an average daily sunshine period of 7.6 hours [15–17]. Although abundant solar energy is available in Pakistan, no such study exists to highlight solar thermal potential using PTC technology. Therefore, in this research work, a small

parabolic trough collector (PTC) is proposed for solar assisted dehumidification applications, which required hot water temperature ranging from 70–90°C. Hence, the small and efficient PTC makes it favorable for roof-mounted hot water applications. Keeping in view the specific application of solar-assisted dehumidification, a group of four small PTC systems is designed with a concentration ratio as low as 9.93 to produce around 3 kW of thermal energy. Furthermore, the developed polished stainless steel PTC system is analyzed numerically and experimentally under a wide range of operating conditions for dehumidification application.

2. Thermal Model

Thermal design of the proposed system is based on calculations of trough focal point, rim angle, and concentration ratio. Furthermore, heat characteristics of the evacuated tube receiver are also calculated in terms of heat loss coefficient, radiation coefficient, overall heat transfer coefficient, and useful heat gain.

The optical efficiency of the receiver is calculated as [18]

$$\eta_o = \rho\tau\alpha\gamma[(1 - A_f\tan(\theta))\cos(\theta)]. \tag{1}$$

2.1. Receiver. In the current study, an evacuated tube is used so that negligible convection losses are encountered. Thermal resistance model of the PTC is given in Figure 1. The heat loss coefficient of the receiver is determined by [18]

$$U_{rl} = \left[\frac{A_r}{\left(h_w + h_{(r,c-a)}\right)A_c} + \frac{1}{h_{r,(r-c)}}\right]^{-1}. \tag{2}$$

Radiation coefficient from the absorber to glass cover is obtained from [18]

$$h_{(r,c-a)} = \frac{\sigma\left(T_r^2 + T_c^2\right)(T_r + T_c)}{\left((1/\varepsilon_r) + (A_r/A_c)(1/\varepsilon_c - 1)\right)}. \tag{3}$$

Overall heat transfer coefficient is calculated by considering tubes outer and inner diameter as [9]

$$U_o = \left[\frac{1}{U_{rl}} + \frac{D_o}{(h_{fi}D_i)} + \frac{D_o\ln(D_o/D_i)}{2k}\right]^{-1}. \tag{4}$$

Nusselt number is calculated from the standard pipe flow equation as [19]

$$Nu = 0.023(Re)^{0.8}(Pr)^{0.4}. \tag{5}$$

The collector efficiency factor is given by [18]

$$F' = \frac{1/U_{rl}}{[(1/U_{rl}) + (D_o/h_{fi}D_i) + ((D_o/2k)\ln(D_o/D_i))]}. \tag{6}$$

Collector heat removal factor is the ratio of actual useful energy gain to maximum energy gain as [18]

$$F_r = \frac{\dot{m}c_{pw}}{A_r U_{rl}}\left[1 - \exp\left(-\frac{U_{rl}F'A_r}{\dot{m}c_{pw}}\right)\right]. \tag{7}$$

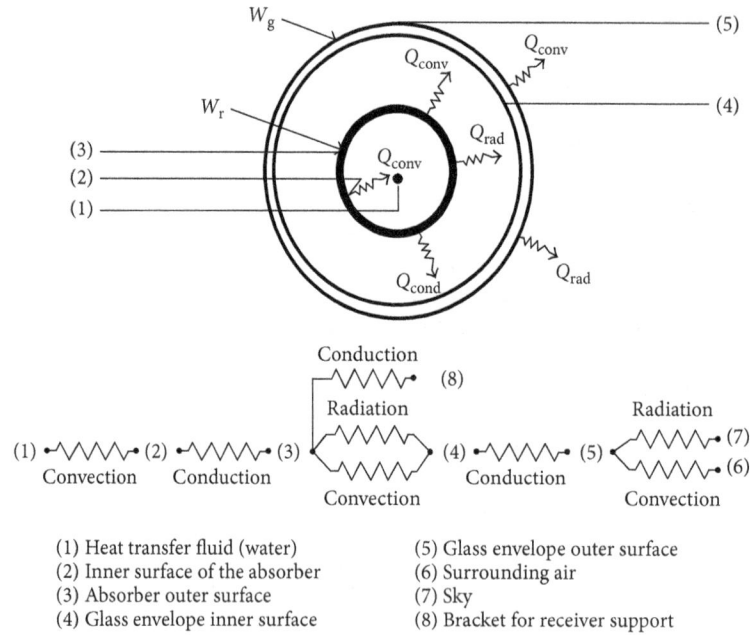

FIGURE 1: Thermal resistance model of the PTC.

(1) Heat transfer fluid (water)
(2) Inner surface of the absorber
(3) Absorber outer surface
(4) Glass envelope inner surface

(5) Glass envelope outer surface
(6) Surrounding air
(7) Sky
(8) Bracket for receiver support

2.2. Useful Heat Gain. Useful energy delivered from PTC is obtained through efficiency factor of the receiver along with incidence angle modifier determined by applying energy balance to the receiver [20]:

$$Q_u = F_r[I_b \eta_o A_a - A_r U_{rl}(T_{in} - T_a)]. \tag{8}$$

Here, to determine the dependency of the incidence angle modifier (IAM) on an angle of incidence, correlation is developed. The correlation is obtained through the application of polynomial curve fitting on experimental data as presented in Figure 2(b).

$$K_\theta = -(5.0^*10^{-6})\theta^3 + (2^*10^{-4})\theta^2 - (4^*10^{-3})\theta + 1.001, \tag{9}$$

whereas the instantaneous thermal efficiency of PTC is found by [21]

$$\eta = \frac{Q_u}{I_b A_a}. \tag{10}$$

Finally, outlet temperature of the PTC is found by [18]

$$T_o = T_{in} + \frac{Q_u}{\dot{m}c_{pw}}. \tag{11}$$

The above design calculations of the system are performed through the development of a mathematical model in the engineering equation solver (EES) [22] due to its coupling compatibility with TRNSYS [23]. The flow chart of the thermodynamic model is presented in the Figure 3.

2.3. Data Reduction. By using various equations and correlations, data is reduced for analysis and graphical representation. The heat energy gain of the PTC is defined as

$$Q_u = \dot{m}c_p(T_o - T_{in}). \tag{12}$$

Therefore, the instantaneous efficiency of PTC in each case can be calculated by [20]

$$\eta = \frac{Q_u}{A_a I_b} = \frac{\dot{m}c_p(T_o - T_{in})}{A_a I_b}. \tag{13}$$

However, thermal efficiency related to the parameters from data points is given by [18]

$$\eta = F_r\left[\eta_o - \frac{U_{rl}}{C}\left(\frac{T_{in} - T_a}{I_b}\right)\right] = a + bT^*, \tag{14}$$

where

$$T^* = \frac{T_{in} - T_a}{I_b}, \tag{15}$$

$$a = F_r \eta_o, \tag{16}$$

$$b = \left(-\frac{F_r U_{rl}}{C}\right). \tag{17}$$

Moreover, incidence angle modifier (IAM) is determined as a ratio of instantaneous thermal efficiency at an angle of incidence to peak thermal efficiency as [24, 25]

$$\text{IAM} = \frac{\eta(T_{in} = T_a)}{F_r(\eta_o)}. \tag{18}$$

Collector time constant is the time needed by the PTC to change the temperature of the working fluid, 63.2% of its steady-state value when a step change in the incident radiation occurs [26].

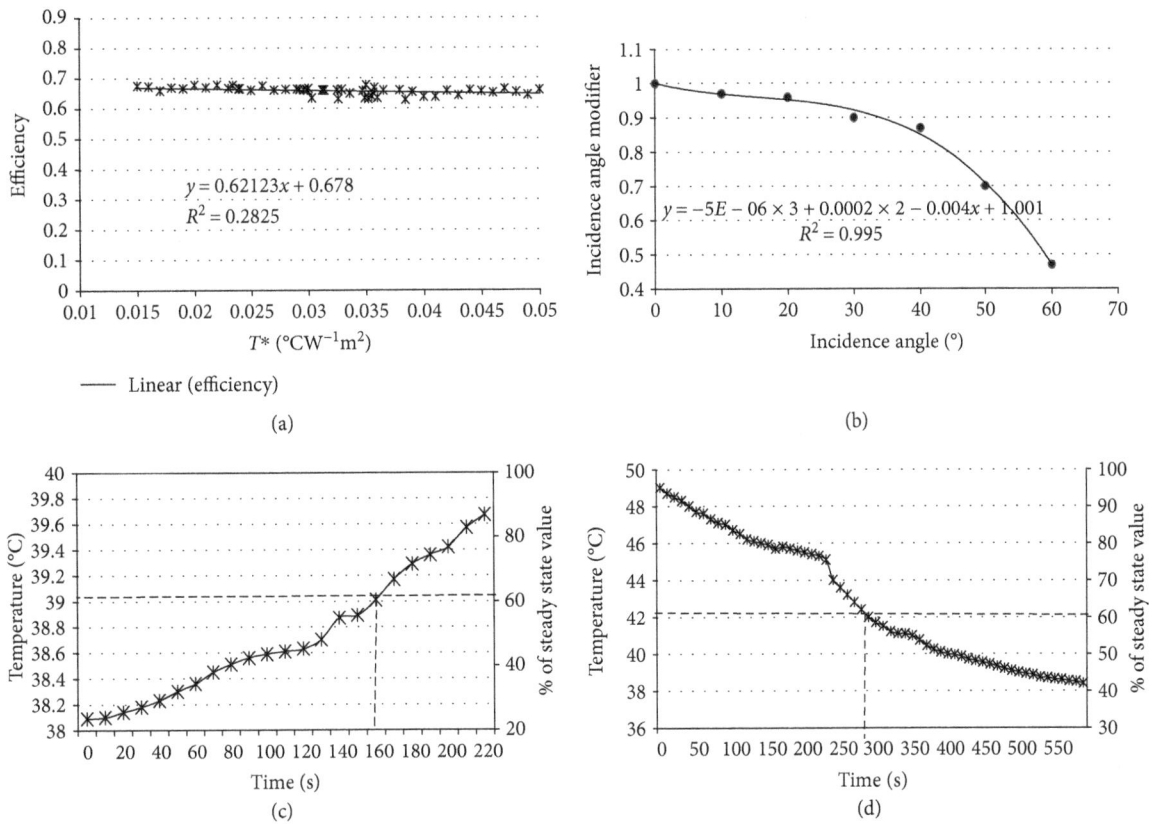

FIGURE 2: Thermal performance according to ASHRAE standard. (a) Efficiency test, (b) incidence angle modifier, (c) PTC heating time constant, and (d) PTC cooling time constant.

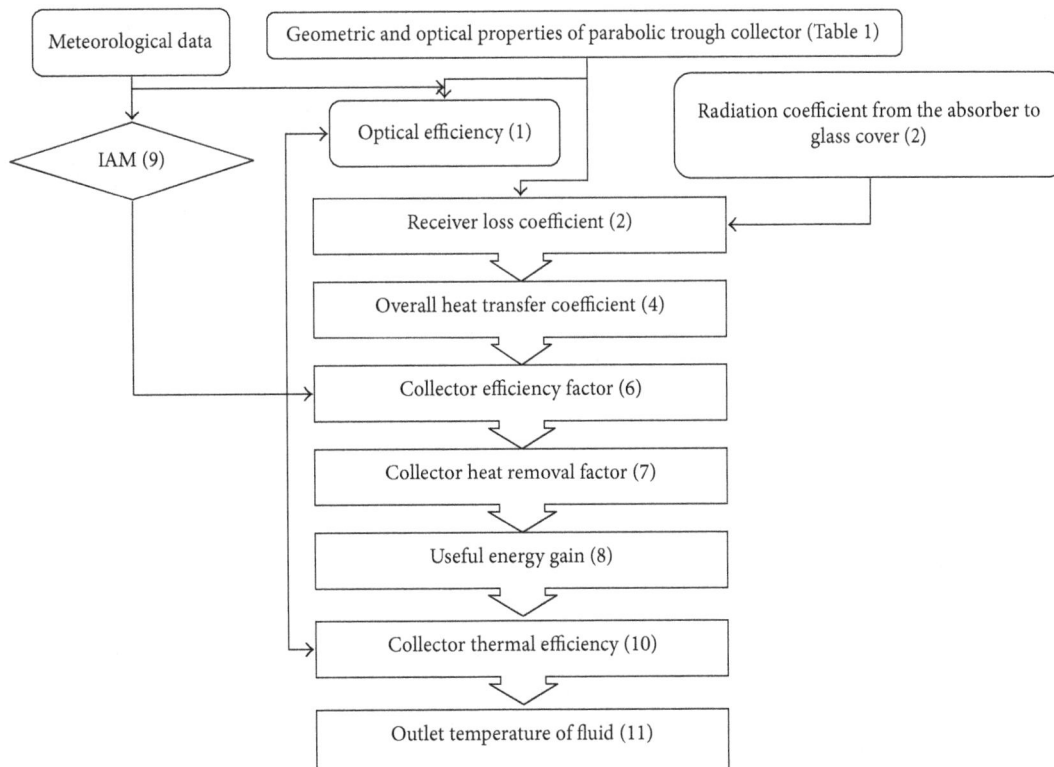

FIGURE 3: Flowchart of the thermodynamic model.

FIGURE 4: Roof-mounted PTCs with accessories for hot water production.

The time constant test for thermal collectors represents the thermal inertia of the receiver or its heat capacity.

The time constant for heating and cooling is given by [27]

$$\frac{T_o - T_{in}}{T_{oi} - T_{in}} = \frac{1}{e} = 0.367. \tag{19}$$

Finally, root mean square error value is determined by [28]

$$\text{RMSE} = \sqrt{\frac{\sum_{i=1}^{n} \left(X_{exp} - X_{model} \right)^2}{n}}. \tag{20}$$

The solar fraction of the PTC system is defined as

$$\text{SF} = \frac{Q_{sol}}{Q_{total}}, \tag{21}$$

where Q_{sol} is the solar energy gain from PTCs and Q_{total} is the sum of solar energy as well as the auxiliary energy from the electric heaters.

Moreover, the dehumidification share (DS) is the product of solar fraction and total dehumidification effect produced given as

$$\text{DS} = \text{SF}(D_e), \tag{22}$$

where D_e is the total dehumidification effect of the desiccant wheel.

3. Material and Methods

This section comprises of the experimental setup, measurement procedure, and correlations used in the current study.

3.1. Experimental Setup. Experimental structure is fabricated for real-time analysis, based on the design parameters obtained through EES model processing. The setup consists of an array of four PTCs, working fluid circuit, and sensing and measuring instruments as presented in Figure 4.

The polished stainless steel sheet of $2 \times 1 \, m^2$ is used as a reflector medium for each trough. The sheet is properly bent to achieve the desired designed concentration ratio of 9.93 and focal point 0.210 m. In total, four PTC units in series are used when analyzing the transient and real-time system performance. The evacuated tube receiver consists of a copper tube 2 m long with outer and inner diameters of 0.21 m and 0.19 m, respectively, with the selective coating. Single

axis E-W tracking was accomplished by solar tracking kit consisting of sunlight sensor and solar tracker linear actuator. The absorber tube is covered by borosilicate glass with a thickness of 0.004 m.

The other major component of the experimental setup is the solid desiccant dehumidifier as shown in Figure 5. The design parameters and overall system specifications of PTCs and solid desiccant wheel are given in Table 1. The hot water stratified storage tank with polyurethane insulation equipped with an electric heater is used for water circulation. The setpoint temperature for the electric heater is 80°C. The hot water is drawn from the top of the tank and supplied to the heating coil. The ambient air at stage 1 passes through the heating coil and after regenerating the desiccant wheel leaves the system at stage 2. Similarly, the humid air enters the dehumidification system at stage 3, dehumidified and leaves the system at stage 4.

3.2. Measurement Procedure. The experiments are conducted at RERDC (Renewable Energy Research and Development Center) located in Taxila (latitude 33.7370°N and longitude 72.7994°E). Figure 5 shows the schematic of experimental setup with measuring and sensing instruments. In the current study, various climate, inlet, and outlet parameters are continuously measured with a time interval of 10 minutes. The parameters include direct solar radiation, ambient temperature, wind velocity, water temperature, air humidity, and mass flow rate.

Operating water temperatures are measured by using K type thermocouples with a sensitivity of 0.01°C. Refrigerated bath circulator model WCR-P12 is used for calibration, with standard thermocouple PT100, calibration range −20°C to 120°C, and accuracy of ±0.01°C. Digital flow transducer S8011R was used to measure mass flow rate in the range of 50–1000 kg/h, whereas climate data including wind velocity and solar radiations are measured through hot wire anemometer and pyrheliometer model TBS-2-2 with a spectral range of 280–3000 nm and sensitivity of $9.876 \, \mu V/Wm^{-2}$. Moreover, the temperature and humidity of the air are measured by using Pro-dual sensors (KLK-100). Hot wire anemometer is used to measure the velocity of the air in the duct. Experiments were conducted from 9:00 AM to 6:00 PM in each case with a fixed mass flow rate of 480 kg/hr., average wind velocity below 1 m/s, ambient temperature within ±3% of 30°C, and absolute humidity of 16 g/kg within ±3%. Climate data was recorded in the Jinzhou Sunshine Science Data collector unit model TRM-Zs1.

3.3. Transient Analysis. Transient analysis is performed by coupling the EES model with TRNSYS through Type 66a. In the current study, the same EES model of PTC with some modifications is used, as discussed in the previous section. Though the TESS library for TRNSYS also contains linear parabolic concentrator model, the EES model is preferred due to certain limitations of Type 536. As intercept efficiency in Type 536 is declared as a parameter which remains constant throughout simulations, while in the developed model, Eq. 8 is used for useful heat gain for dependency of all variables on transient inputs and its greater accuracy during

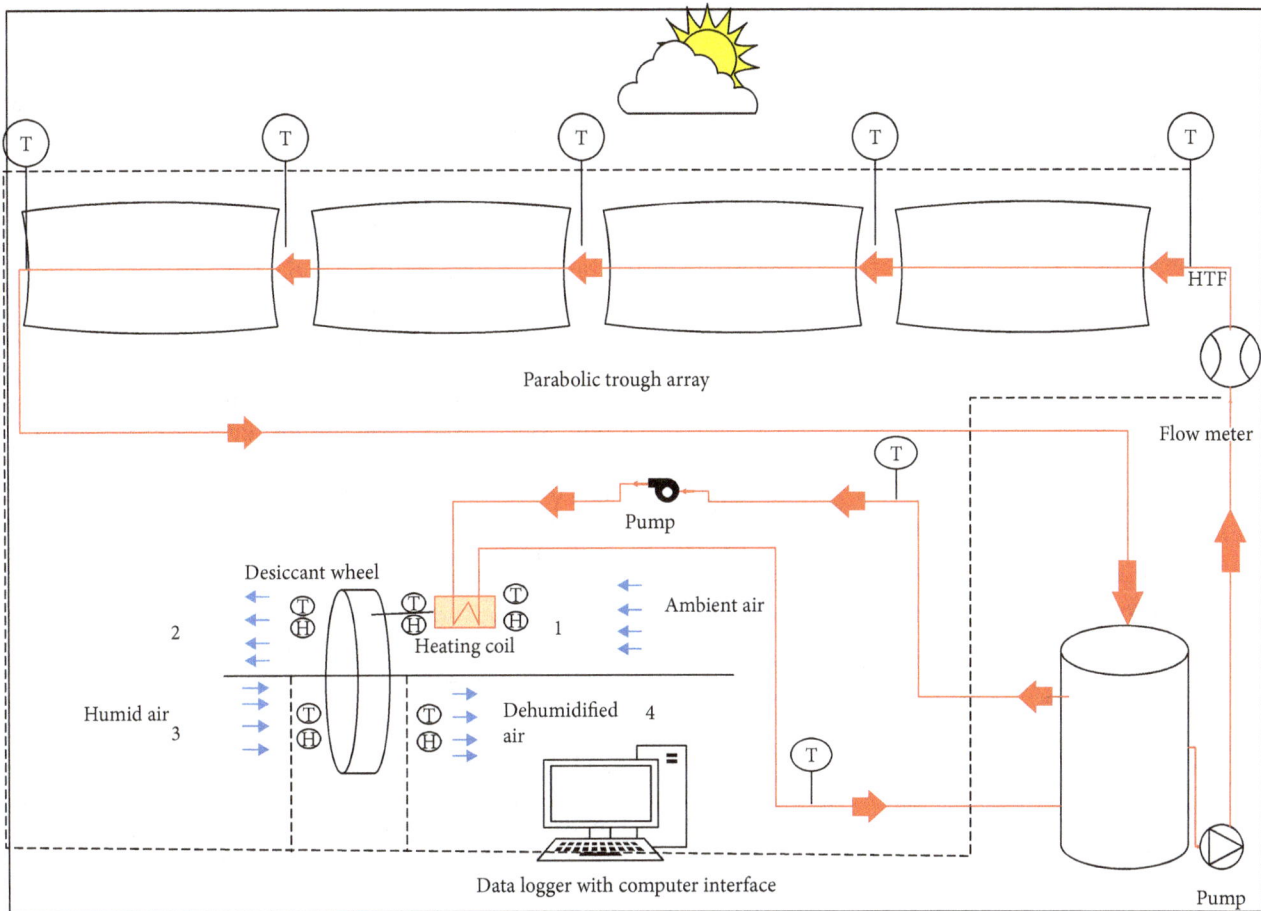

FIGURE 5: Schematic of experimental setup with sensing and measuring instruments.

model validation. Various components (types) are used to generate and process input data along with some outputs. The hourly climate data such as ambient temperature, radiations, and incidence angle based on TMY files are incorporated through Type 15. In addition, the time schedule is implemented from 9:00 AM to 6:00 PM. Equation 1 in the simulation studio is used for conversion of direct solar radiations from kJ/hr to W/m^2 and to calculate IAM values for respective incidence angles. Major components include Type 683 for the solid desiccant wheel model, Type 667 for heat recovery, and Type 4b for a stratified storage tank. The detailed description of the components can be found in the standard TRNSYS manual [23]. Furthermore, monthly useful energy gain is determined by applying an integrator that integrates hourly into monthly outcomes. The output performance parameters, for example, water outlet temperature, useful heat gain, and system efficiency are extracted from the model by using various output components such as printer and plotters as shown in Figure 6. The annual transient simulations are performed by using TMY data for the climate conditions of Taxila, Pakistan. Furthermore, potential assessment for various other cities is also executed with the help of TMY data files [23]. The simulation interval is one hour and executed by using successive substitution solution method with tolerance convergence of 0.0001.

The transient analysis follows almost similar approach as described in the experimental setup with some slight modifications and control strategies. The cold water from the base of the stratified tank is circulated through a pump to the PTC array. The model then calculates thermal energy gain and outlet temperatures based on the given climate data. A differential controller governs the working of the pump with a predefined temperature band of 10°C. Tempering valve restricts the temperature of the outlet stream of supplied water from exceeding the set point temperature. Due to the intermittent nature of the solar energy, an auxiliary gas heater is added in the flow stream. The hot water passes through the heating coils and after exchanging the heat, it returned to the hot water tank again. The regeneration air after passing through the heat recovery wheel absorbs this heat for achieving the desired regeneration temperature. This hot air hence regenerates the desiccant wheel.

4. Results and Discussion

Initially, the thermal performance of PTC is evaluated with different experimental tests according to ASHRAE 93-1986 standard [26] for procedural validation. Afterwards, further detailed experimental analysis categorized in two cases is performed, in terms of temperature variation, useful heat gain,

TABLE 1: Parameters for parabolic trough collector and desiccant wheel.

Parameter	Value
Aperture length (m)	2.1
Aperture width (m)	1.067
Focal distance (m)	0.211
Aperture area (m^2)	2.077
Concentration ratio	9.93
Rim angle (°)	108
Reflector sheet thickness (m)	0.0005
Reflectance of surface	0.89
Transmittance of glass	0.90
Length of receiver (m)	2.0
Absorptance of the receiver	>93%
Outer glass tube diameter (m)	0.123
Outer glass tube thickness (m)	0.002
Inner tube diameter (m)	0.003
Inner tube thickness (m)	0.0013
Conductivity of tube (W/m K)	380
Emittance of tube	<7%
Desiccant wheel	
Regeneration rage of desiccant wheel	60–120°C
Rotation speed	20 RPH
Effectiveness	55% at 80°C
Diameter of the wheel	350 mm
Wheel length	200 mm
Design capacity	800 kg/hr

and thermal efficiency of PTC integrated with a solid desiccant dehumidifier under a fixed mass flow rate of 480 kg/hr.

It can be observed that thermal efficiency η is a linear function of T^* as shown in Figure 2. The slope and the intercept of the η linear fit are represented as "a," "b," and given by (16) and (17), respectively.

To acquire the best fit straight line, the least square fitting technique is used through the set of efficiency points. The PTC thermal efficiency is determined as

$$\eta = 0.678 - 0.6213T^*. \qquad (23)$$

Heat losses of the collector are steady when compared to the temperature of the working fluid. The curve shows that when the inlet fluid temperature is nearby the ambient temperature, the collector efficiency is around 67% which is in agreement with published data comparable with other PTCs efficiency designed and tested for medium temperature ranges [29]. However, η in the present work is closer to the smallest concentration ratio collector present in the literature by Coccia et al. [30].

Similarly, Figure 2(b) represents the incidence angle modifier variation with respect to the incidence angle. As the angle of incidence increases, the incidence angle modifier decreases. A third order polynomial equation is obtained for IAM, by applying curve fitting for its further

application in model validation and transient analysis as given in (9). The determination coefficient (R^2) for curve fitting is 0.99.

The PTC heating and cooling time constant are 160 s and 270 s, respectively, as presented by Figures 2(c) and 2(d), respectively.

4.1. Experimental Analysis

4.1.1. Case 1: Sunny Day. In the current work, the experimental analysis of the PTC integrated with the solid desiccant dehumidifier (SDD) is determined and discussed as follows.

Variation of water inlet temperature and outlet temperature along with temperature difference is shown in Figure 7(a). The increase in incident radiations positively effects the outlet temperature of the water. Variation of inlet temperature during the day is from 72°C to 77°C. The inlet and outlet temperature of the working fluid progressively increases from 9:00 AM and reaches its peak around 2:30 PM. Similarly, working fluid temperature difference at inlet and outlet also exhibits the same trend of progressive increment till 2:00 PM. The maximum value of the outlet temperature is 84°C at 2:30 PM. Furthermore, maximum and minimum temperature difference achieved is 5.52°C and 1.5°C, respectively. The experimental instantaneous thermal efficiency of PTC along with corresponding EES model results are shown in Figure 7(b). Experimental data shows that efficiency (η) bears minimum values during the morning and afternoon.

Whereas the maximum value of η is about 62% at 1:20 PM, the reason for the variation is that the instantaneous efficiency is maximized when direct solar radiations are at its peak and incidence angles are small. Minimum instantaneous efficiency value is 19% at 6:00 PM. Moreover, comparative analysis of experimental data and model results are in good agreement with root mean square error value of about 9%.

Similarly, the hourly energy flows and solar fraction of the system is presented in Figure 7(c). It can be observed that the system thermal demand for dehumidification varies from 2 to 2.83 kW.

Similarly, solar energy gain also varies from 0.62 to 3 kW. Peak solar energy gain occurs corresponding to incident beam radiations of 5 kWh. Moreover, solar fraction also starts increasing in the morning till it reaches one at 12:00 AM; however, it starts decreasing after 3:00 PM.

Finally, dehumidification share of PTC is highlighted in Figure 7(d). It can be observed that dehumidification share of the PTC varies from 1.7 to 6.4 g/kg depending on the availability of solar radiations.

4.1.2. Case 2: Cloudy Day. The outlet temperature of the fluid and temperature difference at inlet and outlet fluctuates along the day corresponding to the intensity of solar radiations, as highlighted in Figure 8(a).

Maximum outlet temperature and temperature difference achieved are 80°C and 4.2°C, respectively. The average outlet temperature of the water is 78°C.

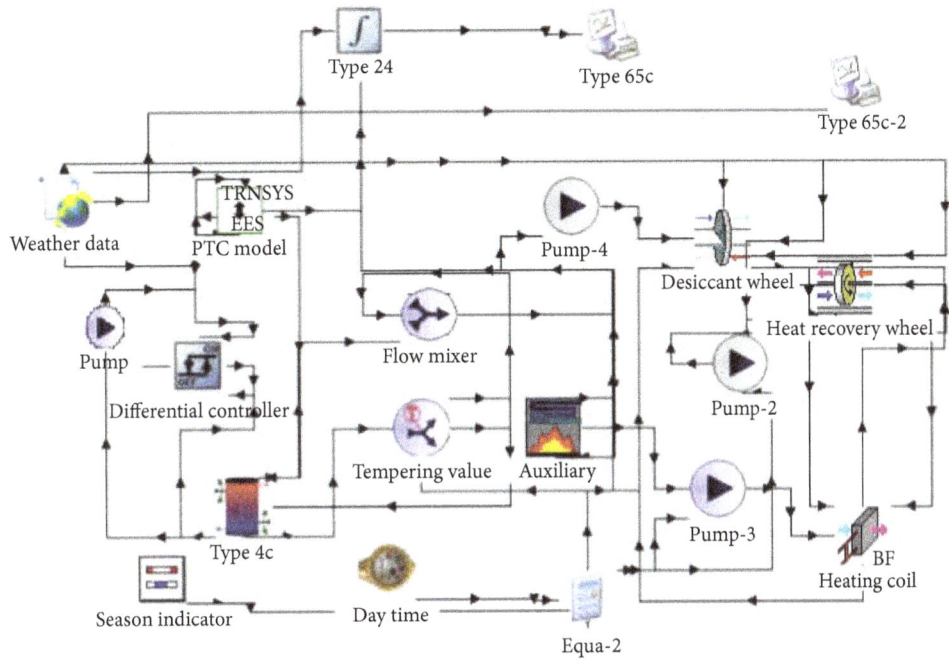

FIGURE 6: Transient simulation model in TRNSYS coupled with EES.

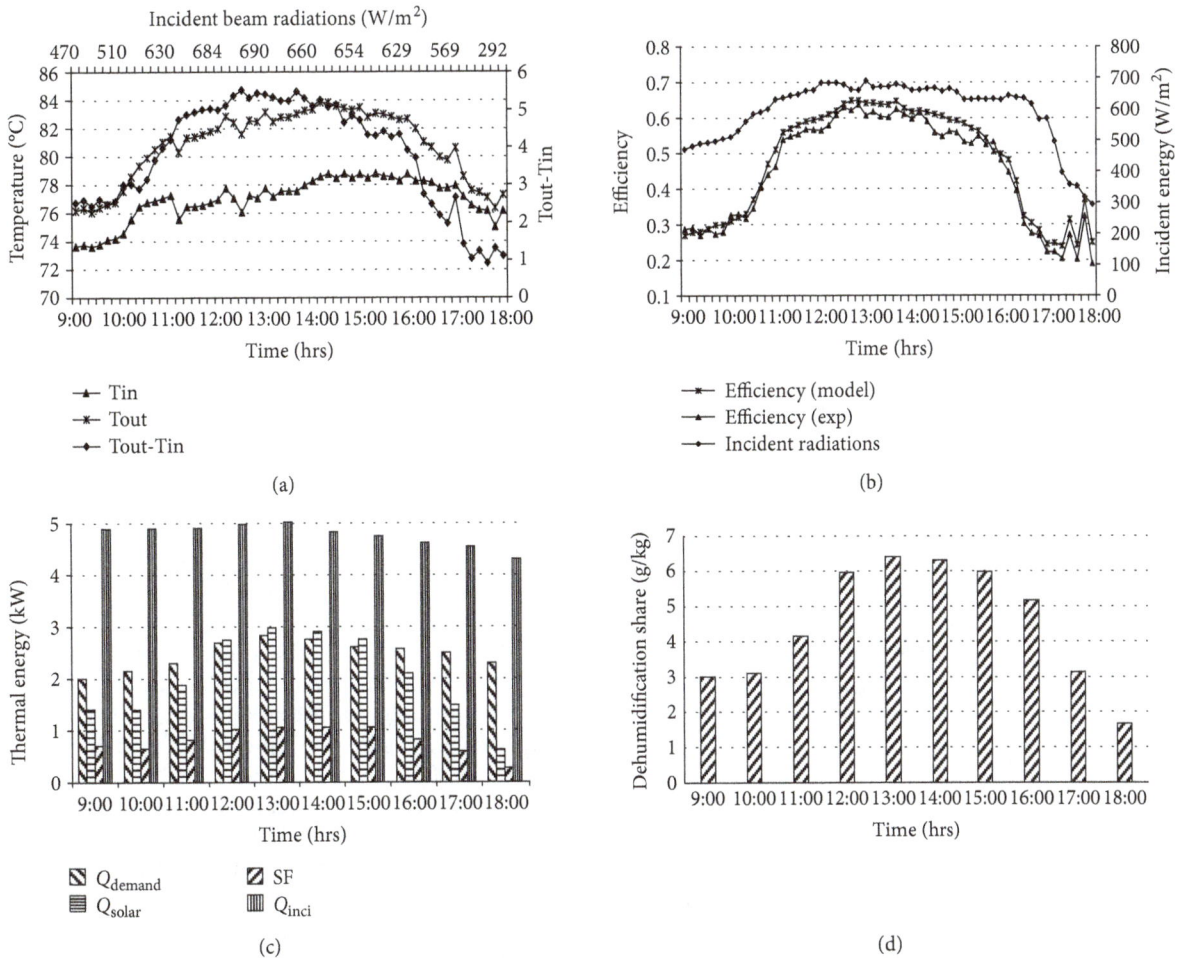

(a)

(b)

(c)

(d)

FIGURE 7: Sunny day. (a)Variation of collector temperatures. (b) Variation of collector efficiency along with model results. (c) Evolution of energy and solar fraction. (d) Dehumidification share of PTC.

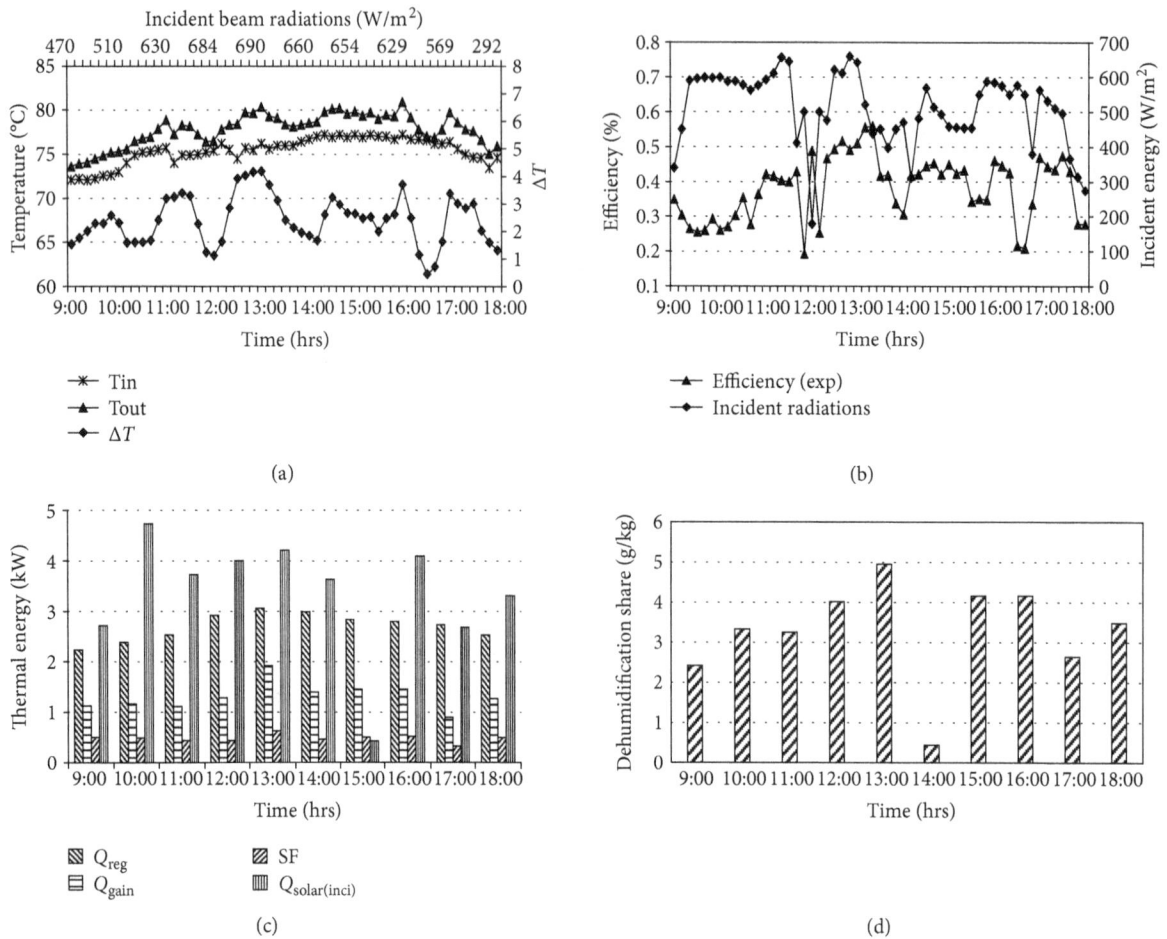

FIGURE 8: Cloudy day. (a) Variation of collector temperatures. (b) Variation of collector efficiency along with model results. (c) Evolution of energy and solar fraction. (d) Dehumidification share of PTC.

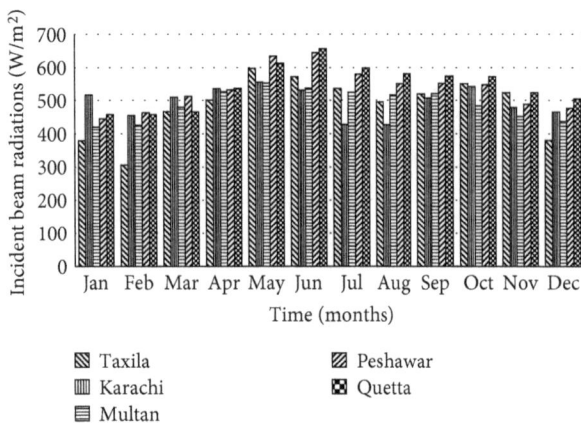

FIGURE 9: Evolution of monthly average direct solar radiations.

Cloudy day is associated with abrupt changes in solar flux between 170 W/m² and 700 W/m² as highlighted in Figure 8(b). The maximum value for incident direct solar radiations is 700 W/m² at 9:20 AM; moreover, the efficiency of PTC during a cloudy day is also presented in Figure 8(c).

The maximum value of efficiency in a cloudy day is up to 56.6% corresponding to 331 W/m² at 1:30 PM.

Similarly, the system energy flows along with solar fraction are presented in Figure 8(c).

The incident solar radiations and hence the solar energy gain constantly fluctuate along the day; however, heat gain is small due to greater incidence angle of the incidence in the morning. Although, the maximum incident energy of 4.72 kWh occurs at 10:00 AM but maximum heat gain of 1.92 kW is achieved corresponding to incident energy of 4.21 kWh at 1:00 PM. So, it is also evident from the comparison that incident direct solar radiations along with incidence angle strongly influence the useful heat gain of PTC.

Moreover, Figure 8(d) presents the dehumidification share of the PTC during a cloudy day. It is evident that it exhibits almost similar trends as solar energy gain in Figure 8(c). The maximum and minimum values for dehumidification share are 5.0 g/kg and 0.6 g/kg, respectively.

4.2. Transient Analysis. The PTC system is further subjected to annual transient analysis and is highlighted from Figures 9–11 by adopting the methodology given in Section 2.3.

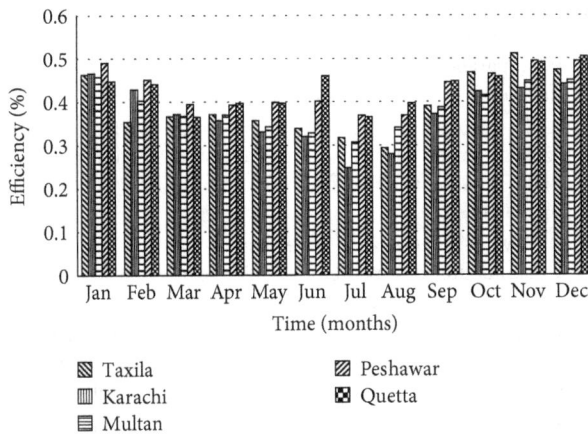

FIGURE 10: Monthly average efficiency variation of PTC.

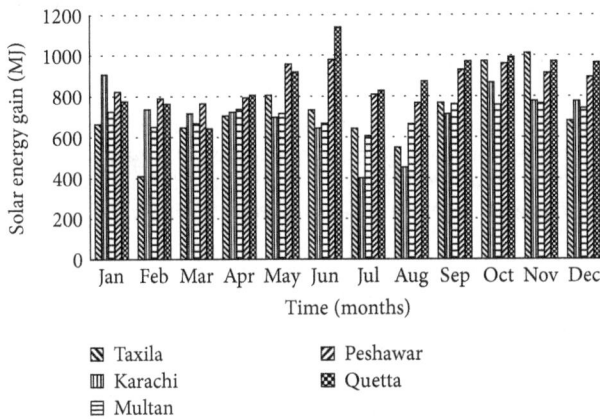

FIGURE 11: Monthly average solar energy gain of integrated PTC.

Monthly average direct solar radiations trends are presented in Figure 9 for five major cities of Pakistan. It is evident from the intensity of solar radiations that considerable potential for solar thermal application exists in the country. Maximum value of direct solar radiations is up to 656 W/m² for Quetta in June. However, minimum monthly average value exists for Taxila in February, that is, 453 W/m².

It can also be observed that monthly average values for efficiency vary from 24 to 51% as shown in Figure 10. Collector efficiency is maximum in Quetta due to better available irradiance while minimum in the humid climate of Karachi.

Finally, monthly average solar energy gain for different climates is presented in Figure 11. It is evident that maximum monthly average useful energy gain of 1140 MJ is achieved in June for Quetta, while minimum value is 408 MJ in February.

5. Conclusions

The experimental investigation of a small-sized parabolic trough collector for the solar dehumidification purpose has been carried out in the current study. The following conclusions have been drawn from the investigation:

(i) The experimental results are compared with those measured from the mathematical model. Various tests conducted according to ASHRAE standard determined that the maximum instantaneous thermal efficiency of the collector is 67%.

(ii) The intercept and the slope determined by the PTC efficiency curve are found to be 0.67 and 0.62, respectively. While the resulted PTC heating and cooling time constants are 160 s and 270 s, respectively.

(iii) Detailed experimental analysis of PTC integrated with the solid desiccant dehumidifier on a sunny and cloudy day revealed maximum instantaneous thermal efficiency of 62% and 56.6%, maximum temperature difference of 5.52°C and 4.1°C, and maximum useful energy gain of 3.0 kW and 1.92 kW, respectively.

(iv) Annual transient analysis in different cities revealed maximum monthly average efficiency up to 52% in Quetta and maximum solar energy gain around 1139 MJ. Finally, the model validation with efficiency parameter shows that a root mean square error value (RMSEV) of 9% exists. Hence, the proposed system is very much suitable for solar thermal-assisted dehumidification and air-conditioning application.

Nomenclature

A_a: Aperture area (m²)
A_c: Glass cover area (m²)
A_r: Receiver area (m²)
A_f: Geometric factor (−)
C: Concentration ratio (−)
c_{pw}: Heat capacity (kJ/kgK)
D_e: Total dehumidification effect (g)
D_i: Tube's inner diameter (m)
D_o: Tube's outer diameter (m)
F_r: Heat removal factor (−)
F: Collector efficiency factor (−)
h_{fi}: Convective heat transfer coefficient inside the tube (W/m²K)
$h_{r,(r-c)}$: Radiation coefficient from receiver to cover (W/m²K)
$h_{r,(c-a)}$: Radiation coefficient from cover to ambient (W/m²K)
h_w: Wind heat transfer coefficient (W/m²K)
I_b: Incident beam radiations (W/m²)
K_θ: Incidence angle modifier
k: Conductivity(W/mK)
$\dot{m}$: Mass flow rate (kg/s)
n: Total number of observations
Nu: Nusselt number (−)
Pr: Prandtl number (−)
Q_{cond}: Heat transfer due to conduction (W)
Q_{conv}: Heat transfer due to convection (W)
Q_{rad}: Heat transfer due to radiation (W)
Q_{sol}: Solar energy (W/m²)

Q_{total}:   Total energy (W)
Q_u: Heat energy gain (W)
R_e: Reynolds number (−)
SF: Solar fraction
T_a: Ambient temperature (K)
T_c: Temperature of glass cover (K)
T_{in}:   Inlet fluid temperature (K)
T_{oi}:   Initial outlet temperature (K)
T_o: Outlet fluid temperature (K)
T_r: Temperature of the receiver tube (K)
U_o: Overall heat transfer coefficient (W/m^{2}K)
U_{rl}:   Receiver loss coefficient (W/m^{2}K)
W_a: Aperture width (m)
W_g: Solar energy absorbed by glass cover (W)
W_r: Solar energy absorbed by the receiver (W)
X_{exp}:   Experimental data value (−)
X_{model}: Model data value (−)
Y: Parabola parameter.

Greek Symbols

ε_c: Emissivity of cover tube (−)
ε_r: Emissivity of receiver tube (−)
η_o: Optical efficiency (−)
Φ: Rim angle (°)
η: Thermal efficiency (−)
σ: Stefan Boltzmann constant (W. m^{-2}. K^{-4})
θ: Angle of incidence (°)
τ: Transmittance of glass cover
α: Absorptance of the receiver
γ: Intercept factor
ρ: Reflectance of the mirror.

Additional Points

Highlights. The highlights of the study are (i) low concentration ratio parabolic trough collector for dehumidification, (ii) thermal model along with experimental validation, (iii) transient analysis of the integrated system in five major cities of Pakistan, (iv) real-time performance investigation on sunny and cloudy days, and (v) dehumidification share of PTC along with solar fraction.

Conflicts of Interest

The authors declare that they have no conflicts of interest.

References

[1] S. Bilgen, "Structure and environmental impact of global energy consumption," *Renewable and Sustainable Energy Reviews*, vol. 38, pp. 890–902, 2014.

[2] A. Fernández-García, E. Zarza, L. Valenzuela, and M. Pérez, "Parabolic-trough solar collectors and their applications," *Renewable and Sustainable Energy Reviews*, vol. 14, no. 7, pp. 1695–1721, 2010.

[3] İ. H. Yılmaz and M. S. Söylemez, "Thermo-mathematical modeling of parabolic trough collector," *Energy Conversion and Management*, vol. 88, pp. 768–784, 2014.

[4] C. Tzivanidis, E. Bellos, D. Korres, K. A. Antonopoulos, and G. Mitsopoulos, "Thermal and optical efficiency investigation of a parabolic trough collector," *Case Studies in Thermal Engineering*, vol. 6, pp. 226–237, 2015.

[5] A. A. Hachicha, I. Rodríguez, R. Capdevila, and A. Oliva, "Heat transfer analysis and numerical simulation of a parabolic trough solar collector," *Applied Energy*, vol. 111, pp. 581–592, 2013.

[6] R. Silva, M. Pérez, and A. Fernández-Garcia, "Modeling and co-simulation of a parabolic trough solar plant for industrial process heat," *Applied Energy*, vol. 106, pp. 287–300, 2013.

[7] O. Behar, A. Khellaf, and K. MohAMmedi, "A novel parabolic trough solar collector model – validation with experimental data and comparison to engineering equation solver (EES)," *Energy Conversion and Management*, vol. 106, pp. 268–281, 2015.

[8] O. García-Valladares and N. Velázquez, "Numerical simulation of parabolic trough solar collector: improvement using counter flow concentric circular heat exchangers," *International Journal of Heat and Mass Transfer*, vol. 52, no. 3-4, pp. 597–609, 2009.

[9] S. A. Kalogirou, "Parabolic trough collectors for industrial process heat in Cyprus," *Energy*, vol. 27, no. 9, pp. 813–830, 2002.

[10] M. Ouagued, A. Khellaf, and L. Loukarfi, "Estimation of the temperature, heat gain and heat loss by solar parabolic trough collector under Algerian climate using different thermal oils," *Energy Conversion and Management*, vol. 75, pp. 191–201, 2013.

[11] A. Valan Arasu and T. Sornakumar, "Design, manufacture and testing of fiberglass reinforced parabola trough for parabolic trough solar collectors," *Solar Energy*, vol. 81, no. 10, pp. 1273–1279, 2007.

[12] D. Kumar and S. Kumar, "Year-round performance assessment of a solar parabolic trough collector under climatic condition of Bhiwani, India: a case study," *Energy Conversion and Management*, vol. 106, pp. 224–234, 2015.

[13] M. Biencinto, L. González, and L. Valenzuela, "A quasi-dynamic simulation model for direct steAM generation in parabolic troughs using TRNSYS," *Applied Energy*, vol. 161, pp. 133–142, 2016.

[14] F. J. Cabrera, A. Fernández-García, R. M. P. Silva, and M. Pérez-García, "Use of parabolic trough solar collectors for solar refrigeration and air-conditioning applications," *Renewable and Sustainable Energy Reviews*, vol. 20, pp. 103–118, 2013.

[15] M. A. Sheikh, "Energy and renewable energy scenario of Pakistan," *Renewable and Sustainable Energy Reviews*, vol. 14, no. 1, pp. 354–363, 2010.

[16] S. Z. Farooqui, "Prospects of renewables penetration in the energy mix of Pakistan," *Renewable and Sustainable Energy Reviews*, vol. 29, pp. 693–700, 2014.

[17] H. B. Khalil and S. J. H. Zaidi, "Energy crisis and potential of solar energy in Pakistan," *Renewable and Sustainable Energy Reviews*, vol. 31, pp. 194–201, 2014.

[18] S. A. Kalogirou, *Solar Energy Engineering: Processes and Systems*, Academic Press, 2013.

[19] R. H. S. Winterton, "Where did the Dittus and Boelter equation come from?," *International Journal of Heat and Mass Transfer*, vol. 41, no. 4-5, pp. 809–810, 1998.

[20] J. A. Duffie and W. A. Beckman, *Solar Engineering of Thermal Processes*, vol. 3, Wiley, New York, 2013.

[21] G. Tiwari and A. Tiwari, "Handbook of Solar Energy: Theory," in *Analysis and Applications*, Springer, 2016.

[22] S. A. Klein and F. Alvarado, *EES: Engineering Equation Solver for the Microsoft Windows Operating System*, F-Chart software, 1992.

[23] Laboratory, U.o.W.-.-M.S.E and S. A. Klein, "TRNSYS, a transient system simulation progrAM," in *Solar Energy Laborataory*, University of Wisconsin–Madison, 1979.

[24] K. S. Reddy, K. Ravi Kumar, and C. S. Ajay, "Experimental investigation of porous disc enhanced receiver for solar parabolic trough collector," *Renewable Energy*, vol. 77, pp. 308–319, 2015.

[25] A. Standard, "Methods of testing to determine the thermal performance of solar collectors," *ANSK*, vol. B198, pp. 1–1977, 1977.

[26] A. Standard, "Methods of testing to determine the thermal performance of solar collectors," in *American Society of Heating*, pp. 93–77, Refrigeration and Air Conditioning Engineers, Inc., New York, NY, USA, 1977.

[27] M. Chafie, M. F. Ben Aissa, S. Bouadila, M. Balghouthi, A. Farhat, and A. Guizani, "Experimental investigation of parabolic trough collector system under Tunisian climate: design, manufacturing and performance assessment," *Applied Thermal Engineering*, vol. 101, pp. 273–283, 2016.

[28] D. Hooper, J. Coughlan, and M. Mullen, "Structural equation modelling: guidelines for determining model fit," *Art*, p. 2, 2008.

[29] G. Coccia, G. Di Nicola, and A. Hidalgo, *Parabolic Trough Collector Prototypes for Low-Temperature Process Heat*, Swizerland, 2016.

[30] G. Coccia, G. Di Nicola, and M. Sotte, "Design, manufacture, and test of a prototype for a parabolic trough collector for industrial process heat," *Renewable Energy*, vol. 74, pp. 727–736, 2015.

8

The Effect of Microcrack Length in Silicon Cells on the Potential Induced Degradation Behavior

Xianfang Gou,[1,2] Xiaoyan Li,[1] Shaoliang Wang,[3] Hao Zhuang⊙,[2] Xixi Huang,[2] and Likai Jiang[2]

[1]Beijing University of Technology, Beijing 100124, China
[2]CECEP Solar Energy Technology (Zhenjiang) Co., Ltd., Zhenjiang 212132, China
[3]Beijing Jiaotong University, Beijing 100044, China

Correspondence should be addressed to Hao Zhuang; zhuanghao@cecsec.cn

Academic Editor: Reyna Natividad-Rangel

The presence of microcracks may lead to loss in the module output power and safety hazard of the module. This paper investigated whether the existed microscopic microcracks in cells will facilitate the PID behavior. Cells with different degrees of microcracks were fabricated into small modules to undergo the simulated PID test. The I-V performance and EL images of the modules were characterized before and after the PID test. The obtained results demonstrate that with the increase in the microcracked area or length, the modules would show a more serious PID behavior. The mechanism of this microcrack length-related degradation under high negative bias was proposed.

1. Introduction

Microcracks refer to the invisible cracks that cannot be easily perceived by the naked eye when a wafer is subjected to mechanical or thermal stress. There are several stages related to the generation of microcracks [1–8]: (i) the cutting process of an ingot or crystal bar due to a local uneven force; (ii) the cell or module fabrication process due to external factors; (iii) improper module installation; and (iv) the power plant operation period due to external factors such as wind or ground subsidence. Since the microcracked silicon wafer is not completely broken apart, microcracks can be detected only through the electroluminescence (EL) test [9]. The presence of microcracks may cause part of the cells to be inactive, leading to the loss in the output power and safety hazard of the module [9].

In a crystalline silicon cell, current is collected from fingers to the busbar and then through the string connector to the output from the junction box. The generated current of a cell is proportional to the cell active area. The inactive area can be judged by whether the current collection from the finger to the busbar is blocked or not. According to the inactive area of the cell, the number of microcracked cells, and the

impact on the output power of modules, microcracks can be divided into three categories: microscopic microcrack, general microcrack, and serious microcrack. Modules with seriously microcracked cells generally need to be replaced in a power station, and those with general microcracks will not affect the power output in the initial stage and will be disposed according to their working condition. Microscopic microcracks generally refer to the microcracks that are single or partial flakes located not at the busbars and basically do not cause failure of the area, and the power degradation of the module with microscopic microcracks should meet the industry standard (i.e., the first-year power degradation less than 2.5%). Therefore, it becomes necessary to develop the means of quantifying the risk of power loss in PV modules with cracked solar cells to ensure their output during the lifetime, and some standards may be discussed and set in the future.

In solar power stations, it is known that modules must be connected in series and parallel to build arrays to meet the load requirements. The connection of single modules in series will produce a high voltage relative to the plane of zero potential (ground). The efficiency of the modules may probably degrade due to this high negative bias under heat and

TABLE 1: Electrical performance of five groups of solar cells before and after the PID test.

Sample	PID test	Voc (V)	Isc (A)	FF (%)	Eta (%)	Rsh (Ω)	Irev2 (A)	Degradation
A	Before	0.64	9.511	73.5	18.38	187.78	0.081	2.61%
	After	0.635	9.425	72.8	17.9	11.03	2.057	
B	Before	0.64	9.575	73.67	18.55	57.83	0.171	32.02%
	After	0.617	8.675	57.37	12.61	0.44	12.277	
C	Before	0.639	9.536	73.58	18.43	57.09	0.4	33.80%
	After	0.576	8.268	62.31	12.2	0.35	12.277	
D	Before	0.639	9.502	73.56	18.36	78.94	0.117	37.53%
	After	0.602	8.505	54.49	11.47	0.33	12.277	
E	Before	0.64	9.449	73.79	18.33	75.15	0.396	49.32%
	After	0.587	8.28	46.52	9.29	0.21	12.28	

humidity, which is known as the potential induced degradation, PID [10, 11]. A number of factors [12–21], such as stacking faults in the silicon wafer, refractive index of the antireflection coating, resistance of the encapsulant material, and design of the power station, have been found and demonstrated to be related with the PID behavior. However, it has never been investigated whether the existed microscopic microcracks in cells will facilitate the PID behavior of modules.

Herein, we fabricated a series of small modules using solar cells with different microcrack lengths. The small modules were then kept in a climate chamber with constant temperature and humidity for the PID simulation test. The I-V curves and EL images of the small modules were measured before and after the PID test. The obtained results demonstrate that with the increase in the microcrack length, the modules would show a more serious PID behavior. Our work reveals the underlying relationship between the microcrack length in cells and PID of modules.

2. Materials and Methods

Conventional cells with different microcrack lengths were selected via EL and divided into 5 groups with 10 cells each group according to the microcrack length: A (0 cm), B (0–0.9 cm), C (1–2 cm), D (4–5 cm), and E (9–10 cm). The length was measured by the maximum length of the cracked area. Then the cells were fabricated into small modules using the normal process and kept in a climate chamber with constant temperature and humidity for the PID simulation test, after which the I-V and EL of the small modules were measured and analyzed.

3. Results and Discussion

The power loss in crystalline silicon-based photovoltaic modules due to microcracks was investigated by Köntges et al. in 2011 [9]. They analyzed the direct impact of microcracks on the module power and the consequences after artificial aging. The approach of artificial aging they adopt was 200 humidity freeze cycles. The main focus of their research is on the degradation of power due to crack propagation after artificial aging.

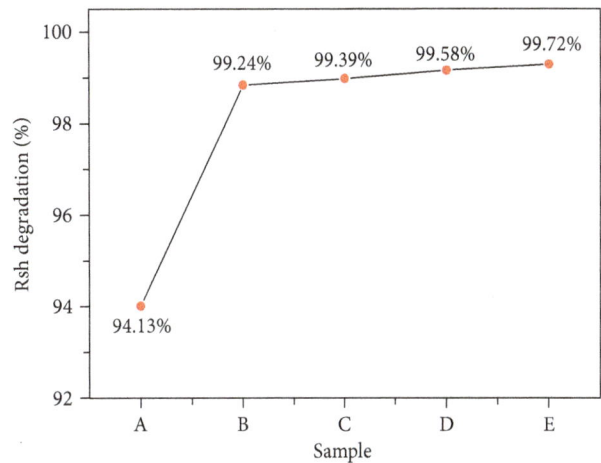

FIGURE 1: Rsh variations of five sample groups with different microcrack lengths after the PID test.

Herein, to investigate the effect of microcrack length in silicon cells on the potential induced degradation behavior, in our work, five groups of cells with different degrees of microcracks were fabricated into small modules. After performing the PID test in an environmental test chamber (85°C, 85% RH), the electrical performance was measured using a Pasan tester. The obtained data are listed in Table 1.

As can be seen from Table 1, after the PID test, degradation was observed for the open-circuit voltage (Voc), short-circuit current (Isc), and fill factor (FF). Figure 1 shows the trend of degradation with microcrack length. It can be clearly noted that with the increase in the microcrack length, a larger degradation would occur. The decrease of the parallel resistance (Rsh) is 94.13%, 99.24%, 99.39%, 99.58%, and 99.72% for groups A to E, respectively. These results demonstrate that the longer the microcrack length, the faster the Rsh degrades after the PID test, which increases the probability of providing the shunt for the current, and the trace current Irev2 is greatly improved after the PID test.

From Table 1, it can be also seen that after the PID test, the module efficiency also exhibits a larger degradation with the increase of the microcrack length. The power degradation of module group A without a microcrack is 2.61% after the test, which meets the <5% standard for IEC 62804. Modules

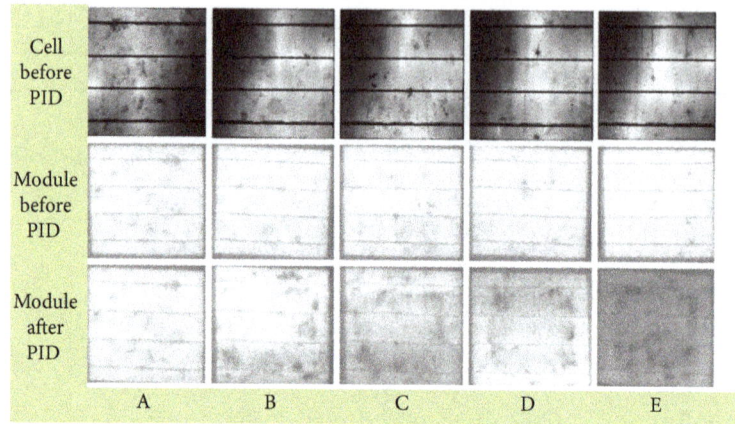

FIGURE 2: EL images of five groups with different microcrack lengths after the PID test.

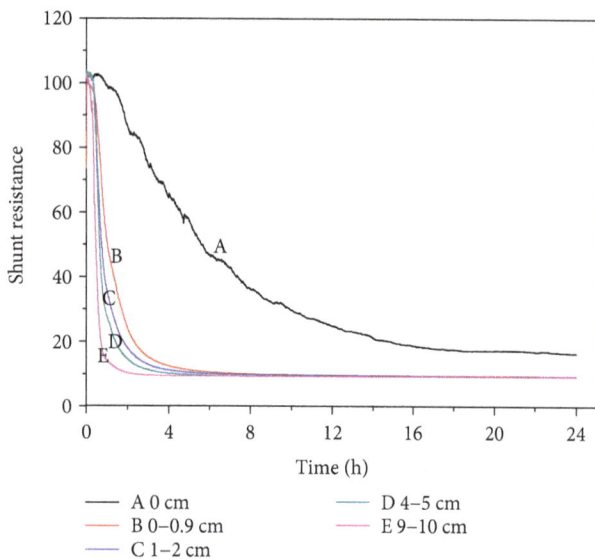

FIGURE 3: Plot of the Rsh versus PID testing time of five groups of solar cells.

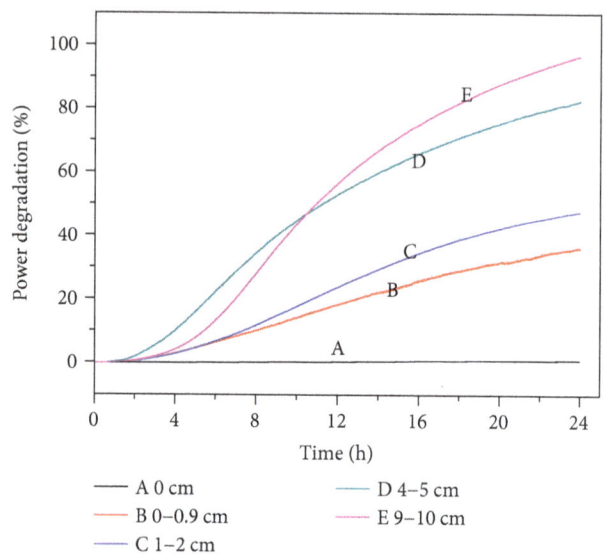

FIGURE 4: Plot of the efficiency degradation versus PID testing time of five groups of solar cells.

of groups B, C, D, and E degraded by 32.02%, 33.80%, 37.53%, and 49.32%, respectively, showing a serious PID phenomenon.

It is shown that with increasing microcrack length, positive charges are easy to gather on the surface of the cell under long-term high bias and high-temperature and humidity conditions. Under the in-built electric field, a large amount of negative charges is attracted to the surface. If a microcrack then exists in the wafer, it can provide a diversion channel for the surface charges, leading to current leakage, which decreases the efficiency of the cell. The larger the microcracked area is, the more leakage occurs, and the greater the efficiency declines.

Figure 2 shows the EL pictures of the cells and small modules before and after the PID test. By comparing these EL pictures, we can find that with the increase in the microcrack length, the EL of the cells after the PID test gradually tarnishes, which is consistent with the degradation trend of modules.

To further verify the obtained results, PIDcon equipment was used to simulate the anti-PID performance of the five sample groups. The parallel resistance change of the samples with test time is shown in Figure 3. It can be seen from the figure that the parallel resistance of sample group A (i.e., without microcrack) first decreased quickly in the initial 12 hours and gradually became steady after that. The Rsh of the B, C, D, and E sample groups decreased rapidly within the initial two hours of the test, especially for group E. After the initial two-hour rapid decrease, it slowly became stable and constant till the end of this test. These results further demonstrate that the increasing microcrack length generally gives a faster decrease rate of parallel resistance after the PID test, indicating a more serious current leakage. Figure 4 shows the PID degradation of modules with different microcrack lengths, and modules are found to degrade less with decreasing microcrack length.

Based on these results, the mechanism of the effect of the wafer microcrack defect on PID is proposed and depicted in

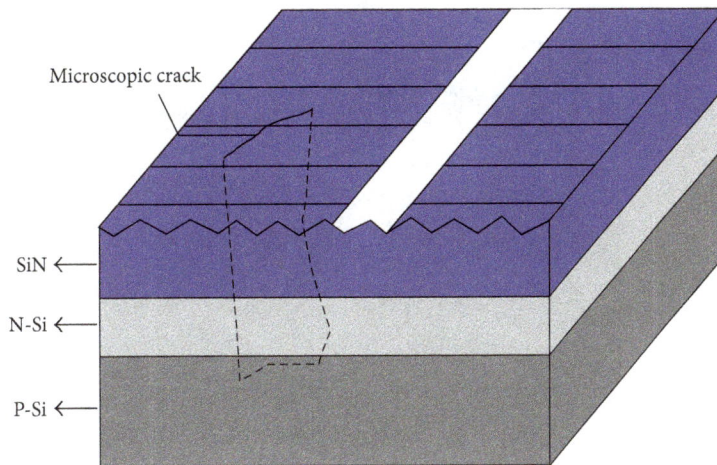

(a)

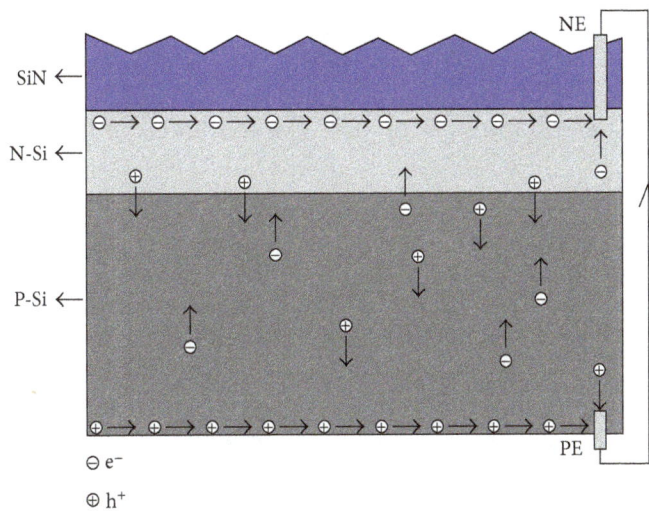

(b)

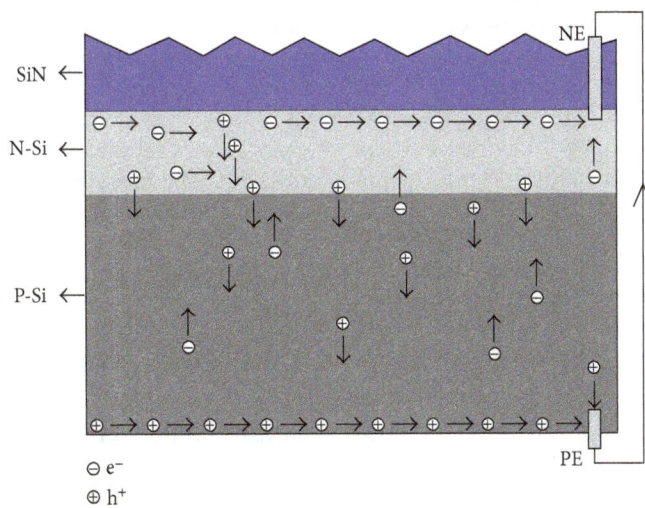

(c)

FIGURE 5: Continued.

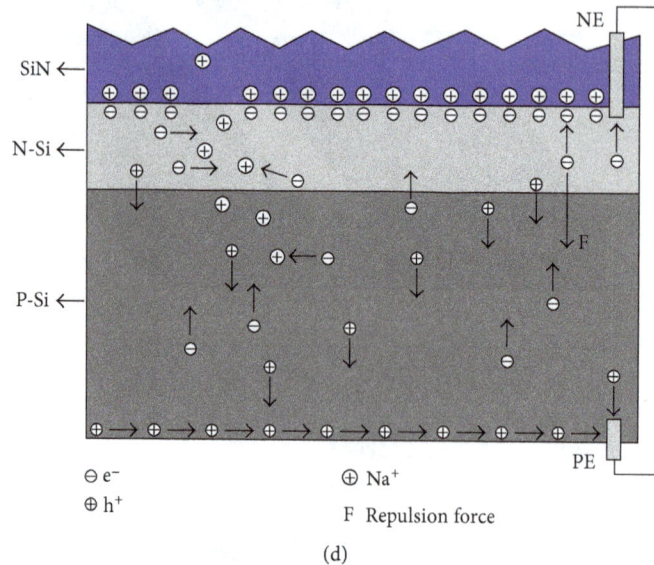

FIGURE 5: Schematic diagrams of the effect of microcracks of silicon wafer on PID: (a) diagram of part of solar cells; (b) conductive diagram of solar cells without microcracks; (c) conductive diagram of solar cells with microcracks; (d) PID conductive diagram of solar cells with microcracks.

Figure 5. The overall crack interface is shown as the dotted area. Figure 5(a) is a scheme of part of the cell, in which the microcrack is located in the cell grid area. The microcrack vertically penetrates through the PN junction. Figure 5(b) is a schematic diagram of the normal crack-free cell. Under light irradiation, photon excites the motion of nonequilibrium carriers in the silicon wafer, and the minorities (i.e., electrons) of the P-type silicon region move to the N-type silicon region. The holes, the minorities of the N-type silicon region, move toward the P-type silicon region and converge through the silver finger to the busbar to generate current. Figure 5(c) is the scheme of charge flow in a microcracked cell. During the lateral or longitudinal movement of electrons and holes, the presence of a microcrack will block the movement of electrons and holes, impeding the transportation of electrons and hence reducing the output current. Due to the limited microcrack area, the degradation in output power is not significant enough to be observed. Figure 5(d) is the scheme of the PID mechanism in a microcracked cell, when the cell is under high temperature/humidity and negative bias; the sodium ions migrate from the glass to the silicon nitride film. Therefore, sodium ions gradually accumulate at the SiN_x/Si interface. In the microcrack-free region, the positive charges of sodium ions will attract a large amount of electrons to the silicon surface, which reduces the convergence of electrons to the silver electrodes.

On the other hand, due to the accumulation of negative charges on the silicon surface, the fixed negative charges repel the electrons moving from the P-type silicon side and simultaneously attract the positive charges, thus reducing the number of electrons and holes. In the microcracked region, while the movement of electrons and holes are hindered, sodium ions are impeded when arriving at the silicon nitride layer and the P- and N-microcracked interface regions. Therefore, the sodium ions are easy to gather at

the edge of the microcracked region to capture the electrons and become the recombination center for the minorities. When more and more sodium ions accumulate in the microcracked region, the collection of current is largely reduced, leading to current leakage.

4. Conclusions

This paper focused on the effect of existing microscopic microcracks in cells on the potential induced degradation behavior. Cells with different degrees of microcrack were fabricated into small modules to undergo a simulated PID test. The I-V performance and EL images of the modules were characterized before and after the PID test. The obtained results indicated that with the increase in the microcrack length, the modules would show a more serious PID behavior. The mechanism of this microcrack length-related degradation under high negative bias was proposed.

Conflicts of Interest

The authors declare that there is no conflict of interests regarding the publication of this paper.

Acknowledgments

This study was financially supported by the research program 13RD1 CECEP (Zhenjiang).

References

[1] V. A. Popovich, A. Yunus, M. Janssen, I. M. Richardson, and I. J. Bennett, "Effect of silicon solar cell processing parameters and crystallinity on mechanical strength," *Solar Energy Materials & Solar Cells*, vol. 95, no. 1, pp. 97–100, 2011.

[2] M. Demant, S. Rein, J. Krisch et al., "Detection and analysis of micro-cracks in multicrystalline silicon wafers during solar cell production," in *Proceedings of the 37th IEEE Photovoltaic Specialists Conference*, Seattle, 2011.

[3] J. I. Mölken, U. A. Yusufoğlu, A. Safiei et al., "Impact of micro-cracks on the degradation of solar cell performance based on two-diode model parameters," *Energy Procedia*, vol. 27, pp. 167–172, 2012.

[4] H. Kim, P. Sungeun, B. Kang et al., "Effect of texturing process involving saw-damage etching on crystalline silicon solar cells," *Applied Surface Science*, vol. 284, pp. 133–137, 2013.

[5] I. Berardone, M. Corrado, and M. Paggi, "A generalized electric model for mono and polycrystalline silicon in the presence of cracks and random defects," *Energy Procedia*, vol. 55, pp. 22–29, 2014.

[6] S. S. Ko, C. S. Liu, and Y. C. Lin, "Optical inspection system with tunable exposure unit for micro-crack detection in solar wafers," *Optik- International Journal for Light and Electron Optics*, vol. 124, no. 19, pp. 4030–4035, 2013.

[7] V. T. Dragišić, "Silicon solar wafers: quality control and improving the mechanical properties," *Procedia Engineering*, vol. 117, pp. 459–464, 2015.

[8] K. O. Davis, M. P. Rodgers, G. Scardera et al., "Manufacturing metrology for c-Si module reliability and durability part II: cell manufacturing," *Renewable and Sustainable Energy Reviews*, vol. 59, pp. 225–252, 2016.

[9] M. Köntges, I. Kunze, S. K. Schröder, X. Breitenmoser, and B. Bjørneklett, "The risk of power loss in crystalline silicon based photovoltaic modules due to micro-cracks," *Solar Energy Materials & Solar Cells*, vol. 95, no. 4, pp. 1131–1137, 2011.

[10] S. Pingel, O. Frank, M. Winkler et al., "Potential induced degradation of solar cells and panels," in *Proceedings of the 35th IEEE Photovoltaic Specialists Conference (PVSC)*, Honolulu, Hawaii, USA, 2010.

[11] V. Fjallstrom, S. PMP, A. Hultqvist et al., "Potential-induced degradation of thin film solar cells," *IEEE Journal of Photovoltaics*, vol. 3, no. 3, pp. 1090–1094, 2013.

[12] H. Nagel, A. Metz, and K. Wangemann, "Crystalline Si solar cells and modules featuring excellent stability against potential-induced degradation," in *Proceedings of the Eu-Pvsec*, Hamburg, Germany, 2011.

[13] W. Herrmann, M. Schweiger, and G. Mathiak, "Potential-induced degradation - comparison of different test methods and low irradiance performance measurements," in *Proceedings of the 27th European Photovoltaic Solar Energy Conference and Exhibition*, Frankfurt, Germany, 2012.

[14] V. Naumann, D. Lausch, and C. Hagendorf, "Sodium decoration of PID-s crystal defects after corona induced degradation of bare silicon solar cells," *Energy Procedia*, vol. 77, pp. 397–401, 2015.

[15] A. Raykov, H. Hahn, K.-H. Stegemann et al., "Towards a root cause model for the potential-induced degradation in crystalline silicon photovoltaic cells and modules," in *Proceedings of the European Photovoltaic Solar Energy Conference and Exhibition*, Paris, France, 2013.

[16] S. Koch, D. Nieschalk, J. Berghold, S. Wendlandt, S. Krauter, and P. Grunow, "Potential induced degradation effects on crystalline silicon cells with various antireflective coatings," in *Proceedings of the European Photovoltaic Solar Energy Conference and Exhibition*, Frankfurt, Germany, 2012.

[17] K. Mishina, A. Ogishi, K. Ueno et al., "Investigation on anti-reflection coating for high resistance to potential induced degradation," *Japanese Journal of Applied Physics*, vol. 53, no. 3S1, article 03CE01, 2014.

[18] S. Koch, C. Seidel, P. Grunow, S. Krauter, and M. Schoppa, "Polarization effects and tests for crystalline silicon cells," in *Proceedings of the 26th European Photovoltaic Solar Energy Conference and Exhibition*, Hamburg, Germany, 2011.

[19] V. Naumann, D. LAUSCH, A. Hähnel et al., "Explanation of potential-induced degradation of the shunting type by Na decoration of stacking faults in Si solar cells," *Solar Energy Materials & Solar Cells*, vol. 120, pp. 383–389, 2014.

[20] S. Jonai, K. Hara, Y. Tsutsui, H. Nakahama, and A. Masuda, "Relationship between cross-linking conditions of ethylene vinyl acetate and potential induced degradation for crystalline silicon photovoltaic modules," *Japanese Journal of Applied Physics*, vol. 54, no. 8S1, article 08KG1, 2015.

[21] S. Koch, J. Berghold, O. Okoroafor, S. Krauter, and P. Grunow, "Encapsulation influence on the potential induced degradation of crystalline silicon cells with selective emitter structures," in *Proceedings of the European Photovoltaic Solar Energy Conference and Exhibition*, Frankfurt, Germany, 2012.

9

Antibacterial Activity of Ag-Doped TiO$_2$ and Ag-Doped ZnO Nanoparticles

Gebretinsae Yeabyo Nigussie ®,[1] Gebrekidan Mebrahtu Tesfamariam,[1] Berhanu Menasbo Tegegne,[1] Yemane Araya Weldemichel,[1] Tesfakiros Woldu Gebreab,[2] Desta Gebremedhin Gebrehiwot,[1] and Gebru Equar Gebremichel[3]

[1]Department of Chemistry, College of Natural and Computational Sciences, Mekelle University, Mekelle, Ethiopia
[2]Department of Physics, College of Natural and Computational Sciences, Mekelle University, Mekelle, Ethiopia
[3]Department of Biology, College of Natural and Computational Sciences, Mekelle University, Mekelle, Ethiopia

Correspondence should be addressed to Gebretinsae Yeabyo Nigussie; g.tinsae21@gmail.com

Academic Editor: P. Davide Cozzoli

We report in this paper antibacterial activity of Ag-doped TiO$_2$ and Ag-doped ZnO nanoparticles (NPs) under visible light irradiation synthesized by using a sol-gel method. Structural, morphological, and basic optical properties of these samples were investigated using X-ray diffraction (XRD), scanning electron microscopy (SEM), energy dispersive X-ray (EDX) spectrum, and UV-Vis reflectance. Room temperature X-ray diffraction analysis revealed that Ag-doped TiO$_2$ has both rutile and anatase phases, but TiO$_2$ NPs only have the anatase phase. In both ZnO and Ag-doped ZnO NPs, the hexagonal wurtzite structure was observed. The morphologies of TiO$_2$ and ZnO were influenced by doping with Ag, as shown from the SEM images. EDX confirms that the samples are composed of Zn, Ti, Ag, and O elements. UV-Vis reflectance results show decreased band gap energy of Ag-doped TiO$_2$ and Ag-doped ZnO NPs in comparison to that of TiO$_2$ and ZnO. Pathogenic bacteria, such as *Staphylococcus aureus*, *Pseudomonas aeruginosa*, and *Escherichia coli*, were used to assess the antibacterial activity of the synthesized materials. The reduction in the viability of all the three bacteria to zero using Ag-doped ZnO occurred at 60 μg/mL of culture, while Ag-doped TiO$_2$ showed zero viability at 80 μg/mL. Doping of Ag on ZnO and TiO$_2$ plays a vital role in the increased antibacterial activity performance.

1. Introduction

Currently, nanosized materials are the most advanced type of materials, both in scientific knowledge and in commercial applications. Inorganic nanoparticles (NPs), such as silver, copper, titanium, and zinc, are the most interesting NPs due to their applications and positive impact on pathogenic microorganisms [1–4]. NPs have been studied for many years because of their size-dependent physical and chemical properties. Among NPs, great attention has been shifted to nanooxides [5–7]. NPs have attracted great interest due to their special or specific properties and selectivity, especially in pharmaceutical and biological applications [8]. In laboratory tests with NPs, different microorganisms have been eliminated within minutes of contact with the NPs [9–12]. The application of NPs on bacteria is very important since NPs have a tendency to be in the lowest level and directly enter the food chain of the ecosystem [13, 14].

Recently, special attention has been given to TiO$_2$ and ZnO NPs due to their unique optical, electrical, and chemical properties. TiO$_2$ is a tremendous photocatalyst, which is widely used for antibacterial activity due to its high photosensitivity, high efficiency, nontoxic nature, strong oxidizing power, relative cheapness, and chemical stability [15]. ZnO is also a promising photocatalyst and plays a pivotal role in antibacterial activity. In line with this, it is low cost, biocompatible, highly catalytic, and environmental friendly [16, 17].

In order to enhance the photocatalytic activity, intensive interdisciplinary researches have been made on TiO_2 [18–20] and ZnO [21–24]. It is known that photocatalytic activity of NPs depends upon their crystalline structure [19], doping [25], surface area [26], and hydroxyl group [19]. Currently, different researchers are engaged in improving the efficiency of photocatalysts by using metal dopants like Ag which is the most effective due to its high stability and good electrical/thermal conductivity. Furthermore, Ag doping on the surface of metal oxides employed to enhance photocatalytic activity by preventing fast e^--h^+ recombination processes [23, 24, 27, 28]; also, this mechanism could lead to the generation of good antibacterial properties. In this work, Ag-doped TiO_2 and Ag-doped ZnO were synthesized using the sol-gel method, and their antibacterial activity was carefully investigated and discussed.

2. Experimental

2.1. Materials and Methods.
High-purity (AR grade) titanium tetrachloride ($TiCl_4$) (Merck, EtOH 99%), silver nitrate ($AgNO_3$), zinc nitrate ($ZnNO_3$), hydrochloric acid (HCl), deionized water (DI), and sodium hydroxide (NaOH) were used as precursors for the preparation of NPs and were acquired from Sigma-Aldrich.

2.2. Synthesis of Nanoparticles.
TiO_2 NPs were synthesized using an acid-catalyzed sol-gel process, as described elsewhere [29]. First, 1 mL of $TiCl_4$ was slowly added dropwise to 10 mL EtOH (Merck, 99.8%) under vigorous stirring. A large amount of HCl gas was evolved, and a yellowish solution was formed. The gel was subjected to an oven, and it was dried at 100°C for 24 h. Finally, the white TiO_2 powder was obtained. To obtain nanocrystalline particles, TiO_2 was annealed at 450°C for 4 hours.

Ag-doped TiO_2 was synthesized using1 mL $TiCl_4$ (Merck, 99%) slowly added dropwise to 10 mL EtOH (Merck, 99.8%) under vigorous stirring. $AgNO_3$ was mixed gently with 0.5 mL deionized water (DI). After gelation, the NPs were left to dry at 100°C for 24 hours. In addition, the amorphous TiO_2 transformed to a crystalline structure using a furnace at 450°C for 4 hours.

First, 0.5 M $ZnNO_3$ was dissolved in ethanol, and the reaction mixture was kept under constant stirring for 1 hour until $ZnNO_3$ was completely dissolved. Similarly, in another reaction vessel, 0.5 M of NaOH was dissolved under constant stirring for 1 hour. Then, 0.5 M NaOH was added dropwise to the $ZnNO_3$ solution under vigorous stirring for 45 minutes. The reaction mixture was left for 2 hours after NaOH was completely added. Then, the solution was centrifuged for 10 minutes, the precipitate was dried in an oven at approximately 80°C, and $Zn(OH)_2$ was completely converted into ZnO.

Ag-doped ZnO nanopowder was synthesized using a 0.5 M concentration of $AgNO_3$, which was added to the zinc solution before the NaOH solution, and then, the Ag-doped ZnO NPs were obtained from the $Ag(OH)_2$ precipitate. The powder was calcined at a heating rate of 2°C/min in an air atmosphere for 4 hours at 450°C.

2.3. Characterizations of Ag-TiO_2 and Ag-ZnO Nanoparticles.
The phase components of the powder were analyzed using PANalytical X'pert PRO with monochromatic Cu Kα radiation, and the samples were scanned over a range of 20–80°. SEM (JEOL JSM-5610 analysis station SEM) was utilized to examine the shape and cross-sectional morphology. The energy dispersive X-ray (EDX) spectrum was recorded with JEOL JSM-5610 SEM equipped with EDX. The UV-Vis diffuse reflectance spectra were measured using a PerkinElmer Lambda 35 spectrometer which was operated at a wavelength range of 200–800 nm.

2.4. Bacterial Strain and Culture Situations.
In this trial, three different characteristic bacterial pathogens were nominated: Staphylococcus aureus (S. aureus, ATCC29737/NCIM-2901), Pseudomonas aeruginosa (P. aeruginosa, PA220), and Escherichia coli (E. coli, ATCC 25922). These were grown in 61748 LB Broth (Luria Bertani Broth) medium in a humidified incubator at 37°C with constant agitation overnight. The microorganisms were cultured on a nutrient agar plate for 24 hours and grown aerobically at 37°C.

2.4.1. Antibacterial Activity of Ag-TiO_2 and Ag-ZnO Nanoparticles

(1) Bacterial Cell Growth and Viability Resting. The bacterial growth rate was observed in the presence of Ag-TiO_2 and Ag-ZnO NPs, on the bacteria. S. aureus, P. aeruginosa, and E. coli cultures were gathered in the midexponential growth phase, and the cells were collected using centrifugation at 3000 rpm for 10 min. Directly, the bacterial pellet was washed three times with saline water to clean the pellet of medium constituents. Then, 10 μL of the cell suspensions was mixed with seven different concentrations of Ag-TiO_2 and Ag-ZnO NPs (0 μg/mL, 10 μg/mL, 20 μg/mL, 40 μg/mL, 60 μg/mL, 80 μg/mL, and 100 μg/mL) and incubated at 37°C for 2 h with slight shaking. The mixture was then transferred to 5 mL tubes containing 2 mL MH medium, and the tubes were incubated on a rotary shaker at 180 rpm and 37°C. The antibacterial activity was tested by using 107 CFU/mL and then incubated with concentrations of Ag-TiO_2 and Ag-ZnO for 4 h. Next, 20 μL of a successive 106-fold dilution of each bacterial suspension in sterile deionized water was spread onto agar plates to grow for 24 h at 37°C.

$$Survival\% = \frac{Colony\ number\ of\ treated\ bacteria}{Colony\ number\ of\ control\ bacteria} \times 100. \quad (1)$$

3. Results and Discussion

3.1. Material Characterization

3.1.1. X-Ray Diffraction.
The XRD pattern of pure TiO_2 and Ag-doped TiO_2 samples is shown in Figure 1. It reveals that the unmodified TiO_2 contains only the anatase phase, and its diffraction peaks are well matching with those of the standard anatase phase of TiO_2 Joint Committee on Powder Diffraction Standard (JCPDS card number 01-071-1167), while Ag-doped TiO_2 exhibited both the anatase and rutile phases (JCPDS card number 4-783). In Figure 1(a, b), the (101)

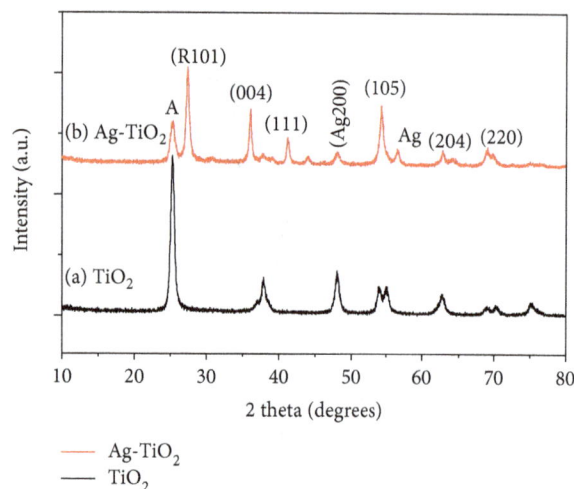

FIGURE 1: XRD patterns of (a) TiO$_2$ and (b) Ag-doped TiO$_2$ nanoparticles.

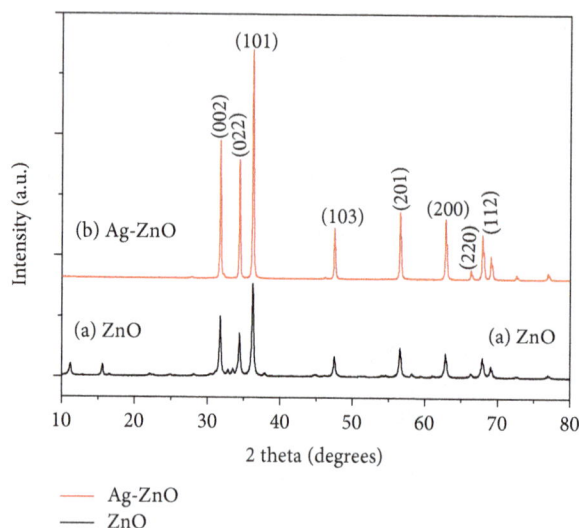

FIGURE 2: XRD patterns of (a) ZnO and (b) Ag-doped ZnO NPs.

diffraction peak appeared to be attributed to both TiO$_2$ and Ag-doped TiO$_2$. Xu et al. [30] reported that a mixture of anatase and rutile phases of Ag-doped TiO$_2$ possesses greater photocatalytic activity than the pure anatase phase of TiO$_2$ under UV light. This is evidenced by the presence of Ag in the diffractogram of the Ag-doped samples with the relative intensity of the stronger Ag peaks, (111) at 2θ of 42.500° and (200) at 2θ of 48.500°. Furthermore, this result agrees with that of other reports [31], in which calcination is used to increase the crystallinity of TiO$_2$ by decreasing the e$^-$-h$^+$ pair recombination and enhance the photocatalytic activity as well.

Figure 2(a, b) represents the XRD pattern of pure and Ag-doped ZnO collected over a 2θ range of 20°–80° using Cu Kα radiation.

The XRD pattern of the ZnO NP (Figure 2) shows its characteristic peaks with a hexagonal wurtzite structure. The broadening of the graph illustrates the nanometer range of the particle. Our data is in agreement with that of JCPDS card number 36-1451 [1]. The strongest peaks observed at 2θ values of 31.791°, 34.421°, 36.252°, 47.511°, 56.602°, 62.862°, 67.961°, and 69.000° correspond to the lattice planes (100), (002), (101), (110), (103), (112), (201), and (200), respectively. The XRD pattern of Ag-doped ZnO NPs (Figure 2(b)) shows the characteristic peak of Ag at 2θ of 38.110°, 44.270°, and 64.420° (JCPDS card number 04-0783). No shift was observed for ZnO peaks, confirming the surface doping or segregation of Ag nanoclusters on the grain boundaries of ZnO NPs [32].

3.1.2. SEM Analysis. The morphologies of the samples were observed by SEM. Figures 3(a)–3(d) show the morphologies of the samples and Ag distribution in the as-synthesized samples. It can be seen that the surface of Ag in Figures 3(b) and 3(d) is smooth. The nanostructure of the sample as revealed by SEM shows a significant difference in the undoped and doped states. Large elongated grains together with a granular structure show evidence of mixed phases which is in accordance with the XRD results

of the corresponding gel powder (Figure 2). These results are in accordance with the values found in the literature [33]. Furthermore, the compositions of the samples were determined with energy dispersive X-ray spectroscopy (EDX) (Figure 4(a, b)). All of the peaks on the curves are recognized to be of Zn, Ag, Ti, O, Cu, C, and Na elements, and no peaks of other elements are observed. The Cu, C, and Na signals might be from the precursors and sample holders. Therefore, it is concluded that the as-synthesized samples are composed of Zn, Ti, Ag, and O elements, which is in good agreement with the above XRD result.

3.1.3. Optical Properties. As shown in Figure 4, Ag-TiO$_2$ and Ag-ZnO exhibited atypical absorption characteristics due to a band gap change in the range 400–550 nm in the visible region caused by the surface plasmon band characteristics of silver; it was further confirmed that Ag was effectively deposited on the surface of TiO$_2$ and ZnO [34]. The shift in absorption spectra provides some evidence of the interaction between Ag and TiO$_2$ or ZnO, which is also in agreement with the XRD patterns.

3.2. Antibacterial Activity of the Nanoparticles. The antibacterial activities of Ag-TiO$_2$ and Ag-ZnO were examined against gram-positive S. aureus and gram-negative P. aeruginosa and E.coli bacteria, as shown in Figure 5. The Ag-TiO$_2$ and Ag-ZnO NPs at concentrations of 60 µg/mL of culture were toxic to the three different bacteria tested Figure 5. However, 40 µg/mL concentrations of Ag-ZnO NPs eliminated 100% P. aeruginosa cells, whereas 15% and 12% viabilities of S. aureus and E. coli were obtained, respectively. In Ag-doped NPs at 60 µg/mL of culture, 0% viability in the case of P. aeruginosa and S. aureus was observed, while in E. coli, viabilities were observed. Therefore, here it was determined that 60 µg/mL Ag-doped NP concentration of bacterial culture (0.2 OD at 600 nm) is the optimal concentration for eliminating the bacteria.

When we compare the antibacterial activity of Ag-ZnO NPs to that of Ag-TiO$_2$ NPs, it was observed that Ag-ZnO

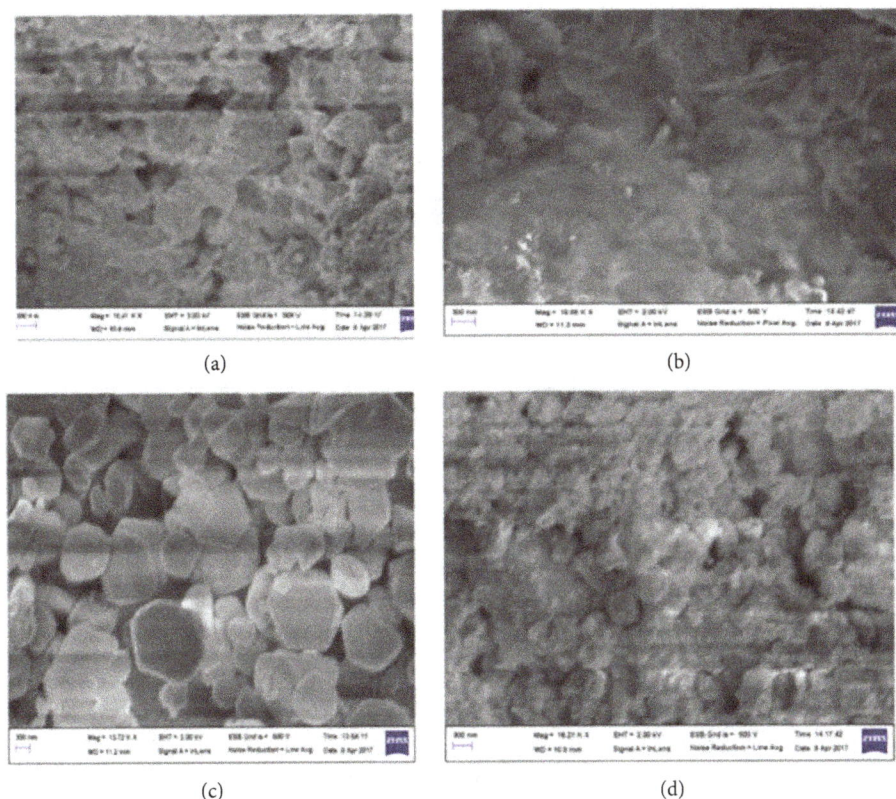

FIGURE 3: SEM images of (a) pure TiO_2, (b) Ag-TiO_2, (c) pure ZnO, and (d) Ag-ZnO NPs.

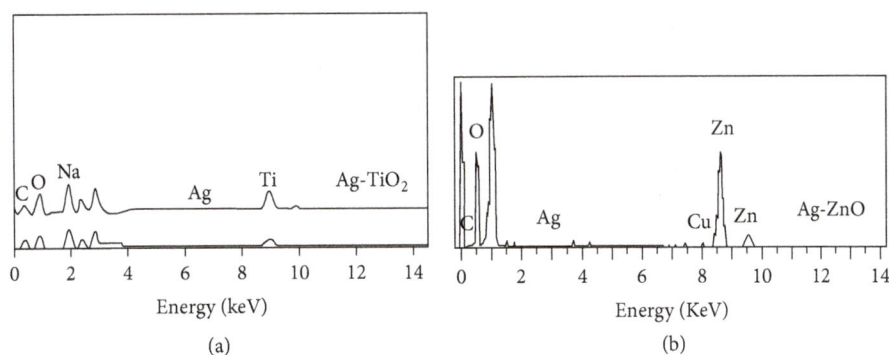

FIGURE 4: EDX spectrum of (a) Ag-TiO_2 and (b) Ag-ZnO.

NPs are highly efficient. A significant antibacterial activity was observed on gram-negative bacteria in both doped NPs. It may be that gram-positive bacteria have a stronger molecular network in the cell wall than gram-negative bacteria and silver ions may enter the cell walls of gram-positive bacteria [29]. The percent viability of bacteria was exponentially reduced as the concentration of Ag doping increased in the TiO_2 and ZnO matrix. The observed results of this study, along with those of a previous study [35], demonstrate that doping with metal or metal oxide on the surface of TiO_2 and ZnO NPs increases the e^--h^+ charge separation by reducing the band gap energy and leads to a delay in the recombination and an increase in the antibacterial activity Figure 6.

4. Conclusion

In summary, Ag-TiO_2 and Ag-ZnO were prepared via the sol-gel method. The synthesized NPs were characterized using XRD, SEM, EDX, and UV-Vis. This study may provide new insights into the design and preparation of nanomaterials and the enhancement of antibacterial activity. In comparison to other materials, the antibacterial results were more favorable for assays conducted with the same species. Moreover, the low antibacterial activities of pure TiO_2 and ZnO were significantly improved by the incorporation of silver. The synthesized Ag-TiO_2 and Ag-ZnO NPs with high thermal stability and strong antibacterial activity are expected to serve in applications in the

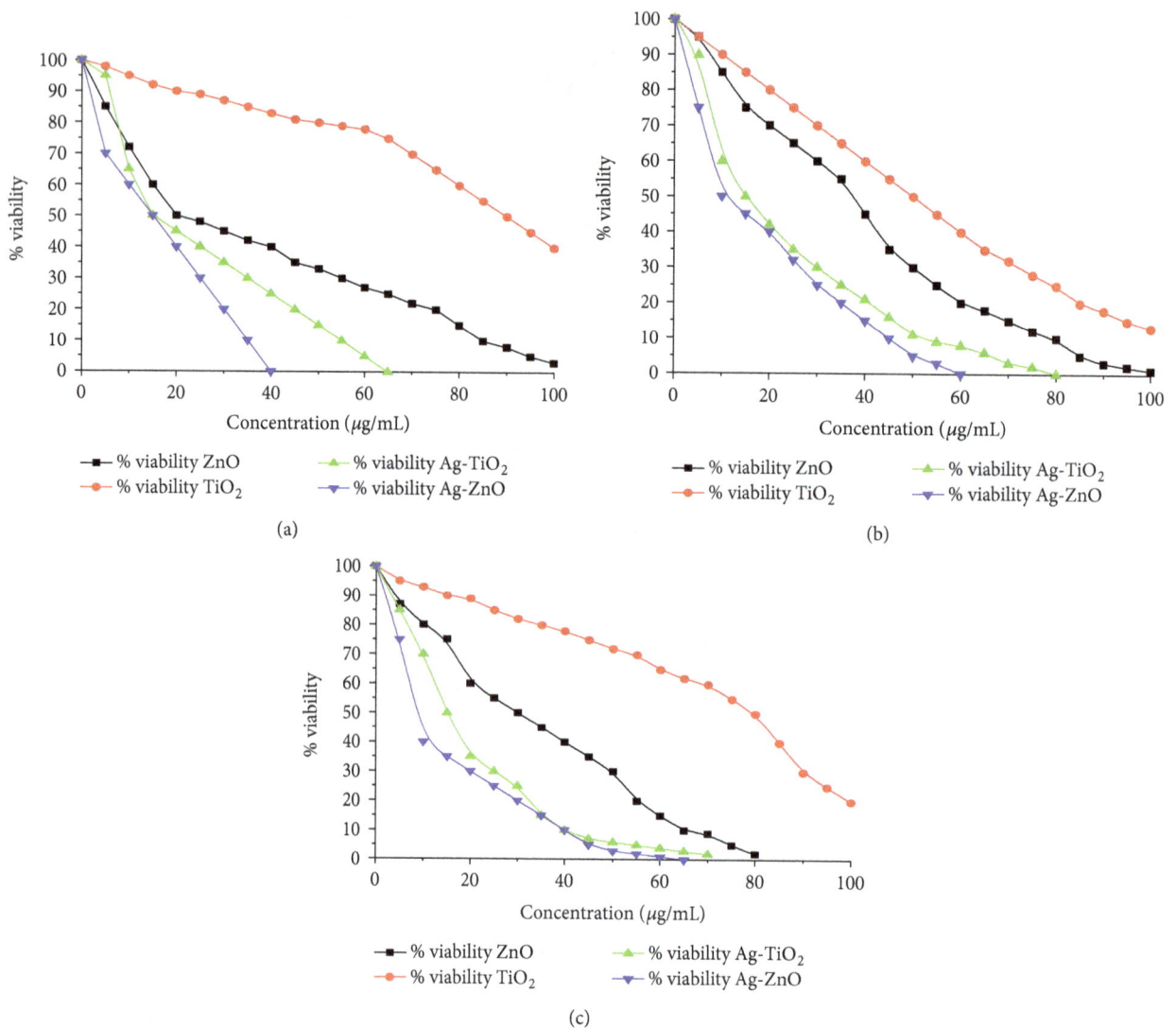

FIGURE 5: Viability of (a) *P. aeruginosa*, (b) *S. aureus*, and (c) *E. coli* with respect to the concentration of NPs in μg/mL.

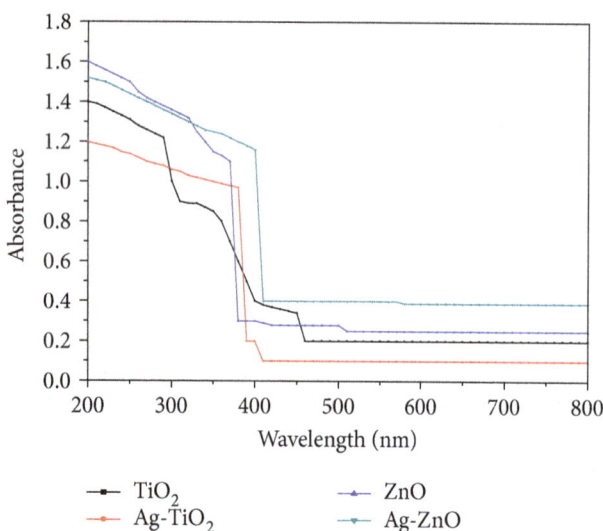

FIGURE 6: UV-Vis diffuse reflectance spectra of the TiO$_2$, ZnO, Ag-TiO$_2$, and Ag-ZnO.

pharmaceutical and nanocomposite fields. Our work provided a possible way to develop nanomaterials with very attractive properties to be applied in antibacterial activities.

Conflicts of Interest

In this, the authors wish to confirm that there are no conflicts of interest associated with this publication.

Acknowledgments

The authors would like to acknowledge Mekelle University for their financial support from the recurrent budget (Grant no. CRPO/CNCS/016/08). They are also grateful to Osmania University, India, and TIGP-SCST for their provision of XRD and SEM analyses.

References

[1] S. Prabhu and E. K. Poulose, "Silver nanoparticles: mechanism of antimicrobial action, synthesis, medical applications, and

toxicity effects," *International Nano Letters*, vol. 2, no. 1, p. 32, 2012.

[2] X. Wan, T. Wang, Y. Dong, and D. He, "Development and application of TiO$_2$ nanoparticles coupled with silver halide," *Journal of Nanomaterials*, vol. 2014, Article ID 908785, 5 pages, 2014.

[3] S. Chen, Y. Guo, S. Chen, Z. Ge, H. Yang, and J. Tang, "Fabrication of Cu/TiO$_2$ nanocomposite: toward an enhanced antibacterial performance in the absence of light," *Materials Letters*, vol. 83, pp. 154–157, 2012.

[4] V. Mohammad, A. Umar, and Y. B. Hahn, "ZnO nanoparticles: growth, properties, and applications," in *Metal Oxide Nanostructures and their Applications*, American Scientific Publishers, New York, NY, USA, 2010.

[5] C. B. Murray, C. R. Kagan, and M. G. Bawendi, "Synthesis and characterization of monodisperse nanocrystals and close-packed nanocrystal assemblies," *Annual Review of Materials Research*, vol. 30, no. 1, pp. 545–610, 2000.

[6] L. Cermenati, D. Dondi, M. Fagnoni, and A. Albini, "Titanium dioxide photocatalysis of adamantane," *Tetrahedron*, vol. 59, no. 34, pp. 6409–6414, 2003.

[7] R. Kumar, D. Rana, A. Umar, P. Sharma, S. Chauhan, and M. S. Chauhan, "Ag-doped ZnO nanoellipsoids: potential scaffold for photocatalytic and sensing applications," *Talanta*, vol. 137, pp. 204–213, 2015.

[8] P. Li, J. Li, C. Wu, Q. Wu, and J. Li, "Synergistic antibacterial effects of β-lactam antibiotic combined with silver nanoparticles," *Nanotechnology*, vol. 16, no. 9, pp. 1912–1917, 2005.

[9] T. Sungkaworn, W. Triampo, P. Nalakarn et al., "The effects of TiO$_2$ nanoparticles on tumor cell colonies: fractal dimension and morphology properties," *International Journal of Biomedical Science*, vol. 2, no. 1, pp. 67–74, 2007.

[10] G. Nagaraju, Udayabhanu, Shivaraj et al., "Electrochemical heavy metal detection, photocatalytic, photoluminescence, biodiesel production and antibacterial activities of Ag–ZnO nanomaterial," *Materials Research Bulletin*, vol. 94, pp. 54–63, 2017.

[11] Y. Xiang, J. Li, X. Liu et al., "Construction of poly(lactic-*co*-glycolic acid)/ZnO nanorods/Ag nanoparticles hybrid coating on Ti implants for enhanced antibacterial activity and biocompatibility," *Materials Science and Engineering: C*, vol. 79, pp. 629–637, 2017.

[12] S. M. Lam, J. A. Quek, and J. C. Sin, "Mechanistic investigation of visible light responsive Ag/ZnO micro/nanoflowers for enhanced photocatalytic performance and antibacterial activity," *Journal of Photochemistry and Photobiology A: Chemistry*, vol. 353, pp. 171–184, 2018.

[13] J. D. Fortner, D. Y. Lyon, C. M. Sayes et al., "C$_{60}$ in water: nanocrystal formation and microbial response," *Environmental Science & Technology*, vol. 39, no. 11, pp. 4307–4316, 2005.

[14] M. Herrera, P. Carrión, P. Baca, J. Liébana, and A. Castillo, "In vitro antibacterial activity of glass-ionomer cements," *Microbios*, vol. 104, no. 409, pp. 141–148, 2001.

[15] A. Kösemen, Z. Alpaslan Kösemen, B. Canimkubey et al., "Fe doped TiO$_2$ thin film as electron selective layer for inverted solar cells," *Solar Energy*, vol. 132, pp. 511–517, 2016.

[16] H. Morkoç and Ü. Özgür, *Zinc Oxide: Fundamentals, Materials and Device Technology*, Wiley-VCH, Weinheim, Germany, 2009.

[17] J. Jayabharathi, C. Karunakaran, V. Kalaiarasi, and P. Ramanathan, "Nano ZnO, Cu-doped ZnO, and Ag-doped ZnO assisted generation of light from imidazole," *Journal of Photochemistry and Photobiology A: Chemistry*, vol. 295, pp. 1–10, 2014.

[18] L. Elsellami, F. Dappozze, A. Houas, and C. Guillard, "Effect of Ag$^+$ reduction on the photocatalytic activity of Ag-doped TiO$_2$," *Superlattices and Microstructures*, vol. 109, pp. 511–518, 2017.

[19] S. Krejčíková, L. Matějová, K. Kočí et al., "Preparation and characterization of Ag-doped crystalline titania for photocatalysis applications," *Applied Catalysis B: Environmental*, vol. 111-112, no. 112, pp. 119–125, 2012.

[20] K. Ubonchonlakate, L. Sikong, and F. Saito, "Photocatalytic disinfection of *P. aeruginosa* bacterial Ag-doped TiO2 film," *Procedia Engineering*, vol. 32, pp. 656–662, 2012.

[21] Y. Zhang, X. Gao, L. Zhi et al., "The synergetic antibacterial activity of Ag islands on ZnO (Ag/ZnO) heterostructure nanoparticles and its mode of action," *Journal of Inorganic Biochemistry*, vol. 130, pp. 74–83, 2014.

[22] Y. Al-Hadeethi, A. Umar, A. A. Ibrahim et al., "Synthesis, characterization and acetone gas sensing applications of Ag-doped ZnO nanoneedles," *Ceramics International*, vol. 43, no. 9, pp. 6765–6770, 2017.

[23] O. Bechambi, M. Chalbi, W. Najjar, and S. Sayadi, "Photocatalytic activity of ZnO doped with Ag on the degradation of endocrine disrupting under UV irradiation and the investigation of its antibacterial activity," *Applied Surface Science*, vol. 347, pp. 414–420, 2015.

[24] M. Elango, M. Deepa, R. Subramanian, and A. Mohamed Musthafa, "Synthesis, characterization of polyindole/Ag—ZnO nanocomposites and its antibacterial activity," *Journal of Alloys and Compounds*, vol. 696, pp. 391–401, 2017.

[25] R. Kumar, J. Rashid, and M. A. Barakat, "Zero valent Ag deposited TiO$_2$ for the efficient photocatalysis of methylene blue under UV-C light irradiation," *Colloids and Interface Science Communications*, vol. 5, pp. 1–4, 2015.

[26] S. Angkaew and P. Limsuwan, "Preparation of silver-titanium dioxide core-shell (Ag@TiO$_2$) nanoparticles: effect of Ti-Ag mole ratio," *Procedia Engineering*, vol. 32, pp. 649–655, 2012.

[27] A. Hastir, N. Kohli, and R. C. Singh, "Ag doped ZnO nanowires as highly sensitive ethanol gas sensor," *Materials Today: Proceedings*, vol. 4, no. 9, pp. 9476–9480, 2017.

[28] M. Jakob, H. Levanon, and P. V. Kamat, "Charge distribution between UV-irradiated TiO$_2$ and gold nanoparticles: determination of shift in the Fermi level," *Nano Letters*, vol. 3, no. 3, pp. 353–358, 2003.

[29] K. Gupta, R. P. Singh, A. Pandey, and A. Pandey, "Photocatalytic antibacterial performance of TiO$_2$ and Ag-doped TiO$_2$ against *S. aureus. P. aeruginosa* and *E. coli*," *Beilstein Journal of Nanotechnology*, vol. 4, pp. 345–351, 2013.

[30] H. Xu, H. Li, C. Wu, J. Chu, Y. Yan, and H. Shu, "Preparation, characterization and photocatalytic activity of transition metal-loaded BiVO$_4$," *Materials Science and Engineering: B*, vol. 147, no. 1, pp. 52–56, 2008.

[31] X. You, F. Chen, and J. J. Zhang, "Effects of calcination on the physical and photocatalytic properties of TiO$_2$ powders prepared by sol–gel template method," *Journal of Sol-Gel Science and Technology*, vol. 34, no. 2, pp. 181–187, 2005.

[32] N. Y. Jamil, S. A. Najim, A. M. Muhammed, and V. M. Rogoz, "Preparation, structural and optical characterization of ZnO/Ag thin film by CVD," *Proceedings of the International*

Conference Nanomaterials: Applications and Properties, vol. 3, no. 2, pp. 32–45, 2014.

[33] B. Cheng, Y. Le, and J. Yu, "Preparation and enhanced photocatalytic activity of Ag@TiO$_2$ core–shell nanocomposite nanowires," *Journal of Hazardous Materials*, vol. 177, no. 1–3, pp. 971–977, 2010.

[34] C. Chen, Y. Zheng, Y. Zhan, X. Lin, Q. Zheng, and K. Wei, "Enhanced Raman scattering and photocatalytic activity of Ag/ZnO heterojunction nanocrystals," *Dalton Transactions*, vol. 40, no. 37, pp. 9566–9570, 2011.

[35] S. Banerjee, J. Gopal, P. Muraleedharan, A. K. Tyagi, and B. Raj, "Physics and chemistry of photocatalytic titanium dioxide: visualization of bactericidal activity using atomic force microscopy," *Current Science*, vol. 90, no. 10, pp. 1378–1383, 2006.

10

The Effect of Irradiation with a 405nm Blue-Violet Laser on the Bacterial Adhesion on the Osteosynthetic Biomaterials

Chika Terada[ID],[1] **Takahiro Imamura**,[1] **Tomoko Ohshima**,[2] **Nobuko Maeda**,[2] **Seiko Tatehara**,[1] **Reiko Tokuyama-Toda**,[1] **Shigeo Yamachika**,[1] **Nagataka Toyoda**,[1] and **Kazuhito Satomura**[ID][1]

[1]*Department of Oral Medicine and Stomatology, School of Dental Medicine, Tsurumi University, 2-1-3 Tsurumi, Tsurumi-ku, Yokohama 230-8501, Japan*
[2]*Department of Oral Microbiology, School of Dental Medicine, Tsurumi University, 2-1-3 Tsurumi, Tsurumi-ku, Yokohama 230-8501, Japan*

Correspondence should be addressed to Kazuhito Satomura; satomura-k@tsurumi-u.ac.jp

Academic Editor: Maria da Graça P. Neves

Delayed postoperative infection is known as a major complication after bone surgeries using osteosynthetic biomaterial such as titanium (Ti) and bioresorbable organic materials. However, the precise cause of this type of infection is still unclear and no effective prevention has been established. The purpose of this study is to investigate the effect of irradiation with a 405 nm blue-violet laser on the bacteria adhered on the Ti and hydroxyapatite-poly-L-lactic acid- (HA-PLLA) based material surfaces and to verify the possibility of its clinical application to prevent the delayed postoperative infection after bone surgeries using osteosynthetic biomaterial. The suspension of *Staphylococcus aureus* FDA 209P was delivered onto the surface of disks composed of Ti or HA-PLLA. Bacterial adhesion on each disk was observed using a scanning electron microscope (SEM). After thorough washing with distilled water, the growth of bacteria attached to the material surfaces was examined with an alamar blue-based redox indicator. Moreover, a bactericidal effect of 405 nm blue-violet laser irradiation on residual bacteria on both materials was investigated using colony-forming assay. As a result, there was no significant difference in the bacterial adhesion between Ti and HA-PLLA materials. In contrast, 45 J/cm^2 of irradiation with 405 nm blue-violet laser inhibited the bacterial growth at approximately 93% on Ti disks and at approximately 99% on HA-PLLA disks. This study clearly demonstrated the possibility that the irradiation with a 405 nm blue-violet laser is useful as an alternative management strategy for the prevention of delayed postoperative infection after bone surgeries using osteosynthetic biomaterials.

1. Introduction

Several types of biomaterials including titanium (Ti) and bioresorbable materials such as poly-L-lactic acid (PLLA) are used for osteosynthesis in the fields of orthopedics and craniomaxillofacial surgery [1, 2]. Ti is a traditional material and most commonly used for osteosynthesis because of its high biocompatibility and superior physical properties [3, 4]. In contrast, PLLA is one of representative bioresorbable materials, which is hydrolyzed by moisture in the environment, resulting in a decrease in the molecular weight and then ultimate decomposition of the material into carbon dioxide and water. For this property,

PLLA has been used as a source reagent of bioresorbable osteosynthetic materials since the 1990s [5, 6]. In 2006, new biomaterial for osteosynthesis was developed, which was composed of incalescent/unsintered hydroxyapatite (u-HA) particles and PLLA, and has been commonly used due to (1) its better osteoconductivity [7–10], (2) no need for secondary surgery to remove the materials [11, 12], (3) no restriction to bone growth in young patients [13–15], and (4) no elution of metal ions that could act as allergens [16–19].

Delayed postoperative infection, one of major and serious complications after bone surgeries using osteosynthetic materials, often follows a protracted course as a persistent complication [20–23], resulting in the eventual removal of materials followed by the failure of bone tissue regeneration/wound healing [22, 24–26]. Particularly for bioresorbable materials, a removal surgery due to postoperative infection completely negates the beneficial property of the materials. Therefore, it is very important to prevent the delayed postoperative infection after bone surgeries using osteosynthetic materials. However, the precise cause of this type of infection is still unclear, and effective prevention has not yet been established.

Although a variety of bacteria are known to cause the postoperative infections, *Staphylococcus aureus* is the most responsible one in surgical site infections. In addition, methicillin-resistant *S. aureus* is one of the offending microorganisms of lethal infections due to its resistant characteristics to β-lactam antibiotics. One of the alternative therapies to antibiotics, the bactericidal effect of blue light against *S. aureus*, has been verified in previous studies. Guffey and Wilborn reported that the irradiation with a 405 nm blue-violet laser had a bactericidal effect against *S. aureus* and *Pseudomonas aeruginosa* [27]. Other studies also reported that visible light is effective against *Porphyromonas gingivalis*, *Fusobacterium nucleatum*, *Staphylococcus aureus*, *Streptococcus mutans*, and *Escherichia coli* [28, 29]. The results of our previous study also confirmed that the irradiation with a 405 nm blue-violet laser had a bactericidal effect against *P. gingivalis* which is a major periodontopathogenic microorganism as well as *Prevotella intermedia*, and even against *Candida albicans* which is a major responsible fungus causing candidiasis [30]. These results demonstrate the possibility that the irradiation with a 405 nm blue-violet laser is capable of eliminating a variety of microorganisms adhered to the surface of osteosynthetic materials and preventing effectively the delayed postoperative infection.

Consequently, in the present study, we attempted to explore the possibility that the irradiation with 405 nm blue-violet laser had a bactericidal effect on *S. aureus* adhered to the surface of Ti and HA-containing PLLA (HA-PLLA) materials and that the irradiation could be effective and useful for the prevention of postoperative infections after bone surgeries using osteosynthetic materials.

2. Materials and Methods

2.1. Preparation of Samples. The Ti sample disks (10 mm in diameter, 2 mm in thickness; Daido Bunseki Research Inc., Nagoya, Japan) were made of commercially available pure Ti (ISO 5832/2, Grade 4A). The surfaces of the disks were barrel-polished and anodized after acid treatment to remove the oxide layer with a diluted mixture of nitric acid and hydrofluoric acid, which is the same processing procedure used in clinical practice. The disks were then ultrasonically degreased in ethanol and then deionized water (DW) for 10 min each. The HA-PLLA sample disks (10 mm in diameter, 2 mm in thickness) which was composed of a mix of PLLA and particulate bioresorbable u-HA (Super Fixsorb®) were supplied by Takiron Co. Ltd., Osaka, Japan.

The surface topographies (i.e., contact angle and surface roughness) of the two materials were examined. Contact

angles were measured by the sessile drop method using a measurement device (Kyowa Interface Science Co. Ltd. DM-300). Surface roughness average (Ra) was measured using the SURFCOM 550A surface roughness tester (Tokyo Seimitsu Co., Ltd.) with a 2 μm radius tip at three separate points on each disk. In addition, the diffuse transmittance of disks was measured at normal incidence (0°) and the reflectance of disks was measured at the angle of incidence of 8° in the 200~2500 nm region using a UV-3100PC (Shimadzu Corporation, Kyoto, Japan). The surface texture of disks of each material was observed using a scanning electron microscope (SEM; JSM 5600LV, JEOL Ltd., Tokyo, Japan) at an accelerating voltage of 10–15 kV.

2.2. Observation of Bacterial Adhesion with SEM. *S. aureus* FDA 209P was precultured aerobically in 3 mL of brain heart infusion (BHI) medium (Becton Dickinson and Company, Sparks, MD, USA) supplemented with 0.5% yeast extract, 0.05% L-cysteine, and 0.025% resazurin overnight at 37°C. Thereafter, 100 μL of culture solution was inoculated to 10 mL of BHI and cultured aerobically for 5 h with agitation at 37°C.

Each disk was sterilized and placed into a well of a 48-well microtiter plate. Then, 100 μL of bacterial suspension adjusted to 1×10^4 colony-forming units (CFU)/mL and 1 mL of BHI medium were added to each well and cultured aerobically at 37°C for 3, 6, 24, or 48 h. Three disks were prepared for each condition of each material. The disks were rinsed twice with DW to remove nonadherent bacteria, further washed in a vortex with DW for 30 s to dislodge bacteria adhered poorly to the material surfaces, and fixed in 2.5% glutaraldehyde in phosphate-buffered saline for 12 h at 4°C. The disks were dehydrated using a series of graded concentrations of ethanol (50%, 60%, 70%, 80%, 90%, and absolute alcohol), freeze-dried using an ID-2 freeze dryer (EIKO Engineering K.K. Ltd., Tokyo, Japan), and coated with gold using the SC-701AT Quick Auto Coater (Sanyu Denshi, Tokyo, Japan). Four areas were arbitrarily selected and observed with a SEM at an accelerating voltage of 10–15 kV.

2.3. Observation of the Growth of Residual Bacteria with Redox Indicator and SEM. For monitoring the growth of residual bacteria on the material surfaces, a modified redox indicator assay (alamarBlue®; Bio-Rad Laboratories, Hercules, CA, USA) was performed. The redox indicator is a nonfluorescent blue oxidation-reduction pigment which is transformed into red pigment when reduced by bacterial proliferation. The absorbance of this red pigment was measured using a multidetection reader (LabSystems Multiskan®; MultiSoft, Helsinki, Finland) at a wavelength of 570 nm.

After incubation in bacterial solution for 3 h at 37°C, the disks of the Ti and HA-PLLA materials were rinsed twice with DW to remove nonadherent bacteria and further washed in a vortex with DW for 30 s to dislodge bacteria adhered poorly to the material surfaces. Aside from this, some disks that were only rinsed twice with DW for 30 s were prepared as a control group. Thereafter, the disks of each material were transferred into 5 mL of BHI medium containing 500 μL of redox indicator (as 10% of the sample volume)

TABLE 1: Roughness and contact angles (CA) of Ti and HA-PLLA surface.

	Ti	HA-PLLA
Ra (μm)	0.45 ± 0.005	0.3 ± 0.05
Rz (μm)	7.63 ± 1.02	3.32 ± 0.59
CA (°)	67.42 ± 5.28	88.62 ± 1.04

Values are expressed as mean ± SD. Ra: calculated average roughness; Rz: maximum height.

TABLE 2: Transmittance and reflectance of Ti and HA-PLLA surface.

	Ti	HA-PLLA
Transmittance (%)	0.0 ± 0.0	0.9 ± 0.02
Reflectance (%)	51.29 ± 0.55	89.83 ± 0.04

Values are expressed as mean ± SD.

and incubated for 120 min at 37°C. The absorbance was measured using an automated multidetection reader at a wavelength of 570 nm. To examine the growth of residual bacteria, some other disks of each material were cultured for an additional 24 h and observed with SEM at an accelerating voltage of 10–15 kV.

2.4. Effect of Irradiation with a 405 nm Blue-Violet Laser on Bacteria Adhered on Material Surfaces. One hundred μL of bacterial suspensions at 1×10^4 CFU/mL and 1 mL of BHI medium were inoculated onto the surfaces of the Ti and HA-PLLA disks in 24-well culture plates for 3 hours at 37°C. After incubation, each disk was washed with DW and excess water on the surface was absorbed. Thereafter, the disks were irradiated with 405 nm of blue-violet laser in a moisture chamber at 100% relative humidity, under constant output power of 0.2 W ($176 \, mW/cm^2$) for various irradiation times of 0 s (as a control), 180 s ($27 \, J/cm^2$), and 300 s ($45 \, J/cm^2$) using a laser-emitting device equipped with bundling of 20 fibers coupled with 405 nm laser diodes (Ushio Inc., Tokyo, Japan). The 405 nm single-wavelength emission of the laser from the present device was verified and confirmed using a spectrophotometer (USB2000 Miniature Fiber Optic Spectrometer and OceanView Spectroscopy Software, Ocean Optics Inc., FL, USA) (data not shown). The laser output was maintained at a stable wattage, which was measured before every irradiation by a laser power meter and a sensor (Orion/TH P/N 1Z01801: 188784, 3A-P-SH-V1 P/N 1Z02622: 187487; Ophir Optronics Solutions Ltd., Jerusalem, Israel). After irradiation, the samples were stamped on tryptic soy agar plates. After incubation at 37°C for 12 hours, the numbers of bacterial colonies on the plates were counted.

To eliminate possible thermal effects on the material surfaces due to the absorption of irradiation, changes in surface temperature during irradiation were measured with a thermocouple and monitoring device (TCS620 NTC Thermistor; Wavelength Electronics Inc., Bozeman, MT, USA).

2.5. Statistical Analysis. For multiple-group comparisons, data were analyzed by one-way analysis of variance. The significance of individual differences was evaluated using the Mann-Whitney U test. A probability value (p) of <0.05 was considered statistically significant.

3. Results

3.1. Morphological Characteristics of the Material Surfaces. The differences in the morphological characteristics were noted between Ti and HA-PLLA. The Ra values of Ti and HA-PLLA were 0.30 ± 0.05 and 0.45 ± 0.005, and the contact

angles of each material were 67.4 ± 5.28 and 88.8 ± 1.04, respectively (Table 1). Transmittance and reflectance of each material in the wavelength of 405 nm are presented in Table 2. SEM images also showed obvious differences in surface topology between the two materials. The surface of Ti was relatively smooth and micropores were found in some places, while that of HA-PLLA was comparatively rougher with interspersed white particulates that seemed to be HA crystals (Figure 1).

3.2. Bacterial Adhesion. Although there were differences in morphological characteristics/surface topology between Ti and HA-PLLA, no differences were noted in the bacterial adhesion between the two materials. At 3 hours after inoculation of the bacterial suspension, only a few bacteria were noted to adhere on the surface of both materials (Figures 2(a) and 2(b)). After incubation for 24 hours, small colonies or the bacterial clumps consisting of a few bacteria were observed in some places over the surface of both materials (Figures 2(c) and 2(d)). After incubation for 48 hours, much more amount of bacteria and larger colonies were formed on the entire surface of both materials (Figures 2(e) and 2(f)).

3.3. Growth of Residual Bacteria after Washing. The growth of residual bacteria on the surface of each material was examined with the redox indicator, alamarBlue® reagent. No obvious difference in bacterial growth was noted until 2–3 hours after 30 s washing with DW between the vortex group and the control group in both materials. In contrast, at 4–5 hours after 30 s washing with DW, more bacterial growth was observed on HA-PLLA compared with Ti in both groups (Figure 3). SEM observation showed the bacterial growth and colony formation at 24 hours after vortex washing on both materials (Figures 4(a) and 4(d)). These results indicated that some few bacteria remained on the surface of both materials even after thorough washing and were capable of growing with time.

3.4. Effect of Irradiation with a 405 nm Blue-Violet Laser on Bacteria Adhered to Material Surfaces. The effects of irradiation with a 405 nm blue-violet laser on bacteria adhered to the surfaces of both materials were analyzed by counting the numbers of colonies formed on tryptic soy agar plates. Obvious differences in the amount of viable bacteria were observed in the control group (without irradiation) between both materials (Figures 5(a) and 5(b)). In the Ti group, irradiation for 180 s (Figure 5(c)) inhibited the colony formation at approximately 76%, while irradiation for 300 s (Figure 5(e)) inhibited at approximately 93%. In the

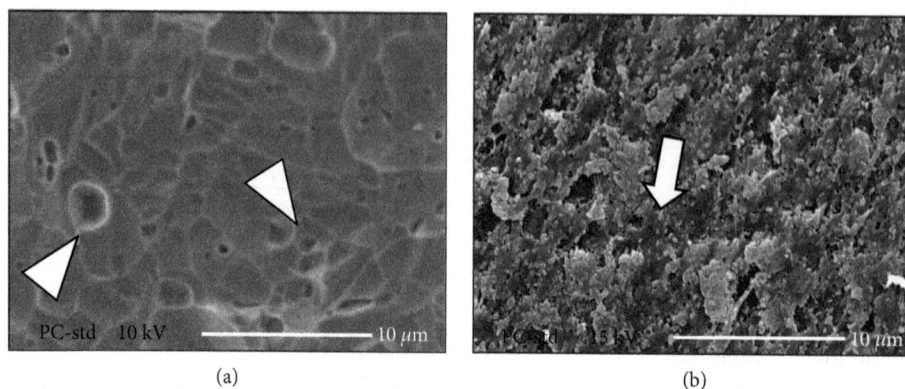

(a)   (b)

Figure 1: Morphological characteristics of the surface of TI (a) and HA-PLLA (b). SEM images show the surface differences between two materials. The surface of Ti appeared relatively smooth, and micropores (arrowheads) were found in some places on Ti. The surface of HA-PLLA appeared relatively rough, and interspersed white particulates, which seemed to be HA crystals (arrow), were visible in some places.

HA-PLLA group, irradiation for 180 s (Figure 5(d)) and 300 s (Figure 5(f)) inhibited the colony formation at approximately 90% and 99%, respectively. Interestingly, the 405 nm blue-violet laser exerted a more bactericidal effect on HA-PLLA than Ti. There was no significant difference in the control group (without irradiation) between both materials (Figure 6). Irradiation for 300 s caused an increase in surface temperature of HA-PLLA from 20°C to 23°C and from 20°C to 30°C of the Ti (data not shown).

4. Discussion

Delayed postoperative infection after bone surgeries using osteosynthetic materials composed of Ti or bioresorbable materials is one of the thorny problems that remain to be solved. However, the precise cause of this type of infection is still unclear and effective prevention measures have not yet been established. The findings of the present study confirmed that S. aureus, which is the most responsible bacterium for postoperative infection, could adhere easily to the surface of either material composed of Ti or HA-PLLA and continue to survive. It was also confirmed that once the surface of osteosynthetic material becomes contaminated, a certain amount of the bacteria could remain lodged on the surface of material even after vigorous washing, which then led to colony formation on the material surface. Even with scrupulous attention during surgery, it is difficult to prevent entirely any contaminations by some few bacteria and to remove all contaminating bacteria by washing with physiological saline. This fact indicates the possibility that some cases of inflammation after surgery using osteosynthetic materials could be due to the growth of trace amounts of bacteria which attached to the material surface in surgery and were introduced into the body. Further studies are necessary, however, to elucidate the precise mechanisms of the establishment of postoperative infection after bone surgeries using osteosynthetic biomaterials.

In fact, there are a considerable number of reported cases of delayed infection or inflammatory abscess formation after surgical treatment using bioresorbable osteosynthetic materials such as HA-PLLA [31–33]. So far, such delayed postoperative infections have conventionally been considered to result from the inflammation associated with the decomposition of the material, that is, a foreign body reaction or defective absorption of the crystal component [26, 34, 35]. However, the results of the present study indicate another possibility that delayed postoperative infections are due to delayed onset of inflammation caused by a small number of contaminating microorganisms adhered to the surfaces of biomaterials. This concept could be expected to give some new suggestion to solve this clinical problem.

Judging from the results of the present study, in order to prevent postoperative infection after implantation of osteosynthetic materials, it seems useful to reduce the amount of microorganisms introduced into the living body just before wound closure. So, we examined the usefulness of 405 nm blue-violet laser irradiation, which has been shown to have bactericidal effects, in sterilization of the microorganisms adhered to osteothynthetic biomaterials. Guffey and Wilborn reported that the irradiation with a 405 nm blue-violet laser at 15 J/cm^2 resulted in a 90–95% mortality rate to a solution of S. aureus [27], while Maclean et al. reported a 100% mortality rate with a 405 nm blue-violet laser at 36 J/cm^2 [36]. Several other studies showed that titanium oxide could exert an antimicrobial effect when irradiated with ultraviolet or short-wavelength visible light [37–39]. In the present study as in previous studies, the irradiation with a 405 nm blue-violet laser exerted a bactericidal effect on S. aureus attached to the surface of biomaterials composed of Ti. This bactericidal effect of the 405 nm blue-violet laser was observed to rise with increasing irradiation energy, in accordance with the findings of previous studies [30, 36, 40]. Imamura et al. confirmed the significant bactericidal effect on P. gingivalis with irradiation at 0.2 W for 300 s [30]. When exposed to irradiation at 0.2 W for 300 s, at an energy density of 45 J/cm^2 of disk, the bactericidal effect was approximately 93% on Ti and 99% on HA-PLLA, which is considered to be sufficient to sterilize the residual amount of S. aureus on biomaterials after washing. Moreover, the irradiation time of 300 s is considered to be within the acceptable range in practice and the potential burden to the patients would be minimal, even when the irradiation is performed during surgery.

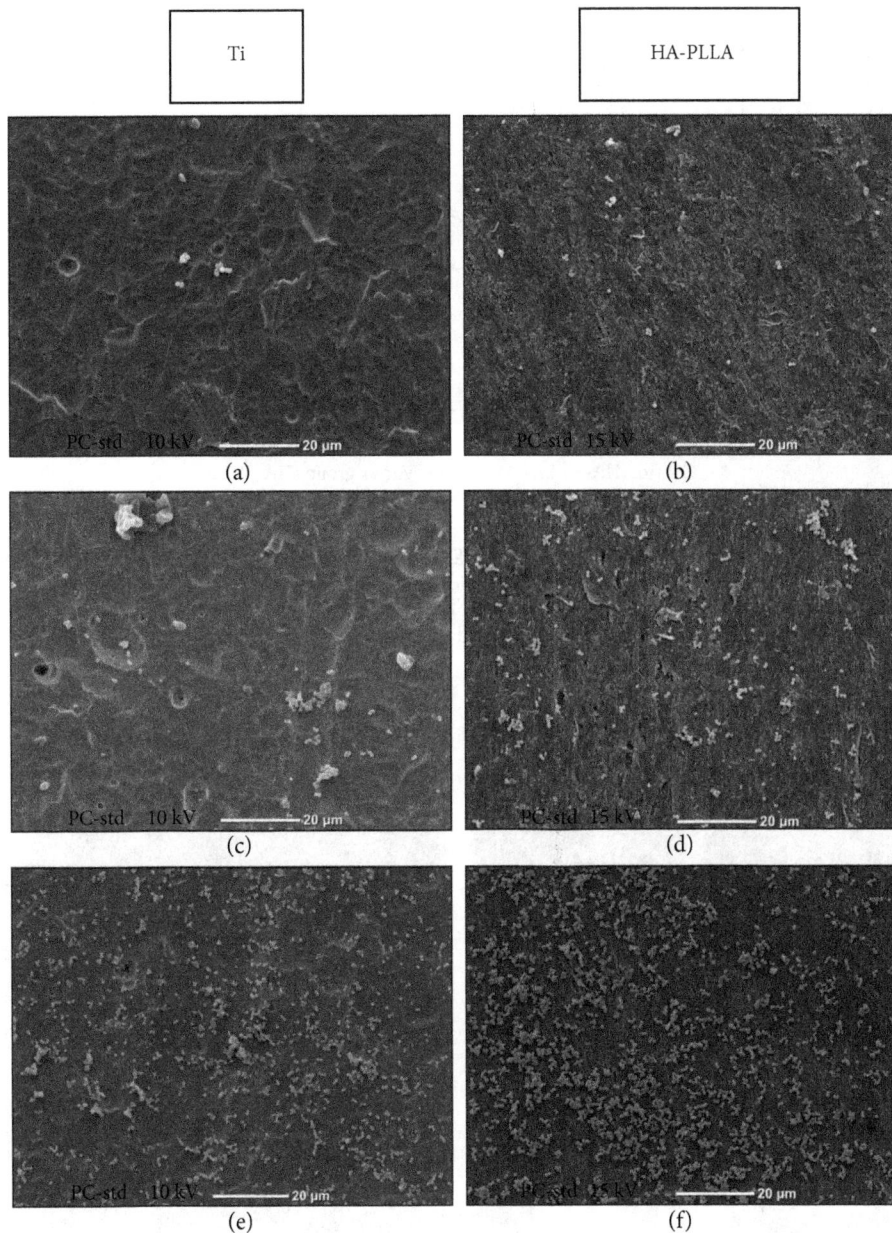

FIGURE 2: Bacterial adhesion of on the surface of Ti (a, c, e) and HA-PLLA (b, d, f) at 3 h (a, b), 24 h (c, d), and 48 h (e, f) after the inoculation of bacterial suspension. SEM images show that more and larger bacterial colonies were formed with the passage of incubation time. No obvious differences were noted in the bacterial adhesion between the two materials.

Interestingly, although there was no significant difference in the bacterial adhesion and the growth rate between two materials, more bactericidal effect was observed on HA-PLLA than Ti. Although the real reasons or precise mechanisms of the difference in bactericidal effect between two materials are still unclear, several possibilities might be involved. One of the possibilities is the difference in color and reflectance between the two materials. Since HA-PLLA has a higher reflectance than Ti as the properties of the material surface, more reflection of 405 nm blue-violet laser light does occur on the surface. Because of this, the bacteria adhered to the surface of HA-PLLA could be more irradiated

with laser light even in the same irradiation time compared with Ti, and the bactericidal effect would be more enhanced. Another possibility is the influence of morphological characteristics of materials. The calculated surface roughness (Ra) and maximum height (Rz) was larger in Ti compared with HA-PLLA. In addition, SEM showed some micropores scattered on Ti. These differences might affect the efficiency of irradiation to bacterial cells lodged on materials; that is, some bacteria adhered on shady areas of the surface of Ti might be much less irradiated compared with those on HA-PLLA. More interestingly, a slightly more bacterial growth was observed on the surface of HA-PLLA compared with Ti at

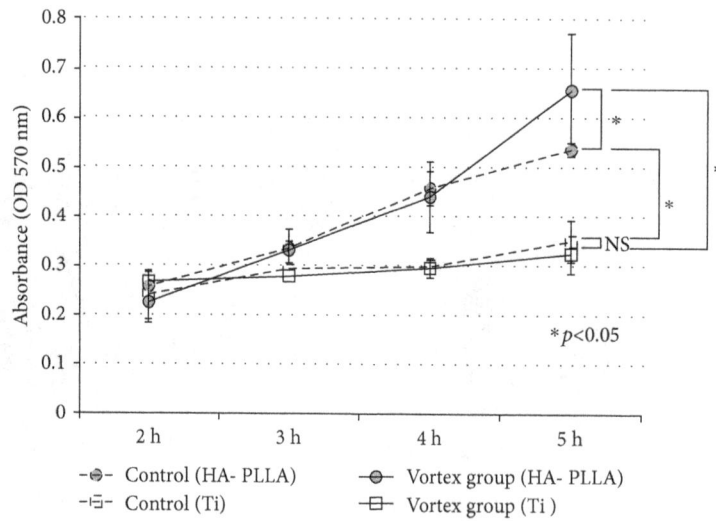

FIGURE 3: The growth of residual bacteria on Ti and HA-PLLA. No significant difference was detected between vortex and control groups in both materials until 2–3 hours after 30 s washing with DW. In contrast, at 4–5 hours of incubation after 30 s washing with DW, more bacterial growth was observed on HA-PLLA compared with Ti in both groups. The error bars represent the standard deviation. Values are mean ± SD from 3 samples per group. *$p < 0.05$. Circle shows HA-PLLA and square shows Ti as control.

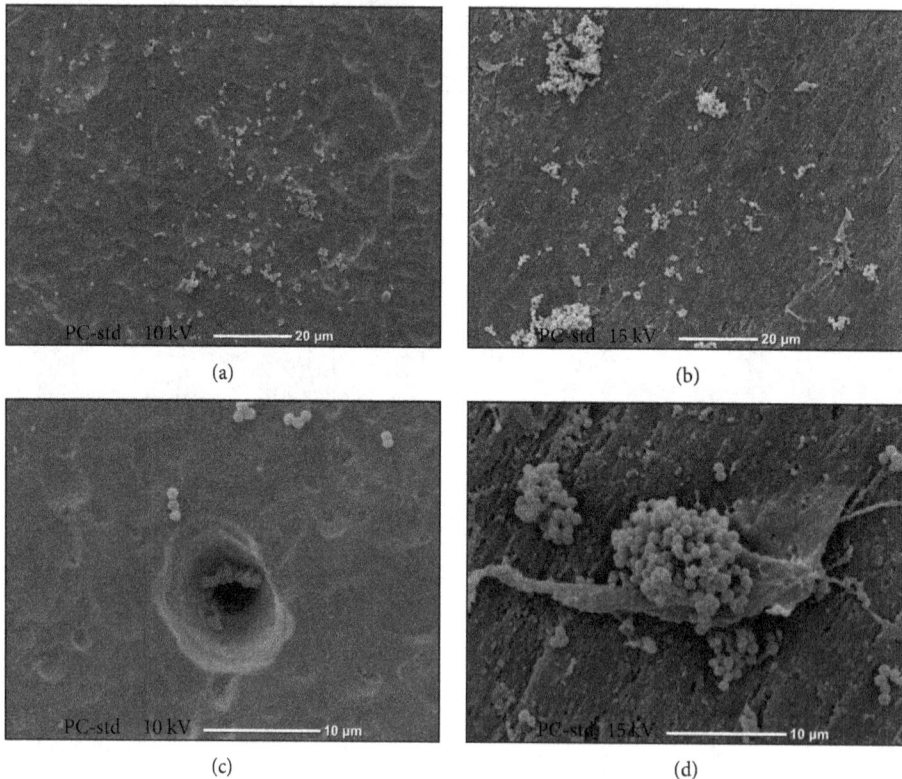

FIGURE 4: The SEM images of the surface of Ti (a, c) and HA-PLLA (b, d) at 24 hours of incubation after vortex washing. The images show that some few bacteria remained on the surface of both materials even after thorough washing and were capable of growing with time.

4–5 hours of culture after vortex washing, despite of no significant differences in bacterial adhesion between both materials. This phenomenon might come from a modest antimicrobial activity of photocatalytic titanium oxide layer covering the surface of Ti, because the investigation was not performed in complete darkness.

In clinical application of the sanitization or disinfection method, their influence on neighboring tissue must be carefully considered in regards to the potential damage of living cells and tissues. Judging from the fact that the wavelength of the 405 nm blue-violet laser used in this study is within the range of visible light and the

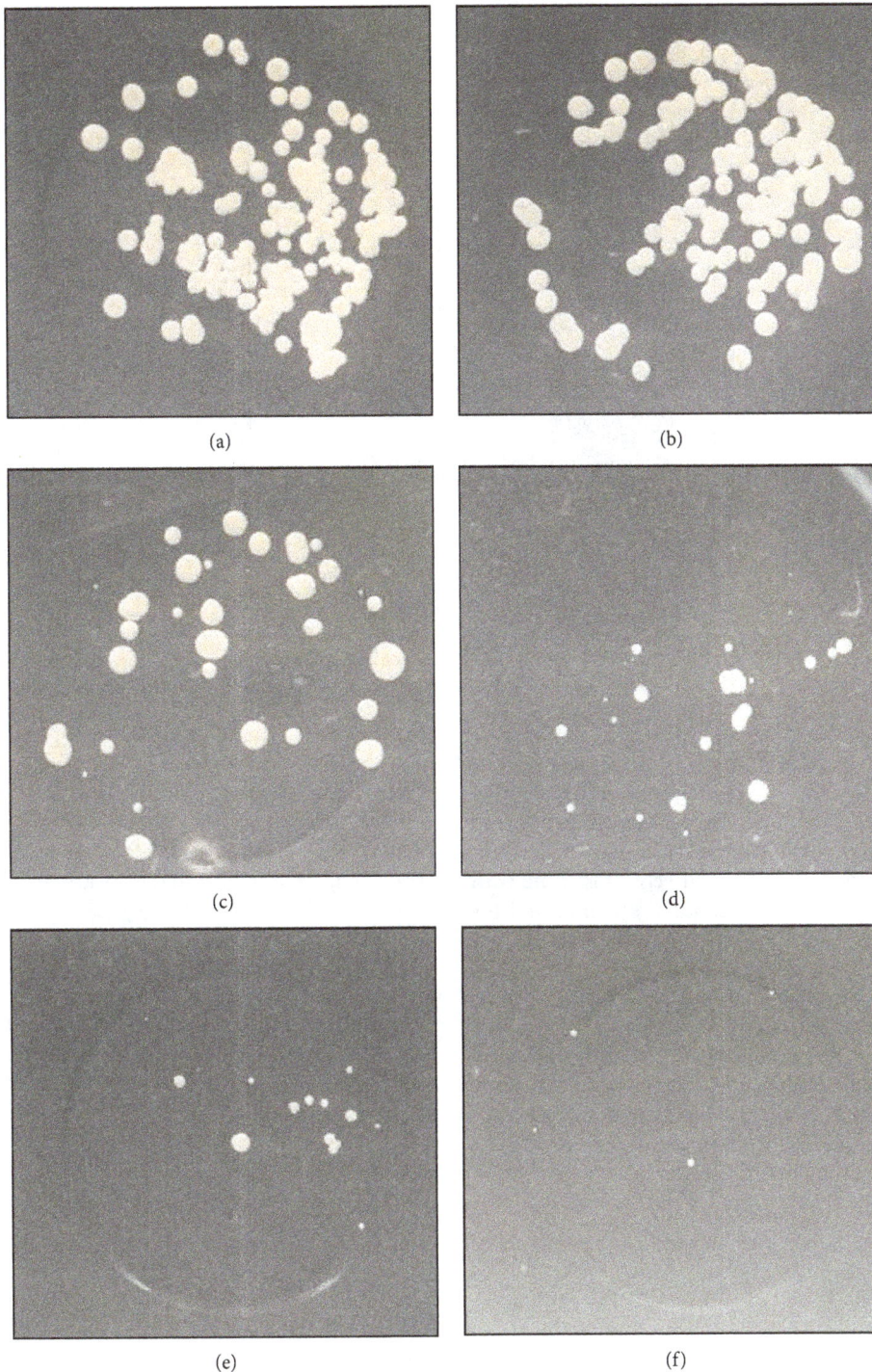

FIGURE 5: Photographs of tryptic soy agar plates in the stamp assay. The effects of irradiation with a 405 nm blue-violet laser on bacteria adhered to the surface of Ti (a, c, e) and HA-PLLA (b, d, f) were analyzed by counting the numbers of colonies formed on tryptic soy agar plates. Compared with the nonirradiated control plates (a, b), colony formation was inhibited by the irradiation for 180 s (c, d) and 300 s (e, f) in both materials.

irradiation time is at most 300 s, this irradiation is considered to be harmless and noninjurious to neighboring tissues. In addition, monitoring the thermal effect of irradiation indicated that the surface temperature of each material increased to at most 30°C in Ti even after the irradiation for 300 s, which suggests that this procedure is harmless to neighboring tissues and also influential in bacterial growth. For these reasons, the irradiation with a 405 nm blue-violet laser is considered to be a clinically applicable and promising measure to prevent the delayed postoperative infection after bone surgeries using osteosynthetic materials.

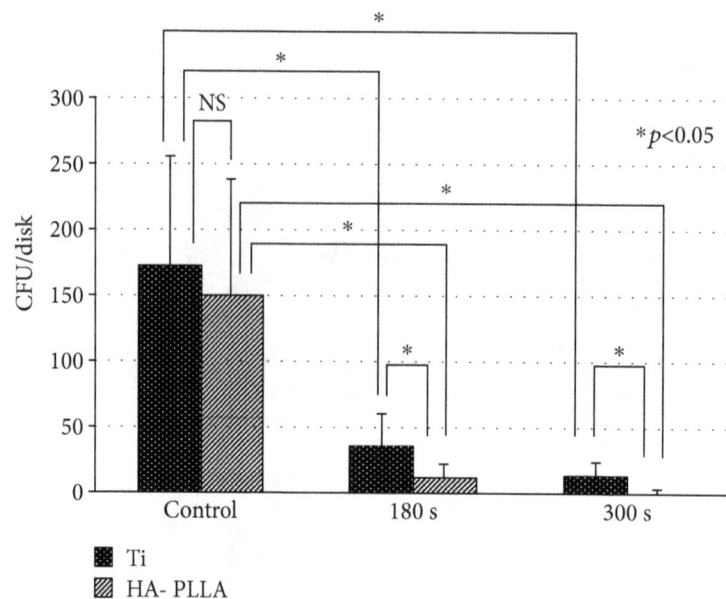

FIGURE 6: Effect of irradiation with a 405 nm blue-violet laser on the growth of bacteria adhered to the surface of Ti and HA-PLLA. Irradiation with the 405 nm blue-violet laser for 180 s and 300 s significantly inhibited the bacterial growth. Values are mean ± SD from 20 samples per group. $^*p < 0.05$.

This study is also the first to examine the bactericidal effect of 405 nm blue-violet laser light on bacteria adhered to surfaces of osteosynthetic biomaterials and to confirm the usefulness of irradiation for a relatively short time with clinical application in mind. Although the mechanisms underlying the bactericidal effect of 405 nm blue-violet laser light are still under investigation, porphyrin, a metabolite present in bacteria, might be involved. Endogenously produced porphyrins could act as photosensitizers and generate reactive oxygen/free radicals under light irradiation, leading to the reduction in bacterial growth and viability, and eventual cell death [41, 42]. Further investigations on intracellular photosensitizers including porphyrins will be capable of resolving this issue in the near future.

5. Conclusion

The present study clearly demonstrated that the irradiation with a 405 nm blue-violet laser effectively reduces the number of microorganisms adhered to osteosynthetic biomaterials and introduced into the living body in bone surgeries. These findings strongly suggest the potential usefulness of the 405 nm blue-violet laser irradiation as an alternative strategy for prevention of postoperative infection after implantation of biomaterials.

Disclosure

The funders had no role in the study design, date collection and analysis, decision to publish, or preparation of the manuscript.

Conflicts of Interest

The authors declare that there is no other conflict of interest regarding the publication of this article.

Acknowledgments

The authors received a financial support regarding this study from Takiron Co. Ltd., Osaka, Japan. This work was supported by Grants-in-Aid for Scientific Research (nos. 25463102, 26463028, 16K11770, 17K17156, and 17H04411) from the Ministry of Education, Culture, Sports, Science, and Technology of Japan. In addition, the authors wish to thank Ushio Inc. for the development of a 405 nm laser-emitting device.

References

[1] K. E. Zakhary and J. S. Thakker, "Emerging biomaterials in trauma," *Oral and Maxillofacial Surgery Clinics of North America*, vol. 29, no. 1, pp. 51–62, 2017.

[2] S. Agarwal, A. Gupta, M. Grevious, and R. R. Reid, "Use of resorbable implants for mandibular fixation: a systematic review," *The Journal of Craniofacial Surgery*, vol. 20, no. 2, pp. 331–339, 2009.

[3] M. S. Gilardino, E. Chen, and S. P. Bartlett, "Choice of internal rigid fixation materials in the treatment of facial fractures," *Craniomaxillofacial Trauma and Reconstruction*, vol. 2, no. 1, pp. 049–060, 2009.

[4] M. Niinomi, "Mechanical properties of biomedical titanium alloys," *Materials Science and Engineering: A*, vol. 243, no. 1-2, pp. 231–236, 1998.

[5] R. Suuronen, P. E. Haers, C. Lindqvist, and H. F. Sailer, "Update on bioresorbable plates in maxillofacial surgery," *Facial Plastic Surgery*, vol. 15, no. 1, pp. 61–72, 1999.

[6] Y. Shikinami and M. Okuno, "Bioresorbable devices made of forged composites of hydroxyapatite (HA) particles and poly-L-lactide (PLLA): part I. Basic characteristics," *Biomaterials*, vol. 20, no. 9, pp. 859–877, 1999.

[7] Y. Shikinami and M. Okuno, "Bioresorbable devices made of forged composites of hydroxyapatite (HA) particles and poly L-lactide (PLLA). Part II: practical properties of miniscrews and miniplates," *Biomaterials*, vol. 22, no. 23, pp. 3197–3211, 2001.

[8] K. L. Gerlach, "In-vivo and clinical evaluations of poly (L-lactide) plates and screws for use in maxillofacial traumatology," *Clinical Materials*, vol. 13, no. 1-4, pp. 21–28, 1993.

[9] R. Suuronen, P. Laine, T. Pohjonen, and C. Lindqvist, "Sagittal ramus osteotomies fixed with biodegradable screws: a preliminary report," *Journal of Oral and Maxillofacial Surgery*, vol. 52, no. 7, pp. 715–720, 1994.

[10] J. Tams, F. R. rozema, R. R. M. Bos, J. L. N. Roodenburg, P. G. J. Nikkels, and A. Vermey, "Poly (L-lactide) bone plates and screws for internal fixation of mandibular swing osteotomies," *International Journal of Oral and Maxillofacial Surgery*, vol. 25, no. 1, pp. 20–24, 1996.

[11] S. Park, J. H. Kim, I. H. Kim et al., "Evaluation of poly (lactic-co-glycolic acid) plate and screw system for bone fixation," *Journal of Craniofacial Surgery*, vol. 24, no. 3, pp. 1021–1025, 2013.

[12] N. B. van Bakelen, B. D. A. Boermans, G. J. Buijs et al., "Comparison of the long-term skeletal stability between a biodegradable and a titanium fixation system following BSSO advancement—a cohort study based on a multicenter randomised controlled trial," *British Journal of Oral and Maxillofacial Surgery*, vol. 52, no. 8, pp. 721–8, 2014.

[13] M. J. Imola, D. D. Hamlar, W. Shao, K. Chowdhury, and S. Tatum, "Resorbable plate fixation in pediatric craniofacial surgery: long-term outcome," *Archives of Facial Plastic Surgery*, vol. 3, no. 2, pp. 79–90, 2001.

[14] F. C. Senel, U. S. Tekin, and M. Imamoglu, "Treatment of a mandibular fracture with biodegradable plate in an infant: report of a case," *Oral Surgery, Oral Medicine, Oral Pathology, Oral Radiology, and Endodontology*, vol. 101, no. 4, pp. 448–450, 2006.

[15] J. An, P. Jia, Y. Zhang, X. Gong, X. Han, and Y. He, "Application of biodegradable plates for treating pediatric mandibular fractures," *Journal of Cranio-Maxillofacial Surgery*, vol. 43, no. 4, pp. 515–520, 2015.

[16] P. A. Lalor, P. A. Revell, A. B. Gray, S. Wright, G. T. Railton, and M. A. Freeman, "Sensitivity to titanium. A cause of implant failure?," *The Journal of Bone and Joint Surgery. British volume*, vol. 73-B, no. 1, pp. 25–28, 1991.

[17] K. Bessho, K. Fujimura, and T. Iizuka, "Experimental long-term study of titanium ions eluted from pure titanium miniplates," *Journal of Biomedical Materials Research*, vol. 29, no. 7, pp. 901–904, 1995.

[18] Y. K. Kim, H. H. Yeo, and S. C. Lim, "Tissue response to titanium plates: a transmitted electron microscopic study," *Journal of Oral and Maxillofacial Surgery*, vol. 55, no. 4, pp. 322–326, 1997.

[19] J. Acero, J. Calderon, J. I. Salmeron, J. J. Verdaguer, C. Concejo, and M. L. Somacarrera, "The behaviour of titanium as a biomaterial: microscopy study of plates and surrounding tissues in facial osteosynthesis," *Journal of Cranio-Maxillofacial Surgery*, vol. 27, no. 2, pp. 117–123, 1999.

[20] H. B. Lee, J. S. Oh, S. G. Kim et al., "Comparison of titanium and biodegradable miniplates for fixation of mandibular fractures," *Journal of Oral and Maxillofacial Surgery*, vol. 68, no. 9, pp. 2065–9, 2010.

[21] K. Bhatt, A. Roychoudhury, O. Bhutia, A. Trikha, A. Seith, and R. M. Pandey, "Equivalence randomized controlled trial of bioresorbable versus titanium miniplates in treatment of mandibular fracture: a pilot study," *Journal of Oral and Maxillofacial Surgery*, vol. 68, no. 8, pp. 1842–8, 2010.

[22] G. J. Buijs, N. B. van Bakelen, J. Jansma et al., "A randomized clinical trial of biodegradable and titanium fixation systems in maxillofacial surgery," *Journal of Dental Research*, vol. 91, no. 3, pp. 299–304, 2012.

[23] R. M. Laughlin, M. S. Block, R. Wilk, R. B. Malloy, and J. N. Kent, "Resorbable plates for the fixation of mandibular fractures: a prospective study," *Journal of Oral and Maxillofacial Surgery*, vol. 65, no. 1, pp. 89–96, 2007.

[24] Z. Pan and P. M. Patil, "Titanium osteosynthesis hardware in maxillofacial trauma surgery: to remove or remain? A retrospective study," *European Journal of Trauma and Emergency Surgery*, vol. 40, no. 5, pp. 587–591, 2014.

[25] L. K. Cheung, L. K. Chow, and W. K. Chiu, "A randomized controlled trial of resorbable versus titanium fixation for orthognathic surgery," *Oral Surgery, Oral Medicine, Oral Pathology, Oral Radiology, and Endodontology*, vol. 98, no. 4, pp. 386–397, 2004.

[26] C. Alpha, F. O'Ryan, A. Silva, and D. Poor, "The incidence of postoperative wound healing problems following sagittal ramus osteotomies stabilized with miniplates and monocortical screws," *Journal of Oral and Maxillofacial Surgery*, vol. 64, no. 4, pp. 659–668, 2006.

[27] J. S. Guffey and J. Wilborn, "In vitro bactericidal effects of 405-nm and 470-nm blue light," *Photomedicine and Laser Surgery*, vol. 24, no. 6, pp. 684–688, 2006.

[28] N. T. A. de Sousa, M. F. Santos, R. C. Gomes, H. E. Brandino, R. Martinez, and R. R. de Jesus Guirro, "Blue laser inhibits bacterial growth of *Staphylococcus aureus*, *Escherichia coli*, and *Pseudomonas aeruginosa*," *Photomedicine and Laser Surgery*, vol. 33, no. 5, pp. 278–282, 2015.

[29] M. Hessling, B. Spellerberg, and K. Hoenes, "Photoinactivation of bacteria by endogenous photosensitizers and exposure to visible light of different wavelengths – a review on existing data," *FEMS Microbiology Letters*, vol. 364, no. 2, article fnw270, 2017.

[30] T. Imamura, S. Tatehara, Y. Takebe et al., "Antibacterial and antifungal effect of 405 nm monochromatic laser on endodontopathogenic microorganisms," *International Journal of Photoenergy*, vol. 2014, Article ID 387215, 7 pages, 2014.

[31] R. B. Bell and C. S. Kindsfater, "The use of biodegradable plates and screws to stabilize facial fractures," *Journal of Oral and Maxillofacial Surgery*, vol. 64, no. 1, pp. 31–39, 2006.

[32] C. A. Landes, A. Ballon, and C. Roth, "Maxillary and mandibular osteosyntheses with PLGA and P (L/DL) LA implants: a 5-year inpatient biocompatibility and degradation experience,"

Plastic and Reconstructive Surgery, vol. 117, no. 7, pp. 2347–2360, 2006.

[33] T. A. Turvey, W. P. Proffit, and C. Phillips, "Biodegradable fixation for craniomaxillofacial surgery: a 10-year experience involving 761 operations and 745 patients," *International Journal of Oral and Maxillofacial Surgery*, vol. 40, no. 3, pp. 244–249, 2011.

[34] E. J. Bergsma, F. R. Rozema, R. R. M. Bos, and W. C. D. Bruijn, "Foreign body reactions to resorbable poly(l-lactide) bone plates and screws used for the fixation of unstable zygomatic fractures," *Journal of Oral and Maxillofacial Surgery*, vol. 51, no. 6, pp. 666–670, 1993.

[35] J. E. Bergsma, W. C. de Bruijn, F. R. Rozema, R. R. Bos, and G. Boering, "Late degradation tissue response to poly (L-lactide) bone plates and screws," *Biomaterials*, vol. 16, no. 1, pp. 25–31, 1995.

[36] M. Maclean, S. J. MacGregor, J. G. Anderson, and G. Woolsey, "Inactivation of bacterial pathogens following exposure to light from a 405-nanometer light-emitting diode array," *Applied and Environmental Microbiology*, vol. 75, no. 7, pp. 1932–7, 2009.

[37] K. Sunada, T. Watanabe, and K. Hashimoto, "Studies on photokilling of bacteria on TiO2 thin film," *Journal of Photochemistry and Photobiology A: Chemistry*, vol. 156, no. 1-3, pp. 227–233, 2003.

[38] G. Villatte, C. Massard, S. Descamps, Y. Sibaud, C. Forestier, and K. O. Awitor, "Photoactive TiO_2 antibacterial coating on surgical external fixation pins for clinical application," *International Journal of Nanomedicine*, vol. 10, no. 1, pp. 3367–3375, 2015.

[39] C. H. Kim, E. S. Lee, S. M. Kang, E. de Josselin de Jong, and B. I. Kim, "Bactericidal effect of the photocatalystic reaction of titanium dioxide using visible wavelengths on *Streptococcus mutans* biofilm," *Photodiagnosis and Photodynamic Therapy*, vol. 18, pp. 279–283, 2017.

[40] M. D. Barneck, N. L. R. Rhodes, M. de la Presa et al., "Violet 405-nm light: a novel therapeutic agent against common pathogenic bacteria," *Journal of Surgical Research*, vol. 206, no. 2, pp. 316–324, 2016.

[41] R. Lubart, A. Lipovski, Y. Nitzan, and H. Friedmann, "A possible mechanism for the bactericidal effect of visible light," *Laser Therapy*, vol. 20, no. 1, pp. 17–22, 2011.

[42] M. Maclean, S. J. MacGregor, J. G. Anderson, and G. Woolsey, "High-intensity narrow-spectrum light inactivation and wavelength sensitivity of *Staphylococcus aureus*," *FEMS Microbiology Letters*, vol. 285, no. 2, pp. 227–232, 2008.

11

A Parametric Investigation on Energy-Saving Effect of Solar Building Based on Double Phase Change Material Layer Wallboard

Xiaoxiao Tong [1,2] **and Xingyao Xiong** [3]

[1]*Horticulture & Landscape College, Hunan Agricultural University, Changsha 410128, China*
[2]*School of Architecture & Design, China University of Mining and Technology, Xuzhou 221116, China*
[3]*Institute of Vegetables and Flowers, Chinese Academy of Agricultural Sciences, Beijing 100081, China*

Correspondence should be addressed to Xingyao Xiong; xiongxingyao@126.com

Academic Editor: Ben Xu

In order to further understand the thermal performance of the double phase change material (PCM) layer wallboard, the wallboard model was established and a comprehensively numerical parametric investigation was carried out. The variation laws of inner wall temperature rise and the heat flux transferred under different phase transition temperatures and thermal conductivities are presented in detail. The main results show that the temperature of the inside wall for case 2 can be reduced by about 1.5 K further compared to that for case 1. About 83% of the heat transferred from the outside is absorbed by the PCM layer in case 2. Reducing the phase transition temperature of the PCM layer can decrease the inside wall temperature to a certain extent in the period of high temperature. The utilization of double PCM layers shows much more performance compared to that of the single PCM layer case, and the temperature of the inside wall can be reduced by 2 K further.

1. Introduction

Energy demand has been increasing quickly with the development of economy. And the conventional fossil energy sources such as oil, coal, and gas are limited. Their use leads to climate changes and environmental pollution [1]. Building energy consumption has become a serious problem due to a large amount of energy that is consumed by the heating, ventilation, and air conditioning system of buildings every day. According to [2], about 40% of the world's total energy was used for buildings and more than 30% of the primary energy consumed in buildings is for the heating and air conditioning system. Therefore, some energy-saving and environment-friendly techniques have been investigated in recent years. Thermal energy storage techniques used in buildings to decrease the energy consumption were considered an effective way [3, 4]. Thermal energy storage can be divided into sensible heat storage, latent heat storage, and chemical energy storage. And among them, the latent heat storage has

received considerable attention in comparison with the other two methods attributed to the obvious advantages of latent heat storage using phase change material (PCM), like high energy storage density and narrow operating temperature range [5, 6]. Furthermore, PCM can store and release a large amount of latent heat during the process of melting and solidifying in its narrow phase transition range [7, 8].

In the past two decades, researches on the application of PCM in building energy conservation can be divided into two categories. One is combining the PCM with the active air conditional system where the PCM system serves as the heat source or cold source of the air conditioning system to increase the refrigerating efficiency or the heat efficiency [9, 10]. For instance, Tyagi et al. [11] designed and experimentally studied the thermal performance of a PCM-based building thermal management system for cool energy storage. The other is the usage in the passive heat insulation and preservation system, comprising the combination of PCM and building materials to obtain novel energy

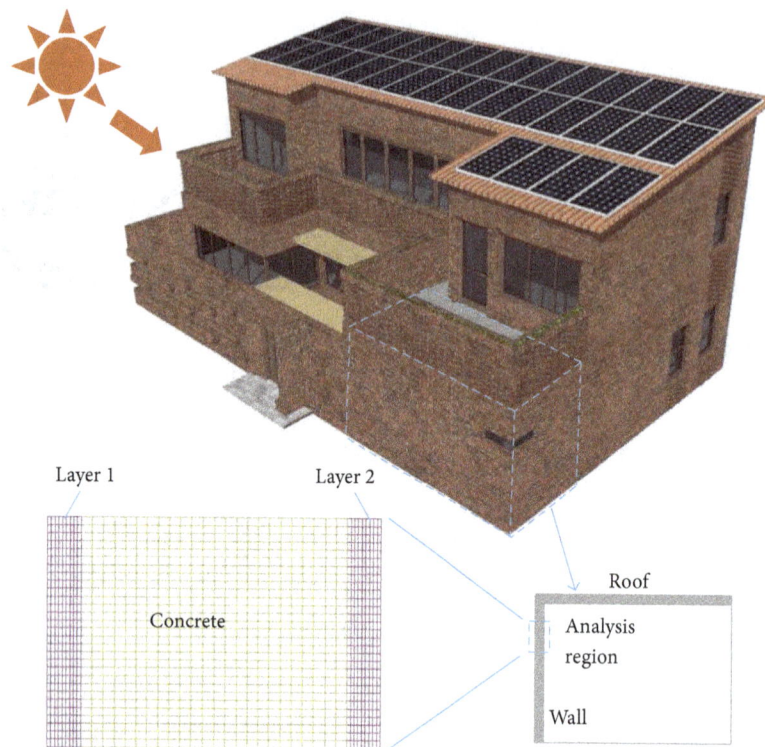

FIGURE 1: Schematics of the resident house and enlargement of the analysis region with structure mesh.

conservation building materials and directly inserting shape-stabilized PCM into the enclosure structure of the building [12–14]. The first category of PCM application needs to be considered early in the design process of the air conditioning system and also needs later maintenance. However, the second one is concentrated on the design of new building material, like coating material with phase change function, insulation wallboard, and brick with PCM encapsulated in it, and is a way to enhance the ability of the building itself to adapt to the climate. Therefore, it is widely of concern to scholars. Li et al. [15] compared the thermal performance of lightweight buildings with and without a PCM layer attached to the inside wallboard and found that the energy consumption to maintain comfortable temperature can be reduced by 40–70%. Ramakrishnan et al. [16] numerically investigated the thermal control effect of building fabrics integrated with PCM under extreme heatwave periods. The result shows that the indoor heat stress risks can be reduced effectively without the function of an air conditioner. Meanwhile, Thiele et al. [17] constructed a numerical model based on a modified admittance model to evaluate the thermal performance of building envelops integrated with PCM whose result turns to agree well with that of the existing finite element simulations. Zhu et al. [18, 19] put forward a new structure of wallboards with double shape-stabilized PCM, proposed a related simplified dynamic model, and then used it to analyze energy performance of office building under different conditions. However, the model and related analysis are concentrated on the whole system and overall efficiency. The heat transfer process and the influence of the PCM parameters on the heat transfer law are also quite important for the actual design and need to be further understood.

In this paper, in order to further explore the heat transfer law and thermal performance of the double PCM layer wallboard put forward by Zhu et al. [18] under different conditions, the wallboard model was established and a comprehensively parametric numerical investigation was carried out. The variation law of temperature rise at the inner side of the wall and the heat flux transferred under different phase transition temperatures, thermal conductivities, and arrangements of PCMs are presented and discussed in detail in the following sections.

2. Model and Methodology

2.1. Model Description. Figure 1 shows the schematics of the resident house and enlargement of the analysis region with structure mesh. The performance of the wall determines the economic and energy-saving efficiency of the whole building to a great extent. Therefore, the wallboard is the key research object. The structure of the wallboard is presented in the enlarged view clearly. As shown in Figure 2, three cases of wallboard with different layer combinations were designed and compared. Case 1 represents the convection wallboard with insulation material only. The insulation layer outside is replaced by the PCM layer in case 2, and both insulation layers are replaced by PCM layers in case 3. The dimensions and thermo-physical properties of these two layers and the concrete can be seen in Table 1.

2.2. Numerical Simulation. With the development of computer technology, numerical study as an effective means of research involving design, analysis, and optimization is being developed quickly. In this exploration, commercial

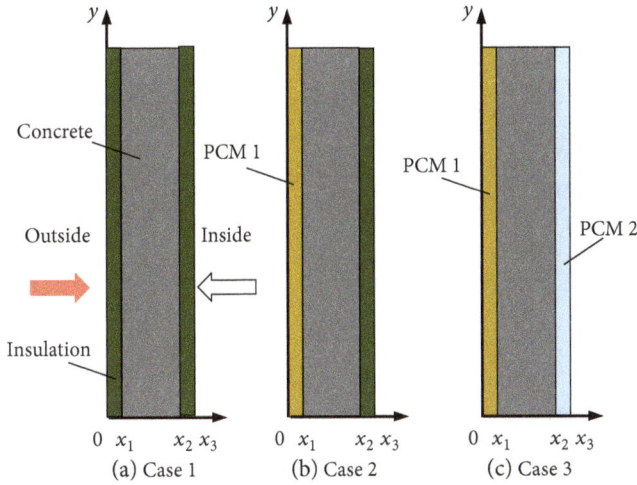

FIGURE 2: Schematics of the building wallboard with/without PCM.

computational fluid dynamics package, FLUENT 14.0, was utilized. The mesh density and the computational parameters such as time step and number of iterations per time step were evaluated by checking the dependency of the total heat transfer flux on models with various mesh quantities and different computational parameters. The time step was set to 10 seconds, and the number of iterations per time step was 50. The pressure-based 1st-order implicit algorithm for this unsteady problem was considered. Some assumptions were made in the following simulation work. The specific heat, the phase transition temperature, and the thermal conductivity of PCMs were constant. Also, the PCMs utilized were isotropic and homogenous. The volume change of the PCM during phase transition was ignored.

The energy conservation equation for the concrete region can be presented as follows:

$$\frac{\partial}{\partial \tau}\left(\rho_c c_{pc} t_c\right) = \nabla \cdot \left(\lambda_c \nabla t_c\right), \tag{1}$$

where ρ_c, c_{pc}, and λ_c are the density, heat capacity, and thermal conductivity of the concrete, respectively. The enthalpy-porosity model was adopted to model the phase changing process in this work. The liquid fraction is computed at each iteration, based on an enthalpy balance. The energy equation of PCM can be expressed as follows [20]:

$$\rho \frac{\partial H}{\partial \tau} = k \nabla^2 H + S, \quad H = H_0 + \Delta H,$$

$$H_0 = H_{ref} + \int_{T_{ref}}^{T} C_{P_{pcm}} dT, \quad C_{P_{pcm}} = \begin{cases} C_{Ps_{pcm}}, & T < T_m, \\ C_{Pl_{pcm}}, & T > T_m, \end{cases}$$

$$\Delta H = \beta \gamma, \quad \beta = \begin{cases} 0, & T < T_m, \\ 1, & T > T_m, \end{cases} \tag{2}$$

where H represents the total enthalpy of PCM, H_0 is the sensible enthalpy, ΔH is the latent heat, β is the liquid fraction, and T_m is the phase transition temperature.

$$q_s''(t) = \begin{cases} q_{s,max}'' \cos\left(\dfrac{\pi t}{43200} - \pi\right), & 6:00\,am \le t \le 6:00\,pm, \\ 0, & 6:00\,pm < t \le 6:00\,am, \end{cases} \tag{3}$$

$$q_{ave}'' = \frac{\int q_s''(t)dt}{12 \times 3600} = \frac{\int q_{s,max}'' \cos((\pi t/43200) - \pi)dt}{12 \times 3600}$$
$$= \frac{q_{s,max}'' \int \cos((\pi t/43200) - \pi)dt}{12 \times 3600}. \tag{4}$$

The external boundary condition is based on the total solar radiation of the Xuzhou area, which can be seen in Figure 3 [21]. The average solar radiation (q_{ave}'') of the Xuzhou area in June is about 385.8 W/m^2, and the maximum daily solar radiation can be observed through (4). Boundary conditions at the top and bottom of the model are thermal isolation. The sun radiation reaches the left side to heat the wall, and the convection heat transfer exists at the same time to cool the wall. But the heat flux and radiation cannot appear in the boundary condition at the same time to complete the numerical solution. Therefore, after simplification, the left boundary condition is time-dependent temperature boundary, which is described as follows:

$$x = 0,$$
$$-k\frac{\partial T}{\partial n} = T(\tau) = \Delta T_{max} + T_{ini}\sin\left(\frac{\pi t}{60000}\right). \tag{5}$$

The right side boundary conditions are mixed boundary conditions. The right side of the wall heats the air inside by radiation and convection synchronously, which can be described as follows:

$$x = x_3,$$
$$-k_i\frac{\partial T}{\partial n} = h_i(T_{wi} - T_{ai}) + \varepsilon\sigma\left(T_{wi}^4 - T_{ai}^4\right). \tag{6}$$

A parametric study was undertaken to investigate the influence of phase transition temperature of PCM and thermal conductivities of PCM and the thermal control effect of the double PCM layers compared to other two cases.

3. Result and Discussion

The aim of this work was to investigate a special kind of wallboard with two PCM layers attached both sides of the concrete wall for heat insulation and energy saving. Temperature variation of the outside wall with time in summer is exhibited in Figure 4. As the figure shows, the temperature of the outside wall increases to 325 K linearly with a relatively high rate of rise before 10:00 am, and the tendency of temperature rising gradually reduces from 10:00 am to 02:00 pm. The highest temperature of the outside wall in one solar day is up to about 338 K at 02:00 pm. After 02:00 pm, the temperature of the outside wall gradually decreases to 330 K (at 06:00 pm). This variation trend of the outside wall

TABLE 1: Thermo-physical properties of the wallboard materials [18].

Materials	c_P (kJ/kg·°C)	λ (W/m·°C)	ρ (kg/m³)	L (kJ/kg)	T_m (°C)	D (m)
Concrete	0.8	2.1	2400	—	—	0.24
Insulation	2.0	0.2	850	—	—	0.03
PCM 1/PCM 2	2.0	0.1~2	850	200	26–29	0.03

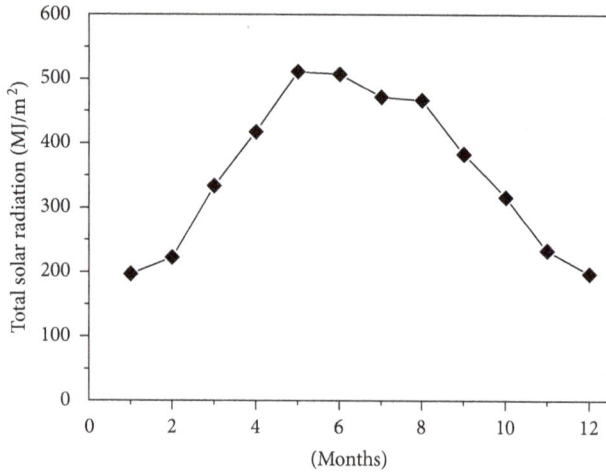

FIGURE 3: Total solar radiation of the Xuzhou area in different months.

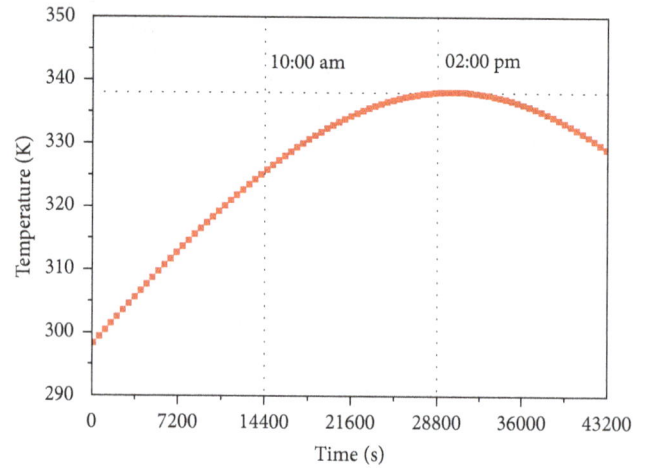

FIGURE 4: Temperature variation of the outside wall at different moments in summer.

temperature can reflect well the actual temperature condition of a solar day in the Xuzhou area in summer.

Figure 5 presents the temperature variation and the heat flux of the inside wall for different cases. As shown in Figure 5(a), the temperature of case 1 and case 2 has little change before 10:00 am, keeping a temperature of 299 K, and after 10:00 am, the temperature of case 1 gradually increases to 300 K, while that of case 2 has only little variation until 02:00 pm. After 02:00 pm, variation of temperature rising of case 1 has a significant improvement and case 2 starts a temperature rising with a similar tendency with that of case 1, and the temperatures of case 1 and case 2 are, respectively, 302 K and 300 K. As the wall is without any heat insulation layer or PCM layer, the temperature of the inside wall rises promptly after 09:00 am and finally rises to 312 K at 06:00 pm. In summary, the function of the insulation layer and PCM layer can both greatly retard the velocity of temperature diffusion, and the inside wall temperature can be decreased by more than 10 K. Case 2 has a certain advantage over case 1, which is attributed to the phase change endothermic behavior of the PCM layer. When the outside insulation layer was replaced by the PCM layer, the temperature of the inside wall can be reduced by about 1.5 K further. As shown in Figure 5(b), the heat flux is positive in the morning, indicating that the wall absorbs the air heat in the room, for the reason that the temperature of external air is set larger than the initial temperature of the wall in the simulation process. The heat flux of case 2 begins to turn negative, and the heat began to go through the wall completely until about 04:00 pm. It can be concluded from area C and area A in Figure 5(b) that the heat transferred into the indoor can be reduced about

98% by the function of the PCM layer and insulation layer. It can also be deduced that about 83% of the heat transferred from the outside is absorbed by the PCM layer through comparing area B and area C.

Figure 6 shows the temperature contours of the wallboard at 06:00 pm for three different cases obviously. The temperature distribution exhibits uniform gradient in the concrete wall case, in which the double layers are replaced by the concrete. It can be seen from case 1 and case 2 that the overall average temperature of the concrete wall in case 2 is obviously lower than that in case 1 for the reason that PCM can absorb large amount of latent heat under lower and stable temperature region, the heat transfer driving force and temperature difference are relatively weak, and less heat gets across the border to the concrete wall.

3.1. The Effect of Phase Transition Temperature of the PCM Layer. Figure 7 presents temperature variation and heat flux of the inside wall for case 2 under different phase transition temperatures (from 299.15 K to 302.15 K). As shown in Figure 7(a), the temperature under different phase transition times is identical before 10:00 am and after 04:00 pm. In the middle range of the solar day, the temperature difference under different phase transition temperatures increases firstly and then decreases. When time goes after 02:00 pm, the temperature difference under different phase transition temperatures decreases in contrast. The temperature of the inside wall under different phase transition temperatures is identical again at 04:00 pm, and the final identical temperature is about 301 K, increasing about 2.7 K. As shown in Figure 7(b), heat flux of the inside wall is also identical under

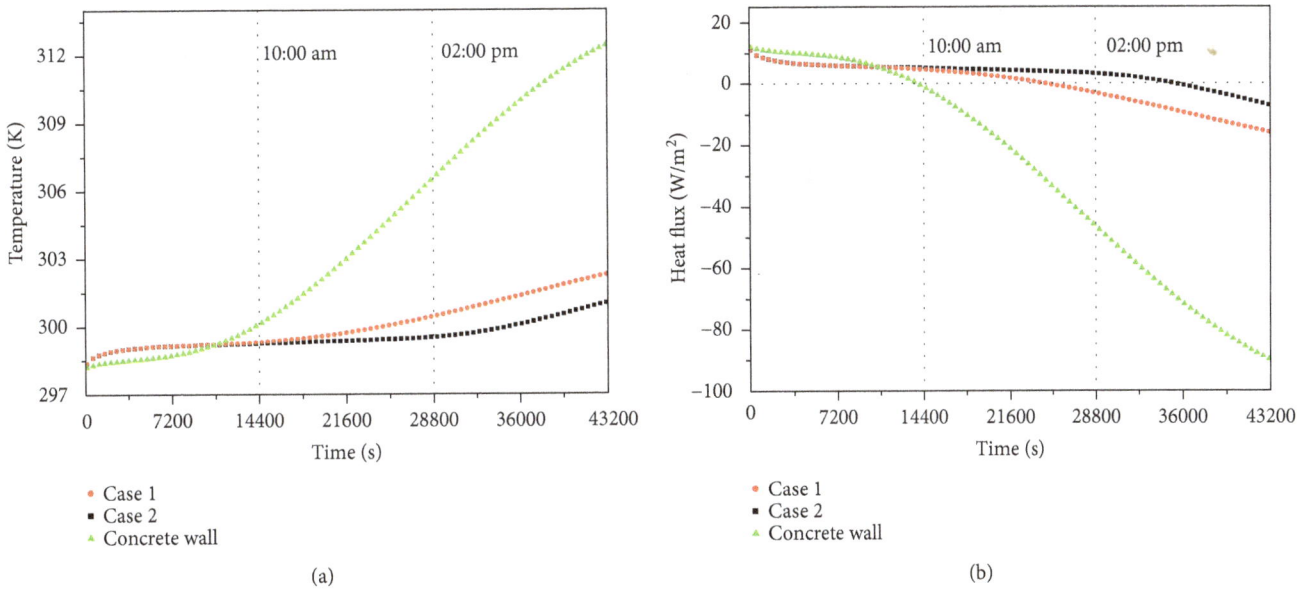

(a)

(b)

FIGURE 5: Temperature variation and heat flux of the inside wall under different cases.

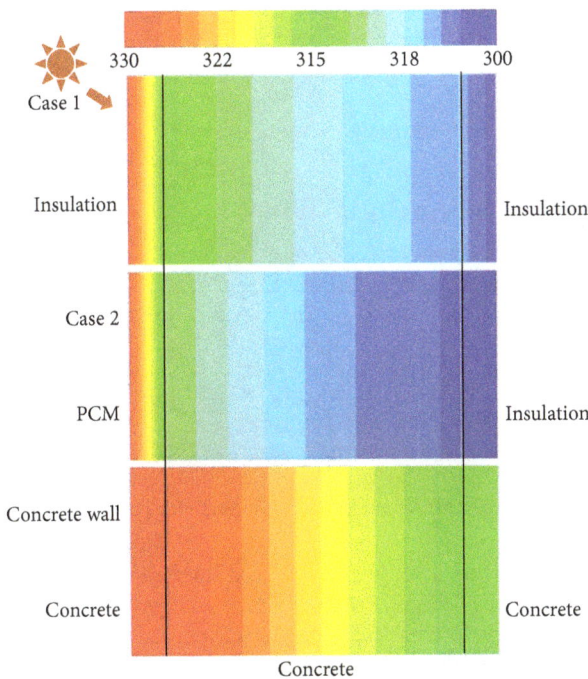

FIGURE 6: Temperature contour of the wallboard at 06:00 pm under different cases.

different transition temperatures before 10:00 am. When time is over 10:00 am, heat flux of the inside wall is higher under lower phase transition temperature, which is opposite to temperature. Heat flux of the inside wall under different phase transition temperatures is identical at the end. It can also be seen that heat flux decreases from 11.2 W/m² to −7 W/m². Actually, heat flux decreases from 11.2 W/m² to 0 W/m², and then it increases to 7 W/m² with the opposite heat transfer direction. In summary, reducing the phase transition temperature of the PCM layer can decrease the inside wall temperature to a certain degree in the middle period of

a solar day; however, the heat transferred into the indoor in the whole daytime is almost not affected by the phase transition temperature.

3.2. The Effect of Thermal Conductivity of PCM. Figure 8 exhibits the temperature variation and heat flux of the inside wall for case 2 under different thermal conductivities of PCM. As shown in Figure 8(a), the temperature of the inside wall has little change before 10:00 am, keeping a temperature of 299 K, though the thermal conductivity of PCM is changed. After 10:00 am, the temperature of the inside wall gradually increases and the final temperatures are 303.2 K, 305.4 K, and 307.4 K while the thermal conductivities of PCM are 0.4 W/(m·K), 0.8 W/(m·K), and 2 W/(m·K), respectively. As the thermal conductivity of PCM reaches 0.2 W/(m·K), the rising trend of the inside wall temperature becomes obvious after 02:00 pm, and the final temperature rises to 301 K. However, the temperature of the inside wall has little change during the whole solar day when the thermal conductivity of PCM is as low as 0.1 W/(m·K). It is obvious that decreasing the thermal conductivity of the PCM layer is beneficial to heat insulation and energy saving. Less heat can be transferred to the indoor. As shown in Figure 8(b), heat flux is almost stable around 5 W/m² and flows toward the outside before 10:00 am. And after 10:00 am, heat flux gradually reverses its direction and reaches about 22.5 W/m², 37.9 W/m², and 52.3 W/m² at 06:00 pm as the thermal conductivities of PCM are 0.4 W/(m·K), 0.8 W/(m·K), and 2 W/(m·K), respectively. Figure 9 shows the phase change ratio of the PCM layer for case 2 under different thermal conductivities. It can be found that when the thermal conductivity is 2 W/(m·K), the PCM melts entirely in almost 7200 s, while it takes 4.5 times longer to melt the PCM layer with a thermal conductivity of 0.1 W/(m·K).

3.3. The Effect of Double PCM Layers. In order to further increase the energy-saving capacity of the wallboard, the

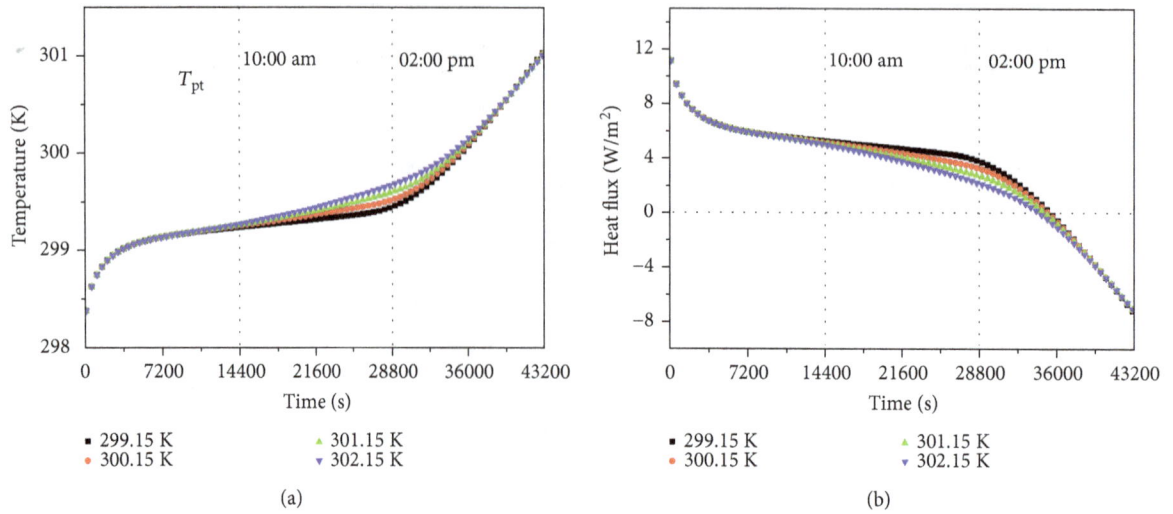

FIGURE 7: Temperature variation and heat flux of the inside wall for case 2 under different phase transition temperatures.

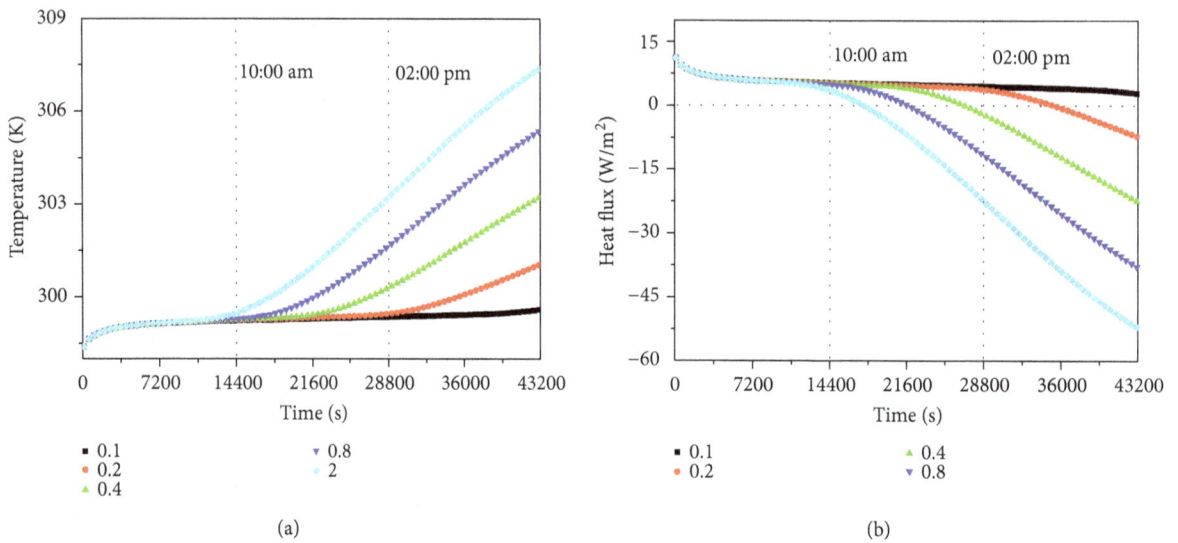

FIGURE 8: Temperature variation and heat flux of the inside wall for case 2 under different thermal conductivities of PCM.

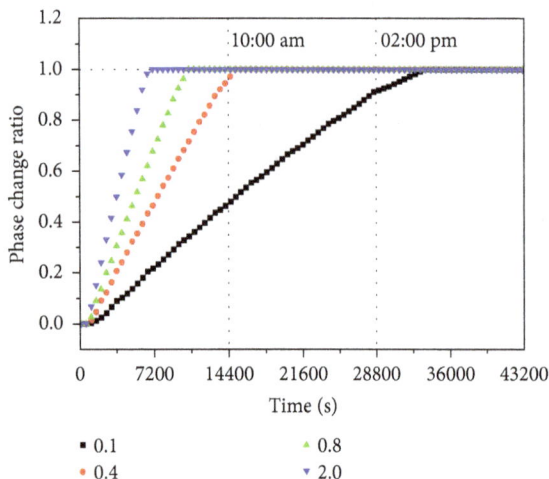

FIGURE 9: Phase change ratio of the PCM layer for case 2 under different thermal conductivities.

right insulation layer is also replaced by the PCM layer, and the PCM has the same thermal physical property as the left layer. Figure 10 presents the temperature variation and heat flux of the inside wall for case 3. It can be observed that the temperature curve for case 3 increases first and becomes stable almost the whole day. The temperature can be stabilized at about 299.15 K. The wallboard with double PCM layers shows much better thermal performance compared to the single PCM layer case, and the temperature of the inside wall can be reduced by 2 K further. The inner wall almost can exclude the interference from external environment. As shown in Figure 10(b), the heat flux for case 3 is positive in the whole daytime. The heat outside cannot be transferred to the indoor, which is the reason why the temperature of the inside wall can be stable. Figure 11 illustrates the phase change ratio of each PCM layer for case 3. The phase change ratio of PCM 1 is on the rise before 12:00 pm, but that of PCM 2 remains constant until 02:00 pm. It can be seen that PCM 1 just takes 21600 s to melt totally. However, the phase

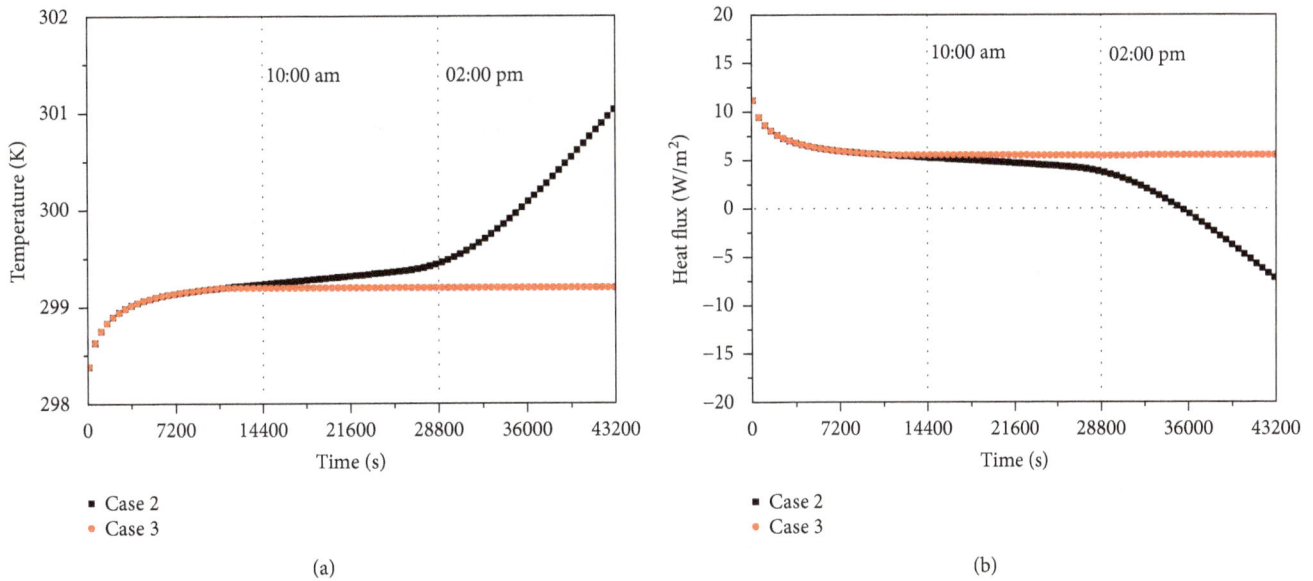

(a)

(b)

FIGURE 10: Temperature variation and heat flux of the inside wall for case 3.

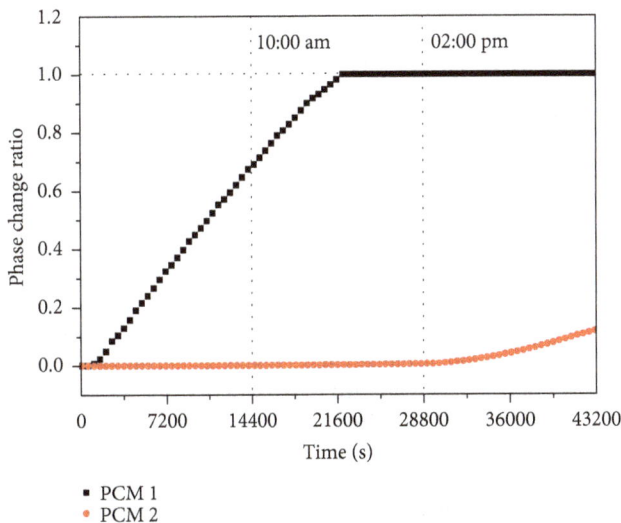

FIGURE 11: Phase change ratio of each PCM layer for case 3.

change ratio of PCM 2 is still under 0.2 at the end of the solar day.

4. Conclusion

In order to further understand the heat transfer law and thermal performance of the double PCM layer wallboard under different conditions, a comprehensively parametric numerical investigation was carried out. The variation law of temperature rise at the inner side of the wall and the heat flux flowed through under different phase transition temperatures, thermal conductivities, and arrangements of PCMs are presented and discussed in detail. The main conclusions can thus be summarized as follows:

(1) The function of the insulation layer and PCM layer can both greatly retard the velocity of temperature

diffusion, and the inside wall temperature can be decreased by more than 10 K. About 83% of the heat transferred from the outside is absorbed by the PCM layer in case 2.

(2) Reducing the phase transition temperature of the PCM layer can decrease the inside wall temperature to a certain degree in the period of high temperature. Increasing the thermal conductivity of the PCM layer is not beneficial to heat insulation and energy saving. More heat can be transferred to the indoor easily.

(3) The utilization of the double PCM layer shows much more performance compared to that of the single PCM layer case, and the temperature of the inside wall can be reduced by 2 K further.

Nomenclature

T: Temperature (K)
ΔT: Temperature difference (K)
c_p: Specific heat (J/(kg·K))
H: Enthalpy (J kg^{-1})
ΔH: Latent heat of PCM (J kg^{-1})
ΔT: Temperature increase (K)
k: Thermal conductivity (W/(m·K))
q: Heat flux (W/m^2)
β: Liquid volume fraction
τ: Time (s)
ρ: Density (kg/m^3)
ε: Emissivity
σ: Stefan-Boltzmann constant.

Subscripts

w: Wall
wi: Wall inside
a: Air

ai: Air inside
ini: Initial
c: Concrete
ave: Average
ref: Reference
m: Melting.

Acronyms

PCM: Phase change material.

Conflicts of Interest

The authors declare that there is no conflict of interest regarding the publication of this paper.

Acknowledgments

This work was supported by the National Natural Science Foundation of China (no. 51778611).

References

[1] T. Qian, J. Li, X. Min, Y. Deng, W. Guan, and L. Ning, "Diatomite: a promising natural candidate as carrier material for low, middle and high temperature phase change material," *Energy Conversion and Management*, vol. 98, pp. 34–45, 2015.

[2] Z. Chen, M. Qin, and J. Yang, "Synthesis and characteristics of hygroscopic phase change material: composite microencapsulated phase change material (MPCM) and diatomite," *Energy and Buildings*, vol. 106, pp. 175–182, 2015.

[3] S. Karaman, A. Karaipekli, A. Sarı, and A. Biçer, "Polyethylene glycol (PEG)/diatomite composite as a novel form-stable phase change material for thermal energy storage," *Solar Energy Materials and Solar Cells*, vol. 95, no. 7, pp. 1647–1653, 2011.

[4] B. Xu and Z. Li, "Performance of novel thermal energy storage engineered cementitious composites incorporating a paraffin/diatomite composite phase change material," *Applied Energy*, vol. 121, pp. 114–122, 2014.

[5] S.-G. Jeong, J. Jeon, O. Chung, S. Kim, and S. Kim, "Evaluation of PCM/diatomite composites using exfoliated graphite nanoplatelets (xGnP) to improve thermal properties," *Journal of Thermal Analysis and Calorimetry*, vol. 114, no. 2, pp. 689–698, 2013.

[6] M. Li, H. Kao, Z. Wu, and J. Tan, "Study on preparation and thermal property of binary fatty acid and the binary fatty acids/diatomite composite phase change materials," *Applied Energy*, vol. 88, no. 5, pp. 1606–1612, 2011.

[7] X. Fu, Z. Liu, Y. Xiao, J. Wang, and J. Lei, "Preparation and properties of lauric acid/diatomite composites as novel form-stable phase change materials for thermal energy storage," *Energy and Buildings*, vol. 104, pp. 244–249, 2015.

[8] A. Sarı and A. Karaipekli, "Fatty acid esters-based composite phase change materials for thermal energy storage in buildings," *Applied Thermal Engineering*, vol. 37, pp. 208–216, 2012.

[9] F. Souayfane, F. Fardoun, and P.-H. Biwole, "Phase change materials (PCM) for cooling applications in buildings: a review," *Energy and Buildings*, vol. 129, pp. 396–431, 2016.

[10] S. Ramakrishnan, X. Wang, M. Alam, J. Sanjayan, and J. Wilson, "Parametric analysis for performance enhancement of phase change materials in naturally ventilated buildings," *Energy and Buildings*, vol. 124, pp. 35–45, 2016.

[11] V. V. Tyagi, D. Buddhi, R. Kothari, and S. K. Tyagi, "Phase change material (PCM) based thermal management system for cool energy storage application in building: an experimental study," *Energy and Buildings*, vol. 51, pp. 248–254, 2012.

[12] D. Feldman, D. Banu, and D. W. Hawes, "Development and application of organic phase change mixtures in thermal storage gypsum wallboard," *Solar Energy Materials and Solar Cells*, vol. 36, no. 2, pp. 147–157, 1995.

[13] M. Chung and J. Park, "An experimental study on the thermal performance of phase-change material and wood-plastic composites for building roofs," *Energies*, vol. 10, no. 2, p. 195, 2017.

[14] J. Lei, K. Kumarasamy, K. T. Zingre, J. Yang, M. P. Wan, and E.-H. Yang, "Cool colored coating and phase change materials as complementary cooling strategies for building cooling load reduction in tropics," *Applied Energy*, vol. 190, pp. 57–63, 2017.

[15] Y. Li, Y. Wang, X. Meng, W. Zhang, and E. Long, "Research on thermal performance improvement of lightweight buildings by integrating with phase change material under different climate conditions," *Science and Technology for the Built Environment*, vol. 23, no. 2, pp. 285–295, 2016.

[16] S. Ramakrishnan, X. Wang, J. Sanjayan, and J. Wilson, "Thermal performance of buildings integrated with phase change materials to reduce heat stress risks during extreme heatwave events," *Applied Energy*, vol. 194, pp. 410–421, 2017.

[17] A. M. Thiele, R. S. Liggett, G. Sant, and L. Pilon, "Simple thermal evaluation of building envelopes containing phase change materials using a modified admittance method," *Energy and Buildings*, vol. 145, pp. 238–250, 2017.

[18] N. Zhu, P. Hu, and L. Xu, "A simplified dynamic model of double layers shape-stabilized phase change materials wallboards," *Energy and Buildings*, vol. 67, pp. 508–516, 2013.

[19] N. Zhu, P. Liu, F. Liu, P. Hu, and M. Wu, "Energy performance of double shape-stabilized phase change materials wallboards in office building," *Applied Thermal Engineering*, vol. 105, pp. 180–188, 2016.

[20] Z. Rao, S. Wang, and G. Zhang, "Simulation and experiment of thermal energy management with phase change material for ageing LiFePO$_4$ power battery," *Energy Conversion and Management*, vol. 52, no. 12, pp. 3408–3414, 2011.

[21] L. Zhang, J. Liu, and R. Zhang, "Assessment of solar radiation condition in Xuzhou area," *Acta Agriculturae Jiangxi*, vol. 22, no. 3, pp. 114–117, 2010.

12

Experimental Study of a Novel Direct-Expansion Variable Frequency Finned Solar/Air-Assisted Heat Pump Water Heater

Jing Qin,[1] **Jie Ji**[iD]**,**[1] **Wenzhu Huang,**[1] **Hong Qin,**[2] **Mawufemo Modjinou,**[1] **and Guiqiang Li**[iD][1]

[1]*Department of Thermal Science and Energy Engineering, University of Science and Technology of China, No. 96 Jinzhai Road, Hefei, Anhui, China*
[2]*Guangdong University of Technology, Guangzhou, China*

Correspondence should be addressed to Jie Ji; jijie@ustc.edu.cn

Academic Editor: Francesco Riganti Fulginei

A novel direct expansion variable frequency finned solar/air-assisted heat pump water heater was fabricated and tested in the enthalpy difference lab with a solar simulator. A solar/air source evaporator-collector with an automatic lifting glass cover plate was installed on the system. The system could be operated in three modes, namely, air, solar, and dual modes. The effects of the ambient temperature, solar irradiation, compressor frequency, and operating mode on the performance of this system were studied in this paper. The experimental results show that the ambient temperature, solar irradiation, and operating mode almost have no effect on the energy consumption of the compressor. When the ambient temperature and the solar irradiation were increased, the COP was found to increase with decreasing heating time. Also, when the compressor frequency was increased, an increase in the energy consumption of the compressor and the heat gain of the evaporator were noted with a decrease in the heating time.

1. Introduction

Water heating consumes nearly 20% of the total energy consumption for an average family [1]. Solar energy is a clean, inexhaustible, and abundant energy resource [2]. The development of an affordable and effective clean energy technologies such as a combination of solar energy and heat pump can have huge long-term benefits, which is the concept of solar-assisted heat pump (SAHP) [3].

The concept of direct expansion solar-assisted heat pump (DX-SAHP) was first presented by Sporn and Ambrose [4]. While the "indirect-expansion SAHP" system has an intermediate heat exchanger between the solar water circuit and a water circuit, a DX-SAHP uses a two-phase solar collector to function directly as an evaporator. The evaporator-collector configuration favourably reduces the number of components in use, which can lower the cost of the system and avoid the nighttime freeze-up problem of a traditional water collector [5].

For the SAHP system, the match between different components is critical. Liu et al. [6] found that the mass flow rate of refrigerant can match well with the thermal load of the evaporator by means of compressor frequency modulation. In this case, the compressor frequency should be adjusted according to the ambient conditions. Chaturvedi et al. [7] showed that the coefficient of performance (COP) of the system can be enhanced extensively by lowering the compressor frequency when the ambient temperatures are higher. Moreno-Rodríguez et al. [8] showed that when the condenser water flow rate is lower than 0.114 kg/s, a higher COP can be obtained by reducing the compressor frequency. However, when the water flow rate is over 0.174 kg/s, lowing the compressor frequency will drastically lead to a lower COP.

SAHP is an effective way for utilizing solar energy, but solar energy is an intermittent energy which changes greatly depending on time and weather [9]. For this reason, supplementary heat source is required to ensure the continuous and reliable operation of the SAHP. Air source heat pump

(ASHP) can operate more reliably by absorbing heat from the ambient air [10]. However, the application of single ASHP is also dependent on the weather conditions. Especially in winter when the evaporating temperature decreases, there is a significant drop in the heating capacity and energy efficiency of the system [11] in good shape. Since SAHP and ASHP have their respective advantages and disadvantages, many researchers combined SAHP and ASHP to improve the reliability of the system. The current dual source heat pump system is made up of different heat exchangers to utilize different heat sources, which would involve more complicated equipment and cause higher cost and other related issues [12–14]. A dual source evaporator-collector, which is made up of finned copper tubes with selective absorption coating is fabricated in this study to improve the performance of the dual source heat pump system. The dual source evaporator-collector configuration could effectively reduce the number of components in use and absorb more heat from solar and air than flat plate evaporator through the fins.

Glazed and unglazed flat plate solar collectors are two major collector types mostly used in SAHP systems [15]. When the temperature of the collector system is higher than the ambient temperature, the use of glazed solar collector will reduce heat loss [16], while, when the temperature of the collector is lower than the ambient temperature, the use of unglazed solar collector produces high heat efficiency [17]. Therefore, an automatic lifting glass cover plate, which could automatically cover and uncover the evaporator-collector, is expected to enhance the performance of SAHP system.

Therefore, there are three approaches to enhance the performance of a solar-assisted heat pump water heater (SAHPWH) system: (1) variable frequency compressor; (2) a dual source evaporator-collector made up of finned copper tubes with selective absorption coating; and (3) an automatic lifting glass cover plate. However, there is no system in published experimental studies that applied all three components above to improve the performance of SAHPWH system. In this paper, a novel DX-SAHPWH system that applied all three components above was fabricated and tested experimentally.

Most studies of SAHP were implemented in outdoor environment where constant solar radiation and ambient temperature are hard to maintain [18]. In this study, the experiment was carried out under constant and controlled solar irradiation and ambient temperature in the enthalpy difference lab with a solar simulator to produce better results.

In this paper, a novel DX-SAHPWH system was experimentally investigated and tested in the enthalpy difference lab with a solar simulator. The effects of the ambient temperature, solar irradiation, compressor frequency, and the operating modes on the performance of this system were studied.

2. System Description and Test Apparatus

2.1. System Description. Figure 1 shows the schematic diagram of the DX-SAHPWH system. This system mainly consists of a solar/air source horizontal finned evaporator-collector, a R410a hermetic rotary DC inverter compressor, a plate-type heat exchanger condenser, a 150 L pressure bearing structure water-storage tank, and an electronic expansion valve.

In this system, the refrigerant absorbs energy from the air and solar; and gets vaporized in the evaporator-collector. Then, it is compressed in the compressor and becomes superheated vapor of high temperature and pressure. The vapor enters the condenser and gets condensed into liquid. The heat rejected in this procedure is absorbed by water-storage tank. Therefore, the water is heated. After this, the refrigerant passes through the electronic expansion valve where it expands irreversibly and adiabatically. Finally, the refrigerant flows back to the evaporator-collector and gets vaporized again, the cycle to continues.

As shown in Figure 2, the evaporator-collector is made up of finned copper tubes with selective absorption coating to increase the area for absorbing solar and air energy, its main parameters are shown in Table 1. A fan is also installed at the back of the evaporator-collector. Besides, an automatic lifting glass cover plate, which can be used to cover and uncover the evaporator-collector automatically, was fabricated on the top of the evaporator-collector, as shown in Figure 3.

Based on the operating conditions of the glass cover plate and fan, this system could be operated in three modes, namely, air, solar, and dual modes, although the system could absorb energy from both air and solar in three working modes, as shown in Table 2.

2.2. Test Apparatus. To study the performance of the DX-SAHPWH system, the system has been tested in the enthalpy difference lab with a solar simulator. The enthalpy difference lab can maintain constant ambient temperature and humidity by operating the air handling units (AHU), as shown in Figure 4. The AHU consists of coolers, heaters, humidifier, and fan, which offer cooling capacity, heating capacity, controllable humidity, and air cycle in the lab, respectively. The solar simulator was then configured in the lab to simulate solar irradiation on the surface of the evaporator-collector. As the spectrum distribution satisfies the national class B level standard, and the heterogeneity and instability of the solar simulator are under 5%, the solar simulator effectively simulated the solar irradiation for the study. Its luminous area is $2 \times 2\,m^2$, and the adjustable range of the irradiance is from 0 to $1200\,W/m^2$. The solar simulator was set parallel to the collectors. Hence, the influence of solar irradiation on this system can be studied.

Ambient dry bulb and wet bulb temperature of the enthalpy difference lab was measured using platinum resistance thermometer (with grade A accuracy). The temperatures of water in the storage tank, at the inlet and outlet of the evaporator-collector, the compressor, the condenser, and the electronic expansion valve were measured using T-type copper-constantan thermocouples. Also, the refrigerant pressure in evaporator-collector and condenser was measured using Huba pressure sensor. The solar irradiance and the energy consumption of the system were also measured and recorded during the experiments. A digital power meter (YOKOGAWA WT230) was used to measure the energy consumption of the system. The experimental data was recorded automatically for every 6 seconds by a data logger (Agilent 34970A). The details of the sensors and measurement instruments are shown in Table 3.

(a)

(b)

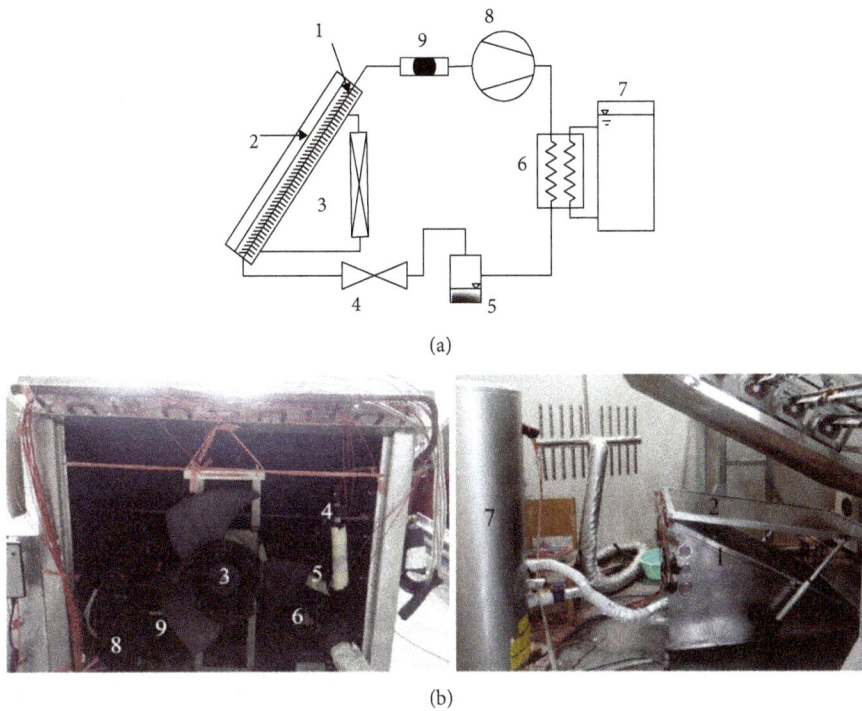

FIGURE 1: Configuration diagram of the DX-SAHPWH system. (a) Schematic. (b) Photo. 1: solar/air source horizontal finned evaporator-collector; 2: automatic lifting glass cover plate; 3: fan; 4: electronic expansion valve; 5: accumulator; 6: condenser; 7: water tank; 8: variable frequency compressor; 9: strainer.

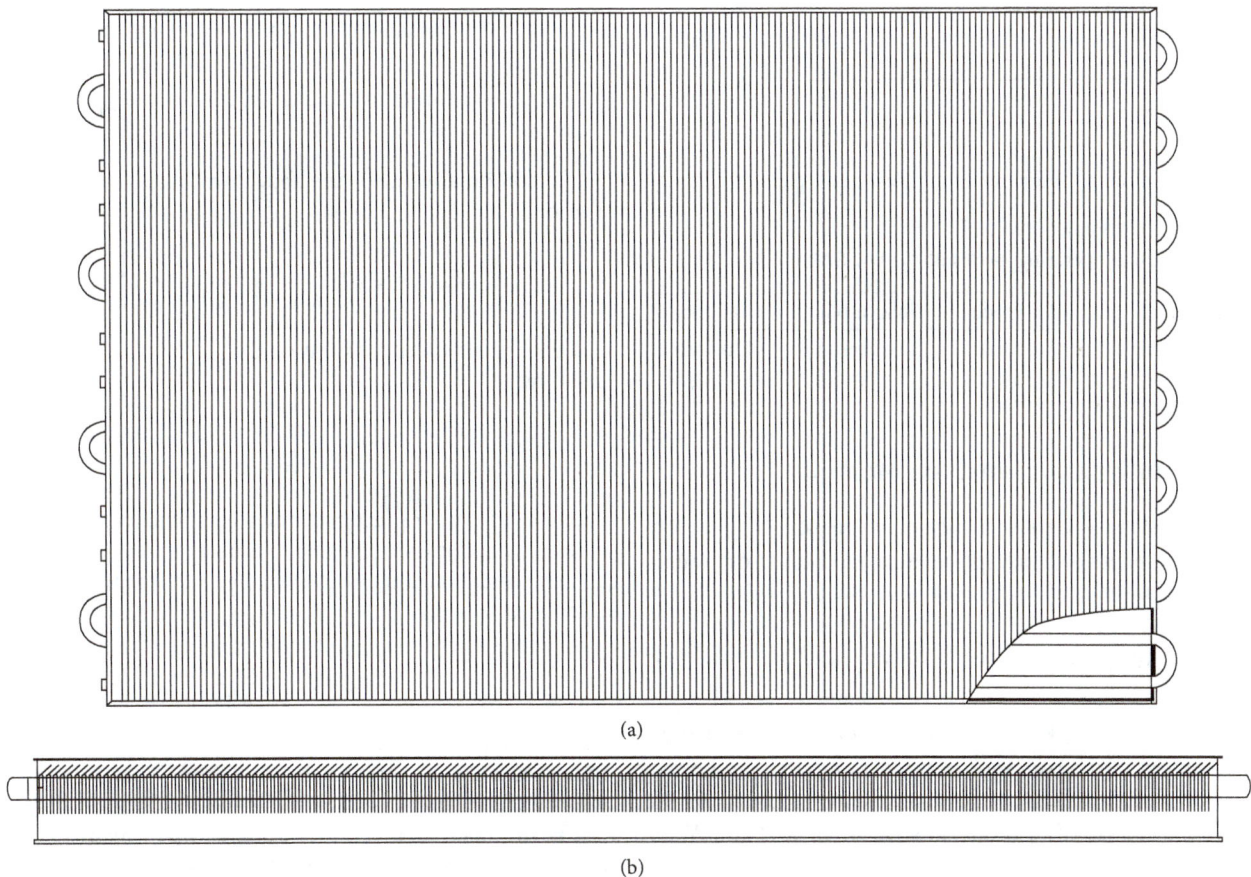

(a)

(b)

FIGURE 2: Diagram of the evaporator-collector. (a) Front view. (b) Lateral view.

TABLE 1: Parameters of evaporator-collector.

Length	1475 mm
Width	800 mm
Solar absorptivity of selective absorption coating	0.95
Transmittance of glass plate	0.85
External diameter of copper tube	9.52 mm
Thickness of copper tube	0.35 mm
Number of tube rows	16
Thickness of fin	0.35 mm
Punching diameter of fin	10 mm
Height of fin	15 mm
Vertical pitch of fin	3 mm
Number of fins	208
Angle of fin and copper tube	28°

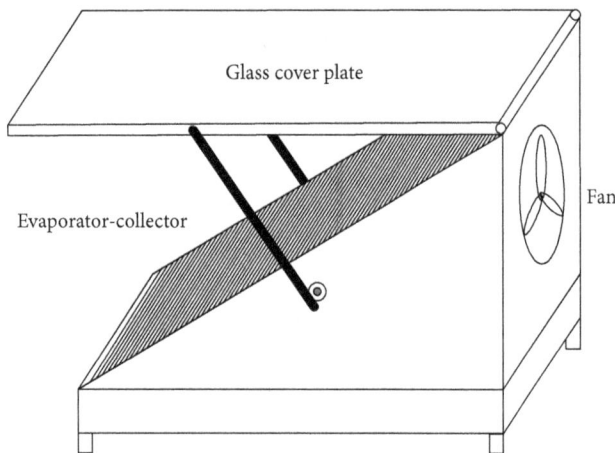

FIGURE 3: Automatic lifting glass cover plate.

TABLE 2: Three operating modes of the DX-SAHPWH system.

Component	Air mode	Solar mode	Dual mode
Glass cover plate	Uncover	Cover	Uncover
Fan	Fast	Off	Slow

3. Thermodynamic Analysis Methods

For this system, the total energy consumption W_{sys} is the sum of the energy consumption of the compressor, water pump, fan, and electric controller.

$$W_{sys} = W_{com} + W_{pump} + W_{fan} + W_{ele}, \quad (1)$$

where $W_{com}, W_{pump}, W_{fan}$, and W_{ele} represent the energy consumption of the compressor, water pump, fan, and electric controller, respectively. In this system, W_{pump} is 60 W and W_{ele} is 10 W. W_{fan} is 0 W in solar mode and 100 W in dual and air modes.

For the compressor, the refrigerant gas temperatures (T_d, T_s) and pressures (P_d, P_s) at the discharge and suction ends of the compressor, respectively, are related by the compression index γ as follows:

$$\frac{T_d}{T_s} = \left(\frac{P_d}{P_s}\right)^{\gamma - 1/\gamma}. \quad (2)$$

The rotating speed n of the compressor is influenced by the inverter frequency.

$$n = \frac{120 \times f \times (1 - s)}{p}, \quad (3)$$

where p is motor magnetic series and s is motor transfer ratio.

The mass flow rate m_r of the refrigerant is given by

$$m_r = \frac{\eta_v n V_d}{60 v_s}, \quad (4)$$

where η_v is the volumetric efficiency, V_d is the displacement volume, and v_s is the specific volume of refrigerant gas at compressor suction.

According to the first law of thermodynamics, the heat absorbed by the evaporator-collector Q_e should be equal to heat gain of the refrigerant.

$$Q_e = m_r(h_{r,out} - h_{r,in}), \quad (5)$$

where $h_{r,out}$ is refrigerant enthalpy leaving the evaporator-collector and $h_{r,in}$ is refrigerant enthalpy entering the evaporator-collector.

The theoretical power consumption W_{th} is given by

$$W_{th} = \eta_v V_d P_s \frac{n}{60} \frac{\gamma}{\gamma - 1} \left[\left(\frac{P_d}{P_s}\right)^{\gamma - 1/\gamma} - 1\right]. \quad (6)$$

In combination with the above two formulas, W_{th} could also be

$$W_{th} = \eta_v V_d P_s \frac{n}{60} \frac{\gamma}{\gamma - 1} \left[\left(\frac{T_d}{T_s}\right) - 1\right]. \quad (7)$$

The input power consumption W_{com} of the compressor is given by

$$W_{com} = \frac{W_{th}}{\eta_i \eta_m \eta_{mo}}, \quad (8)$$

where η_i, η_m, and η_{mo} are the indicated efficiency, mechanical efficiency, and motor efficiency, respectively.

The instantaneous heat exchange rate of this system is calculated by

$$Q_{sys} = c_w m_w \frac{dT_w}{dt}, \quad (9)$$

where c_w represents the specific heat capacity of water, m_w is the gross mass of water in the tank, and T_w stands for the water temperature.

The fundamental and predominant performance evaluation approach for the DX-SAHPWH system is coefficient of the performance (COP). It is defined as the ratio of heat exchange rate of the system Q_{sys} to the energy consumption

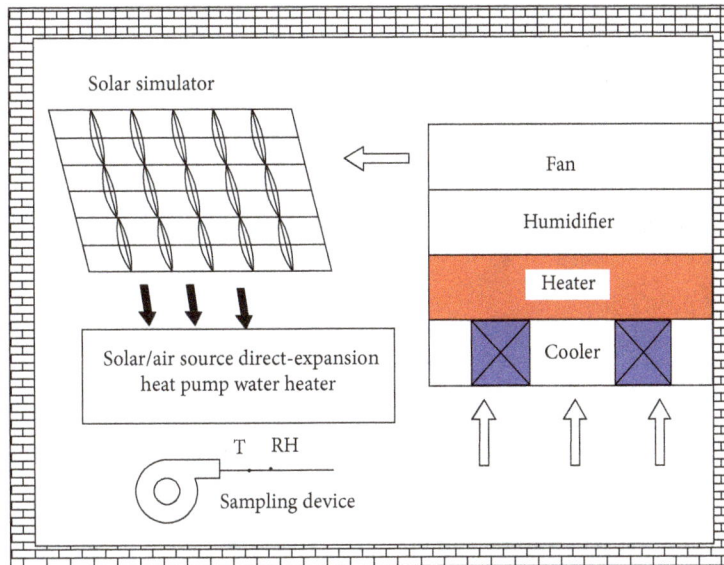

(a)

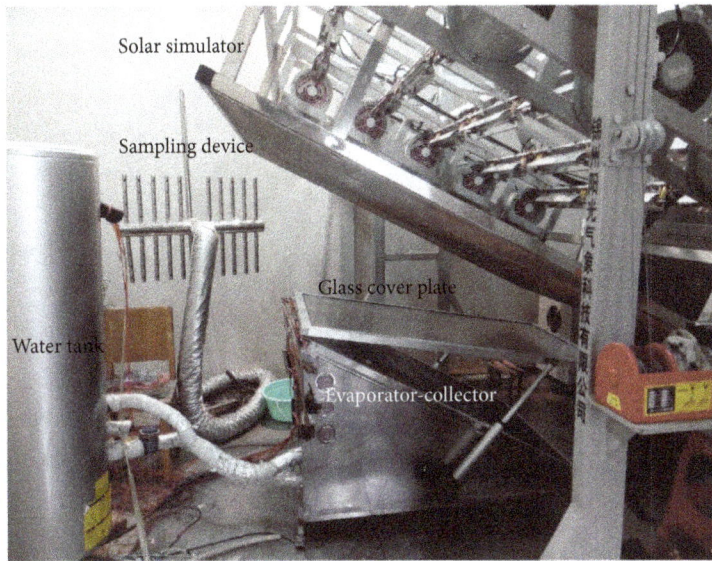

(b)

FIGURE 4: Configuration diagram of the enthalpy difference lab. (a) Schematic. (b) Photo.

TABLE 3: Characteristics of sensors and measurement instruments.

Parameter	Range	Accuracy	Measuring apparatus
Temperature (°C)	−100 to 100	±0.2	T-type thermocouple
Pressure (MPa)	0–3.04	0.3%	Huba pressure sensor
Irradiance (W/m^2)	0–2000	2%	Pyranometer
Power (kW)	0–12	1.2%	YOKOGAWA power sensor WT230

of the system W_{sys}. This description is mathematically stated as

$$\text{COP} = \frac{Q_{sys}}{W_{sys}}. \tag{10}$$

Based on (1), (9), and (10), instantaneous COP can be expressed by

$$\text{COP}_i = \frac{c_w m_w (dT_w/dt)}{W_{com} + W_{pump} + W_{fan} + W_{ele}}. \tag{11}$$

The average COP with the water temperature rising from the initial value T_{ini} to the final value T_{fin} is

$$\text{COP}_m = \frac{\int Q_{sys}}{\int W_{sys}} = \frac{c_w m_w (T_{fin} - T_{ini})}{\int (W_{com} + W_{pump} + W_{fan} + W_{ele})}. \quad (12)$$

Since the heat absorbed by the evaporator-collector includes convectional and radiant heat exchanging, Q_e can also be expressed by

$$Q_e = A(K_{ei}\Delta T_m + \alpha_e R), \quad (13)$$

where A is the heat exchanging area of the evaporator-collector. $i = 1, 2, 3$, and K_{e1}, K_{e2}, K_{e3} stands for convectional heat exchanging coefficient of the evaporator-collector in solar, dual, and air mode, respectively. According to the difference of wind speed in each mode, K_{e1}, K_{e2}, K_{e3} is different, and $K_{e1} \leq K_{e2} < K_{e3}$. ΔT_m is the mean temperature difference of convectional heat exchanging; α_e is the absorptivity of the evaporator-collector; R is the intensity of solar irradiation.

4. Experiment Results and Discussion

4.1. Effect of the Solar Irradiation on Heat Performance of the System. According to GB/T 23137-2008, the nominal working conditions of heat pump water heater were chosen as the test conditions, which was set to an ambient temperature of 20°C, an ambient relative humidity of 59%, and the water temperature rising from 15°C to 55°C. To investigate the effect of the solar irradiation on heat performance of this system, the experiments were taken under above conditions and the solar irradiation set at 500 W/m² and 900 W/m². The results are shown in Figures 5–9.

As illustrated in Figure 5, W_{com} increases significantly with the rising water temperature. The reason is that when T_w increases, T_d, T_s, P_d, and P_s increase as well with a decrease in η_m and the increment of T_d is significantly larger than the increment of T_s, so when P_s and T_d/T_s both is increased, according to (7) and (8), W_{com} will increase significantly. Besides when the solar irradiation increases from 500 W/m² to 900 W/m², P_s is found to be almost the same, so was W_{com}. And when the operating mode is set different, P_s shows no significant difference, so W_{com} is almost the same.

As shown in Figure 6, when solar irradiation increases from 500 W/m² to 900 W/m², the heat absorbed from solar irradiation by the evaporator-collector increases, so Q_e increases. Q_e decreases with the rising of the water temperature. The reason is that with the rising of the water temperature, the evaporator temperature increases, which means the mean temperature difference of convectional heat exchanging ΔT_m decreases, thus the heat absorbed from air by the evaporator decreases.

As shown in Figure 7, when the solar irradiation is increased from 500 W/m² to 900 W/m², W_{com} is almost the same and Q_e is higher, so COP_i is higher. And with the rising of the water temperature, Q_e decreases and W_{com} increased, so COP_i decrease. And there are three reasons for high COP_i when T_w is low. First, when T_w is low, P_s and T_d/T_s are small, according to (7) and (8), W_{com} is small. Second, the refrigerant temperature in the evaporator is very low when

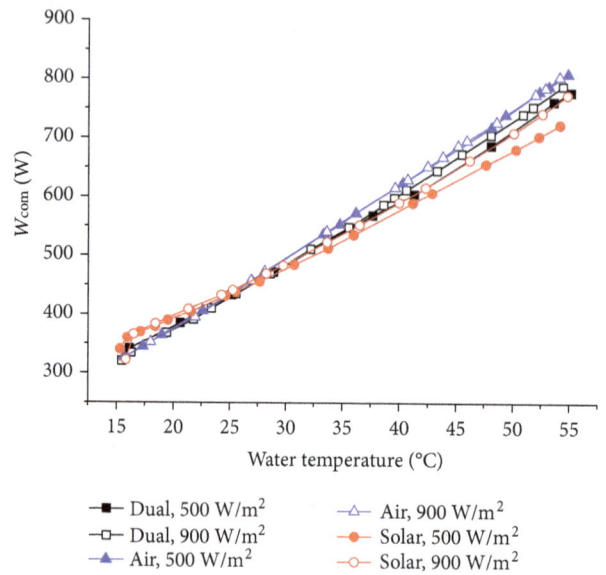

FIGURE 5: Effect of the solar irradiation on W_{com}.

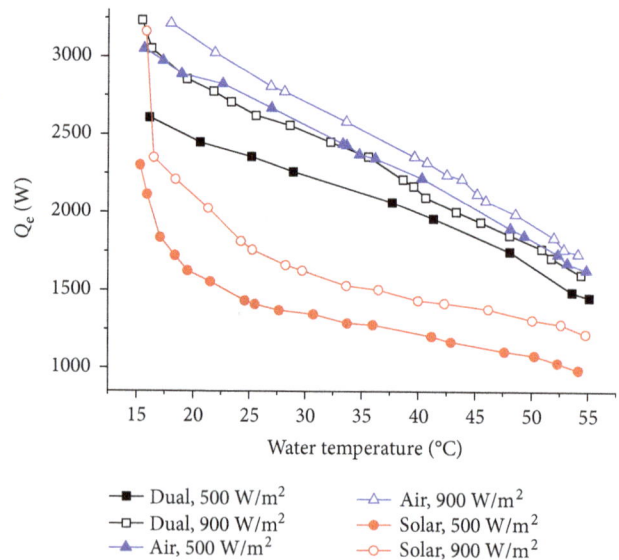

FIGURE 6: Effect of the solar irradiation on Q_e.

T_w is low, so the refrigerant could absorb more energy from air. Finally, before the system is turned on, the solar simulator needs to be adjusted for 50 min or so, and the evaporator-collector could absorb a lot of energy from the solar simulator during this period. After the system is turned on, the energy absorbed by the evaporator-collector will be transferred to the water. Therefore, when T_w are low, W_{com} is low and Q_e is very high, and so COP_i is found to be very high.

As described in Figures 8 and 9, COP_m is higher and heating time for the water temperature rising from 15°C to 55°C is less when the solar irradiation is higher, because the evaporator-collector can absorb more heat from solar irradiation under higher solar irradiation, which will speed up the heating process and increases the average COP.

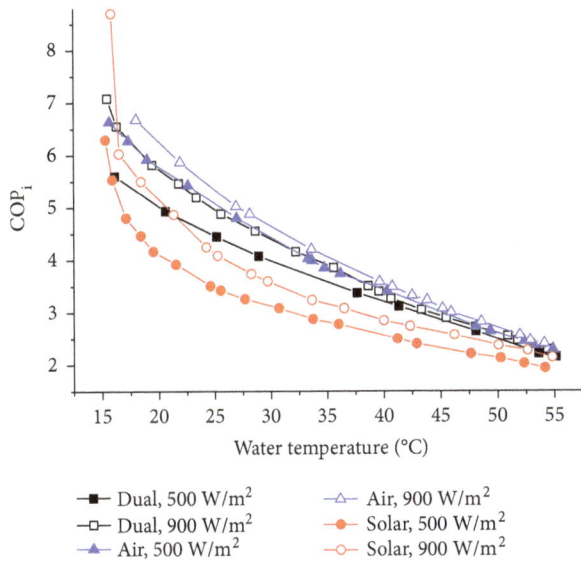

FIGURE 7: Effect of the solar irradiation on COP_i.

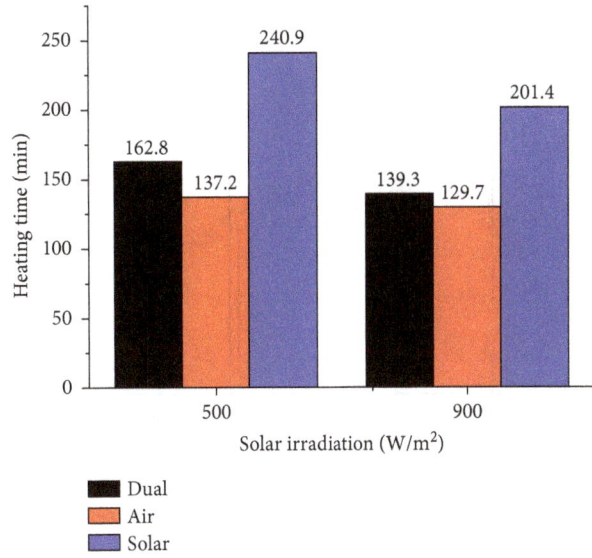

FIGURE 9: Effect of the solar irradiation on heating time.

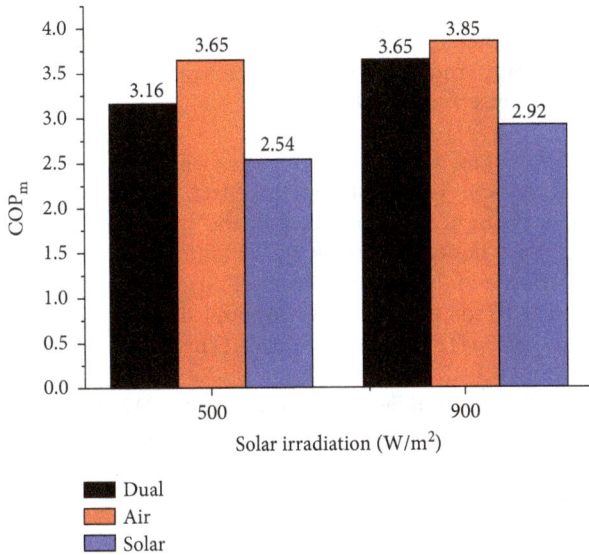

FIGURE 8: Effect of the solar irradiation on COP_m.

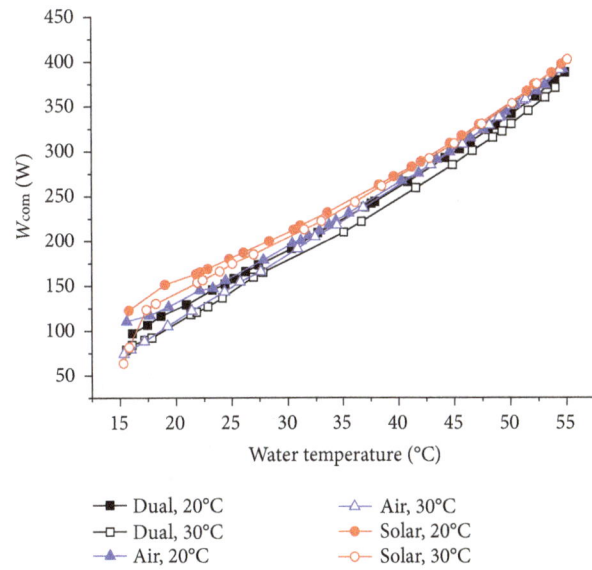

FIGURE 10: Effect of the ambient temperature on W_{com}.

When the solar irradiation increases from 500 W/m² to 900 W/m², COP_m increases by 0.20, 0.49, and 0.38 in air, dual, and solar modes, respectively. And the heating time decreases by 7.5 min, 23.5 min, and 39.5 min in air, dual, and solar modes, respectively.

4.2. Effect of the Ambient Temperature on Heat Performance of the System. To investigate the effect of the ambient temperature on heat performance of the system, experiments were done under the following conditions of water temperature rising from 15°C to 55°C, ambient relative humidity of 59%, the solar irradiation of 900 W/m²,the compressor frequency of 30 Hz, and the ambient temperature of 20°C and 30°C. Results are shown in Figures 10–12 and Table 4.

As described in Figure 10, when the ambient temperature rises from 20°C to 30°C, the increment of P_s can be neglected, so W_{com} is almost the same.

As presented in Figure 11, when the ambient temperature is higher, the mean temperature difference of convectional heat exchanging ΔT_m increases, so the energy absorbed from air by the evaporator-collector increases, which leads to a higher Q_e.

As shown in Figure 12, when the ambient temperature rises from 20°C to 30°C, W_{com} is almost the same, Q_e is higher, so COP_i is higher. It is noteworthy that after the water temperature rises to 52°C, COP_i decreases below 1.00 in air mode, which means that the performance of this system is not better than electric water heater. The reason is that the heat exchange rate of the evaporator-collector is extremely

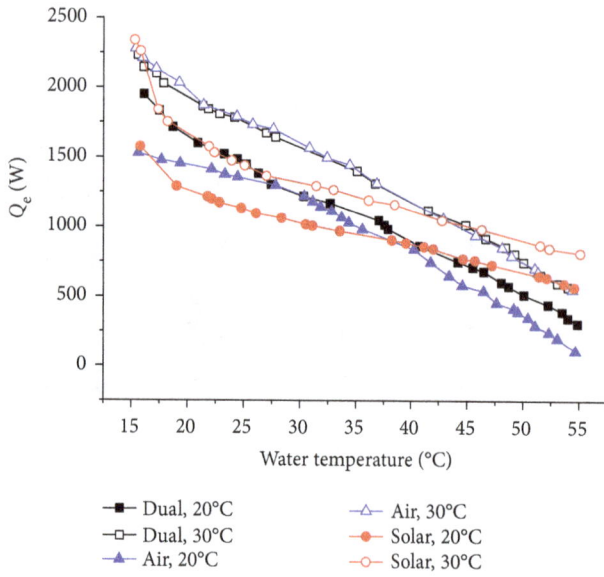

FIGURE 11: Effect of the ambient temperature on Q_e.

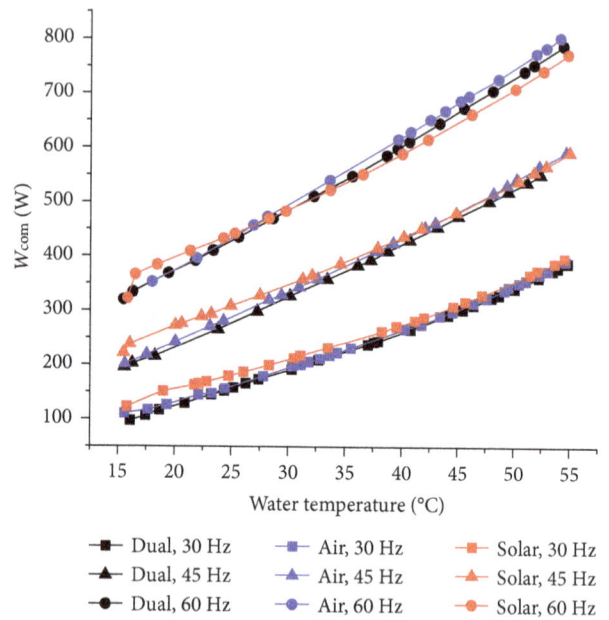

FIGURE 12: Effect of the ambient temperature on COP_i.

TABLE 4: Effect of the ambient temperature on average COP and heating time.

Ambient temperature (°C)	Average COP			Heating time (min)		
	Air	Solar	Dual	Air	Solar	Dual
20	2.23	3.35	2.65	382.2	336.6	328.8
30	3.40	4.15	3.48	260	276.3	262.2

low and the energy consumption of the compressor is too high after the water temperature rises to 52°C.

As shown in Table 4, when the ambient temperature is higher, COP_m is higher and heating time is less. Because when the ambient temperature is higher, the mean temperature difference of convectional heat exchanging ΔT_m

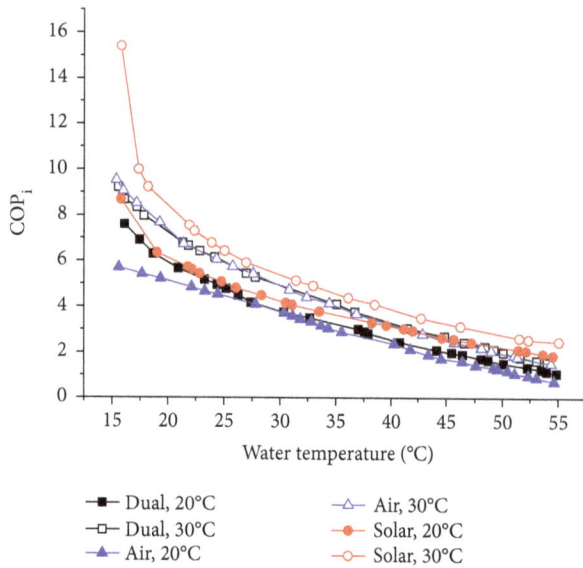

FIGURE 13: Effect of the compressor frequency on W_{com}.

increases; therefore, the evaporator-collector could absorb more energy from air, which will accelerate the heating process and raise average COP.

4.3. Effect of the Compressor Frequency on Heat Performance of the System.

To investigate the effect of the compressor frequency on heat performance of the system, experiments were done under the following conditions of an ambient temperature of 20°C, the ambient relative humidity of 59%, the solar irradiation of 900 W/m², and the compressor frequency of 30 Hz, 45 Hz, and 60 Hz. The results are shown in Figures 13–17.

As illustrated in Figure 13, according to (3) and (7), when the compressor frequency increases, the rotating speed of compressor n increases, so W_{com} increases. For the system working in different operating modes, P_s is almost the same, so W_{com} is almost same.

As presented in Figure 14, Q_e decreases with the increment of T_w and increases with the rising of the compressor frequency. According to (3), (4), and (5), when the compressor frequency f increases, the rotating speed n of the compressor increases, then the mass flow rate of the refrigerant m_r increases, so Q_e increases.

As shown in Figure 15, since the compressor frequency could both affect W_{com} and Q_e, the changing of COP_i is more complicated under different compressor frequencies. In air mode, COP_i is higher when the compressor frequency is higher. In solar mode, COP_i is higher when the compressor frequency is lower. And the difference of COP_i gradually decreased to 0 after the water temperature increases to 45°C. In dual mode, before the water temperature rises to 23°C, COP_i is higher when the compressor frequency is lower and, afterwards, COP_i is higher when the compressor frequency is higher.

As shown in Figures 16 and 17, heating time for the water temperature rising from 15°C to 55°C is less when the

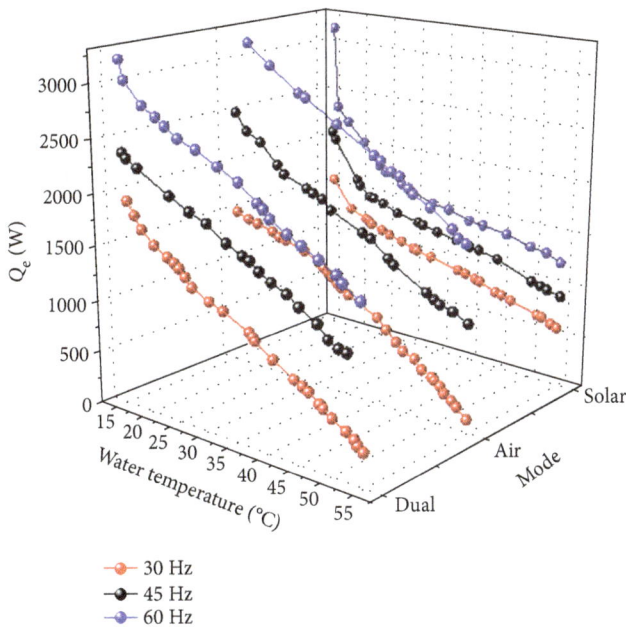

FIGURE 14: Effect of the compressor frequency on Q_e.

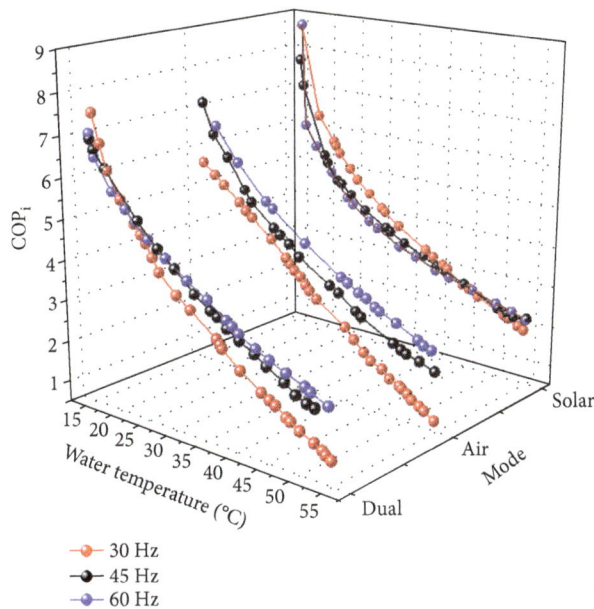

FIGURE 15: Effect of the compressor frequency on COP_i.

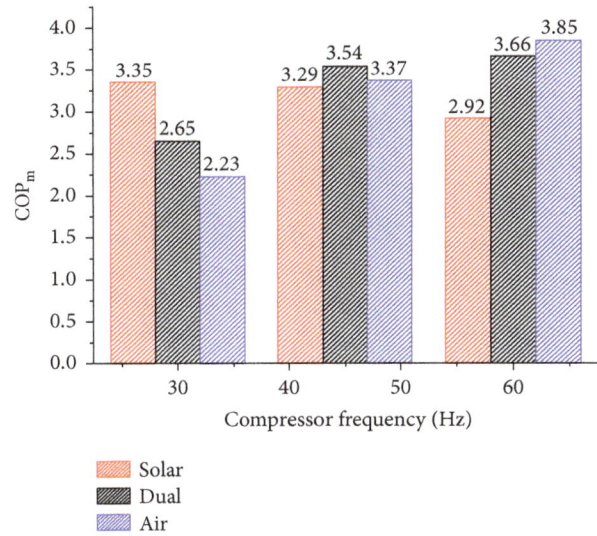

FIGURE 16: Effect of the compressor frequency on COP_m.

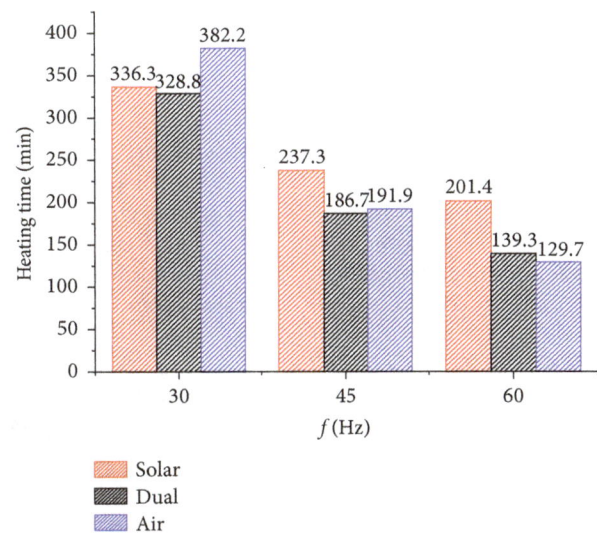

FIGURE 17: Effect of the compressor frequency on heating time.

compressor frequency is higher. Because when the compressor frequency is higher, Q_e is higher, which will speed up the heating process. In air and dual modes, COP_m is higher when the compressor frequency is higher. So it is better to set the compressor frequency at 60 Hz for a higher COP_m and less heating time under above conditions. In solar mode, COP_m is lower when the compressor frequency is higher, so it is preferable to set the compressor frequency at 30 Hz for a higher COP_m at the cost of longer heating time.

4.4. Comparison of Three Modes. Several experiments were tested in order to provide an operating strategy about the

operating mode and the compressor frequency under certain ambient conditions, results are shown in Table 5.

It is generally accepted that when the average COP is higher and heating time is less, the performance of heat pump water heater system is better. As shown in Table 5, under conditions that the solar irradiation is 900 W/m² and the ambient temperature is 20°C and 30°C, in air and dual modes, it is better for this system to be operated at compressor frequency of 60 Hz for a relatively higher average COP and less heating time. And in solar mode, it is preferable to set the compressor frequency at 30 Hz for a higher average COP at the cost of longer heating time.

Under the conditions that the solar irradiation is 900 W/m² and the ambient temperature is 30°C, the system reaches the largest average COP of 4.15 in solar mode with the compressor frequency at 30 Hz. While under above conditions, the system shows better performance in air mode with the

TABLE 5: Average COP and heating time under different ambient and operation conditions.

| Ambient conditions | | Compressor frequency | Average COP | | | Heating time (min) | | |
Ambient temperature	Solar irradiation		Air	Dual	Solar	Air	Dual	Solar
20°C	900 W/m²	30 Hz	2.23	2.65	3.35	382.2	328.8	336.3
		45 Hz	3.37	2.54	3.29	191.9	186.7	237.3
		60 Hz	3.85	3.66	2.92	129.7	139.3	201.4
30°C	900 W/m²	30 Hz	3.40	3.48	4.15	260.0	262.2	276.3
		45 Hz	4.12	3.96	3.88	147.9	167.2	200.9
		60 Hz	4.07	3.97	3.61	122.8	128.6	161.7

compressor frequency at 60 Hz, because the heating time is much less and the average COP is only a little lower than 4.15.

As shown in Figure 15, when the ambient temperature, the solar irradiation, and the water temperature are different, the corresponding operating mode and the compressor frequency for a largest instantaneous COP are different. Thus, the operating mode and the compressor frequency should be chosen according to the ambient temperature, the solar irradiation, and the water temperature.

5. Conclusion

A novel DX-SAHPWH system was fabricated and tested experimentally in the enthalpy difference lab with a solar simulator. The effects of ambient temperature and solar irradiation on the performance of this system were analyzed. And the performance of the system in three operating modes with different compressor frequencies was compared by the paper.

(1) The ambient temperature, solar irradiation, and operating mode have almost no effect on the energy consumption of the compressor. When the ambient temperature and the solar irradiation are higher, Q_e and COP_i are higher as well.

(2) With the rising of the water temperature, W_{sys} increases while Q_e and COP_i decrease. When the compressor frequency increases, W_{sys} and Q_e both increase.

(3) It is therefore recommended to operate this system in air and solar modes with the compressor frequency at 60 Hz under an ambient temperature condition of 20°C to 30°C and an average solar irradiation of 900 W/m².

(4) In order to enhance the performance of this system, the operating mode and the compressor frequency should be chosen according to the water and ambient temperature, and solar irradiation.

Nomenclature

Symbols

W: Energy consumption, W

Q: Heat exchange rate, W
c: Specific heat capacity, J/(kg°C)
m: Mass, kg
T: Temperature, °C
t: Time, min
η: Efficiency
COP_i: Instantaneous coefficient of the performance
COP_m: Average coefficient of the performance
K: Convectional heat exchanging coefficient, J/(m²°C)
$\triangle T_m$: Mean temperature difference, °C
α: Absorptivity
R: Intensity of solar irradiation, W/m²
P: Pressure
γ: Compression index
V_d: Displacement volume of compressor
f: Inverter frequency of compressor
n: Rotating speed of compressor
p: Motor magnetic series of compressor
s: Motor transfer ratio of compressor
m_r: Mass flow rate of refrigerant
$h_{r,out}$: Refrigerant enthalpy leaving the evaporator-collector
$h_{r,in}$: Refrigerant enthalpy entering the evaporator-collector
v_s: Specific volume of refrigerant gas at compressor suction.

Subscripts

sys: System
com: Compressor
pump: Water pump
fan: Fan
ele: Electric controller
w: Water
e: Evaporator
d: Discharge end of compressor
s: Suction end of compressor
th: Theoretical
v: Volumetric
i: Indicated
m: Mechanical
mo: Motor
ini: Initial
fin: Final
1: Solar mode
2: Dual mode
3: Air mode.

Conflicts of Interest

The authors declare that they have no conflicts of interest.

Acknowledgments

The work was supported by the National Science and Technology Support Program of China (no. 2015BAA02B03), Dong Guan Innovative Research Team Program (no. 2014607101008), and National Natural Science Foundation of China (no. 51378483).

References

[1] L. Yang, H. Yuan, J.-W. Peng, and C.-L. Zhang, "Performance modeling of air cycle heat pump water heater in cold climate," *Renewable Energy*, vol. 87, Part 3, pp. 1067–1075, 2016.

[2] S. Singh, M. Singh, and S. C. Kaushik, "A review on optimization techniques for sizing of solar-wind hybrid energy systems," *International Journal of Green Energy*, vol. 13, no. 15, pp. 1564–1578, 2016.

[3] Z. M. Amin and M. N. A. Hawlader, "Analysis of solar desalination system using heat pump," *Renewable Energy*, vol. 74, pp. 116–123, 2015.

[4] P. Sporn and E. R. Ambrose, "The heat pump and solar energy," in *Proceeding of the world symposium on applied solar energy, 1955 November 1–5, 1955*in Phoenix, Ariz.

[5] T. T. Chow, G. Pei, K. F. Fong, Z. Lin, A. L. S. Chan, and M. He, "Modeling and application of direct-expansion solar-assisted heat pump for water heating in subtropical Hong Kong," *Applied Energy*, vol. 87, no. 2, pp. 643–649, 2010.

[6] X. Q. Kong, D. Zhang, Y. Li, and Q. M. Yang, "Thermal performance analysis of a direct-expansion solar-assisted heat pump water heater," *Energy*, vol. 36, no. 12, pp. 6830–6838, 2011.

[7] M. Mohanraj, S. Jayaraj, and C. Muraleedharan, "Modeling of a direct expansion solar assisted heat pump using artificial neural networks," *International Journal of Green Energy*, vol. 5, no. 6, pp. 520–532, 2008.

[8] A. Moreno-Rodríguez, A. González-Gil, M. Izquierdo, and N. Garcia-Hernando, "Theoretical model and experimental validation of a direct-expansion solar assisted heat pump for domestic hot water applications," *Energy*, vol. 45, no. 1, pp. 704–715, 2012.

[9] W. Huang, J. Ji, N. Xu, and G. Li, "Frosting characteristics and heating performance of a direct-expansion solar-assisted heat pump for space heating under frosting conditions," *Applied Energy*, vol. 171, pp. 656–666, 2016.

[10] M. Mohanraj, S. Jayaraj, and C. Muraleedharan, "Exergy assessment of a direct expansion solar-assisted heat pump working with R22 and R407C/LPG mixture," *International Journal of Green Energy*, vol. 7, no. 1, pp. 65–83, 2010.

[11] X. Sun, J. Wu, Y. Dai, and R. Wang, "Experimental study on roll-bond collector/evaporator with optimized-channel used in direct expansion solar assisted heat pump water heating system," *Applied Thermal Engineering*, vol. 66, no. 1-2, pp. 571–579, 2014.

[12] L. Keliang, J. Jie, C. Tin-tai et al., "Performance study of a photovoltaic solar assisted heat pump with variable-frequency compressor – a case study in Tibet," *Renewable Energy*, vol. 34, no. 12, pp. 2680–2687, 2009.

[13] S. K. Chaturvedi, D. T. Chen, and A. Kheireddine, "Thermal performance of a variable capacity direct expansion solar-assisted heat pump," *Energy Conversion and Management*, vol. 39, no. 3-4, pp. 181–191, 1998.

[14] L. Zhao, T. J. Zhang, Q. Zhang, and G. L. Ding, "Influence of two systematic parameters on the geothermal heat pump system operation," *Renewable Energy*, vol. 28, no. 1, pp. 35–43, 2003.

[15] B. J. Huang and J. P. Chyng, "Integral-type solar-assisted heat pump water heater," *Renewable Energy*, vol. 16, no. 1–4, pp. 731–734, 1999.

[16] X. Sun, Y. Dai, V. Novakovic, J. Wu, and R. Wang, "Performance comparison of direct expansion solar-assisted heat pump and conventional air source heat pump for domestic hot water," *Energy Procedia*, vol. 70, pp. 394–401, 2015.

[17] T. Changqing, S. Wenxing, and W. Sen, "Research on two-stage compression variable frequency air source heat pump in cold regions," *Acta Energiae Solaris Sinica*, vol. 25, pp. 388–393, 2004.

[18] J. Ji, G. Pei, T.-t. Chow et al., "Performance of multi-functional domestic heat-pump system," *Applied Energy*, vol. 80, no. 3, pp. 307–326, 2005.

13

Treatment of a Textile Effluent by Electrochemical Oxidation and Coupled System Electooxidation–*Salix babylonica*

Alejandra Sánchez-Sánchez,[1] Moisés Tejocote-Pérez,[2] Rosa María Fuentes-Rivas,[3] Ivonne Linares-Hernández,[1] Verónica Martínez-Miranda,[1] and Reyna María Guadalupe Fonseca-Montes de Oca ⓘ[1]

[1]*Centro Interamericano de Recursos del Agua, Universidad Autónoma del Estado de México, Carretera Toluca-Atlacomulco, Km 14.5 Unidad San Cayetano, 50200 Toluca, MEX, Mexico*
[2]*Centro de Investigación en Ciencias Biológicas Aplicadas, Universidad Autónoma del Estado de México, Carretera Toluca-Atlacomulco, Km 14.5 Unidad San Cayetano, 50200 Toluca, MEX, Mexico*
[3]*Facultad de Geografía, Universidad Autónoma del Estado de México, Cerro de Coatepec s/n, Ciudad Universitaria, 50110 Toluca, MEX, Mexico*

Correspondence should be addressed to Reyna María Guadalupe Fonseca-Montes de Oca; mgfonsecam@uaemex.mx

Academic Editor: Carlos A Martínez-Huitle

The removal of pollutants from textile wastewater via electrochemical oxidation and a coupled system electrooxidation—*Salix babylonica*, using boron-doped diamond electrodes was evaluated. Under optimal conditions of pH 5.23 and 3.5 mA·cm^{-2} of current density, the electrochemical method yields an effective reduction of chemical oxygen demand by 41.95%, biochemical oxygen demand by 83.33%, color by 60.83%, and turbidity by 26.53% at 300 minutes of treatment. The raw and treated wastewater was characterized by infrared spectroscopy to confirm the degradation of pollutants. The wastewater was oxidized at 15-minute intervals for one hour and was placed in contact with willow plants for 15 days. The coupled system yielded a reduction of the chemical oxygen demand by 14%, color by 85%, and turbidity by 93%. The best efficiency for the coupled system was achieved at 60 minutes, at which time the plants achieved more biomass and photosynthetic pigments.

1. Introduction

The textile industry is one of the greatest generators of liquid effluent pollutants due to the high quantities of water used in the dyeing processes. The chemical composition involves a wide range of pollutants: inorganic compounds, polymers, and organic products [1–3]. Treatment of textile dye effluent is difficult and ineffective with conventional processes because many synthetic dyes are very stable in light and high temperature, and they are also nonbiodegradable. Moreover, partial oxidation or reduction can generate very toxic by-products [4–6].

Advanced oxidation processes (AOPs) have emerged as potentially powerful methods that can transform recalcitrant pollutants into harmless substances. AOPs rely on the generation of very reactive free radicals and very powerful oxidants, such as the hydroxyl radical, HO• (redox potential $= 2.8$ V) [7, 8]. These radicals react rapidly with most organic compounds, either by addition to a double bond or by the abstraction of a hydrogen atom from organic molecules [9, 10].

The resulting organic radicals, then, react with oxygen to initiate a series of degradative oxidation reactions that lead to products, such as CO_2 and H_2O [1, 11]. Electrochemical oxidation is carried out by indirect and/or direct anodic reactions in which oxygen is transferred from the solvent (water) to the product to be oxidized [12]. The main characteristic of this treatment is that it uses electrical energy as a vector for environmental decontamination [13]. During direct anodic oxidation, pollutants are initially adsorbed on the

TABLE 1: Percentage of removal performance of color, turbidity, nitrogen, and COD by electrochemical or electrochemical with other electrochemical or biological procedures.

Treatment	Sample	Optimal operating conditions	Removal performance	References
Vertical-flow constructed wetlands with planted *Phragmites australis*	Acid Blue 113 (AB113), Basic Red 46 (BR46)	7 mg·L^{-1} (AB113); 208 mg·L^{-1} (BR46) 48 h (AB113); 96 h (BR46) contact times	Nitrate nitrogen (NO$_3$-N) 85–100%	Hussein and Scholz (2018)
Electrochemical oxidation	Malachite green oxalate (MG)	pH = 3; stainless steel cathode Boron-doped diamond (BDD) anode 32 mA·cm^{-2} current density Na$_2$SO$_4$ supporting electrolyte	COD 98% (60 min) COD 91% (180 min)	Guenfoud et al. [1]
Electrocoagulation-phytoremediation (*Myriophyllum aquaticum*)	Industrial wastewater	pH = 8; Fe electrode; 45.45 A·m^{-2} current density	COD 94%, color 97% Turbidity 98%	Cano-Rodríguez et al. [35]
Electrochemical Fenton (EF) Chemical Fenton (CF)	Textile wastewater	H$_2$O$_2$ 1978 mg·L^{-1} 350 mA electrical current	COD 70.6% (EF) COD 72.9% (CF) (60 min)	Eslami et al. [36]
Electrochemical oxidation	Textile dyehouse	pH = 1; 8 mA·cm^{-2} current density Boron-doped diamond (BDD) anode 0.25 M HClO$_4$ supporting electrolyte	Color 100% Mineralization 85% (180 min)	Tsantaki et al. [37]
Electrochemical	Dyestuff effluent	Boron-doped diamond (BDD) anode; pH = 10; 60°C, 40 mA·cm^{-2} current density	COD 100%, color 100% (15 h)	Martínez-Huitle et al. [15]
Combined electrochemical, microbial, and photocatalytic	Procion blue dye	RuO$_x$-TiO$_x$ catalytic anode 2 A and 30 V power source	COD 80% to 95%	Basha et al. [16]
Electrochemical	Novacron Deep Red C-D (NDRCD) Novacron Orange C-RN (NOCRN)	pH = 3; 170 A·m^{-2} current density Graphite carbon as anode and cathode Potential: +1.0 to −0.4 V (NDRCD) and +0.5 to −0.2 V (NOCRN); 300.15 K NaCl (7 g·L^{-1}) supporting electrolyte	Color 99% (NDRCD); 97% (NOCRN) COD 88% (NDRCD); 82% (NOCRN)	Kariyajjanavar et al. [6]

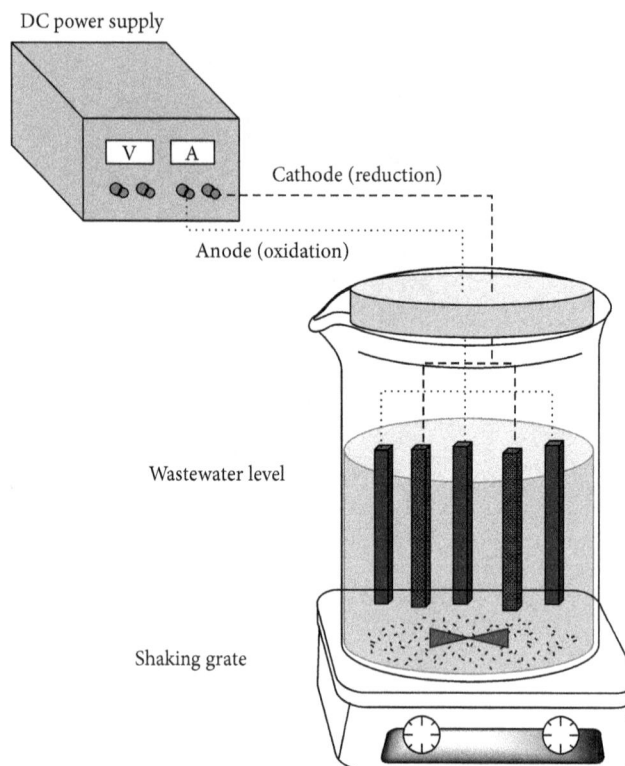

FIGURE 1: A schematic diagram of the electrochemical reactor.

surface of the anode, where the anodic electron transfer reaction degrades them [6]. In indirect anodic oxidation, strong oxidants, such as hypochlorite, chlorine, ozone, or hydrogen peroxide, are electrochemically generated.

The pollutants are degraded via the oxidation reactions with these strong oxidants [11]. Boron-doped diamond (BDD) thin films are electrode materials that possess several technologically important characteristics, including an inert surface with low adsorption properties, an acceptable conductivity, and remarkable corrosion stability even in strongly acidic media and extremely high O_2 evolution overvoltage [14, 15].

On the other hand, biotechnology continues to be used to solve environmental problems [16–18]. Phytoremediation (PR) is a green technology that uses plant systems for the remediation and restoration of contaminated sites [19]. PR's advantages are solar energy dependence and an esthetically pleasant method of treatment [20]. Plants have inbuilt enzymatic characteristics that are capable of degrading complex structures, and they can be used for cleaning contaminated sites [17].

Plants, however, remove pollutants predominantly via adsorption, accumulation, and subsequent enzyme-mediated degradation [20]. Therefore, plants are considered organisms with complex metabolic activity when referring to the assimilation of toxic substances. Plant species that have different growth forms have been proposed for the treatment of textile effluents, for instance, *Glandularia pulchella*, *Phragmites australis*, *Tagetes patula*, *Alternanthera philoxeroides*, *Eichhornia crassipes*, *Nasturtium officinale*,

TABLE 2: Experimental design of the electrooxidation process.

Experiment	pH	Current density
1	5.23	3.5
2	5.23	7
3	5.23	10
4	7	3.5
5	7	7
6	7	10
7	10	3.5
8	10	7
9	10	10

Hydrocotyle vulgaris, *Petunia grandiflora*, and *Gaillardia grandiflora* [17, 18, 21–24].

Another alternative is to use species of fast-growing woody plants with high biomass production and high genetic variability [25–27]. Trees from the Salicaceae family with the genera *Salix* and *Populus* are suitable candidates for this purpose [28–30]. Willows (*Salix* spp.) have several characteristics that make them ideal plant species for PR application, including easy propagation and cultivation, a large amount of biomass, a deep root system, a high transpiration rate, tolerance to hypoxic conditions, and high metal accumulation capability [30, 31].

Salix babylonica has been used to solve the problems associated with aquifers contaminated with ethanol-blended gasoline [32] and studies of the biotransformation and metabolic response of cyanide and dieldrin [33, 34]. In recent years, several authors have described dye removal by electrochemical (EC) or coupled electrochemical with other chemical, electrochemical, or biological procedures (Table 1). However, no studies have been conducted on the implementation of coupled electrochemical oxidation-phytoremediation with weeping willow in the remediation of textile effluents, and because it is an introduced species, is noninvasive, and is widely distributed in Mexico, the aim of this study was to evaluate the removal of the pollutants of a textile effluent using an electrochemical oxidation process and to compare their performance with *Salix babylonica*.

2. Materials and Methods

2.1. Wastewater Sampling. A textile wastewater sample was collected from a textile industry whose business is the dyeing and washing of denim garments in Almoloya del Río, State of Mexico, Mexico. The wastewater that this industry discharges does not receive any treatment and is discharged into the sewage system, so it is necessary to give it some kind of treatment to improve its quality. The textile wastewater sample was placed in plastic containers and transported to the laboratory, where it was refrigerated at 4°C for analysis and for conducting the electrochemical oxidation and coupled system electrooxidation-*Salix babylonica*.

2.2. Electrochemical Reactor. In this study, a batch electrochemical reactor was used. The reactor contained five vertical parallel electrodes of BDD (titanium/BDD) that CONDIAS DIACHEM manufactured, two as cathodes and three as anodes. Each electrode was 20.5 cm long and 2.5 cm wide, resulting in an area of 102.5 cm² for each electrode and a total anodic area of 307.5 cm². A schematic diagram of the electrochemical reactor is shown in Figure 1. The tests were carried out in a 1 L cylindrical reactor. The reactor was operated at different pH values (5.23, 7, and 10). A current density power supply provided 1, 2, and 3 A and 5–6.75 V, corresponding to a current density of 3.5, 7, and 10 mA·cm⁻².

The experiment design used included the two factors of pH and current density. The levels of each of the factors are listed in Table 2. Different aliquots were taken, and the chemical oxygen demand (COD), biochemical oxygen demand (BOD₅), color, turbidity, and conductivity were analyzed. The boron-doped diamond electrodes (BDD) were cleaned for one hour in Na_2SO_4 (0.03 M) after each experiment to remove adsorbed molecules at the electrode surface, and then they were rinsed with distilled water.

2.3. Coupled System with Salix babylonica Treatment. For *Salix babylonica* treatment, secondary branches of weeping willows located in five regions near the discharge site were collected based on some defined phenotypic characteristics: intense green color, wide coverage, height greater than 8 meters, absence of pests, and straight shaft. Branch cuttings of 20 cm in length were placed in hydroponics [28] in 1 L containers with 300 mL of distilled water. They

TABLE 3: Physicochemical characterization of textile wastewater.

Parameter	Raw wastewater
pH	5.23
Acidity (mg/L CaCO₃)	962.8
Alkalinity (mg/L CaCO₃)	1000
BOD (mg/L)	1400
BOD/COD	0.7
COD (mg/L)	2022
Color (Pt-Co U)	3000
Chlorides (mg/L Cl⁻)	843.71
EC (mS/cm)	2.811
Hardness (mg/L CaCO₃)	546.2
N-NO₂ (mg/L)	0.848
N-NO₃ (mg/L)	17.28
N-NH₃ (mg/L)	4.72
Phosphorus (mg/L P)	715.1
Sulfates (mg/L SO₄²⁻)	429.5
Turbidity (NTU)	735
TOC (mg/L)	1396.6
TDS (mg/L)	1367
Ca²⁺ (mg/L)	36.537
K⁺ (mg/L)	64.43
Mg²⁺ (mg/L)	17.346
Na⁺ (mg/L)	392.79

were kept at room temperature (19–22°C) for a normal photoperiod (12 h light, 12 h dark). Ten willows per region were placed in jars of 1 L and were then placed in 500 mL of textile wastewater. They remained in contact with wastewater for 15 days, and water aliquots were taken at baseline and at intervals of eight days. Likewise, the development of plants during those time periods was assessed.

3. Methods of Analysis

3.1. Physicochemical Characterization. The characterization of textile wastewater was performed. During both electrochemical and phytoremediation treatment, COD, BOD₅, color, turbidity, pH, and electrolytic conductivity analyses were performed as indicated in the standard methods procedures by the American Public Health Association [38]. In addition, infrared spectroscopy of the raw and treated water was performed.

3.2. Biological Parameters. Once roots and leaves were developed in hydroponics, they were weighed on an analytical balance (BEL Engineering), and the lengths of the plants and roots were measured using a vernier. The numbers of roots and leaves were counted, and the leaf areas and photosynthetic pigments were measured by using the method that Val et al. [39] and Moisés et al. [40] established. These measurements were performed at the beginning of biological treatment and every eight days.

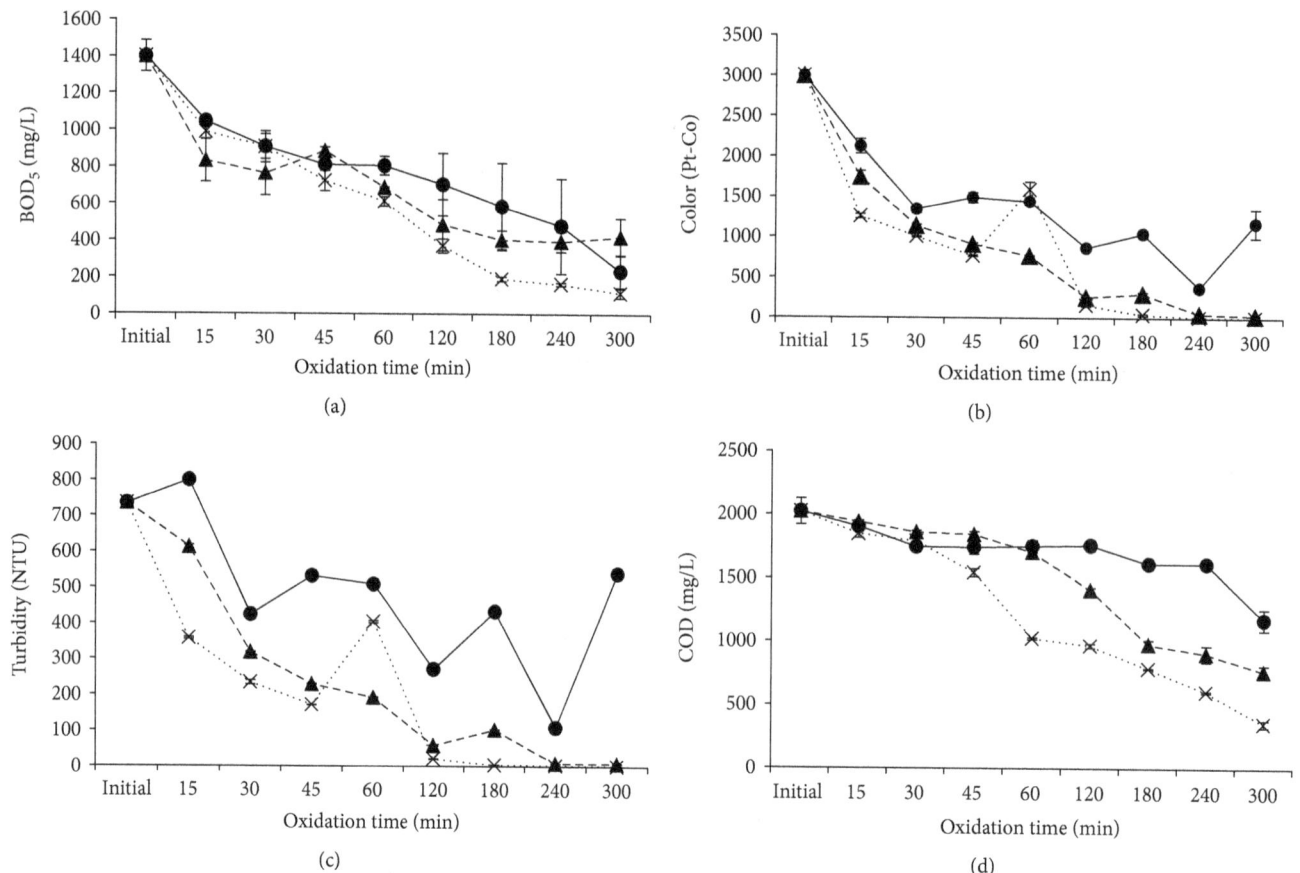

FIGURE 2: Behavior of (a) BOD$_5$, (b) color, (c) turbidity, and (d) COD. Applying three current densities: 3.5 mA·cm^{-2} (●), 7 mA·cm^{-2} (▲), and 10 mA·cm^{-2} (X), at pH 5.23.

4. Results and Discussion

4.1. Wastewater Characterization. Physicochemical chacarcterization of textile wastewater is show in Table 3. The organic parameters indicate for the BOD$_5$ a value of 1400 mg/L. According to Mexican regulation, the allowed limit for discharging wastewater into rivers is 150 mg/L. The COD was 2022 mg/L; in this situation, the BOD/COD ratio (0.7) indicates good biodegradability [9]. The TOC was 1396.6 mg/L, and the color was 3000 Pt-Co U; this high level of color stemmed from the indigo blue dye in the textile effluent. Regarding inorganic matter, different ions contribute to high conductivity (2.811 mS/cm). This parameter could be beneficial to the electrooxidation process because it was not necessary to add any support electrolyte. However, the presence of ions as nitrates, phosphates, and alkalinity could reduce the oxidation speed of organic compounds; on the other hand, chlorides (843.71 mg/L) could improve the indirect organic oxidation.

4.2. Electrooxidation Treatment

4.2.1. Current Density Effect. An important operating variable of the electrochemical process is the current density, which is the input current divided by the surface area of

TABLE 4: Removal efficiencies of different parameters in the electrooxidation process.

Experiment	pH	Current densities	BOD (%)	Color (%)	Turbidity (%)	COD (%)
1	5.23	3.5	83.33	60.83	26.53	41.95
2	5.23	7	69.95	98.88	98.95	62.01
3	5.23	10	91.81	99.81	99.91	82.39
4	7	3.5	58.34	96.95	98.77	47.73
5	7	7	91.15	99.66	99.77	63.14
6	7	10	91.58	99.82	99.90	94.66
7	10	3.5	58.56	92.45	98.09	82.10
8	10	7	94.73	99.01	99.04	98.11
9	10	10	94.38	97.04	93.87	95.03

the electrode [11]. From other variables effective in the electrochemical process is current density as the rate of electrochemical reactions is controlled by this parameter. Further, the performance of electrodes is highly dependent on this parameter. Three different current densities were applied (3.5, 7, and 10 mA·cm^{-2}) to investigate the effect in the oxidation process. All experiments were carried out at pH 5.23 (sample pH), and in all cases, a direct effect of the current densities was observed: If the current increases,

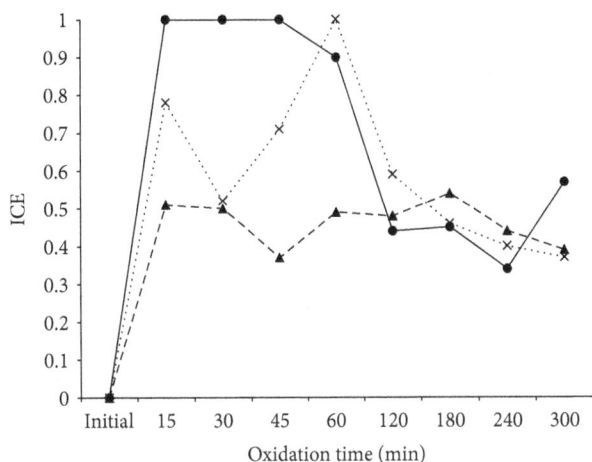

FIGURE 3: Instantaneous current efficiency (ICE) for the anodic oxidation process: 3.5 mA/cm^2 (●), 7.0 mA/cm^2 (▲), and 10 mA/cm^2 (X).

the removal efficiency increases. This could be due to the increased rate of the generation of oxidants, such as hydroxyl radicals and chlorine/hypochlorite at higher current densities [6]. The results at different densities are shown in Figure 2. The best removal efficiency was when 10 mA·cm^{-2} was applied. BOD_5 was reduced considerably from 1400 mg/L to 114 mg/L with 92% of efficiency; COD was 2022 mg/L and was reduced to 356.05 mg/L with 82% of removal efficiency; color was reduced from 3000 Pt-Co U to 5.5 Pt-Co U (99.8% removal efficiency); and initial turbidity was 735 NTU and at the end of the process was 0.65 NTU, achieving 99.99% of removal efficiency. Efficiency was measured during 300 min of treatment time (Table 4).

The instantaneous current efficiency (ICE) for the anodic oxidation was calculated from the values of COD using

$$ICE = FV \frac{COD_i - COD_t}{8I\Delta t}, \quad (1)$$

where F is the Faraday constant (96487 C/mol), V is the volume (L), COD_i and COD_t are the chemical oxygen demand (g/L) at initial time and time t, I is the applied current (A), Δt is the treatment time (s), and 8 is the equivalent mass of oxygen (g·eq^{-1}). The instantaneous current efficiency (ICE) decreased during the electrolysis as wastewater was oxidized. This behavior is shown in Figure 3.

The best ICE percentage was when the lowest current was applied 1 A (3.5 mA·cm^{-2}) in the middle stage of electrooxidation (15–45 min). This may be attributed to the presence of a higher concentration of organics near the electrodes. This indicates that the electrooxidation was under the current control regime at least in the middle stage of electrooxidation. The ICE decreased after 60 min of the electrooxidation process. This may be due to the depletion of the concentration of organics on the electrode surface.

The energy consumption per volume of treated effluent was estimated and expressed in kWh·m^{-3}. The average cell voltage during the electrolysis (cell voltage is reasonably

constant with just some minor oscillations, and for this reason, the average cell voltage was calculated) was measured to calculate the energy consumption by using [15]

$$\text{Energy consumption} = \frac{\Delta E_c \times I \times t}{1000 \times V}, \quad (2)$$

where t is the time of electrolysis (h); ΔE_c (V) and I (A) are the average cell voltage and the electrolysis current, respectively; and V is the sample volume (m^3). According to the results, 5.87 kWh·m^{-3} is required to oxidize the pollutants in the textile wastewater. In another study, a real textile effluent was treated using a BDD anode, applying a current density of 20 mA·cm^{-2}. The energy consumption was 20 kWh·m^{-3} [15].

The specific energy consumption (Ec) in kWh·(kg COD)$^{-1}$ removed was determined according to [37]

$$Ec = \frac{UIt/60}{(COD_0 - COD)V}, \quad (3)$$

where U is the mean applied voltage (V), I is the current (A), t is the treatment time (min), V is the liquid volume (L), and COD_0 and COD are the COD values (g O_2 L^{-1}) at times 0 and t. The results showed that 21.87 kWh·(kg COD)$^{-1}$ was required in the electrooxidation process. In a previous work, 95 kWh·(kg COD)$^{-1}$ was applied for the same COD removal [37].

4.2.2. pH Effect. The studies were performed at three different initial pH values (5.23, 7, and 10) to investigate their effects as depicted in Figure 4. The current density applied in these experiments was 3.5 mA·cm^{-2}. At alkaline pH (10), the best efficiencies were achieved: COD (85.9%), color (99.6%), BOD_5 (70.3%), and turbidity (99.7%). However, an addition of NaOH was required to adjust the pH, and this could be a disadvantage in the oxidation process.

The pH solution was an important factor for wastewater treatment. In anodic oxidation, many reports exist on the influence of pH solution, but the results are diverse and even contradictory due to different organic structures and electrode materials [1]. In an acidic solution, the degradation process of azo dyes is higher than in a basic solution, as in acidic solutions, chlorides are reduced to free chlorine, which is a dominant oxidizing agent [6]. During all experiments, the initial pH decreased during the treatment time (2.3–2.75). This could be attributed to the fragmentation of organic matter into carboxylic acids, carbonic acid, and ions as by-products of mineralization. Figure 5 shows the behavior of the conductivity during the treatment time; it increased at the end of the process probably as a result of the mineralization in the electrochemical oxidation process.

4.2.3. Degradation Mechanism. Previous research studies [41, 42] indicated that the oxidation of organics with concomitant oxygen evolution assumes that both organic oxidation and oxygen evolution take place on a BDD anode surface via the intermediation of hydroxyl radicals generated from the reaction with water shown in

$$BDD + H_2O \rightarrow BDD(OH\bullet) + H^+ + e^- \quad (4)$$

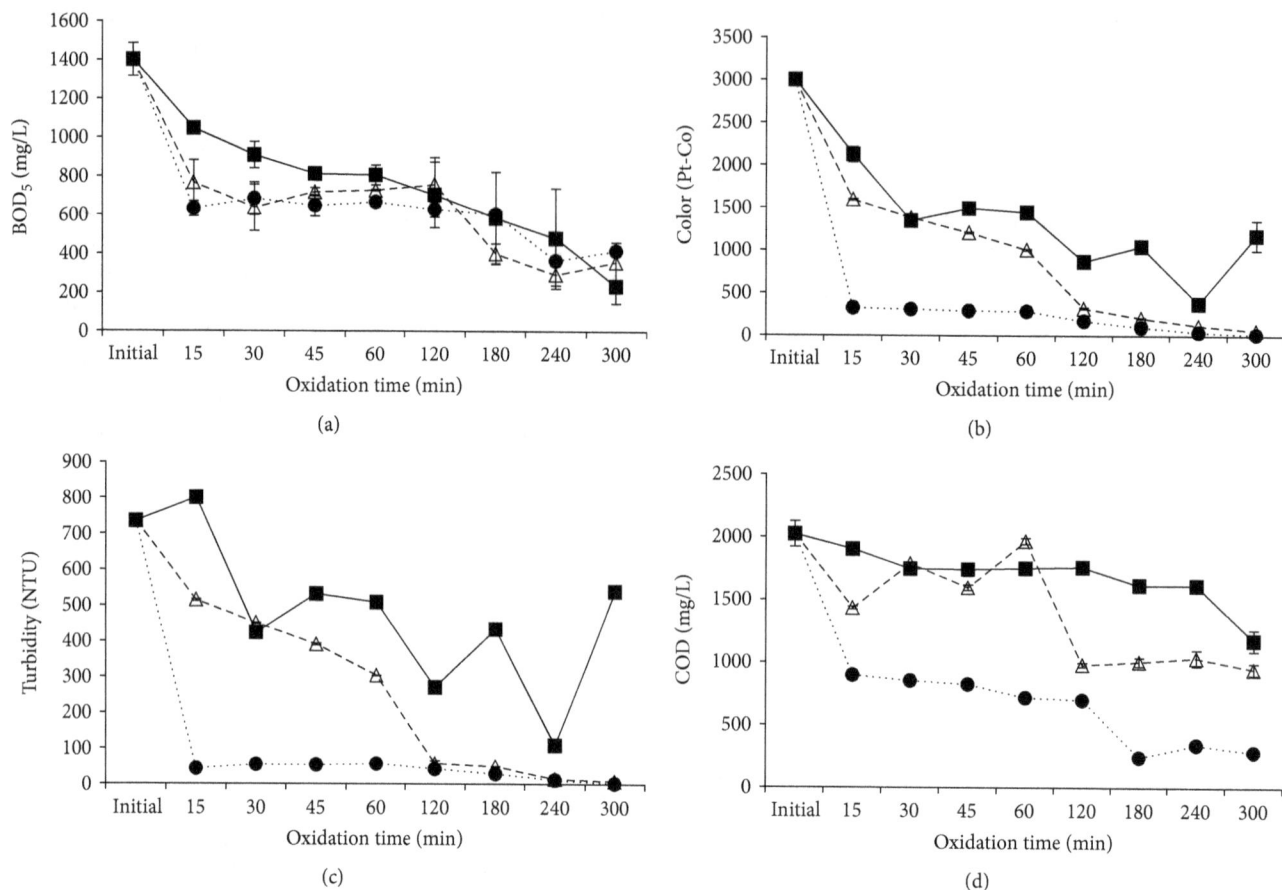

FIGURE 4: Behavior of (a) BOD$_5$, (b) color, (c) turbidity, and (d) COD. Applying 3.5 mA·cm^{-2}, at pH 5.23 (■), 7 (Δ), and 10 (●).

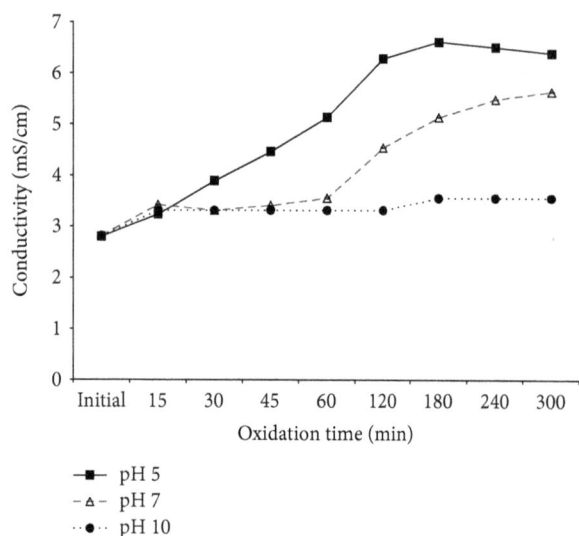

FIGURE 5: Conductivity behavior during the treatment time at three different pH values.

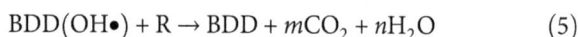

$$BDD(OH\bullet) + R \rightarrow BDD + mCO_2 + nH_2O \qquad (5)$$

Reaction (4) is in competition with the side reaction of hydroxyl radical conversion to O_2 without any participation of the anode surface as indicated in

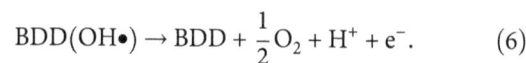

$$BDD(OH\bullet) \rightarrow BDD + \frac{1}{2}O_2 + H^+ + e^-. \qquad (6)$$

Textile wastewater was analyzed via infrared spectroscopy before and after the electrochemical oxidation process, and the spectra are shown in Figure 6. The principal functional groups found in the aqueous solution of dye were –NH– (3305 cm^{-1}), the C–H aromatic bond (2910 and 2845 cm^{-1}), –NH$_3^+$ (2340 cm^{-1}), aromatic –C=C– (1614 cm^{-1}), sulfoxides (1101 and 1022 cm^{-1}), and C–CO–C in ketones (611 cm^{-1}). The spectra of oxidized water showed that the intensity of corresponding bands to sulfoxides and secondary amines diminished after treatment, whereas the bands of R–COOH and O–C=O increased. In accordance with the above, the proposed dye degradation mechanism is shown in Figure 6.

4.3. Phytoremediation with Salix babylonica

4.3.1. Textile Wastewater. After oxidation treatment, the oxidized water was placed in contact with plants for 8 and 15 days, as shown in Figure 7. Parameters of the COD, color, and turbidity were minimally reduced. However, at 15 days of contact time, a visible reduction in color and turbidity was noted. According to the results, the plants assimilated better with the pollutants in the raw water than in the oxidized water due to the structural changes that the compounds suffered with the electrooxidation treatment. The coupled system (electrooxidation + phytoremediation)

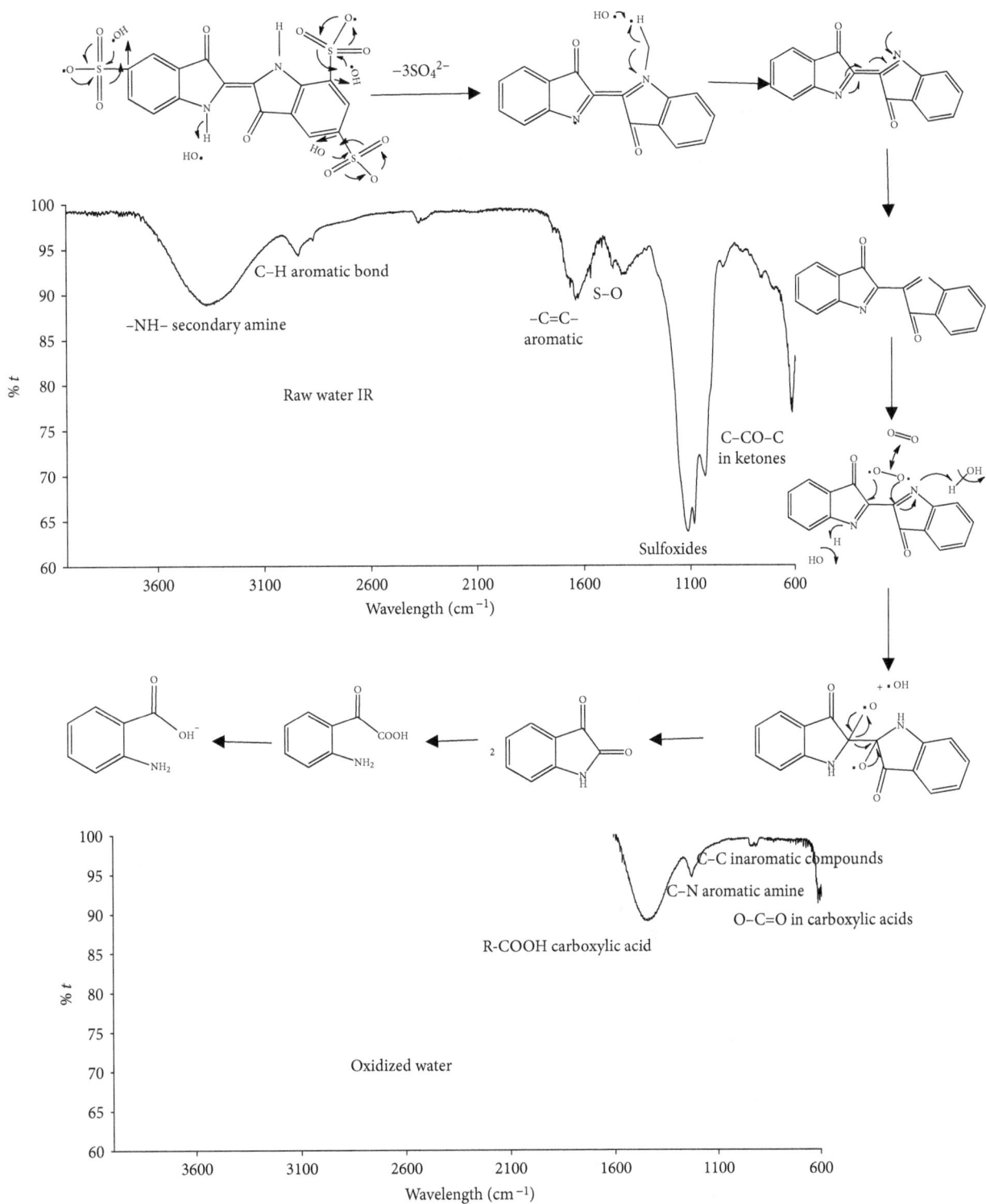

FIGURE 6: Diagram of the degradation mechanism proposed. Within the figure are shown the infrared spectra of raw water and oxidized water.

yielded a reduction of the COD by 14%, color by 85%, and turbidity by 93%.

4.3.2. Salix babylonica Biomass. The willow biomass tolerance was analyzed by using Minitab 15.1.20 statistic program analysis of variance (ANOVA) to find significant differences between treatments. As shown in Figure 8(a), significant differences were found in the leaf numbers, leaf areas, and root numbers among the plants that were in contact with oxidized water for different amounts of time ($P < 0.05$; $F = 8.20$).

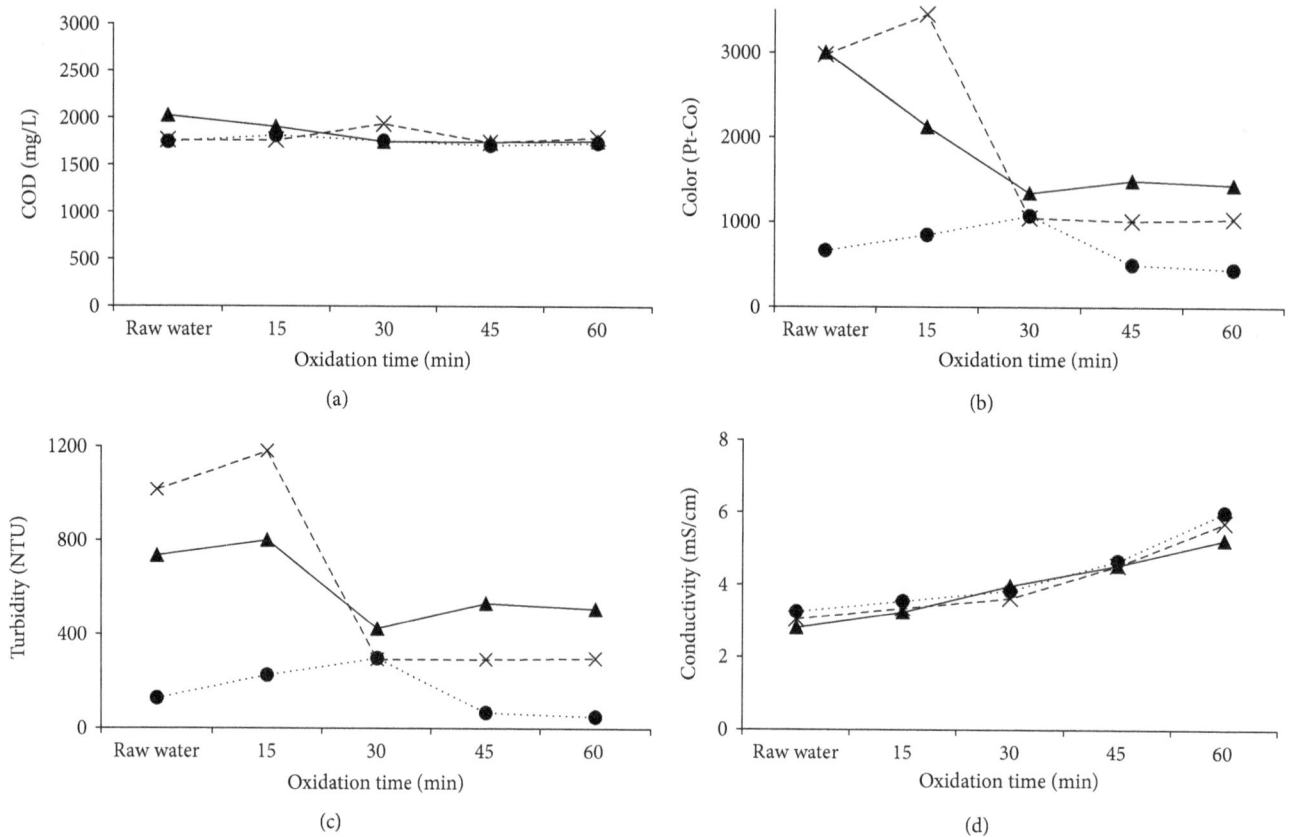

FIGURE 7: Behavior of (a) COD, (b) color, (c) turbidity, and (d) conductivity, with *Salix babylonica* contact at initial time (▲), 8 days (×), and 15 days (●).

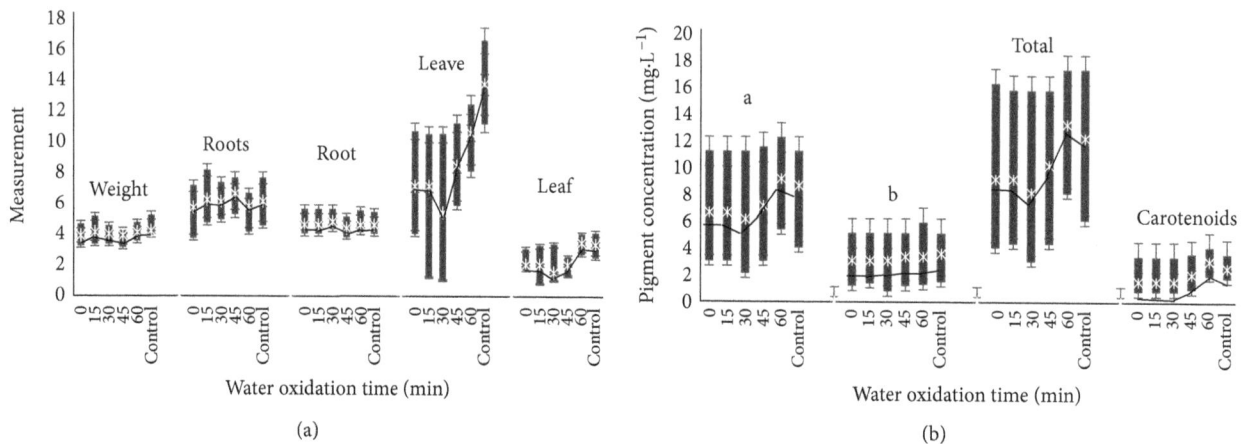

FIGURE 8: Willow tolerance (a) biomass parameters: weight (g), root number, root length (cm), leaf number, and leaf area (cm^2); (b) pigment concentration (mg·L^{-1}) at different oxidized water contact times.

Willow plants in contact with oxidized water for 60 minutes reached a biomass close to that of the control plants. The same behavior was observed in the pigment concentration as shown in Figure 8(b). Willow plants tend to lose leaves in a stressful environment, but the root system and photosynthetic metabolism remain.

With respect to the contact time, willow plants reduced their photosynthetic metabolism and lost leaves at eight days

of contact time, but after this time, such plants recovered their photosynthetic metabolism to some extent as shown in Figure 9. This could be because willow plants became adapted to the new environmental conditions. The mechanism by which *Salix babylonica* decreases color and pollutant concentration is unknown, but an increase in the concentration of chlorophylls indicates that the plant is photosynthesizing and thus absorbing nutrients from wastewater.

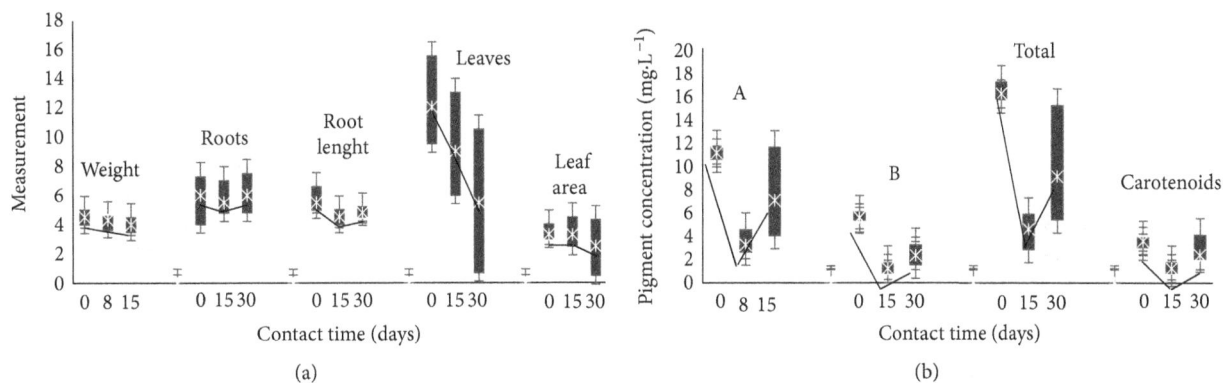

FIGURE 9: Willow tolerance (a) biomass parameters: weight (g), root number, root length (cm), leaf number, and leaf area (cm^2); (b) pigment concentration (mg·L^{-1}) at different contact times.

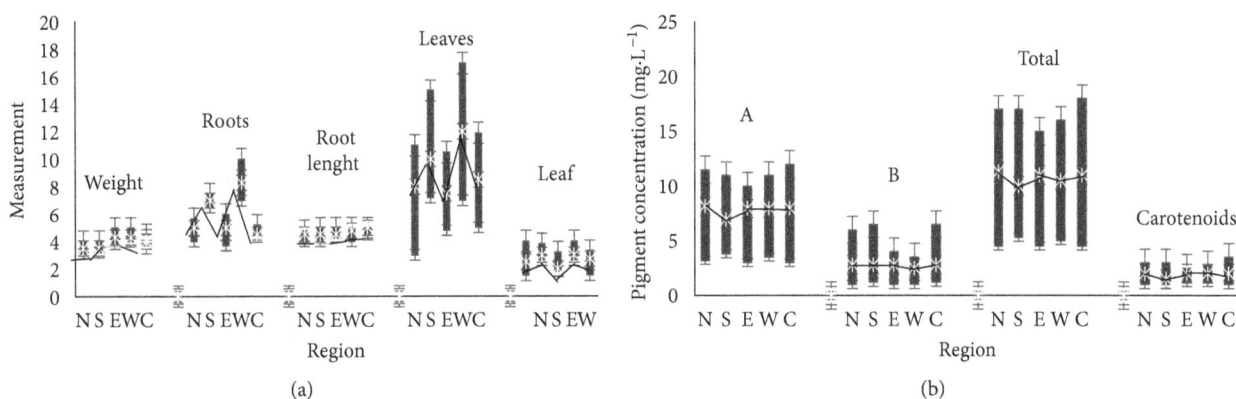

FIGURE 10: Willow tolerance (a) biomass parameters: weight (g), root number, root length (cm), leaf number, and leaf area (cm^2); (b) pigment concentration (mg·L^{-1}) from different regions: north (N), south (S), east (E), west (W), and center (C).

Furthermore, the adsorption of contaminants in plant roots has been documented [20]. The results may indicate that the willow phenotype of the western region is characterized by very dense foliage and root system, by genetics, or by environmental influence, but in terms of photosynthetic metabolism, they are the same as the willows of the other regions (Figure 10).

5. Conclusions

Textile wastewater composition was favorable for carrying out electrochemical oxidation due to the high salt content. All experiments were carried out at the original pH (5.23), and it was determined that if the current density was increased, the removal efficiency increased. However, the current efficiency decreased during this process. For this reason, the lower current density was chosen (3.5 mA/cm^{-2}) as optimal. The infrared spectroscopy of the wastewater before and after electrooxidation showed a degradation of dye. The proposed degradation mechanism showed carboxylic acids and sulfates as degradation products. In the coupled system, a reduction of the COD was decreased by 14%, color by 85%, and turbidity by 93%. The biomass and pigment of

willow *Salix babylonica* demonstrated that this species has the ability to adapt to adverse conditions very quickly.

Conflicts of Interest

The authors declare that they have no conflicts of interest.

Acknowledgments

The authors thank Consejo Nacional de Ciencia y Tecnología Project 219743 and the scholarship 622274 for support during the development of this work.

References

[1] F. Guenfoud, M. Mokhtari, and H. Akrout, "Electrochemical degradation of malachite green with BDD electrodes: effect of electrochemical parameters," *Diamond and Related Materials*, vol. 46, pp. 8–14, 2014.

[2] T. Robinson, G. McMullan, R. Marchant, and P. Nigam, "Remediation of dyes in textile effluent: a critical review on current treatment technologies with a proposed alternative," *Bioresource Technology*, vol. 77, no. 3, pp. 247–255, 2001.

[3] R. G. Saratale, G. D. Saratale, J. S. Chang, and S. P. Govindwar, "Bacterial decolorization and degradation of azo dyes: a review," *Journal of the Taiwan Institute of Chemical Engineers*, vol. 42, no. 1, pp. 138–157, 2011.

[4] I. Bazin, A. Ibn Hadj Hassine, Y. Haj Hamouda et al., "Estrogenic and anti-estrogenic activity of 23 commercial textile dyes," *Ecotoxicology and Environmental Safety*, vol. 85, pp. 131–136, 2012.

[5] V. Buscio, M. Crespi, and C. Gutiérrez-Bouzán, "Sustainable dyeing of denim using indigo dye recovered with polyvinylidene difluoride ultrafiltration membranes," *Journal of Cleaner Production*, vol. 91, pp. 201–207, 2015.

[6] P. Kariyajjanavar, N. Jogttappa, and Y. A. Nayaka, "Studies on degradation of reactive textile dyes solution by electrochemical method," *Journal of Hazardous Materials*, vol. 190, no. 1-3, pp. 952–961, 2011.

[7] A. Asghar, A. A. Abdul Raman, and W. M. A. Wan Daud, "Advanced oxidation processes for *in-situ* production of hydrogen peroxide/hydroxyl radical for textile wastewater treatment: a review," *Journal of Cleaner Production*, vol. 87, pp. 826–838, 2015.

[8] S. Hussain, S. Shaikh, and M. Farooqui, "COD reduction of waste water streams of active pharmaceutical ingredient – atenolol manufacturing unit by advanced oxidation-Fenton process," *Journal of Saudi Chemical Society*, vol. 17, no. 2, pp. 199–202, 2013.

[9] J. L. De Morais and P. P. Zamora, "Use of advanced oxidation processes to improve the biodegradability of mature landfill leachates," *Journal of Hazardous Materials*, vol. 123, no. 1–3, pp. 181–186, 2005.

[10] A. R. Ribeiro, O. C. Nunes, M. F. R. Pereira, and A. M. T. Silva, "An overview on the advanced oxidation processes applied for the treatment of water pollutants defined in the recently launched Directive 2013/39/EU," *Environment International*, vol. 75, pp. 33–51, 2015.

[11] P. Asaithambi and M. Matheswaran, "Electrochemical treatment of simulated sugar industrial effluent: optimization and modeling using a response surface methodology," *Arabian Journal of Chemistry*, vol. 9, Supplement 2, pp. S981–S987, 2016.

[12] C. A. Martínez-Huitle and E. Brillas, "Decontamination of wastewaters containing synthetic organic dyes by electrochemical methods: a general review," *Applied Catalysis B: Environmental*, vol. 87, no. 3-4, pp. 105–145, 2009.

[13] S. A. Alves, T. C. R. Ferreira, N. S. Sabatini et al., "A comparative study of the electrochemical oxidation of the herbicide tebuthiuron using boron-doped diamond electrodes," *Chemosphere*, vol. 88, no. 2, pp. 155–160, 2012.

[14] E. Butrón, M. E. Juárez, M. Solis, M. Teutli, I. González, and J. L. Nava, "Electrochemical incineration of indigo textile dye in filter-press-type FM01-LC electrochemical cell using BDD electrodes," *Electrochimica Acta*, vol. 52, no. 24, pp. 6888–6894, 2007.

[15] C. A. Martínez-Huitle, E. V. Dos Santos, D. M. De Araújo, and M. Panizza, "Applicability of diamond electrode/anode to the electrochemical treatment of a real textile effluent," *Journal of Electroanalytical Chemistry*, vol. 674, pp. 103–107, 2012.

[16] C. A. Basha, K. V. Selvakumar, H. J. Prabhu, P. Sivashanmugam, and C. W. Lee, "Degradation studies for textile reactive dye by combined electrochemical, microbial and photocatalytic methods," *Separation and Purification Technology*, vol. 79, no. 3, pp. 303–309, 2011.

[17] A. V. Patil and J. P. Jadhav, "Evaluation of phytoremediation potential of *Tagetes patula* L. for the degradation of textile dye Reactive Blue 160 and assessment of the toxicity of degraded metabolites by cytogenotoxicity," *Chemosphere*, vol. 92, no. 2, pp. 225–232, 2013.

[18] N. R. Rane, V. V. Chandanshive, A. D. Watharkar et al., "Phytoremediation of sulfonated Remazol Red dye and textile effluents by *Alternanthera philoxeroides*: an anatomical, enzymatic and pilot scale study," *Water Research*, vol. 83, pp. 271–281, 2015.

[19] A. N. Kabra, R. V. Khandare, T. R. Waghmode, and S. P. Govindwar, "Phytoremediation of textile effluent and mixture of structurally different dyes by *Glandularia pulchella* (Sweet) Tronc," *Chemosphere*, vol. 87, no. 3, pp. 265–272, 2012.

[20] R. V. Khandare and S. P. Govindwar, "Phytoremediation of textile dyes and effluents: current scenario and future prospects," *Biotechnology Advances*, vol. 33, no. 8, pp. 1697–1714, 2015.

[21] L. C. Davies, C. C. Carias, J. M. Novais, and S. Martins-Dias, "Phytoremediation of textile effluents containing azo dye by using *Phragmites australis* in a vertical flow intermittent feeding constructed wetland," *Ecological Engineering*, vol. 25, no. 5, pp. 594–605, 2005.

[22] S. Torbati, A. R. Khataee, and A. Movafeghi, "Application of watercress (*Nasturtium officinale* R. Br.) for biotreatment of a textile dye: investigation of some physiological responses and effects of operational parameters," *Chemical Engineering Research and Design*, vol. 92, no. 10, pp. 1934–1941, 2014.

[23] F. Vafaei, A. Movafeghi, A. R. Khataee, M. Zarei, and S. Y. Salehi Lisar, "Potential of *Hydrocotyle vulgaris* for phytoremediation of a textile dye: inducing antioxidant response in roots and leaves," *Ecotoxicology and Environmental Safety*, vol. 93, pp. 128–134, 2013.

[24] A. D. Watharkar and J. P. Jadhav, "Detoxification and decolorization of a simulated textile dye mixture by phytoremediation using *Petunia grandiflora* and, *Gailardia grandiflora*: a plant-plant consortial strategy," *Ecotoxicology and Environmental Safety*, vol. 103, pp. 1–8, 2014.

[25] M. N. Dos Santos Utmazian, G. Wieshammer, R. Vega, and W. W. Wenzel, "Hydroponic screening for metal resistance and accumulation of cadmium and zinc in twenty clones of willows and poplars," *Environmental Pollution*, vol. 148, no. 1, pp. 155–165, 2007.

[26] K. Drzewiecka, M. Mleczek, M. Gąsecka, Z. Magdziak, and P. Goliński, "Changes in *Salix viminalis* L. cv. 'Cannabina' morphology and physiology in response to nickel ions – hydroponic investigations," *Journal of Hazardous Materials*, vol. 217-218, pp. 429–438, 2012.

[27] W. Yang, Z. Ding, F. Zhao et al., "Comparison of manganese tolerance and accumulation among 24 *Salix* clones in a hydroponic experiment: application for phytoremediation," *Journal of Geochemical Exploration*, vol. 149, pp. 1–7, 2015.

[28] C. Cosio, P. Vollenweider, and C. Keller, "Localization and effects of cadmium in leaves of a cadmium-tolerant willow (*Salix viminalis* L.): I. macrolocalization and phytotoxic effects of cadmium," *Environmental and Experimental Botany*, vol. 58, no. 1–3, pp. 64–74, 2006.

[29] A. Evlard, K. Sergeant, B. Printz et al., "A multiple-level study of metal tolerance in *Salix fragilis* and *Salix aurita* clones," *Journal of Proteomics*, vol. 101, pp. 113–129, 2014.

[30] M. Vaculík, C. Konlechner, I. Langer et al., "Root anatomy and element distribution vary between two *Salix caprea* isolates

with different Cd accumulation capacities," *Environmental Pollution*, vol. 163, pp. 117–126, 2012.

[31] M. Mleczek, M. Łukaszewski, Z. Kaczmarek, I. Rissmann, and P. Golinski, "Efficiency of selected heavy metals accumulation by *Salix viminalis* roots," *Environmental and Experimental Botany*, vol. 65, no. 1, pp. 48–53, 2009.

[32] H. X. Corseuil and F. N. Moreno, "Phytoremediation potential of willow trees for aquifers contaminated with ethanol-blended gasoline," *Water Research*, vol. 35, no. 12, pp. 3013–3017, 2001.

[33] S. V. Skaates, A. Ramaswami, and L. G. Anderson, "Transport and fate of dieldrin in poplar and willow trees analyzed by SPME," *Chemosphere*, vol. 61, no. 1, pp. 85–91, 2005.

[34] X.-Z. Yu, J. D. Gu, and S. Liu, "Biotransformation and metabolic response of cyanide in weeping willows," *Journal of Hazardous Materials*, vol. 147, no. 3, pp. 838–844, 2007.

[35] C. T. Cano-Rodríguez, G. Roa-Morales, A. Amaya-Chávez, R. A. Valdés-Arias, C. E. Barrera-Díaz, and P. Balderas-Hernández, "Tolerance of *Myriophyllum aquaticum* to exposure of industrial wastewater pretreatment with electrocoagulation and their efficiency in the removal of pollutants," *Journal of Environmental Biology*, vol. 35, no. 1, pp. 127–136, 2014.

[36] A. Eslami, M. Moradi, F. Ghanbari, and F. Mehdipour, "Decolorization and COD removal from real textile wastewater by chemical and electrochemical Fenton processes: a comparative study," *Journal of Environmental Health Science and Engineering*, vol. 11, no. 1, p. 31, 2013.

[37] E. Tsantaki, T. Velegraki, A. Katsaounis, and D. Mantzavinos, "Anodic oxidation of textile dyehouse effluents on boron-doped diamond electrode," *Journal of Hazardous Materials*, vol. 207-208, pp. 91–96, 2012.

[38] APHA and AWWA, *Standard Methods for Examination of Water and Wastewater*, American Public Health Association y Water Pollution Control Federation, Washington, DC, USA, 21st edition, 2005.

[39] J. Val, L. Heras, and E. Monge, "New equations for the determination of photosynthetic pigments in acetone," *Annals Aula Dei*, vol. 17, no. 3, pp. 231–238, 1985.

[40] T.-P. Moisés, B.-H. Patricia, C. E. Barrera-Díaz, R.-M. Gabriela, and R. Natividad-Rangel, "Treatment of industrial effluents by a continuous system: electrocoagulation – activated sludge," *Bioresource Technology*, vol. 101, no. 20, pp. 7761–7766, 2010.

[41] A. Kapałka, G. Fóti, and C. Comninellis, "Investigations of electrochemical oxygen transfer reaction on boron-doped diamond electrodes," *Electrochimica Acta*, vol. 53, no. 4, pp. 1954–1961, 2007.

[42] I. Linares-Hernández, C. Barrera-Díaz, B. Bilyeu, P. Juárez-GarcíaRojas, and E. Campos-Medina, "A combined electrocoagulation–electrooxidation treatment for industrial wastewater," *Journal of Hazardous Materials*, vol. 175, no. 1-3, pp. 688–694, 2010.

14

Pretreatment of Real Wastewater from the Chocolate Manufacturing Industry through an Integrated Process of Electrocoagulation and Sand Filtration

Marco A. García-Morales ⓘ,[1] **Julio César González Juárez,**[1] **Sonia Martínez-Gallegos,**[1] **Gabriela Roa-Morales,**[2] **Ever Peralta,**[3] **Eduardo Martin del Campo López,**[2] **Carlos Barrera-Díaz,**[2] **Verónica Martínez Miranda,**[2] **and Teresa Torres Blancas**[4]

[1]*Instituto Nacional de México, Instituto Tecnológico de Toluca, Av. Tecnológico s/n, Col. Agrícola Buenavista, 52149 Toluca, MEX, Mexico*

[2]*Facultad de Química, Paseo Colón s/n, Residencial Colón, Universidad Autónoma del Estado de Mexico (UAEMéx), 50120 Toluca de Lerdo, MEX, Mexico*

[3]*Universidad del Mar, Campus Puerto Angel, Ciudad Universitaria s/n, 70902 Puerto Angel, OAX, Mexico*

[4]*Instituto de Química, Carretera Toluca-Atlacomulco Km 14.5, Universidad Nacional Autónoma de México and Centro Conjunto de Investigación en Química Sustentable UAEM-UNAM, 50200 Toluca, MEX, Mexico*

Correspondence should be addressed to Marco A. García-Morales; magm0904@hotmail.com

Academic Editor: Mark van Der Auweraer

The purpose of this study was to evaluate the efficiency of removal of suspended solids in terms of turbidity, color, and chemical oxygen demand (COD) when integrating the electrocoagulation process using aluminum sacrificial anodes and the sand filtration process as a pretreatment of wastewater from the chocolate manufacturing plant in Toluca, México. Wastewater from the chocolate manufacturing industry used in this study is classified as nontoxic, but is characterized as having a high content of color (5952 ± 76 Pt-Co), turbidity (1648 ± 49 FAU), and COD (3608 ± 250 mg/L). Therefore, enhanced performance could be achieved by combining pretreatment techniques to increase the efficiencies of the physical, chemical, and biological treatments. In the integrated process, there was a turbidity reduction of $96.1 \pm 0.2\%$ and an increase in dissolved oxygen from 3.8 ± 0.05 mg/L (inlet sand filtration) to 6.05 ± 0.03 mg/L (outlet sand filtration) after 120 min of treatment. These results indicate good water quality necessary for all forms of elemental life. Color and COD removals were $98.2 \pm 0.2\%$ and $39.02 \pm 2.2\%$, respectively, during the electrocoagulation process (0.2915 mA/cm^2 current density and 120 min of treatment). The proposed integrated process could be an attractive alternative of pretreatment of real wastewater to increase water quality of conventional treatments.

1. Introduction

Chocolate has a uniquely attractive taste and might even be beneficial for health. The popularity of this food appears to be mainly due to its potential to arouse sensory pleasure and positive emotions. Chocolates are complex multiphase systems of particulate (sugar, cocoa, and certain milk components) and continuous phases (cocoa butter, milk fat, and emulsifiers) [1]. The industrial chocolate manufacturing process consists of the following steps: cocoa collection, cleaning, fermentation, drying, roasting, grinding, pressing, spraying, and mixing, during which a large amount of water is used [2]. The wastewater in the chocolate manufacturing industry contains no hazardous ingredients, but it has a high content of color, total solids (TS), biochemical oxygen demand (BOD), and chemical oxygen demand (COD) [3].

The selection of treatment method is mainly based on the composition of the wastewater. Various treatment methods like (a) biological process, namely, anaerobic and aerobic; (b) physicochemical treatment, namely, adsorption, membrane process, reverse osmosis, and coagulation/flocculation; and (c) oxidation processes, namely, ozone and Fenton, have been used for the treatment of industrial wastewater [4]. The aerobic process involves the use of free or dissolved oxygen by microorganisms (aerobes) in the conversion of organic wastes to biomass and CO_2. In the anaerobic process, complex organic wastes are degraded into methane, CO_2, and H_2O through three basic steps (hydrolysis and acidogenesis including acetogenesis and methanogenesis) in the absence of oxygen [5]. Although the biological method is widely applied for the treatment of wastewater, too many disadvantages tend to focus on other technologies: the need for longer aeration times, requirement of large land areas, high energy demand, excess sludge production, and microbial inhibition due biomass poisoning [6]. The physicochemical treatment processes are effective for the treatment of industrial wastewater and are quick and compact but are not generally employed due to the associated high chemical and operational costs as well as complex sludge generation [7, 8]. Oxidation processes generate and use mainly hydroxyl radicals to oxidize the organic compounds. HO• has a high oxidation or standard reduction potential (2.8 V) [9]. The main characteristics of HO• are as follows: it is short-lived, it is simply produced, it is a powerful oxidant, it has an electrophilic behavior, it is ubiquitous in nature, it is highly reactive, and it is practically nonselective. It reacts with a wide variety of organic compound classes, producing shorter and simpler organic compounds, or in case of full mineralization [10]. Nevertheless, some researchers have reported that these processes were highly not effective for industrial application [11]. Decolorization through chemical treatment with ozone, Fenton's reagent, and H_2O_2/UV leads to color reduction due to breaking of the conjugation and or bonds in chromophoric groups. In addition, the formation of potentially toxic oxidation intermediates may occur; therefore, these are not preferred solutions [4].

Due to the complexity of chocolate manufacturing plant wastewater, in which pollutants may be suspended, emulsified, or dissolved, electrocoagulation (EC) represents an interesting alternative for water remediation, providing comparable results with even some advanced oxidation processes in the removal of persistent compounds from pharmaceutical and food industrial effluents [12, 13]. Among the advantages of an EC process, the following can be highlighted: nonspecificity, similar treatment for drinking water and wastewater, low dosage of chemical reagents, low operating costs, low sludge production (compared to traditional chemical coagulation), absence of moving parts in the reaction setup, and low power consumption if solar energy is used [14]. From a practical point of view, EC must be considered a parallel mechanism that includes charge neutralization and adsorption. At the beginning of the process, realized ions destabilize the system forming metal hydroxide complexes that aggregate suspended particles (flocs) and adsorb dissolved particles [15, 16]. In addition to the above, at the cathode, gas formation takes place allowing floc floatation. The electrical corrosion of metal in the sacrificial electrode and the formation of hydroxyl anion are the essential reactions in any EC process [17]. The floc formation is a complex process; according to the Derjaguin-Landau-Verwey-Overbeek (DLVO) theory, aggregates depend on interaction forces (van der Waals) and double-layer forces [18].

It is well known that iron and aluminum are the preferred materials to be used as sacrificial anodes. For iron, anode oxidation could lead to either ferrous or ferric ion formation; however, low solubility of Fe^{3+} ions suggests the release of Fe^{2+} ions, which are oxidized to ferric ions due to pH and dissolved oxygen concentration. For aluminum, anode oxidation leads to Al^{3+} ion formation. In both cases, the subsequent formation of hydroxide compounds induces the presence of monomeric and polymeric amorphous species that trap colloidal particles and promote the soluble pollutant adsorption [14]. Both metals are fine as construction materials for sacrificial electrodes, but for economics, iron has a slight advantage: it is nontoxic, meaning it can be used for drinking water, and it has a lower price. Otherwise, there are many studies that report the effectiveness of aluminum anodes in the EC process for emerging contaminants [17, 19, 20].

Sand filters are a natural medium that can be used as a filter for wastewater treatment. It displays two roles: the retention of solids and biomass fixation that could be developed on the granular material and the biodegradation of organic, phosphorus, and nitrogenous pollutants [21–23]. For disinfection of wastewater reuse, the turbidity and suspended solids must be reduced to prevent the hiding of pathogens and organisms that hide behind these solids. Currently, the most widely used process to remove residual TSS (total suspended solids) is treated effluent filtration [24, 25]. The main mechanisms contributing to the removal of suspended solids in sand filters are cast [22, 26]. This has been identified as the major operating mechanism for the removal of suspended solids during filtration of secondary effluent from processes and biological treatments. Perhaps, other mechanisms, such as interception, impact, and adhesion, are operational, although its effects are minor and mostly marked by the action of casting [27–30].

The purpose of this study is to evaluate the efficiency of integrated electrocoagulation and sand filtration processes as a pretreatment of wastewater from the chocolate manufacturing plants in terms of turbidity, color, and chemical oxygen demand (COD).

2. Materials and Methods

2.1. Sampling. The wastewater samples used in this study were collected at the effluent of an industrial chocolate manufacturing plant, preserved, and analyzed according to the standard methods for conventional characterization APHA/AWWA/WEF [31]. The electrocoagulation and filtration were monitored for turbidity, color, and COD, as well as pH variation. A UV-VIS spectrum of the effluent was done on a PerkinElmer Lambda 25 UV/VIS Spectrophotometer (USA). Color and turbidity were monitored at 465 and 860 nm wavelengths, respectively, using a Hach DR/4000U 110 spectrometer. COD was analyzed by the closed reflux colorimetric method (Method 5220 D; according to APHA) [31].

2.2. Electrocoagulation Process.

Electrocoagulation was carried out in a laboratory-scale batch reactor; two rectangular commercial aluminum plates (99.3 wt% Al) served as anode and cathode. The anodic and cathodic active surface area was 343 cm^2 immersed in wastewater with 0.1372 1/cm of the SA/V ratio. A DC power source supplied the system with 0.1 A, corresponding to 0.2915 mA/cm^2 current density which was kept for 120 min. Electrocoagulation was performed without additional electrolyte (750 μS/cm conductivity in wastewater). The electrodes were connected to a digital DC power supply (GW Instek GPR-1820HD, 0–18 V; 0–20 A, China). Twenty-five mL of sample was taken every 30 min during the 2 h electrocoagulation process. The efficiency of EC was evaluated by measuring the turbidity, color, and chemical oxygen demand (COD) [32].

2.3. Filtration Process.

Two horizontal conventional downflow filters (sand filter A and sand filter B) installed in parallel were used for the filtration process; the filter media of each filter included two beds composed of gravel 1/4 × 1/8 (24 cm) and gravel 1/8 × 1/16 (51 cm), each filter measuring 168 cm in length and 15 cm in diameter. The effective size for each filter was 0.71 mm. In the bottom of each filter, gravel particles (3/4 × 1/2) were placed to support upper layers. The filtration process was carried out after completing the electrocoagulation process, as shown in Figure 1. At the filter outlet, turbidity and dissolved oxygen (DO) were monitored. Filtration experiments were performed without recirculation.

The removal efficiency was calculated using

$$Y(\%) = \frac{(y_0 - y)}{y_0}, \tag{1}$$

where Y is the removal efficiency of turbidity/color/COD and y_0 and y correspond to the initial and final values of a determined parameter, respectively.

3. Results and Discussion

Table 1 presents the initial physicochemical parameters of wastewater from the chocolate manufacturing process. The wastewater contained pollutants, which were reflected in high levels of COD, due to ingredients used in chocolate manufacturing such as cocoa bean, chocolate liquor obtained from the broken beans that are ground, cocoa butter obtained from the broken-down cell walls, sugar, and emulsifiers in conjunction with the waste from the processes of cleaning, fermentation, drying, roasting, grinding, pressing, spraying, and mixing [1, 33]. The color and turbidity values obtained are harmful to aquatic life, obstructing light penetration in the water, inhibiting thus the photosynthesis-based biological processes [31]. The pH was about 7.5, which was basically a neutral pH environment. Contaminants all achieved maximum removal in this pH condition [30]. Therefore, the following tests were performed using the raw water without pH adjustment, in agreement with previous studies, where

the wastewater was used directly for electrocoagulation experiments [34].

3.1. Electrocoagulation Process Efficiency.

After 120 min of treatment, the reductions in turbidity, color, and COD for the electrocoagulation process were 87.8 ± 0.6%, 98.2 ± 0.2%, and 39.02 ± 2.2%, respectively (Figure 2). The experiments were repeated three times to verify the reproducibility of the results; in any experiment, the values of the coefficient of variation were no higher than 5%, indicating that recollected data have an statistical acceptance criteria. Zhao et al. carried out EC experiments as a pretreatment applied to wastewater containing oil, grease, and other inorganic contaminants; their results showed a removal of 93.8% in turbidity under the following conditions: 5.56 mA/cm^2 of current density and 30 min of reaction time [34]. Regarding color removal, Ricordel and Djelal obtained a removal of 80% after EC treatment of landfill leachate, although not the same substrate; this had high levels of organic matter, refractory compounds, inorganic contaminants, and color [32]. In addition, an efficiency greater than 32% was obtained for Farhadi et al. by comparing electrocoagulation (1.83 mA/cm^2 of current density and 30 min of time reaction) and an advanced oxidation process during pharmaceutical wastewater treatment [35]. The removal of turbidity, color, and COD is attributed to sweep flocculation [34]. In the EC process, coagulating ions are produced in situ, involving three successive stages: (i) formation of coagulants by electrolytic oxidation of the sacrificial electrode of Al; (ii) destabilization of the contaminants, particulate suspension, and breaking of emulsions; and (iii) aggregation of the destabilized phases to form flocs. Al gets dissolved from the anode, generating corresponding metal ions that almost immediately hydrolyze to polymeric aluminum oxyhydroxides [4, 36]. These polymeric oxyhydroxides are excellent coagulating agents. When aluminum electrodes in the EC process are used as anode and cathode, the main reactions at the anode are as follows:

$$Al \rightarrow Al^{3+} + 3e^- \tag{2}$$

Also, oxygen evolution can compete with aluminum dissolution at the anode via

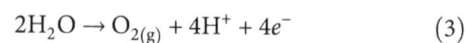

$$2H_2O \rightarrow O_{2(g)} + 4H^+ + 4e^- \tag{3}$$

At the cathode, hydrogen evolution takes place via the following reaction, assisting in the floatation of the flocculated particles out of the water:

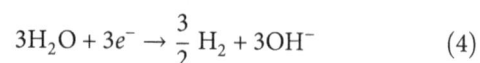

$$3H_2O + 3e^- \rightarrow \frac{3}{2} H_2 + 3OH^- \tag{4}$$

At high pH values, OH$^-$ generated at the cathode during hydrogen evolution may attack the cathode by the following reaction [18, 37]:

$$2Al + 6H_2O + 2OH^- \rightarrow 2Al(OH)_4^- + 3H_{2(g)} \tag{5}$$

Al^{3+} and hydroxyl ions are generated by electrode reactions as shown in (2), (4), and (5) to form various monomeric-polymeric species transformed initially into

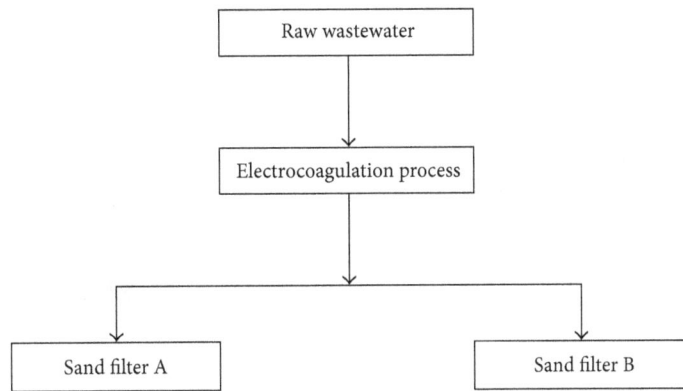

FIGURE 1: Flow chart used in this study.

TABLE 1: Initial and final physicochemical parameters from treated wastewater.

	Raw wastewater			Treated wastewater			% Removal in the integrated process
Parameter	Value	Units	Parameter	Value	Units	
COD	3608 ± 250	mg/L	COD	2200 ± 11	mg/L	39.02 ± 2.2
Color	5952 ± 76	Pt-Co	Color	101 ± 17	Pt-Co	98.2 ± 0.26
Turbidity	1648 ± 49	mg/L	Turbidity	64 ± 5	mg/L	96.1 ± 0.2
pH	7.4 ± 0.06		pH	9.11 ± 0.03		
Conductivity	750 ± 28	μS/cm	Conductivity	520 ± 9	μS/cm	
			Energy consumption	0.32	kWh/m^3	

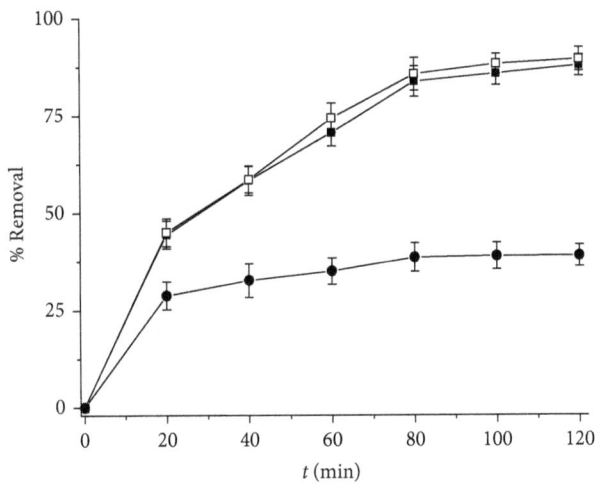

FIGURE 2: Contaminant removal efficiency in wastewater treated after the electrocoagulation process at a current density of 0.2915 mA/cm^2: turbidity (■), color (□), and COD (●).

$Al(OH)_{3(s)}$ and finally polymerized to $Al_n(OH)_{3n}$ ((6) and (7)) in the solution [32, 37–39]:

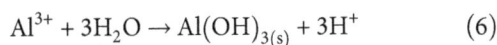

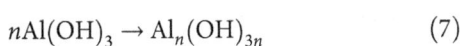

$$Al^{3+} + 3H_2O \rightarrow Al(OH)_{3(s)} + 3H^+ \qquad (6)$$

$$nAl(OH)_3 \rightarrow Al_n(OH)_{3n} \qquad (7)$$

Since the electrocoagulation process is based on removal of the colloidal/particulate COD fraction of wastewater, the

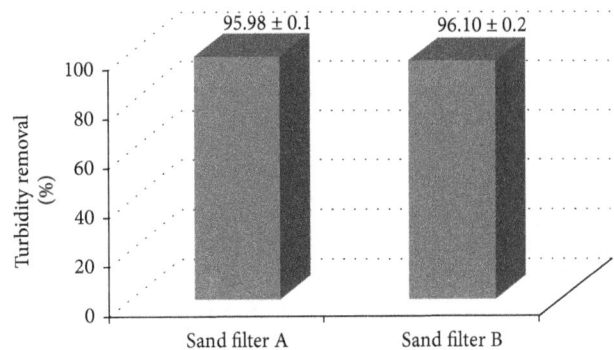

FIGURE 3: Turbidity removal efficiency in treated wastewater.

low efficiency of COD obtained in this process is attributed to the soluble COD fraction in raw wastewater [40].

3.2. Filtration Process Efficiency. As shown in Figure 3, the turbidity after the filtration processes was lower than the initial turbidity, which was 1648 ± 49 FAU in raw wastewater; turbidity removals (%) reached were 95.98 ± 0.1 and 96.10 ± 0.2 for filters A and B, respectively. The turbidity removal was due to the working-in stage (characterized by a rapid decrease in effluent turbidity) and working stage (the effective stage of filtration giving satisfactory effluent quality) [30]. Similar results were obtained by Ramadan, whose results reached 98.05% removal of total suspended solids (TSS) using nonconventional sand filters (TiO_2 was added

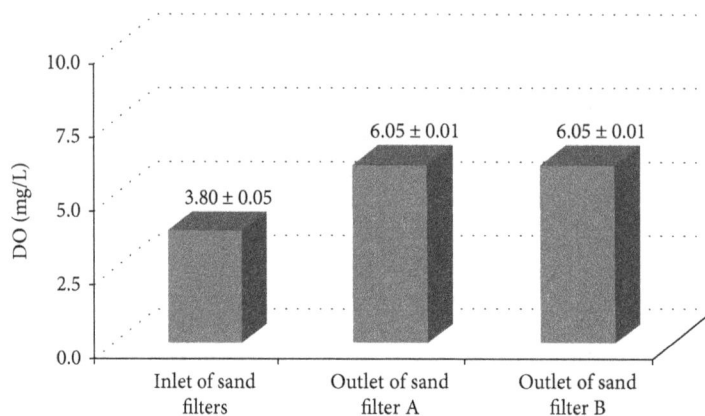

FIGURE 4: Dissolved oxygen in filtration process.

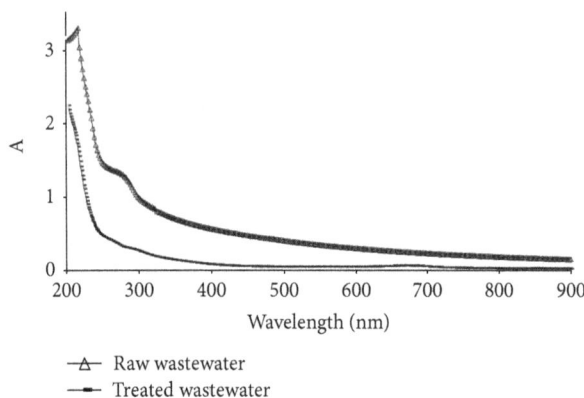

FIGURE 5: UV/VIS spectrum of the raw wastewater and treated wastewater in the electrocoagulation-filtration process at a current density of 0.2915 mA/cm^2.

to the sand filter) in reducing pollutants from wastewater [41]. Finally, Achak et al. obtained results of 90% turbidity removal from olive mill wastewater generated by the olive oil extraction process [42]. After electrocoagulation, sand filtration could remove flocs by attaching them to the sand grain to improve the removal of turbidity from 87.80% in the EC process to 99.10% after filtration obtained an increase by 11.40%; a statistical test was performed, determining that a statistically significant difference exists between both processes (p value = 0.018).

The concentration of dissolved oxygen in the inlet of each sand filter increased from 3.8 ± 0.05 mg/L to 6.05 ± 0.01 mg/L after the filtration process during the 120 min experiment (Figure 4). The oxygenation of sand filters is due to the gaseous exchange between the atmosphere and the interstices of sand in the surface of the filter. These results are good because adequate dissolved oxygen is needed for good water quality and is necessary for all forms of elemental life.

3.3. UV/VIS Spectra of the Treated Wastewater. The spectrum for the raw wastewater (Δ) presented a baseline with one absorbance peak at 290 nm, which was associated with the contaminants in this matrix (3608 ± 250 mg/L COD). For 120 min of reaction, treated wastewater by an electrocoagulation-filtration-coupled process (■) showed the highest efficiency in removing turbidity, color, and COD together with a decrease in the baseline and in the absorbance peaks, as shown in Figure 5. The removal of contaminants (initial DQO = 3608 ± 250 mg/L and final DQO = 2200 ± 11 mg/L) was indicated by the decrease in the absorption band at 290 nm and baseline [43].

4. Conclusions

Removal of contaminants was efficient by integrating electrocoagulation and filtration processes as a pretreatment of wastewater from chocolate manufacturing plants, which was reflected in increased removal percentages of turbidity and color and an increase in dissolved oxygen after the integrated filtration processes.

UV/VIS spectra intensity decreased between raw and treated wastewater, indicating the removal of pollutants in the integrated process.

The EC process removed the colloidal/particulate COD fraction in raw wastewater; however, the remaining COD due to the soluble COD fraction was not removed in the integrated process.

Disclosure

This work was presented as a poster with the title "Wastewater Treatment of Chocolate Manufacture Industry through an Integrated Process of Electrocoagulation and Filtration" at Electrochemical Treatments for Organic Pollutant Degradation in Water and Soils, 2014, ECS and SMEQ Joint International Meeting.

Conflicts of Interest

The authors declare that there is no conflict of interest regarding the publication of this paper.

Acknowledgments

The authors wish to acknowledge the support given by CONACYT and the Tecnológico Nacional de México for the development of this research through Project TecNM 5652.15-P.

References

[1] M. El-kalyoubi, M. F. Khallaf, A. Abdelrashid, and E. M. Mostafa, "Quality characteristics of chocolate – containing some fat replacer," *Annals of Agricultural Science*, vol. 56, no. 2, pp. 89–96, 2011.

[2] P. B. Botelho, M. Galasso, V. Dias et al., "Oxidative stability of functional phytosterol-enriched dark chocolate," *LWT - Food Science and Technology*, vol. 55, no. 2, pp. 444–451, 2014.

[3] M. V. Copetti, B. T. Iamanaka, M. A. Nester, P. Efraim, and M. H. Taniwaki, "Occurrence of ochratoxin A in cocoa by-products and determination of its reduction during chocolate manufacture," *Food Chemistry*, vol. 136, no. 1, pp. 100–104, 2013.

[4] E. Gengec, M. Kobya, E. Demirbas, A. Akyol, and K. Oktor, "Optimization of baker's yeast wastewater using response surface methodology by electrocoagulation," *Desalination*, vol. 286, pp. 200–209, 2012.

[5] Y. J. Chan, M. F. Chong, C. L. Law, and D. G. Hassell, "A review on anaerobic–aerobic treatment of industrial and municipal wastewater," *Chemical Engineering Journal*, vol. 155, no. 1-2, pp. 1–18, 2009.

[6] M. Luan, G. Jing, Y. Piao, D. Liu, and L. Jin, "Treatment of refractory organic pollutants in industrial wastewater by wet air oxidation," *Arabian Journal of Chemistry*, vol. 10, no. 1, pp. S769–S776, 2017.

[7] M. S. Nawaz and M. Ahsan, "Comparison of physico-chemical, advanced oxidation and biological techniques for the textile wastewater treatment," *Alexandria Engineering Journal*, vol. 53, no. 3, pp. 717–722, 2014.

[8] S. Renou, J. G. Givaudan, S. Poulain, F. Dirassouyan, and P. Moulin, "Landfill leachate treatment: review and opportunity," *Journal of Hazardous Materials*, vol. 150, no. 3, pp. 468–493, 2008.

[9] D. Shahidi, R. Roy, and A. Azzouz, "Advances in catalytic oxidation of organic pollutants – prospects for thorough mineralization by natural clay catalysts," *Applied Catalysis B: Environmental*, vol. 174-175, pp. 277–292, 2015.

[10] G. Boczkaj and A. Fernandes, "Wastewater treatment by means of advanced oxidation processes at basic pH conditions: a review," *Chemical Engineering Journal*, vol. 320, pp. 608–633, 2017.

[11] L. Bilińska, M. Gmurek, and S. Ledakowicz, "Comparison between industrial and simulated textile wastewater treatment by AOPs – biodegradability, toxicity and cost assessment," *Chemical Engineering Journal*, vol. 306, pp. 550–559, 2016.

[12] W. Baran, E. Adamek, M. Jajko, and A. Sobczak, "Removal of veterinary antibiotics from wastewater by electrocoagulation," *Chemosphere*, vol. 194, pp. 381–389, 2018.

[13] D. T. Moussa, M. H. El-Naas, M. Nasser, and M. J. Al-Marri, "A comprehensive review of electrocoagulation for water treatment: potentials and challenges," *Journal of Environmental Management*, vol. 186, no. 1, pp. 24–41, 2017.

[14] J. N. Hakizimana, B. Gourich, M. Chafi et al., "Electrocoagulation process in water treatment: a review of electrocoagulation modeling approaches," *Desalination*, vol. 404, pp. 1–21, 2017.

[15] Y. A. Ouaissa, M. Chabani, A. Amrane, and A. Bensmaili, "Removal of tetracycline by electrocoagulation: kinetic and isotherm modeling through adsorption," *Journal of Environmental Chemical Engineering*, vol. 2, no. 1, pp. 177–184, 2014.

[16] D. Valero, J. M. Ortiz, V. García, E. Expósito, V. Montiel, and A. Aldaz, "Electrocoagulation of wastewater from almond industry," *Chemosphere*, vol. 84, no. 9, pp. 1290–1295, 2011.

[17] E. Nariyan, A. Aghababaei, and M. Sillanpää, "Removal of pharmaceutical from water with an electrocoagulation process; effect of various parameters and studies of isotherm and kinetic," *Separation and Purification Technology*, vol. 188, pp. 266–281, 2017.

[18] S. Garcia-Segura, M. M. S. G. Eiband, J. V. de Melo, and C. A. Martínez-Huitle, "Electrocoagulation and advanced electro-coagulation processes: a general review about the fundamentals, emerging applications and its association with other technologies," *Journal of Electroanalytical Chemistry*, vol. 801, pp. 267–299, 2017.

[19] S. Ahmadzadeh, A. Asadipour, M. Pournamdari, B. Behnam, H. Reza Rahimi, and M. Dolatabadi, "Removal of ciprofloxacin from hospital wastewater using electrocoagulation technique by aluminum electrode: optimization and modelling through response surface methodology," *Process Safety and Environmental Protection*, vol. 109, pp. 538–547, 2017.

[20] M. Çırak, "High-temperature electrocoagulation of colloidal calcareo-argillaceous suspension," *Powder Technology*, vol. 328, no. 1, pp. 13–25, 2018.

[21] S. E. Keithley and M. J. Kirisits, "An improved protocol for extracting extracellular polymeric substances from granular filter media," *Water Research*, vol. 129, pp. 419–427, 2018.

[22] Y. Gheraiiri, A. Amrane, Y. Touil, M. Hadj Mahammed, F. Gheraiiri, and I. Baameur, "A comparative study of the addition effect of activated carbon obtained from date stones on the biological filtration efficiency using sand dune bed," *Energy Procedia*, vol. 36, pp. 1175–1183, 2013.

[23] P. Laaksonen, A. Sinkkonen, G. Zaitsev, E. Mäkinen, T. Grönroos, and M. Romantschuk, "Treatment of municipal wastewater in full-scale on-site sand filter reduces BOD efficiently but does not reach requirements for nitrogen and phosphorus removal," *Environmental Science and Pollution Research*, vol. 24, no. 12, pp. 11446–11458, 2017.

[24] S. Mtavangu, A. M. Rugaika, A. Hilonga, and K. N. Njau, "Performance of constructed wetland integrated with sand filters for treating high turbid water for drinking," *Water Practice and Technology*, vol. 12, no. 1, pp. 25–42, 2017.

[25] R. Bauer, H. Dizer, I. Graeber, K.-H. Rosenwinkel, and J. M. López-Pila, "Removal of bacterial fecal indicators, coliphages and enteric adenoviruses from waters with high fecal pollution by slow sand filtration," *Water Research*, vol. 45, no. 2, pp. 439–452, 2011.

[26] M. De Sanctis, G. Del Moro, S. Chimienti, P. Ritelli, C. Levantesi, and C. Di Iaconi, "Removal of pollutants and pathogens by a simplified treatment scheme for municipal wastewater reuse in agriculture," *Science of the Total Environment*, vol. 580, pp. 17–25, 2017.

[27] C. N. Mushila, G. M. Ochieng, F. A. O. Otieno, S. M. Shitote, and C. W. Sitters, "Hydraulic design to optimize the treatment capacity of multi-stage filtration units," *Physics and Chemistry of the Earth, Parts A/B/C*, vol. 92, pp. 85–91, 2016.

[28] E. Bar-Zeev, N. Belkin, B. Liberman, T. Berman, and I. Berman-Frank, "Rapid sand filtration pretreatment for SWRO: microbial maturation dynamics and filtration efficiency of organic matter," *Desalination*, vol. 286, pp. 120–130, 2012.

[29] M. Elbana, F. Ramírez de Cartagena, and J. Puig-Bargués, "Effectiveness of sand media filters for removing turbidity and recovering dissolved oxygen from a reclaimed effluent used for micro-irrigation," *Agricultural Water Management*, vol. 111, pp. 27–33, 2012.

[30] A. Y. Zahrim and N. Hilal, "Treatment of highly concentrated dye solution by coagulation/flocculation–sand filtration and nanofiltration," *Water Resources and Industry*, vol. 3, pp. 23–34, 2013.

[31] APHA, AWWA, WEF, *Standard Methods for Examination of Water and Wastewater*, American Public Health Association, Washington, DC, USA, 22nd edition, 2012.

[32] C. Ricordel and H. Djelal, "Treatment of landfill leachate with high proportion of refractory materials by electrocoagulation: system performances and sludge settling characteristics," *Journal of Environmental Chemical Engineering*, vol. 2, no. 3, pp. 1551–1557, 2014.

[33] M. Gültekin-Özgüven, İ. Berktaş, and B. Özçelik, "Influence of processing conditions on procyanidin profiles and antioxidant capacity of chocolates: optimization of dark chocolate manufacturing by response surface methodology," *LWT - Food Science and Technology*, vol. 66, pp. 252–259, 2016.

[34] S. Zhao, G. Huang, G. Cheng, Y. Wang, and H. Fu, "Hardness, COD and turbidity removals from produced water by electrocoagulation pretreatment prior to reverse osmosis membranes," *Desalination*, vol. 344, pp. 454–462, 2014.

[35] S. Farhadi, B. Aminzadeh, A. Torabian, V. Khatibikamal, and M. Alizadeh Fard, "Comparison of COD removal from pharmaceutical wastewater by electrocoagulation, photo-electrocoagulation, peroxi-electrocoagulation and peroxi-photoelectrocoagulation processes," *Journal of Hazardous Materials*, vol. 219-220, pp. 35–42, 2012.

[36] M. Kobya, M. S. Oncel, E. Demirbas, E. Şık, A. Akyol, and M. Ince, "The application of electrocoagulation process for treatment of the red mud dam wastewater from Bayer's process," *Journal of Environmental Chemical Engineering*, vol. 2, no. 4, pp. 2211–2220, 2014.

[37] S. Tchamango, C. P. Nanseu-Njiki, E. Ngameni, D. Hadjiev, and A. Darchen, "Treatment of dairy effluents by electrocoagulation using aluminium electrodes," *Science of The Total Environment*, vol. 408, no. 4, pp. 947–952, 2010.

[38] M. Mechelhoff, G. H. Kelsall, and N. J. D. Graham, "Electrochemical behaviour of aluminium in electrocoagulation processes," *Chemical Engineering Science*, vol. 95, pp. 301–312, 2013.

[39] M. A. Sandoval, R. Fuentes, J. L. Nava, and I. Rodríguez, "Fluoride removal from drinking water by electrocoagulation in a continuous filter press reactor coupled to a flocculator and clarifier," *Separation and Purification Technology*, vol. 134, pp. 163–170, 2014.

[40] Z. Hu, K. Chandran, B. F. Smets, and D. Grasso, "Evaluation of a rapid physical–chemical method for the determination of extant soluble COD," *Water Research*, vol. 36, no. 3, pp. 617–624, 2002.

[41] M. Ramadan, "Efficiency of new Miswak, titanium dioxide and sand filters in reducing pollutants from wastewater," *Beni-Suef University Journal of Basic and Applied Sciences*, vol. 4, no. 1, pp. 47–51, 2015.

[42] M. Achak, L. Mandi, and N. Ouazzani, "Removal of organic pollutants and nutrients from olive mill wastewater by a sand filter," *Journal of Environmental Management*, vol. 90, no. 8, pp. 2771–2779, 2009.

[43] M. A. García-Morales, G. Roa-Morales, C. Barrera-Díaz, V. Martínez Miranda, P. Balderas Hernández, and T. B. Pavón Silva, "Integrated advanced oxidation process (ozonation) and electrocoagulation treatments for dye removal in denim effluents," *International Journal of Electrochemical Science*, vol. 8, pp. 8752–8763, 2013.

15

The Design and Comparison of Central and Distributed Light Sensored Smart LED Lighting Systems

Mehmet Ali Özçelik⊙

Technical Science, Electric and Energy Department, Gaziantep University, Şehitkamil, 27310 Gaziantep, Turkey

Correspondence should be addressed to Mehmet Ali Özçelik; ozcelik@gantep.edu.tr

Academic Editor: Mahmoud M. El-Nahass

There is a lack of published peer-reviewed research comparing the efficiencies of distributed versus central sensor-controlled LED lighting systems. This research proposes improving the smart illumination of a room with external fenestration using central and distributed light sensors. The optical and electrical measurements of the daylight have been made in the case where the light was not distributed evenly and not sufficient. Test results show that the proposed distributed light sensor illumination system has increased the efficiency by 28% when compared to the proposed central system. It has also been shown that the two tested systems are more cost-effective than common smart illumination systems.

1. Introduction

Energy-efficient resources are essential in today's world [1–3]. Petroleum- and coal-based conventional power plants are continuously spreading harmful gasses such as nitrogen dioxide and carbon dioxide that threaten the environment and human health [4]. The renewable energy resources emerge as an effective solution for increasing the need for clean energy [5]. Among renewable energy resources, solar energy is one of the promising ones providing major benefits, such as being environmentally friendly and being silent [6]. Studies show that the photovoltaic (PV) system can reduce release of one ton of carbon dioxide per kWh of electricity [7]. The modular structure of PV panels enables easy integration to energy-efficient buildings. Because such a modular structure can efficiently integrate electrical, electronic, and mechanical systems [8], the electrical energy used for illumination is equivalent to 20% of the total electrical energy production of the world [9, 10]. For this reason, the energy demand for illumination obtained from renewable energy sources is essential in terms of pollution-free environment. In the USA, the UK, and China, at least 20% of the total electric power production from renewable energy sources has been targeted till 2020. Moreover, it is envisaged that renewables will contribute to over 50% by 2050 in some

countries [11–13]. Light-emitting diode (LED) technologies have significantly lowered down the energy demand needed for illumination. Moreover, they are durable and environmentally friendly [14–17]. Energy efficiency, smart buildings, and green buildings have recently come to the foreground as some striking topics. Furthermore, an illumination control is usually designed according to the needs of spots in the buildings to decrease energy consumption [18, 19]. With LED luminaires, it is possible to control the light output easily and accurately. Additionally, LED luminaires enable the flexible adaptation of a lighting system to its environment [20]. Artificial lighting accounts for a major fraction of global electrical energy consumption. In a typical office building, the energy consumed due to artificial lighting can be up to 40% of the total energy consumption [21]. As the need for the use of energy resources is increasing, the need for illumination control becomes essential.

Automatic or photoelectrically controlled lighting systems in the buildings can significantly reduce the lighting energy consumption down to as low as 50% [22, 23]. Some authors have proposed an illumination model-based method and algorithm for intelligent open-loop lighting control. Specifically, the simulation results were presented using a simplistic virtual room [24, 25]. A single light sensor-driven lighting system was made early by Rubinstein [26] and

Peruffo et al. [27]. Distributed optimization algorithms for lighting control with daylight and occupancy adaptation were proposed in Caicedo and Pandharipande [28] and Lee and Kwon [29], under networking and information exchange constraints. Techniques such as daylight harvesting and automatic dimming control with wireless sensor, illumination balancing, Konnex Association Worldwide Standard for Home and Building Control (KNX), digital addressable lighting interface (DALI) standard, and stochastic hill climbing optimization are applied to lighting control [30, 31]. However, DALI and KNX which are used to add intelligence to buildings have higher costs as much as tens of thousands of dollars for the basic installation [14].

In smart lighting systems, the illumination level of the indoor environment can be determined by means of light sensors [32]. In these systems, the electrical information obtained from the sensors is converted into illumination knowledge [33]. The light information of the environment can be measured by a centrally located sensor or by placing more than one sensor in a distributed manner [34, 35]. By comparing the ambient light value measured by the sensors and the microprocessor system with the desired reference illumination value, the lighting levels of the LED luminaires can be increased or decreased to achieve the desired illumination rate. In this study, the efficiency and cost analysis of two different proposed smart LED architectures with the centralized and distributed sensor structure were performed in the 54 m² classroom having two windows. The solution has been designed keeping in mind that low power, low consumption, and scalability are addressed. To the best of the author's knowledge, it is the first time to compare the central sensor and the distributed sensor smart LED lighting.

2. Method

2.1. The PV System, LED Illuminating System Set on the DC Grid and the Positioning of LED Panels.
As the sunlight hits on PV cells, photovoltage and photocurrent act like a forward diode on a large surface. The current expression resulted from the sunlight hitting on the cell is given.

$$I = I_{PH} - I_S \cdot \left\{ \exp\left[\frac{q}{A \cdot k_B \cdot T} (V + I \cdot R_L) \right] - 1 \right\} - \frac{(V + I \cdot R_S)}{R_{SH}},$$

(1)

where I_{PH} is the photocurrent, I_S is the saturation current, R_L is the load resistance, R_S is the series equivalent circuit resistance, R_{SH} is the parallel equivalent circuit resistance, V is the terminal voltage, I is the load current, A is the

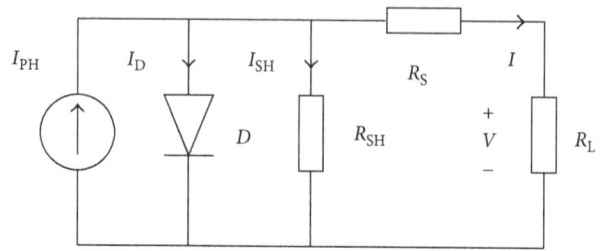

FIGURE 1: Equivalent circuit diagram for solar cell.

TABLE 1: Electrical parameters of the PV and the LED lighting systems.

Maximum power PV system	390 W
Maximum current of PV system	10.83 A
The maximum current and power of each of the LED panel lamp	1 A/36 W
The number of LED lamps	8
The maximum power of the LED illuminating system	288 W

diode ideality factor, k_B is Boltzmann's constant, and T is the temperature of the PV panel, respectively. As it is seen, photocurrent, saturation current, load resistance, series equivalent circuit resistance, parallel equivalent circuit resistance, terminal voltage, load current, diode ideality factor, Boltzman's constant, and temperature of the PV panel are denoted by I_{PH}, I_S, R_L, R_S, R_{SH}, V, I, A, k, and T, respectively.

The equivalent circuit diagram of a solar cell is shown in Figure 1 [36].

In the smart LED lighting system, six PV panels are used and each panel is rated at 12 V and 65 W. PV panels are connected in series and are parallel with PWM charge regulators to obtain 36 V. Table 1 shows the electrical parameters of the PV and the LED lighting systems.

The lumens per watt (lm/W) of the LED lamps used in the lighting systems is 88. The maximum luminous flux is 3200 lumen. The specifications of the room used in the experiment are as follows: (1) The room shape is rectangular having width $a = 6$ m and length $b = 9$ m²). The room height is $h = 3$ m³). The vertical distance between the working plane and the luminaire is 1.85 m. Based on the EN 12464 lighting of indoor work places standard [37], the recommended classroom light level is 300 lux.

The LED distribution curve for the classroom illumination is shown in Figure 2.

Calculations of energy storage capacity are as follows:

$$\text{Maximum daily power consumption (MDPC)} = \text{total LED panels power} \times \text{operation time} = 288 \times 8 = 2304 \text{ Wh,}$$

$$\text{Battery capacity} = \frac{\text{MDPC}}{\text{charge controller efficiency} \times \text{depth discharge} \times \text{system voltage}} = \frac{2304}{0.9 \times 0.75 \times 36} = \cong 100 \text{ Ah.}$$

(2)

When the battery efficiency is 80%, the capacity value must be increased by $100 - 80 = 20\%$. In this case, the

required battery capacity is as follows: the required battery capacity = $100 + 100 \times 0.20 = 120$ Ah.

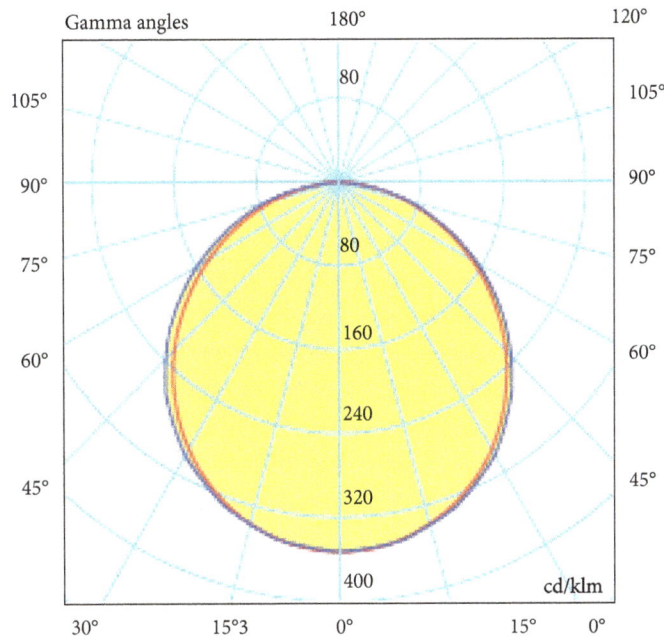

FIGURE 2: Light distribution curve for the classroom [38].

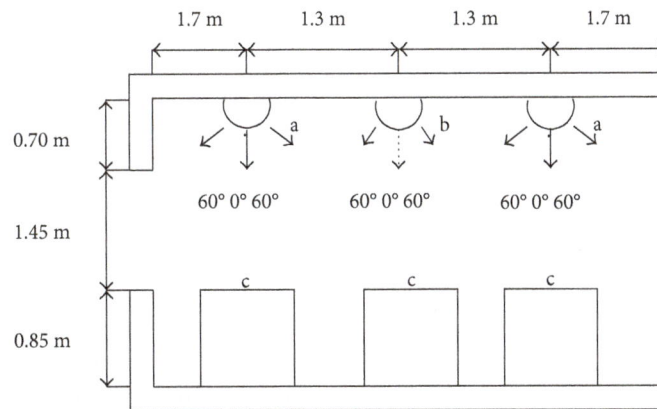

FIGURE 3: Locations of light sensors and illuminance meters: (a) light sensors of the distributed system, (b) light sensor of the central system, and (c) illuminance meters.

The algorithm designed for the illumination control system in this study is a closed loop. In the closed-loop system, the light sensors are installed on the ceiling where changes in illuminance are related to variations in illuminance on the work surface. In areas where either indirect or direct-indirect electrical lighting systems are used, the light sensor should be installed at a place that should not get direct light from luminaires [39]. The placement of the light sensors used in the distributed and central illumination systems together with the illuminance meters are given in Figure 3. The light sensors are placed 60 degrees so that they are not exposed to direct daylight or LED light.

A TEMT6000 Ambient Light Sensor is utilized in this work [10, 40]. This sensor can measure incident illuminance up to 1000 lux with a peak sensitivity at around 580 nm with a spectral sensitivity curve adapted to match the human eye responsivity. The sensor output is an analog current

information, converted to a variable voltage (0–5 V) that is read by the STM32F407 microcontroller through a 10 K resistor. The photocurrent and illuminance of the sensor are presented in Figure 4.

The sensor TEMT6000 is the wide angle of half sensitivity $\varphi = \pm 60°$. Its relative radiant sensitivity and angular displacement are given in Figure 5. This device of the sensor is strictly linear between ≤10 lux and ≥1000 lux, and the typical photocurrent is specified for $50\,\mu A$ (at 100 lx). The photocurrent (I_{PCE}) and illuminance equation of the TEMT6000 is given.

$$\text{Illuminance (lux)} = \frac{1}{2} I_{PCE} \left(\mu A \right). \quad (3)$$

In the system where the ambient light sensors are used, the purpose is to adjust the LED illumination level to the desired one and thus to decrease power loss. In daylight-

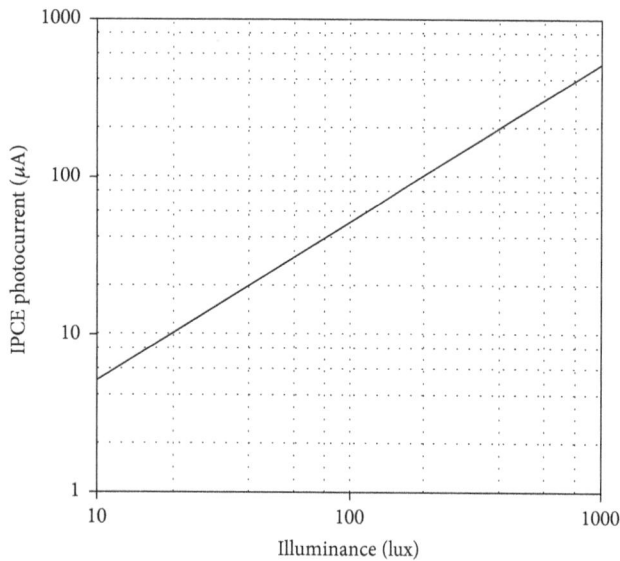

FIGURE 4: The relative photocurrent and illuminance of the TEMT6000 ambient light sensor [40].

exposed buildings and spaces, it can be realized that significant energy saving is achieved, which improves the quality of the visual environment and reduces the demand for electric illumination through the effective employment of automatic lighting controls [41, 42]. A light sensor with a wide spatial sensitivity mounted on the ceiling will have more difficulty during tracking the daylight when it receives more direct light from the window [43]. Also, the presence sensor was used. The reason for using the presence sensor is for the system to cut off energy to save energy and to reduce unnecessary power usage when there is no presence in a given environment for a certain time. The presence sensor's detection distance is 360°, 32 meters.

In the applied central and distributed sensored PV-based smart LED lighting system, an option for switching off the grid is available. The PV-based smart LED illuminating system view is shown in Figure 6. The block charts of the central and the distributed systems are illustrated in Figures 7 and 8, respectively. A direct current (DC) obtained from PV panel's LED lighting system using a DC-level dimming technique has been proposed and compared with distributed (zone) and central light sensor smart control systems. The smart illumination of the $54 \, m^2$ classroom having two windows, each 3.1 m long, has been implemented in a PV-equipped building. The lighting system has been designed for 36 V DC. To obtain this value, 12 V battery and PV panel groups have been used. The smart lighting system consists of 36 W over plaster LED panels designed to work with both PV system and grid. The windows are positioned at one side of the classroom which impede the homogeneous distribution of lighting the evening or in cloudy weathers. Hence, the control of the LED panels is carried out separately for the windows and the walls.

In the PV-based central sensor smart LED illuminating system, the block chart as shown in Figure 7 has been adjusted in such a way that, with the potentiometer

connected to the TEMT6000 ambient light sensor, the ambient reference value can be obtained. The reference value is designed as 300 lux. The maximum rate of this reference for a classroom environment is 350 lux while the minimum value is 250 lux. These values are in the hysteresis range, and these values have been determined by users' preferences [10]. If the light knowledge taken from the ambient light sensor is less than the reference minimum limit, the volume of the light is increased. Similarly, both actions are performed by the proposed algorithm with the PWM method. The sensor light information is sent to the analog input of the STM32F407 microcontroller kit. The PV-based central sensor smart LED illuminating system block chart is shown Figure 7, The PWM control output over the optocoupler brings the illumination level of LED lamps to the required level by adjusting the duty cycle of MOSFETs.

The block chart of the PV-based distributed sensor smart LED illuminating system is shown in Figure 8. In the applied distributed sensored PV-based illuminating system, there exists an option for switching to grid connection.

With potentiometers 1 and 2 connected to the TEMT6000 light sensors, the reference lighting level of the environment can be adjusted. By adjusting PWM values that are produced based on the reference lighting level, thanks to the IRFP460 N-Channel Power MOSFETs, it is possible to control one by one the LED lighting level on the wall column and the window column separately. The reference lighting level depends on the user preference which is recommended as 300 lux for the classroom environment [15].

The right detection of the sensor points is important which affects the whole system performance. For example, when the system is exposed to direct sunlight or LED panel light, wrong feedbacks can be given to the control algorithm. Moreover, the same wrong feedbacks could be caused due to the shadows created by human or object motions [43]. For this reason, the sensors have been positioned close to the ceiling and in the mid points of LED illumination groups. At these points, the sensors will not be exposed to the LED panel, shadow lightings, and direct daylight. If the light knowledge taken from the ambient light sensors is below the reference minimum limit, the volume of the light is increased; in case the light knowledge is over the maximum reference limit, the light volume is decreased. The sensor light knowledge is sent to the analog input of the STM32F407 microcontroller kit. The PWM control output over the optocoupler brings the illumination level of LED lamps to the desired level by adjusting the duty cycle of MOSFETs. In the block structure seen in Figure 8, if there exists solar irradiation or there is redundant power in battery groups, DC power, without requiring conversion into AC, can meet the LED light energy. As shown in Figure 6, the ambient light sensors have been placed near the ceiling of the classroom. The lux voltage information obtained from the ambient sensors is sent to the microcontroller card's 32-bit analog input. The values obtained from here, in relation with the reference value by the algorithm, regulates the duty cycle of both MOS-FETs. Thus, by adjusting the lighting levels of the LED panels, the desired level of the lighting of the environment

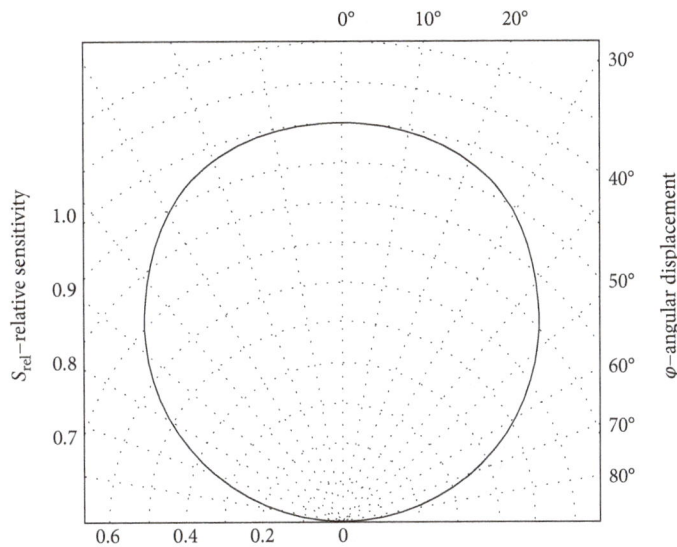

FIGURE 5: Relative radiant sensitivity and angular displacement of TEMT6000 [40].

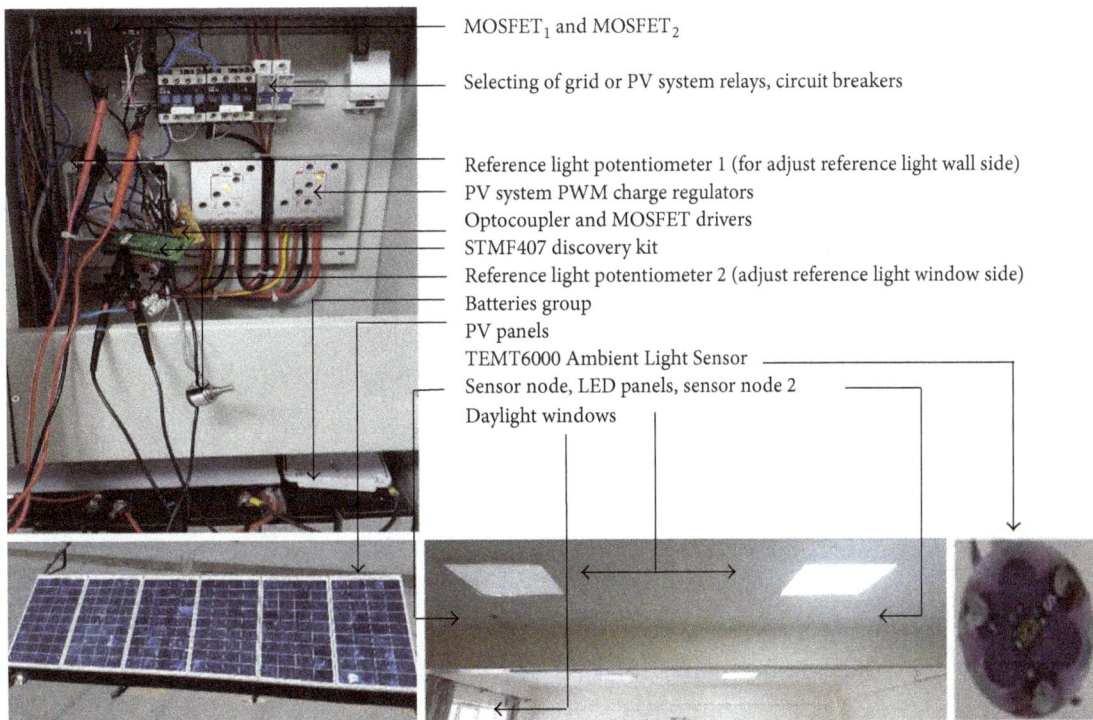

FIGURE 6: PV-based smart LED illuminating system.

is reached. With illumination control, a high level of energy saving has been reached.

2.2. Smart Control Algorithm of Central Sensor and Distributed Sensors LED Lighting System. In the proposed control algorithm, in the light of the voltage information based on the light received from the central or distributed ambient light sensors, it could be adjusted automatically which LED will work at which illumination lux value. The block diagrams and pseudocodes of the proposed algorithm can be seen in Figures 9 and 10, Pseudocodes 1 and 2, respectively. Here the P_s information signifies the presence sensor's 0 or 1 information which is obtained on the human presence in the environment. The user-defined inputs to the control algorithm are room reference light (RRL), presence sensor time delay (P_{std}) and its detection distance, increment (I) value, sampling period, and hysteresis (H). The light information and the active (logic 1) and passive (logic 0) states of the presence sensor signify the nonuser-definable entry information. This nonuser-definable input is the authentic

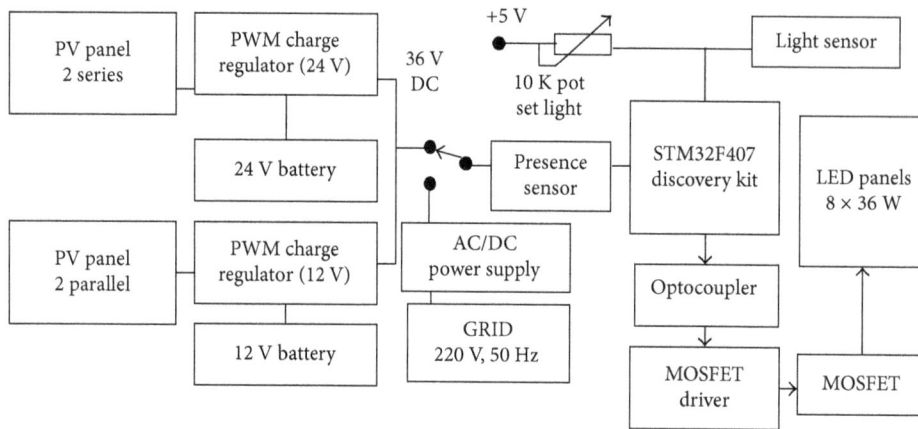

FIGURE 7: PV-based central sensor smart LED lighting system block chart.

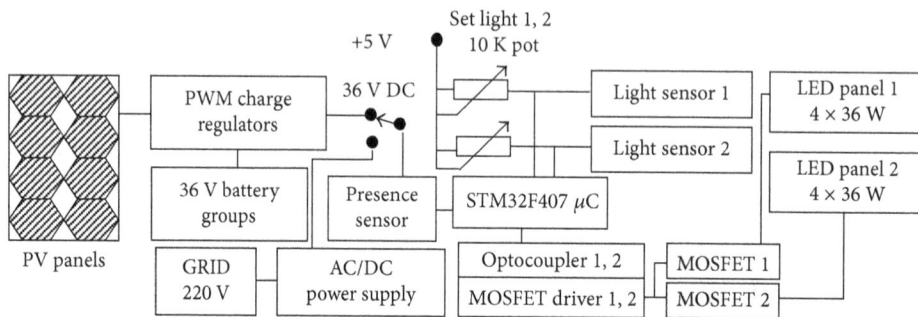

FIGURE 8: PV-based distributed sensor smart LED illuminating system block chart.

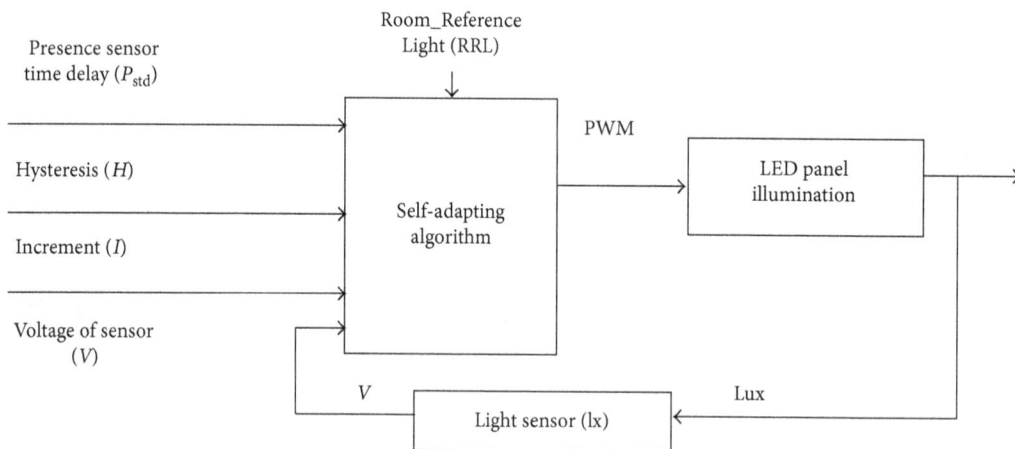

FIGURE 9: The proposed control algorithm central ambient light sensor smart LED system.

information received from the presence sensor. The hysteresis (H) is a user-definable input that is determined at an accepted RRL interval of error 1 and error 2. Value H enables the illumination level to be at a certain interval and precludes unnecessary MOSFET switching losses by not applying duty cycles at minimum changes. Value increment (I) signifies PWM outputs' increase or decrease coefficient. If the value

I in Figures 9 and 10 decreases, then the tracking of the MOSFET for regulating PWM will be slower and the time to reach reference value illumination will be longer. If the I value increases, then the tracking will be faster. However, a large value I will lead to overshoot. The value of I is a user-definable input. In the proposed approach, 500 ms was selected for the sampling period because it is a good

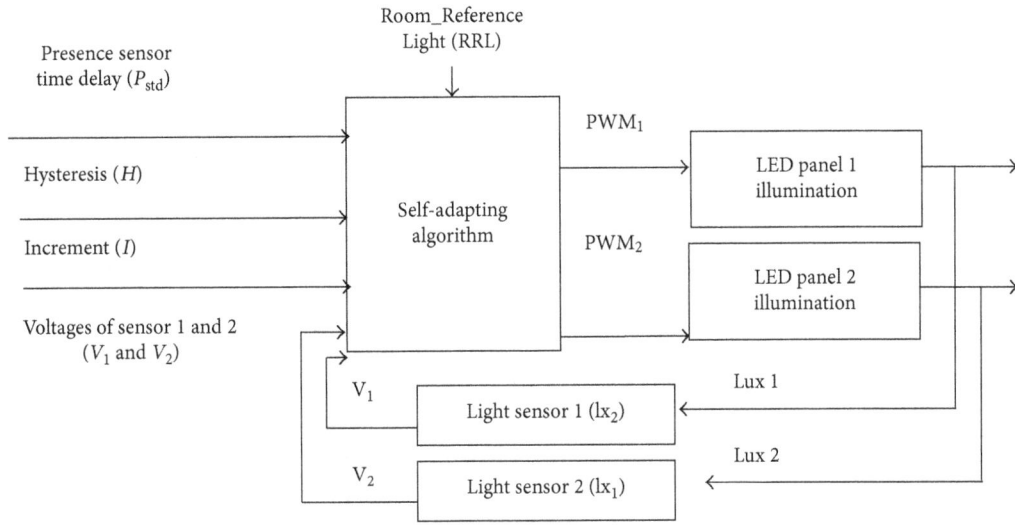

FIGURE 10: The proposed control algorithm-distributed ambient light sensor smart LED system.

trade-off between power and reactivity [14]. The aim of the control algorithm is to minimize error 1 and error 2, as calculated in (4), which should be kept in the hysteresis range. These error signals are calculated

$$\text{Error } 1, 2\,(t_1, t_2) = \text{Room\_Reference\_Light} - \text{Measure\_Light } 1, 2.$$
(4)

The STM32F407 control card minimizes errors 1 and 2 by adjusting the duty cycle value and over the optocouplers composes the duty cycle (PWM_1 and PWM_2) signals of MOSFET 1 and MOSFET 2. This way, the lighting level of the LED groups by the window and wall, which are connected to the MOSFETS' outputs, is controlled. The control card algorithm calculates the PWM values according to

$$\text{PWM}_{1,2}(t_1, t_2) = \begin{cases} \text{Presence\_value } (0 \text{ or } 1)\,[\text{PWM } 1, 2(t_1, t_2) + I] & \text{if error } 1, 2(t_1, t_2) \geq I, \\ \text{Presence\_value } (0 \text{ or } 1)[\text{PWM } 1, 2(t_1, t_2) - I] & \text{if error } 1, 2(t_1, t_2) \leq I. \end{cases}$$
(5)

As it is seen in equation 5, the logic 1 information, which is received from the presence sensor in line with the human presence, will start the PWM signals' creation. The error 1 and 2 values are kept at hysteresis level, and the increase and decrease of the PWM values are provided with value I. The LED panels in the central system are controlled by the MOSFET (Figure 7) PWM signal, and the LED panels in the distributed system are controlled by the $\text{PWM}_{1,2}$ signals of MOSFET 1 and MOSFET 2 (Figure 8). The PWM values of the MOSFETs are produced by the control algorithm according to the reference illumination value. In this way, the LED panels give light at the desired illumination level. All the LED panels in the central sensor illumination system in Figure 9 are controlled by the values obtained from a single light sensor.

The sensor class shown in Figure 9 is located at the center of the ceiling. The information from the sensor is compared with the reference value of 300 lux, and the brightness of all the LED lamps is controlled by a single PWM signal.

In Figure 10, the first sensor gives the brightness information of the four LED lamps on the wall side and the second sensor gives the brightness information of the four LED lamps on the side of windows. The obtained information is compared with the reference value (300 lux), and two PWM signals are obtained. With these PWM signals, the first and second LED lamp groups are controlled separately. The pseudocode of the control algorithm for the central sensor smart lighting system is shown in Pseudocode 1.

When the codes in Pseudocode 1 are examined, the control algorithm starts with the presence sensor being active. With the PWM+I command, the light is increased, and with the PWM−I command, the light of all the lamps is decreased. The pseudocode of the control algorithm for the distributed sensor smart lighting system is shown in Pseudocode 2.

In the proposed smart system algorithm, the illumination level is tried to be kept at a certain value, and in case of minimal changes, the MOSFET PWM has not been regulated. At the same time, the LED lamp power has been regulated to the requested illumination value, which prevents unnecessary power loss and thus brings about an efficient lighting system.

In the distributed sensor LED lighting system, as shown in Pseudocode 2, MOSFET_1 driven by PWM_1 signals controls LED panels by the wall side. On the other hand, MOS-FET_2 driven by PWM_2 signals controls LED panels by the window side.

```
#define Default_Room_Reference_Light
Room_Reference_Light=User_Preference
Light=Light_From_ADC_Lightsensor
Presence_Sensor=Presence_From_Timer
if presence =1                 //presence is sensed
smart_lightingt=1;        // start the saving system
else
smart_lighting=0;
end if
while smart_lighting=1 do
Measure_Light
error=Room_Reference_Light-Measure_Light1
if error>hysteresis then        //adjust lighting level
PWM=PWM-I;
else
PWM=PWM+I;
end if
end while
```

PSEUDOCODE 1: Pseudocode of the control algorithm for the central sensor smart lighting system.

```
#define Default_Room_Reference_Light
Room_Reference_Light=User_Preference
Light1=Light_From_ADC_Lightsensor1
Light2=Light_From_ADC_Lightsensor2
Presence_Sensor=Presence_From_Timer
if presence =1
    smart_lightingt=1;
else
    smart_lighting=0;
end if
while smart_lighting=1 do
Measure_Light1
Measure_Light2
error1=Room_Reference_Light-Measure_Light1
error2=Room_Reference_Light-Measure_Light2
    if error1>hysteresis then
        PWM1=PWM1-I;
    else
        PWM1=PWM1+I;
    end if
    if error2>hysteresis then
        PWM2=PWM2-I;
    else
        PWM2=PWM2+I;
    end if
end while
```

PSEUDOCODE 2: Pseudocode of the control algorithm for the distributed sensor smart lighting system.

In this way, the LED panel groups by the wall and the window are managed by multiple ambient light sensors and distributed smart. Figure 11 illustrates the oscilloscope screenshot of the working system, as the impact of the outer environment lighting level by the window side is higher in comparison with that of the wall side; the LEDs by the window are used with less lighting level whereas the ones

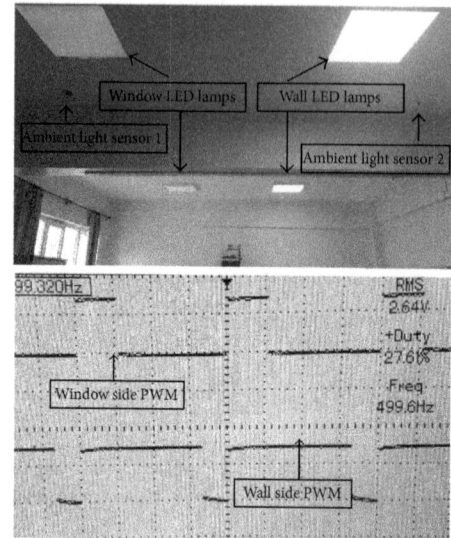

FIGURE 11: Distributed sensored LED control system working example and the oscilloscope image.

by the wall are used at a higher lighting level. As it is noticed from the oscilloscope screenshot, the rate of the window side duty cycle rate is 27.61, but the wall side duty cycle rate is higher. Consequently, these results show that the lighting level of the groups of the LED lamps in the classroom has been efficiently controlled.

In Figure 11, it is targeted that the illumination level on the work surface is 300 lux. The contrast in LED lamps means that the illumination in the window area is higher than the light on the wall side. When the proposed distributed system balances this situation, the brightness levels of the lamps are as shown in Figure 11.

When the distributed sensored smart LED lighting system is controlled with a central single light sensor, it is seen that the LEDs by the window side work with a higher illumination value. Here, the positioning of the single light sensor can be seen in Figure 12.

3. Results and Discussions

3.1. The Comparison of the Proposed Central and Distributed Sensor Lighting System. The central and distributed sensor smart LED lighting system in the DC grid based on the PV system is performed, and the experiment outcomes were compared. Various experiments were performed on the developed smart LED lighting system powered by the DC grid to understand the performance under several environments in classroom conditions.

At the time of implemention of this experiment, the system has been working nonstop for two months and the values obtained belong to the date May 28. During the application, at the hours towards the evening it was noticed that the LED groups by the window side did not illuminate whereas those by the wall side illuminated. For the sunlight illumination measurement by the window side, a DT-1307 solar power meter was utilized (range: 0 to 1.999 W/m²,

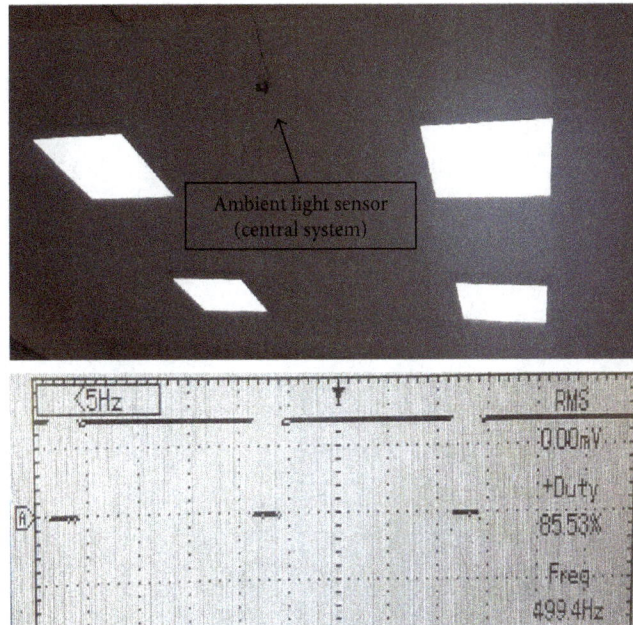

FIGURE 12: Central sensored LED control system working example image.

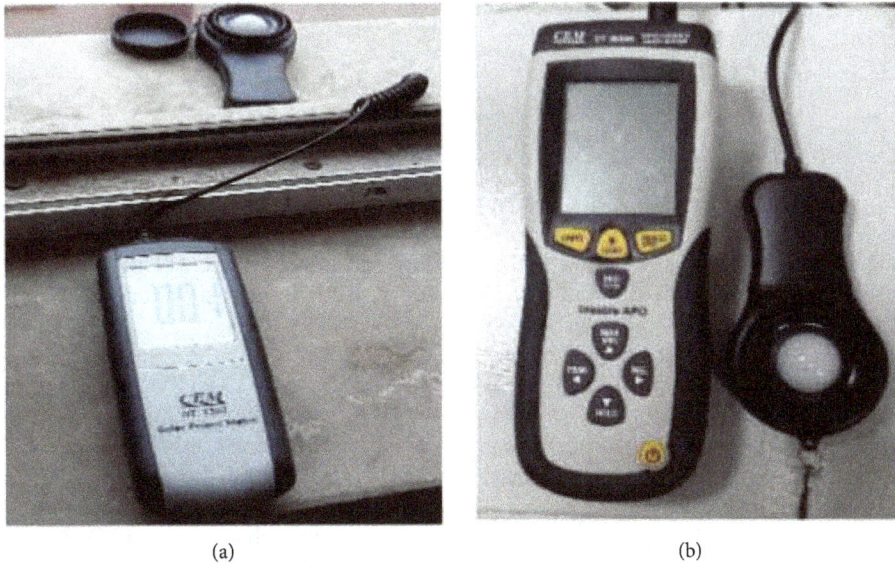

(a)   (b)

FIGURE 13: (a) Solar power meter (W/m^2) and (b) light meter (lux).

resolution: 1 W/m^2, accuracy: ±10 W/m^2), and for the classroom ambient lighting measurement, a DT-8809A professional light meter (max. range: 400,000 lux, resolution: 0.1 lux, accuracy: ±5%, spectral response: CIE photopic, photo detector: one silicon photodiode and spectral response filter) was utilized, as shown in Figure 13.

Depending on the light detected by the photocensors on the ceiling at night time, the workplace task illuminance value, which is 300 lux, has been acquired. In this way, the correlation between the illuminance levels on the ceiling has been provided.

The saved power values can be calculated by (6) [10], where $P_c(t)$ and $P_d(t)$ represent the output power of the proposed central and the proposed distributed sensor control lighting system, respectively [10].

$$P_{\text{saved}}(t) = \Delta P = \int_0^t P_c(t)\mathrm{d}t - \int_0^t P_d(t)\mathrm{d}t. \qquad (6)$$

In the experiment, three cases are conducted to compare the algorithm with both proposed central and distributed light sensors smart lighting systems.

Case 1. Towards evening, the time is 18:00 PM. While the solar power meter by the window shows 8 W/m^2 value, PWM outputs have the values for the window side at 0%

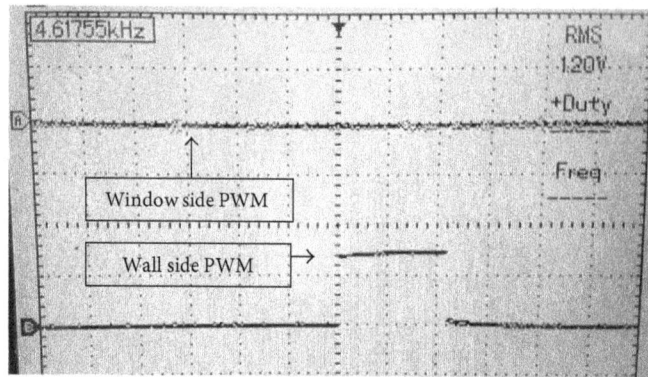

FIGURE 14: PWM signals of the proposed distributed lighting system: case 1.

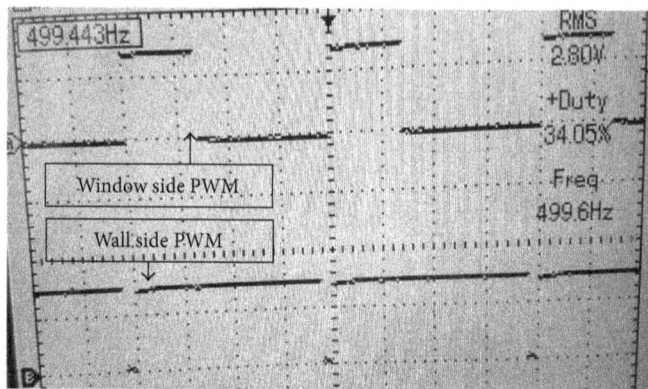

FIGURE 15: PWM signals of the proposed distributed lighting system: case 2.

and the wall side at 25%. Namely, while the LEDs by the window side are illuminating, the LEDs by the wall side of the classroom are functioning at a 25% duty cycle level. When the control system has reached the reference lighting level, the total current value of the LEDs by the wall side is measured as 0.45 A and the current value of the LED lamps by the window side is measured as 0 A. In this case, the power value of the proposed distributed ambient sensor smart LED lighting system is $36 \times 0.45 = 16.2$ W and the power of the central ambient sensor smart LED lighting system is 32.4 W. The difference is $32.4 - 16.2 = 16.2$ W. The oscilloscope screen of the PWM signals of the distributed sensored led lighting system for case 1 is given in Figure 14.

Case 2. The time is 19:30 PM. While the solar power meter shows 1.3 W/m^2, PWM outputs have the following values: for the LEDs by the window side 34.05% and for the LEDs by the wall side 90%. When the control system has reached the reference lighting level, the total current value for the wall side LED panels is measured as 2.3 A and the current value for the window side LED panels is measured 0.6 A. In this case, the power of the proposed distributed ambient sensor smart LED lighting system is $36 (2.3 + 0.6) = 104.4$ W and the power of the central ambient sensor LED lighting system is 165.6 W. The difference value is $165.6 - 104.4 = 61.2$ W.

The oscilloscope screen of PWM signals of the distributed sensor led lighting system for case 2 is given in Figure 15.

Case 3. Time is 19:45 PM, the solar power meter shows 0.1 W/m^2 value, PWM outputs indicate the following values; for the LED panels by the window side 100% and for the LED panels by the wall side 90%. When the control system has reached 304 lux reference lighting level, the total current value of the LED panels by the wall side is measured as 1.99 A, the current value of the LED panels by the window side is 2 A. In this case, while the power of the distributed sensor smart LED lighting system is $36 (1.99 + 2) = 143.64$ W, that of the central sensor smart LED lighting system's is 143.2 W. The difference value between the two is $143.20 - 143.64 = -0.44$ W. Here, it can be seen that the results are almost identical. The oscilloscope screen PWM signals of distributed sensor LED lighting system for case 3 is given in Figure 16.

All the three cases above were applied when it got dark and there was need for artificial illumination. At the earlier hours of the day, artificial illumination was not needed as the classroom had windows. Central and distributed light sensor artificial illumination was applied from 18:00 PM until 20:15 PM (the last hour before the classroom is closed). When all three cases are analyzed, it is seen that the proposed

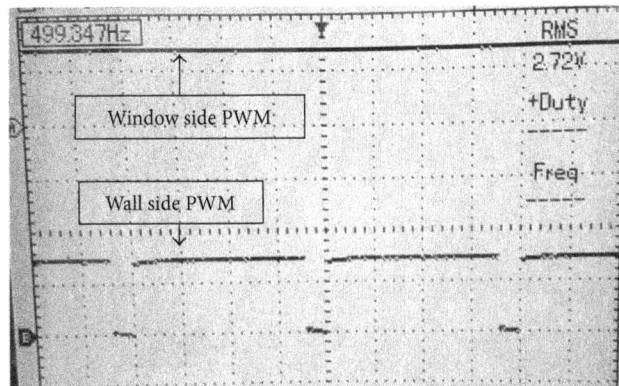

FIGURE 16: PWM signals of the proposed distributed lighting system: case 3.

TABLE 2: The comparison of the proposed distributed sensors lighting with that of the central system.

Time	Solar power (W/m^2)	The total consumed power in distributed sensor smart LED lighting system (W)	The total consumed power in central sensor smart LED lighting system (W)	ΔP (power difference) (W)
18:00	8	16.2	32.4	16.2
18:15	7.1	21.6	43.2	21.6
18:30	6.2	34.5	69.2	34.6
18:45	4.6	44.6	89.2	44.6
19:00	2.9	61.9	123.8	61.9
19:15	2.1	84.9	169.9	84.9
19:30	1.3	104.4	165.6	61.2
19:45	0.1	143.6	143.2	−0.4
20:00	0	146.1	146.1	0
20:15	0	146.1	146.1	0
ΔMean		$\Delta P_{\mathrm{D}} = 80.3$	$\Delta P_{\mathrm{C}} = 112.8$	

distributed light sensor smart illumination has been successful in capturing the desired reference values, and when compared with the proposed central system, it is seen that the proposed distributed system provides more energy saving than the proposed central sensor smart LED lighting system.

In conventional fluorescent lamps, the facility of adjusting PWM is not possible. At the same time, conventional LED illuminating systems function with fixed power. In the proposed central sensored smart LED system, in line with voltage information received from the ambient light sensor stationed in the midpoint of the classroom close to the ceiling, the illumination level of all the LED panels is controlled. With the proposed distributed light sensor lighting system, wall and window LED panels are self-adapting and dimmed separately, and by bringing about an energy-saving situation, it has increased the efficiency of both illumination systems. The comparison of the proposed distributed light sensor smart LED lighting system with that of the central smart light sensor LED system is given in Table 2. By showing the powers consumed by the two systems and their power differences starting towards the evening time, the comparison of the two systems presents us a clearer understanding.

In Table 2, when the power change between the proposed central ambient sensor smart LED illumination system and

the proposed distributed ambient light sensor LED illumination system is compared, it is seen that the highest value is 84.9 W. An efficiency calculation regarding Table 2 can be made as follows:

$$\eta_{\mathrm{efficiency}} = \frac{\Delta P_{\mathrm{C}} - \Delta P_{\mathrm{D}}}{\Delta P_{\mathrm{C}}} = \frac{112.8 - 80.4}{112.8} \times 100 = \%28.7. \quad (7)$$

It is seen that the proposed distributed ambient light sensor LED lighting system is 28.7% more efficient than the proposed central ambient light sensor. The LED lighting system works successfully at the universally determined ambient necessity illumination values. In spring and summer seasons, the use of LED panels and distributed light sensors or central light sensor smart LED lighting system is on the decrease, but in the other seasons its use is higher. The month when the measurement took place is May, which belongs to spring time.

3.2. Cost Analysis. The total setup cost of the central and distributed ambient light sensor smart LED lighting system controller is calculated and listed in Table 3.

TABLE 3: Cost analysis of the developed energy-saving controller.

Item	Unit cost (USD)	Central/distributed system number	Total cost (USD) central/distributed system
STM32F407 discovery kit	25.6	1/1	25.6/25.6
IRFP250 MOSFET	1.4	1/2	1.4/2.8
10 K multiturn pot	4	1/2	4/8
Optocoupler	0.2	1/2	0.2/0.4
5 V power switch regulator	11.2	1/1	11.2/11.2
MOSFET driver	1.9	1/2	1.9/3.8
TEMT6000 Ambient Light Sensor	5.3	1/2	5.3/10.6
Resistor	0.06	1/2	0.06/0.1
100 nF capacitor	0.07	1/2	0.07/0.1
MOSFET cooler	2.1	1/2	2.1/4.3
PCB	0.4	1/1	0.4/0.4
Proposed central/distributed sensor LED lighting system costs			52.2/67.3

TABLE 4: Results of economic analysis for the central and distributed smart led lighting system.

System	kWh unit cost (USD) [44]	Average daily consumption (kWh)	Average total daily consumption for 5 hours	Average monthly consumption (kWh)	Total cost (USD)
The central sensor smart LED lighting	0.12	112.8	564	16.92	1.92
The distributed sensor LED lighting	0.12	80.3	401.5	12	1.44
The normal LED lighting	0.12	288	1440	43.2	5.18

The total costs of the proposed central and distributed system are 52.2 and 67.3 USD, respectively. The system is flexible, meaning it can be expanded.

A monetary payback period for the proposed central-distributed light sensor LED lighting and the normal LED lighting systems is made, as shown in Tables 2 and 4.

When Table 4 is examined, it is seen that the 288 W LED panel-installed power capacity of the central illumination system has an average monthly consumption of 16.92 kWh, and its cost is 1.92 USD. It shows that there is a total cost difference of $5.18 - 1.92 = 3.26$ USD between the central smart LED lighting and the normal LED lighting systems. In Table 4, if the total cost amount for the central smart LED lighting system is divided by the value found, the simple monetary payback period is calculated as 52.2/ $3.26 = 16$ months.

3.3. Visual Comfort. The daylight source can be measured, but it is uncontrollable. Color, distribution, and other features of daylight are highly variable, and the speed of this variability is coincident with many components. This is a characteristic of human nature. In Figure 17, there is a building which is 6 meters south of the working class and there is a light shelf over the roof of the class building which provides indirect daylight diffusion which is one of the best types for both comfort and work. As seen in Figure 17, a daylight of the room reduces energy consumption. The sensors are placed on the ceiling in such a way

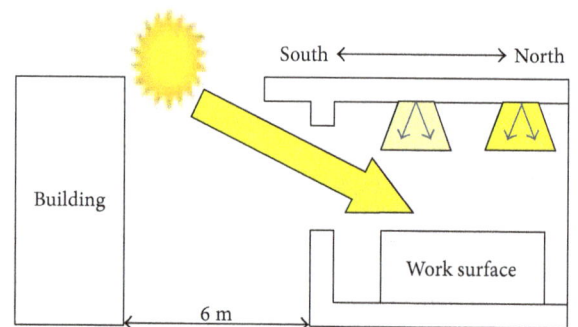

FIGURE 17: The position of the building and artificial lighting controlled with daylight availability.

that they are not affected by direct daylight and the LED lamps are light. A reference 300 lux was targeted for the working surface illumination value.

The color rendering index (CRI) and color temperature (kelvin) are important parameters in achieving visual comfort. The scale range of the color rendering index is 0–100; the natural light has a value of 100 CRI, and it is recommended that the CRI value should be higher than 80. The LED lamps selected for illumination of the class have a color rendering index which is greater than 80, and the color temperature of LED is 4000 K. Visual comfort zones based on color temperature and room illumination level are shown in Figure 18. To achieve better user satisfaction

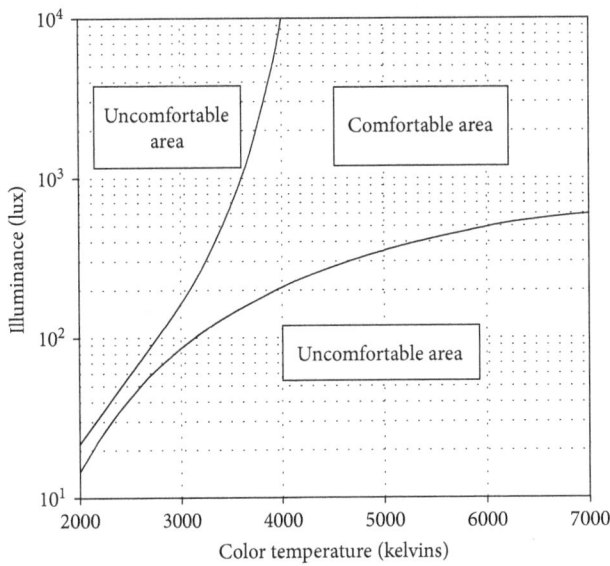

FIGURE 18: Comfort zone of the Kruithof curve [45].

and according to the Kruithof curve given in Figure 18, the intersection of the reference value of the illumination and color temperature of the LED lamp lays in the visual comfort zone.

4. Conclusion

Natural lighting reduces energy consumption of artificial lighting such as LED or fluorescent lighting, and it provides a healthy and comfortable environment. As is known, not much research comparing the distributed and central sensor smart LED lighting systems with respect to energy efficiency has been carried out. In this study, a PV system set up on the building meets the energy demand of the DC grid LED panel structure whose illumination level has been controlled by the proposed self-adapting algorithm based on the information taken from distributed ambient light sensors. The system is composed of the presence sensor, ambient light sensors, PV system, control card, and battery elements. The use of the ambient light sensors and the presence sensor in combination with the user preferences allows the central or the distributed ambient sensor smart to save energy and thus reduce the light intensity. From the measurement taken two hours before sunset, the distributed system energy consumption was 16.2 W while that of the central sensor's stood at 32.4 W. One hour before the sunset, these values stood at 61.92 W and 123.84 W, respectively. When the proposed distributed ambient light sensors and the proposed central ambient light sensor smart LED lighting systems are compared, it is seen that the proposed distributed ambient light sensor smart LED lighting system comes with 28.762% more efficiency. At the same time, both proposed systems have a low consumption rate and building integrated photovoltaic (BIPV) basis which offers an environmentally friendly approach.

Conflicts of Interest

The author declares that there is no conflict of interest regarding the publication of this paper.

References

[1] A. M. Vural, "Contribution of high voltage direct current transmission systems to inter-area oscillation damping: a review," *Renewable and Sustainable Energy Reviews*, vol. 57, pp. 892–915, 2016.

[2] L. Shen, B. He, L. Jiao, X. Song, and X. Zhang, "Research on the development of main policy instruments for improving building energy-efficiency," *Journal of Cleaner Production*, vol. 112, Part 2, pp. 1789–1803, 2016.

[3] M. S. Hossain, N. A. Madlool, N. A. Rahim, J. Selvaraj, A. K. Pandey, and A. F. Khan, "Role of smart grid in renewable energy: an overview," *Renewable and Sustainable Energy Reviews*, vol. 60, pp. 1168–1184, 2016.

[4] R. Smale, M. Hartley, C. Hepburn, J. Ward, and M. Grubb, "The impact of CO_2 emissions trading on firm profits and market prices," *Climate Policy*, vol. 6, no. 1, pp. 31–48, 2006.

[5] S.-H. Kim, I.-T. Kim, A.-S. Choi, and M. K. Sung, "Evaluation of optimized PV power generation and electrical lighting energy savings from the PV blind-integrated daylight responsive dimming system using LED lighting," *Solar Energy*, vol. 107, pp. 746–757, 2014.

[6] S.-H. Yoo, "Simulation for an optimal application of BIPV through parameter variation," *Solar Energy*, vol. 85, no. 7, pp. 1291–1301, 2011.

[7] M. D. Archer and R. Hill, *Clean electricity from photovoltaics*, Imperial College Press, London, 2001.

[8] B. Celik, E. Karatepe, S. Silvestre, N. Gokmen, and A. Chouder, "Analysis of spatial fixed PV arrays configurations to maximize energy harvesting in BIPV applications," *Renewable Energy*, vol. 75, pp. 534–540, 2015.

[9] D. Tran and Y. K. Tan, "Sensorless illumination control of a networked LED-lighting system using feedforward neural network," *IEEE Transactions on Industrial Electronics*, vol. 61, no. 4, pp. 2113–2121, 2014.

[10] I. Chew, V. Kalavally, N. W. Oo, and J. Parkkinen, "Design of an energy-saving controller for an intelligent LED lighting system," *Energy and Buildings*, vol. 120, pp. 1–9, 2016.

[11] I. G. Mason, S. C. Page, and A. G. Williamson, "A 100% renewable electricity generation system for New Zealand utilising hydro, wind, geothermal and biomass resources," *Energy Policy*, vol. 38, no. 8, pp. 3973–3984, 2010.

[12] P. Mallet, P.-O. Granstrom, P. Hallberg, G. Lorenz, and P. Mandatova, "Power to the people!: European perspectives on the future of electric distribution," *IEEE Power and Energy Magazine*, vol. 12, no. 2, pp. 51–64, 2014.

[13] C. K. Lee, S. Li, and S. Y. Hui, "A design methodology for smart LED lighting systems powered by weakly regulated renewable power grids," *IEEE Transactions on Smart Grid*, vol. 2, no. 3, pp. 548–554, 2011.

[14] M. Magno, T. Polonelli, L. Benini, and E. Popovici, "A low cost, highly scalable wireless sensor network solution to achieve smart LED light control for green buildings," *IEEE Sensors Journal*, vol. 15, no. 5, pp. 2963–2973, 2015.

[15] H.-J. Chiu, Y.-K. Lo, J.-T. Chen, S.-J. Cheng, C.-Y. Lin, and S.-C. Mou, "A high-efficiency dimmable LED driver for low-

power lighting applications," *IEEE Transactions on Industrial Electronics*, vol. 57, no. 2, pp. 735–743, 2010.

[16] M. Arias, D. G. Lamar, J. Sebastian, D. Balocco, and A. A. Diallo, "High-efficiency LED driver without electrolytic capacitor for street lighting," *IEEE Transactions on Industry Applications*, vol. 49, no. 1, pp. 127–137, 2013.

[17] S. Li, H. Chen, S.-C. Tan, S. Y. R. Hui, and E. Waffenschmidt, "Power flow analysis and critical design issues of retrofit light-emitting diode (LED) light bulb," *IEEE Transactions on Power Electronics*, vol. 30, no. 7, pp. 3830–3840, 2015.

[18] D. H. W. Li, K. L. Cheung, S. L. Wong, and T. N. T. Lam, "An analysis of energy-efficient light fittings and lighting controls," *Applied Energy*, vol. 87, no. 2, pp. 558–567, 2010.

[19] Y. K. Tan, T. P. Huynh, and Z. Wang, "Smart personal sensor network control for energy saving in DC grid powered LED lighting system," *IEEE Transactions on Smart Grid*, vol. 4, no. 2, pp. 669–676, 2013.

[20] A. Pandharipande and D. Caicedo, "Smart indoor lighting systems with luminaire-based sensing: a review of lighting control approaches," *Energy and Buildings*, vol. 104, pp. 369–377, 2015.

[21] G. Lowry, "Energy saving claims for lighting controls in commercial buildings," *Energy and Buildings*, vol. 133, pp. 489–497, 2016.

[22] F. Rubinstein, G. Ward, and R. Verderber, "Improving the performance of photo-electrically controlled lighting systems," *Journal of the Illuminating Engineering Society*, vol. 18, no. 1, pp. 70–94, 1989.

[23] M. T. Koroglu and K. M. Passino, "Illumination balancing algorithm for smart lights," *IEEE Transactions on Control Systems Technology*, vol. 22, no. 2, pp. 557–567, 2014.

[24] M. Fischer, K. Wu, and P. Agathoklis, "Intelligent illumination model-based lighting control," in *2012 32nd International Conference on Distributed Computing Systems Workshops*, pp. 245–249, Macau, China, June 2012.

[25] G. Boscarino and M. Moallem, "Daylighting control and simulation for LED-based energy-efficient lighting systems," *IEEE Transactions on Industrial Informatics*, vol. 12, no. 1, pp. 301–309, 2016.

[26] F. Rubinstein, "Photoelectric control of equi-illumination lighting systems," *Energy and Buildings*, vol. 6, no. 2, pp. 141–150, 1984.

[27] A. Peruffo, A. Pandharipande, D. Caicedo, and L. Schenato, "Lighting control with distributed wireless sensing and actuation for daylight and occupancy adaptation," *Energy and Buildings*, vol. 97, pp. 13–20, 2015.

[28] D. Caicedo and A. Pandharipande, "Distributed illumination control with local sensing and actuation in networked lighting systems," *IEEE Sensors Journal*, vol. 13, no. 3, pp. 1092–1104, 2013.

[29] S. H. Lee and J. K. Kwon, "Distributed dimming control for LED lighting," *Optics Express*, vol. 21, Supplement 6, pp. A917–A932, 2013.

[30] A. Fernandez-Montes, L. Gonzalez-Abril, J. Ortega, and F. Morente, "A study on saving energy in artificial lighting by making smart use of wireless sensor networks and actuators," *IEEE Network*, vol. 23, no. 6, pp. 16–20, 2009.

[31] J. Vanus, T. Novak, J. Koziorek, J. Konecny, and R. Hrbac, "The proposal model of energy savings of lighting systems in the smart home care," in *12th IFAC conference on programmable devices and embedded systems*, Czech Republic, September 2013.

[32] J. Liu, W. Zhang, X. Chu, and Y. Liu, "Fuzzy logic controller for energy savings in a smart LED lighting system considering lighting comfort and daylight," *Energy and Buildings*, vol. 127, pp. 95–104, 2016.

[33] İ. Kıyak, B. Oral, and V. Topuz, "Smart indoor LED lighting design powered by hybrid renewable energy systems," *Energy and Buildings*, vol. 148, pp. 342–347, 2017.

[34] M. A. Özçelik, "Light sensor control for energy saving in DC grid smart LED lighting system based on PV system," *Journal of Optoelectronics and Advanced Materials*, vol. 18, no. 5-6, pp. 468–474, 2016.

[35] M. A. Özçelik, "The design and implementation of PV-based intelligent distributed sensor LED lighting in daylight exposed room environment," *Sustainable Computing: Informatics and Systems*, vol. 13, pp. 61–69, 2017.

[36] D. K. Gupta, M. Barink, and M. Langelaar, "CPV solar cell modeling and metallization optimization," *Solar Energy*, vol. 159, pp. 868–881, 2018.

[37] "EN 12464-1, light and lighting-lighting of work places part 1: indoor workplaces," EN Standard. September 2017, https://www.en-standard.eu/csn-en-12464-1-light-and-lighting-lighting-of-work-places-part-1-indoor-work-places/.

[38] "LED catalogue," Pelsan. December 2016, http://www.pelsan.com.tr/en-US/catalogues/52212.

[39] J. D. Jennings, F. M. Rubinstein, D. DiBartolomeo, and S. L. Blanc, "Comparison of control options in private offices in an advanced lighting controls testbed," *Journal of the Illuminating Engineering Society*, vol. 29, no. 2, pp. 39–60, 2000.

[40] "TEMT6000 ambient light sensor," Vishay. August 2017, https://www.vishay.com/docs/81579/temt6000.pdf.

[41] A.-S. Choi, K.-D. Song, and Y.-S. Kim, "The characteristics of photosensors and electronic dimming ballasts in daylight responsive dimming systems," *Building and Environment*, vol. 40, no. 1, pp. 39–50, 2005.

[42] S. Ranasinghe and R. Mistrick, "A study of photosensor configuration and performance in a daylighted classroom space," *Journal of the Illuminating Engineering Society*, vol. 32, no. 2, pp. 3–20, 2003.

[43] L. Doulos, A. Tsangrassoulis, and F. V. Topalis, "Multi-criteria decision analysis to select the optimum position and proper field of view of a photosensor," *Energy Conversion and Management*, vol. 86, pp. 1069–1077, 2014.

[44] "Guide for electricity tariff regulations in Turkey, 2016," EMRA. December 2016, http://www.emra.org.tr/en/home.

[45] S. Afshari, S. Mishra, A. Julius, F. Lizarralde, J. D. Wason, and J. T. Wen, "Modeling and control of color tunable lighting systems," *Energy and Buildings*, vol. 68, Part A, pp. 242–253, 2014.

16

LEDs for the Implementation of Advanced Hydrogenation Using Hydrogen Charge-State Control

Chee Mun Chong [ID],[1] Stuart Wenham,[1] Jingjia Ji,[1] Ly Mai,[1] Sisi Wang,[1] Brett Hallam,[1] and Hua Li[2]

[1]*School of Photovoltaic and Renewable Energy Engineering, University of New South Wales, Sydney, NSW, Australia*
[2]*LONGi Lerri Solar, Xi'an, China*

Correspondence should be addressed to Chee Mun Chong; cm.chong@unsw.edu.au

Academic Editor: Chun-Sheng Jiang

Light-induced degradation (LID) of p-type Cz solar cells has plagued the industry for many decades. However, in recent years, new techniques for solving this LID have been developed, with hydrogen passivation of the boron-oxygen defects appearing to be an important contributor to the solution. Advanced hydrogenation approaches involving the control of the charge state for the hydrogen atoms in silicon to enhance their diffusivity and reactivity are developed and evaluated in this work for commercial application using a prototype industrial tool in conjunction with solar cells manufactured on commercial production lines. This prototype tool, unlike the previous successful laser-based laboratory approaches, is based on the use of LEDs for controlling the charge state of the hydrogen atoms. The illumination from the LEDs is also used in this work to passivate process-induced defects and contamination from the respective production lines with significant improvements in both efficiency and stability. The results indicate that the low-cost LED-based industrial tool performs as well as the laser-based laboratory tool for implementing these advanced hydrogen passivation approaches.

1. Introduction

Hydrogen passivation of defects within silicon has been exploited and evaluated for several decades with perhaps more related published papers than most areas of semiconductor physics relating to semiconductor devices. Despite this, some discrepancies remain within published literature relating to areas of key importance such as the involvement of hydrogen in passivating defects such as B-O defects [1–4] and the diffusivity of hydrogen [5]. The theory behind charge states for interstitial hydrogen within silicon is well understood by some authors [6–8] as are the corresponding complications caused in interpreting data relating to the diffusivity of hydrogen and its reactivity [9, 10]. Hydrogen diffusivity is reported to vary by 5 orders of magnitude, often confusing interpretations and corresponding conclusions in the literature. New insights and understanding in this important area are able to be gained for the first time through the

innovative technology developed at UNSW Australia for control of the hydrogen charge state [11, 12] and its use in a range of ways for enhancing mobility and reactivity of hydrogen atoms [13]. This is particularly important for understanding the passivation of defects and recombination mechanisms within silicon solar cells that currently strongly dominate commercial markets.

This new technology from UNSW Australia makes possible for the first time the ability to control the charge state of hydrogen atoms within the silicon wafers, which in turn has facilitated drastic improvements in both the diffusivity (by as much as five orders of magnitude) and reactivity of the hydrogen atoms. Laboratory use of this innovative technology has demonstrated that by controlling the charge state, hydrogen atoms can be used to fix defects and recombination mechanisms within the silicon wafers that are not effectively passivated using conventional hydrogen passivation processes. This manipulation of hydrogen charge states

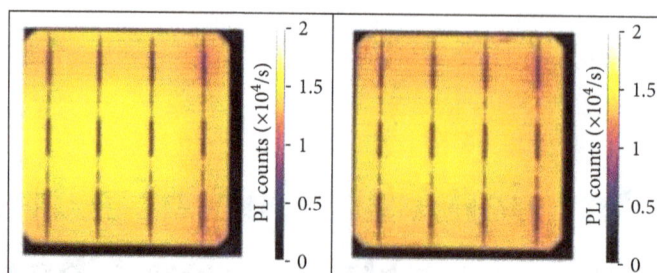

FIGURE 1: Photoluminescence images of two normal production solar cells (group 1) showing relatively uniform PL counts across the cell and only small reductions in the PL response in localised regions.

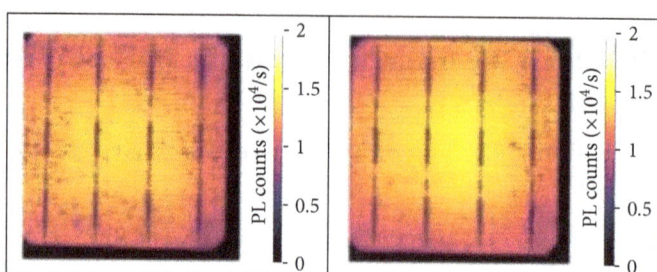

FIGURE 2: Photoluminescence images of two contaminated production solar cells (group 2) with many localised regions of heavily reduced PL response.

can transform the quality of even the worst commercial silicon wafers, into ones with quality equivalent to that of the very best wafers used by the semiconductor industry that can cost as much as one hundred times more [14–16]. This potentially enables the use of significantly lower-cost wafers in photovoltaic (PV) manufacturing without significant performance loss and in fact has simultaneously demonstrated marked improvements in both efficiency and stability.

This technology has been developed and extensively demonstrated in the laboratory through the use of a laser illumination system [17–19] to control the carrier concentrations and hence the hydrogen charge state. As this new technology appears to have significant application in manufacturing, this work aims to study and demonstrate the compatibility of this technology to large area production solar cells using nonlaser illumination sources such as LED or metal halide lights. A new prototype production tool using an array of LED light sources combined with mirror-like reflectors was used to implement the control of the interstitial hydrogen charge states for large area screen-printed solar cells produced on commercial production lines. In particular, this prototype production tool enables the implementation of the hydrogenation technology to solve light-induced degradation (LID) in Czochralski (Cz) wafers and appears to also passivate many other forms of recombination due to contamination and crystallographic imperfections.

This new prototype production tool was specifically used to investigate the impact of the innovative hydrogenation technology on the performance and stability of silicon solar cells that are fabricated from p-type Cz on the production lines of an industry partner, Lerri Solar. These solar cells after treatment were stable and no longer susceptible to LID following the hydrogenation treatment.

2. Experiment

Commercial silicon solar cells from two production lines were obtained from Lerri Solar. These are 156 cm × 156 cm screen-printed p-type monocrystalline silicon solar cells. One group of cells was obtained from the "normal" production line while a second group of cells were from a poorly performing, "contaminated" production line.

Being fabricated from the same Cz silicon wafers, both groups of cells are subject to LID. However, the cells from the contaminated production line also have localised degraded regions of low lifetime that appear to be as a result of contamination introduced during processing. Such contaminated regions are clearly evident in photoluminescence (PL) images [20] such as that shown in Figure 1 with their main impact being the increased ideality factor for the device suggesting that the contamination is creating significant junction recombination in these contaminated regions.

Cells from the normal production line represent our baseline control cells. Figure 2 shows the photoluminescence images [20] of these solar cells with PL counts as the unit. PL images have been deconvolved to negate the effects of photon smearing and improve clarity by the method of Payne et al. [21]. These cells are similar to any other p-type Cz production solar cells and are therefore susceptible to light-induced degradation (LID) caused by the formation of recombination-active boron-oxygen defects. The aim is to use the new LED-based prototype production tool to implement the innovative hydrogenation technology involving the control of the hydrogen charge states on these cells to stabilise them and reduce or eliminate any subsequent LID.

Cells from the contaminated production line are similarly susceptible to LID, and their performance is also further

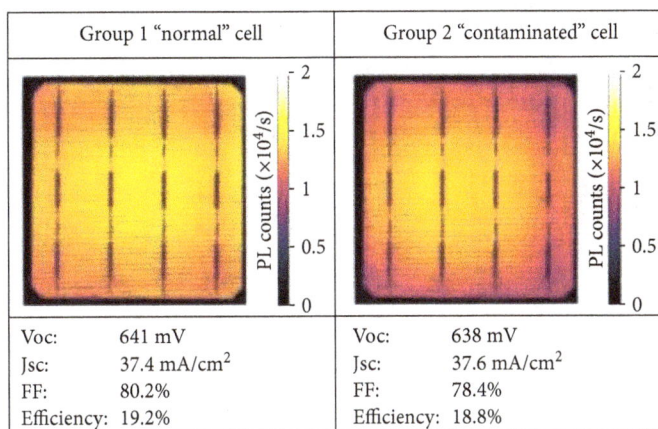

Group 1 "normal" cell	Group 2 "contaminated" cell

Voc:	641 mV	Voc:	638 mV
Jsc:	37.4 mA/cm^2	Jsc:	37.6 mA/cm^2
FF:	80.2%	FF:	78.4%
Efficiency:	19.2%	Efficiency:	18.8%

FIGURE 3: PL image and IV characteristics of a group 1 normal solar cell and a group 2 contaminated solar cell as received.

degraded by the contamination introduced during production. The aim is to use the hydrogenation technology to both passivate the damage caused by the contamination and to stabilise these cells to eliminate their LID. Figure 1 shows the photoluminescence images of these contaminated production solar cells. The localised contamination in these solar cells clearly shows up as dark spots in the open circuit photoluminescence images of Figure 1. Given that the PL response is proportional to carrier lifetime [20], these dark spots indicate areas of lower minority carrier lifetime throughout the solar cell.

Twelve solar cells from both these groups of production cells were put through the following process:

(1) PL images and IV characteristics of cells upon receipt from the industry partner were obtained.

(2) Half the cells were set aside (no treatment) while the other half of the cells were treated in the prototype hydrogenation tool.

(3) All the cells were subjected to a prolonged light-induced degradation (LID) step performed at 35–40°C and 78 mW/cm^2 under halogen lamp illumination for up to 66 hours.

(4) PL images and IV characteristics of the cells following the prolonged LID step were obtained.

The prototype industrial hydrogenation tool is similar in design to many belt firing furnaces except that, in this case, the actual cell temperature is monitored and used as the measurement that determines the power delivered to the furnace. The primary heat source is the LED lights that operate predominantly at wavelengths of 490 nm and 560 nm (measured peak wavelengths) and at about 1.2 W/cm^2 but whereby the LED lights can be pulsed at any required frequency (or mark-space ratio) necessary to reduce the power delivered to the cells to keep the temperature at the desired level. These LEDs were chosen due to availability and being of a sufficiently short wavelength to generate electron-hole pairs. Since the carrier concentrations in the silicon are used for controlling the charge states for the

hydrogen atoms, provided the off-time for the LED lights is significantly less than the minority carrier lifetimes in the wafer, the charge-state control for the hydrogen is not adversely affected by the pulsing. Lenses were used to ensure uniform illumination of the entire solar cell at an intensity of around 12 suns equivalent in relation to the density of photons available to generate electron-hole pairs within the solar cell.

The duration of the hydrogen passivation process is dependent upon the length of the furnace and the corresponding belt speed. For the present work, the duration of the passivation process was about 240 seconds, but with similar results achieved with much shorter times.

The cell PL images were measured using the lab tool R1 developed by BT Imaging. The open circuit voltage (Voc), short circuit current (Jsc), fill factor (FF), and cell efficiency were measured by a h.a.I.m. IV tester (AM 1.5, one sun, 25°C).

3. Results and Discussion

The electrical properties of the solar cells are outlined in Figure 3, showing that group 1 normal cells perform better than group 2 contaminated cells, primarily due to higher fill factors.

The PL image of the group 2 contaminated solar cell shows dark spots across the cell, particularly around the edges of the cell. These represent locally defective areas that are most likely caused by contamination during processing. Such dark spots are not present in group 1 normal solar cells, and the PL images of group 1 normal solar cells are generally uniformly brighter by comparison. These cells have few regions with slightly reduced PL response, indicating only a low level of contamination.

Cells from group 1 and group 2 are light soaked for 48 hours to check their susceptibility to light-induced degradation. These cells showed a decrease in their performance, and the corresponding electrical properties are outlined in Figure 4.

The performance drops by 0.2–0.3% absolute in the solar cells after the 48-hour light soaking. This is most likely

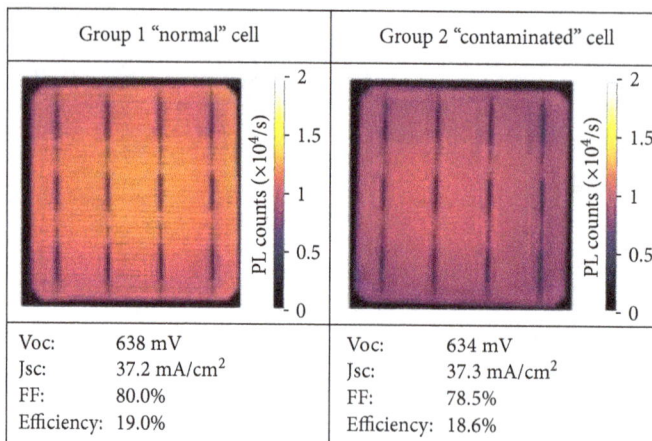

FIGURE 4: PL image and IV characteristics of a group 1 normal cell and group 2 contaminated cell solar cell after light soaking for 48 hours.

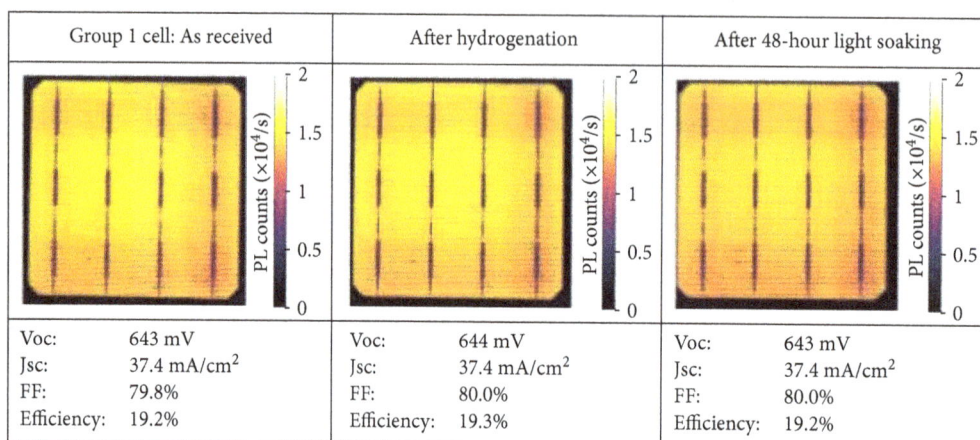

FIGURE 5: Performance of a group 1 normal cell after hydrogenation and then light soaking showing stable lifetimes after light soaking.

caused by the formation of B-O defects in the solar cells under illumination, a common degradation mechanism in p-type Cz silicon [11, 22]. During light soaking, boron and oxygen in the solar cell are believed to bond together to form recombination-active B-O defects, which then degrade the performance of the solar cells [23]. The formation of B-O defects reduces carrier lifetime, which can significantly reduce the Voc. This reduction in lifetime also causes a slight reduction in the Jsc. Due to the injection-level dependence of the recombination caused by B-O defects, a reduction in FF also occurs [18]. This lifetime degradation typically takes 24–48 hours at room temperature to take place. However, if the samples are light soaked for too long at these low temperatures, hydrogen can slowly passivate the B-O defects [24], therefore increasing the lifetime of the wafers, leading to a recovery in performance.

The UNSW hydrogenation technology is then applied to solar cells from both group 1 and group 2, and these cells were then light soaked for 48 hours. This gives a good indication of the impact of the hydrogenation technology on the performance and stability of these solar cells. The electrical properties of these cells are recorded in Figure 5 (group 1) and Figure 6 (group 2), respectively. This increased stability

is due to the hydrogen passivation of the B-O defects, which protects against subsequent degradation. This results in stable electrical parameters after light soaking.

The performance of the solar cell improved marginally after undergoing hydrogenation. Most importantly, the solar cell was stable and did not degrade following exposure to light for 48 hours. These cells appear to be no longer susceptible to LID.

The performance of the group 2 contaminated cell improved by about 0.2–0.3% absolute following hydrogenation. This improvement in electrical performance is most likely due to the passivation of the defects in the solar cell caused by contamination introduced in the cell during production. The localised spots of high recombination evident in the PL images of Figure 1 are not nearly as prominent following the passivation treatment, suggesting that the passivation was having its greatest beneficial impact in these high recombination areas. Although the actual cause of the contamination is not as yet known, the fact that it affects every cell from the affected production line during device fabrication has made it relatively simple to use in-line PL monitoring throughout the line to ascertain the point at which the contamination is occurring.

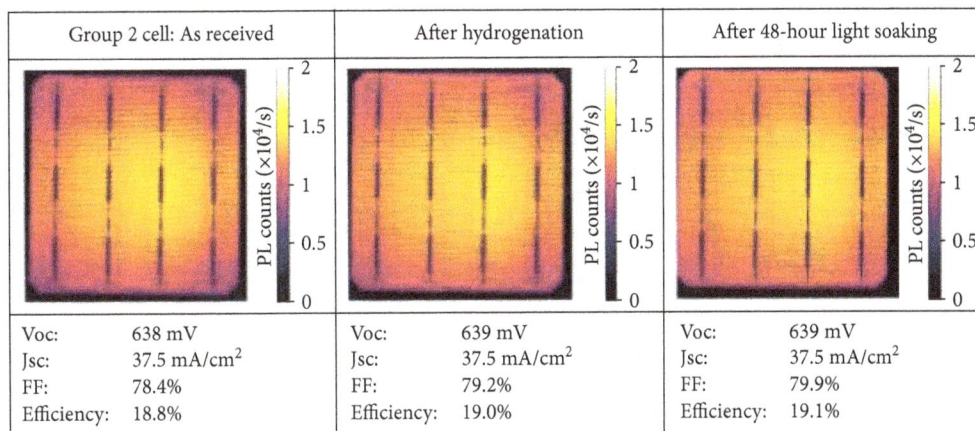

Group 2 cell: As received	After hydrogenation	After 48-hour light soaking
Voc: 638 mV	Voc: 639 mV	Voc: 639 mV
Jsc: 37.5 mA/cm^2	Jsc: 37.5 mA/cm^2	Jsc: 37.5 mA/cm^2
FF: 78.4%	FF: 79.2%	FF: 79.9%
Efficiency: 18.8%	Efficiency: 19.0%	Efficiency: 19.1%

FIGURE 6: Performance of a group 2 contaminated cell after hydrogenation showing stable lifetimes after light soaking.

After light soaking, the passivated solar cell did not degrade, thus demonstrating that the cell is stable and not susceptible to LID. Interestingly, the performance of the solar cell actually increased slightly during light soaking. One possible reason for this is that only some of the defects caused by contamination were passivated during the hydrogenation process, with the remainder continuing to be passivated under the conditions imposed by the light soaking conditions. During the light soaking process which was conducted at slightly elevated temperatures, it is possible that defects caused by contamination near the junction region continue to be passivated and rendered inactive. As a result, the performance of the solar cell improved following light soaking. This suggests that the hydrogenation process needs to be further optimised so as to better passivate the defects caused by contamination introduced during production. In the same way, it is expected that the hydrogenation process is also able to passivate defects in solar cells arising from the use of lower cost, more highly defected silicon wafers, potentially enabling such wafers to lower costs while avoiding significant performance loss.

Interestingly, following hydrogenation and light soaking, the performance levels of the solar cells from group 1 are very similar to the performance of solar cells from group 2, with all appearing to be stable against future LID. This shows that for the latter, the hydrogenation process has passivated both the defects caused by contamination as well as the B-O defects upon exposure to light. Cells from this work from both groups are currently being encapsulated into panels so as to confirm the stability of the solar panels and monitor their performance in the field.

4. Conclusions

Hydrogen passivation of defects and recombination mechanisms within silicon wafers have been demonstrated in the laboratory. Following this treatment, solar cells demonstrated significant improvements in both efficiency and stability. A new prototype industrial tool for the implementation of this technology has been developed using LED lights. Large-area screen-printed p-type Cz silicon production solar cells treated in this prototype industrial tool showed a decrease in the degradation caused by contamination. Following light soaking, performance of these large area production solar cells was stable and did not decrease, indicating that the hydrogenation technology implemented with the LED lights has solved light-induced degradation in these p-type Cz silicon solar cells. Interestingly, following hydrogenation and light soaking, the performance levels of the contaminated solar cells are very similar to the performance levels of the normal solar cells indicating that the hydrogenation process implemented in the LED-based prototype industrial tool has passivated both the defects caused by contamination as well as the boron-oxygen defects upon exposure to light.

Disclosure

The views expressed herein are not necessarily the views of the Australian government, and the Australian government does not accept responsibility for any information or advice contained herein.

Conflicts of Interest

The authors declare that there is no conflict of interest regarding the publication of this paper.

Acknowledgments

This program has been supported by the Australian government through the Australian Research Council (ARC) and the Australian Renewable Energy Agency (ARENA). The authors would like to thank the UK Institution of Engineering and Technology (IET) for their funding support for this work through the A.F. Harvey Engineering Prize.

References

[1] K. Munzer, "Hydrogenated silicon nitride for regeneration of light induced degradation," in *Proceedings of the 24th European Photovoltaic Solar Energy Conference 2009*, pp. 1558–1561, Hamburg, 2009.

[2] S. Wilking, C. Beckh, S. Ebert, A. Herguth, and G. Hahn, "Influence of bound hydrogen states on BO-regeneration

kinetics and consequences for high-speed regeneration processes," *Solar Energy Materials and Solar Cells*, vol. 131, pp. 2–8, 2014.

[3] D. Walter, B. Lim, K. Bothe, V. Voronkov, R. Falster, and J. Schmidt, "Effect of rapid thermal annealing on recombination centres in boron-doped Czochralski-grown silicon," *Applied Physics Letters*, vol. 104, no. 4, article 042111, 2014.

[4] N. Nampalli, B. Hallam, C. Chan, M. Abbott, and S. Wenham, "Evidence for the role of hydrogen in the stabilization of minority carrier lifetime in boron-doped Czochralski silicon," *Applied Physics Letters*, vol. 106, no. 17, article 173501, 2015.

[5] B. Sopori, Y. Zhang, and N. Ravindra, "Silicon device processing in H-ambients: H-diffusion mechanisms and influence on electronic properties," *Journal of Electronic Materials*, vol. 30, no. 12, pp. 1616–1627, 2001.

[6] C. Herring, N. M. Johnson, and C. G. Van de Walle, "Energy levels of isolated interstitial hydrogen in silicon," *Physical Review B*, vol. 64, no. 12, article 125209, 2001.

[7] N. Johnson and C. Herring, "Diffusion of negatively charged hydrogen in silicon," *Physical Review B*, vol. 46, no. 23, pp. 15554–15557, 1992.

[8] C. Seager and R. Anderson, "Two-step debonding of hydrogen from boron acceptors in silicon," *Applied Physics Letters*, vol. 59, no. 5, pp. 585–587, 1991.

[9] D. Mathiot, "Modeling of hydrogen diffusion in *n*- and *p*-type silicon," *Physical Review B*, vol. 40, no. 8, pp. 5867–5870, 1989.

[10] R. Job, W. Fahrner, N. Kazuchits, and A. Ulyashin, "A two-step low-temperature process for a P-N junction formation due to hydrogen enhanced thermal donor formation in P-type Czochralski Silicon," *MRS Proceedings*, vol. 513, no. 1, p. 337, 1998.

[11] B. J. Hallam, P. G. Hamer, S. R. Wenham et al., "Advanced bulk defect passivation for silicon solar cells," *IEEE Journal of Photovoltaics*, vol. 4, no. 1, pp. 88–95, 2014.

[12] P. Hamer, B. Hallam, S. Wenham, and M. Abbott, "Manipulation of hydrogen charge states for passivation of P-type wafers in photovoltaics," *IEEE Journal of Photovoltaics*, vol. 4, no. 5, pp. 1252–1260, 2014.

[13] S. Wenham, S. Wenham, P. Hamer et al., "Advanced hydrogenation of silicon solar cells, 2013," 2014, United States Patent 2015/0111333 A1.

[14] S. Wang, A. Wenham, P. Hamer et al., "Stability of hydrogen passivated UMG silicon with implied open circuit voltages over 700mV," in *2015 IEEE 42nd Photovoltaic Specialist Conference (PVSC)*, New Orleans, LA, USA, 2015.

[15] J. Kraiem, B. Drevet, F. Cocco et al., "High performance solar cells made from 100% UMG silicon obtained via the PHOTO-SIL process," in *2010 35th IEEE Photovoltaic Specialists Conference*, pp. 001427–001431, Honolulu, HI, USA, 2010.

[16] G. del Coso, C. del Canizo, and W. C. Sinke, "The impact of silicon feedstock on the PV module cost," *Solar Energy Materials and Solar Cells*, vol. 94, no. 2, pp. 345–349, 2010.

[17] P. Hamer, S. Wang, B. Hallam et al., "Laser illumination for manipulation of hydrogen charge states in silicon solar cells," *Physica Status Solidi (RRL) - Rapid Research Letters*, vol. 9, no. 2, pp. 111–114, 2015.

[18] B. Hallam, P. Hamer, S. Wang et al., "Advanced hydrogenation of dislocation clusters and boron-oxygen defects in silicon solar cells," *Energy Procedia*, vol. 77, pp. 799–809, 2015.

[19] L. Song, A. Wenham, S. Wang et al., "Laser enhanced hydrogen passivation of silicon wafers," *International Journal of Photoenergy*, vol. 2015, Article ID 193892, 13 pages, 2015.

[20] T. Trupke, R. Bardos, M. Schubert, and W. Warta, "Photoluminescence imaging of silicon wafers," *Applied Physics Letters*, vol. 89, no. 4, article 044107, 2006.

[21] D. N. R. Payne, M. K. Juhl, M. E. Pollard, A. Teal, and D. M. Bagnall, "Evaluating the accuracy of point spread function deconvolutions applied to luminescence images," in *2016 IEEE 43rd Photovoltaic Specialists Conference (PVSC)*, Portland, OR, USA, 2016.

[22] C. Sun, F. E. Rougieux, and D. Macdonald, "A unified approach to modelling the charge state of monatomic hydrogen and other defects in crystalline silicon," *Journal of Applied Physics*, vol. 117, no. 4, article 045702, 2015.

[23] J. H. Brett, E. C. Catherine, C. Ran et al., "Rapid mitigation of carrier-induced degradation in commercial silicon solar cells," *Japanese Journal of Applied Physics*, vol. 56, no. 8S2, article 08MB13, 2017.

[24] H. Brett, B. Jose, P. David et al., "Modelling the long-term behaviour of boron-oxygen defect passivation in the field using typical meteorological year data (TMY2)," in *32nd European Photovoltaic Solar Energy Conference and Exhibition*, pp. 555–559, Munich, Germany, 2016.

17

Thermal Feature of a Modified Solar Phase Change Material Storage Wall System

Chenglong Luo ⓘ,[1] **Lijie Xu ⓘ,**[2] **Jie Ji ⓘ,**[2] **Mengyin Liao,**[1] **and Dan Sun**[1]

[1]*Institute of Energy Research, Jiangxi Academy of Sciences, Nanchang 330096, China*
[2]*Department of Thermal Science and Energy Engineering, University of Science and Technology of China, Hefei 230027, China*

Correspondence should be addressed to Chenglong Luo; xxlong@ustc.edu and Jie Ji; jijie@ustc.edu.cn

Academic Editor: Zhonghao Rao

This work is to study a novel solar PCM storage wall technology, that is, a dual-channel and thermal-insulation-in-the-middle type solar PCM storage wall (MSPCMW) system. The system has the following four independent functions, passive solar heating, heat preservation, heat insulation, and passive cooling, and it can agilely cope with the requirements of climatization of buildings in different seasons throughout the year and is exactly suitable for building in regions characterized by hot summer and cold winter. The present work experimentally analyzes thermal feature of the system working in summer and winter modes, respectively.

1. Introduction

The application of solar energy in buildings to reduce the final energy consumption of conventional energy is an important approach to develop a low-carbon society. The way the phase change material (PCM) provides indirect heat storage is related to energy absorption, which turns into latent heat instead of self-temperature rise. Phase change material (PCM) has strengths of small volume, low temperature, and high heat storage. Therefore, it is a good and efficient heat storage material to be used in building climatization. Hence, the investigation of applying the combination of PCM storage technology and solar energy technology to energy efficient building emerges and receives more and more attention.

Focusing on reducing energy consumption for space heating in building, much work has been done to study and use Trombe wall system which is a high-efficiency simple structure that does not require maintenance [1]. Although this system has been well developed, hurdles remain such as the low annual utilization rate in places with hot and lengthy summer and the often suffering problem of summer overheating [2, 3]. Therefore, its broad application, improvement, and development have been implemented in the decades since it was proposed. Jie et al. [4, 5] made a lot of efforts on investigation of a PV-Trombe wall system. Koyunbaba et al. [6] proposed a BIPV Trombe wall model by computational fluid dynamics (CFD) analysis. The three-dimensional model for the shutter structure of Trombe wall was established in [7], and the comparison with experimental data and an optimal design scheme were conducted. But there are a few research work for the combination of Trombe wall technology and PCM envelope structure.

With regard to the combination of solar energy application technology and PCM envelope structure, current researches and discussions mainly focus on PCM floor [8], PCM wall [9–12], and PCM roof [13, 14]. Soares et al. [15] demonstrated that the approach of combining solar energy utilization technology and PCM envelope structure can effectively reduce the room temperature fluctuation in solar energy building caused by lack of solar energy during the night or uncertain weather, thus enhancing the in-room thermal comfort. However, some large areas of China are characterized by hot summers and cold winters and usually require more than three months of air conditioning cooling. In winter, buildings need heating/heat preservation, while in summer they need heat cooling/insulation. Since solar energy is a thermal energy, it can be used conveniently for building heating in winter, but it can also cause overheating and heavy load of air conditioner in summer. Thus, in regions with hot summers and cold winters, current solar PCM storage technology cannot fully satisfy the application requirements.

Based upon the analysis above, a former work proposed a novel solar PCM storage wall technology that combines Trombe-wall-like technology and phase change material storage technology, that is, a dual-channel and thermal-insulation-in-the-middle type solar PCM storage wall (MSPCMW) system [16]. This system has the following four independent functions: passive heating, heat preservation, heat insulation, and passive cooling. Therefore, it can easily cope with the requirements of different seasons throughout the year when applied to buildings in regions with hot summers and cold winters. To deeply experimentally analyze thermal feature of the system working in summer and winter modes, respectively, temperature variation and distribution of PCM plates, insulated absorbing plate, and air channels are studied by comparison of temperature difference between the monitoring nodes in the present work.

2. Principles of MSPCMW System and Experiment Introduction

MSPCMW system is a combination of Trombe-wall-like technology and phase change material storage technology. A schematic of the system is shown in Figure 1, mainly consisting of a MSPCMW module and a hot-box room with indoor upper and lower vents. The module includes PCM wall, thermal insulation layer, interior and exterior flow channels, heat-absorbing aluminum plate covered by selective absorption coating, indoor upper and lower vents, outdoor upper and lower vents, insulation layer upper and lower vents, glass cover board, and frame. Heat-absorbing aluminum plate and thermal insulation layer are combined as insulated absorbing plate. The structure in which interior and exterior flow channels are separated by thermal insulation layer differs from the single-channel structure of conventional Trombe wall system. The detailed operation modes and functions are as follows:

Summer mode: (a) Heat insulation mode: in summer's daytime, when building needs thermal insulation protection, the indoor and middle layer upper and lower vents are closed, while the outdoor ones are kept opened. Ambient wind pressure together with thermosiphon pressure would form a circular flow between the exterior channel and the outdoor air that takes the solar energy absorbed by aluminum plate back to the environment. Meanwhile, the thermal insulation layer prevents heat conduction into the room, reducing building's absorption of solar energy. (b) Passive cooling mode: when building needs insulation protection, such as summer's night, the indoor upper and lower vents are shut while the middle layer and outdoor ones are kept opened. Under the action of ambient wind pressure, the formed circular flow among the interior channel, the exterior channel, and the outdoor cool air can cool down the PCM wall, reducing indoor air temperature and storing PCM wall's cold energy.

Winter mode: (a) Solar passive heating function: in winter daytime, when building needs heating, the outdoor upper and lower vents of system close, the middle layer ones

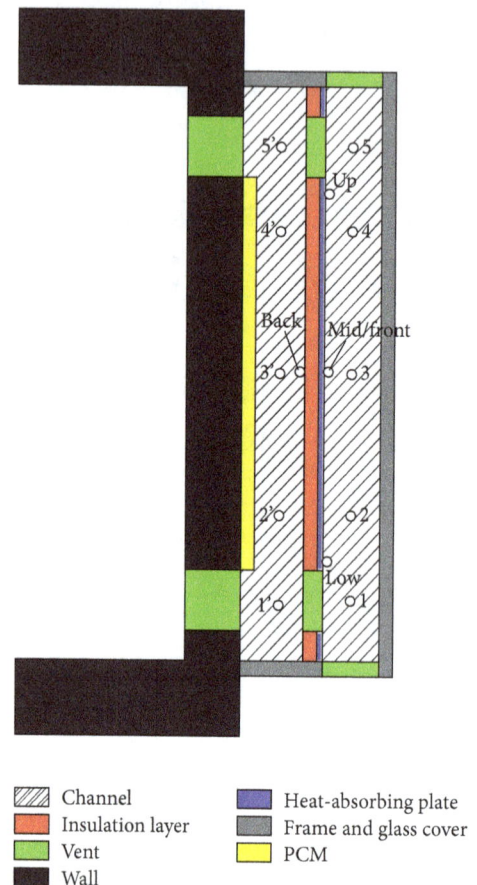

FIGURE 1: The structural principle of the proposed system and the arrangement diagram of thermocouples.

open, and the indoor ones can open/close to implement the interactive adjustment between the indoor temperature rise rate and the stored heat amount of PCM. Aluminum absorbing plate heats up the air of exterior flow channel by absorbing solar radiation irradiated on it. Among the air in the exterior channel, the interior channel, and indoor, the natural circulation due to thermosiphon occurs and induces circular exchange, heating up PCM wall and indoor air, eventually achieving the solar passive heating in building. (b) Heat preservation function: in winter nights, when the building needs heat preservation, the indoor, outdoor, and middle layer upper and lower vents all close, and PCM wall transfers the heat stored during daytime into the room via heat conduction to its neighboring building wall. In addition, the thermal insulation layer composed of insulation material can block the heat loss toward outdoor as much as possible.

The experimental test of the system was carried out on a comparative hot-box test platform located in Hefei City, Anhui Province, characterized by hot summer and cold winter zones. The test system included two hot-box rooms. The experimental room was the hot-box room installed with MSPCMW module, and the other one was the reference room. PCM wall was made by orderly laying and pasting 11 PCM plates on the building's south-facing wall using thermal silicone grease. Each plate measured 0.45 m × 0.3 m × 0.01 m

(L, W, and H). Figure 2 presents the array pattern of PCM plates and monitoring point distribution of thermocouples. The plate was wrapped in aluminum and plated with anticorrosive coating. Its interior components are crystalline hydrate and organic PCM; thus, it benefits from both phase change materials of hydrate and organic matter.

Copper-constantan thermocouple with ice-point compensation (accuracy of ±0.2°C) measures the temperature of experimental system. Five thermocouples are arranged in the interior air channel and the exterior air channel, respectively. Three thermocouples are placed on heat-absorbing aluminum plate, and one is on the back of the insulation layer (shown in Figure 1). There are 5 thermocouples placed on internal surfaces (bonding with the south wall surface) of the middle PCM plates (plates 1, 2, and 3). The locations of measuring points are referred to as circular marks in Figure 2. As shown, three monitoring points are evenly arranged on PCM plate 2 along vertical direction, and the other two are, respectively, positioned in the center of PCM plates 1 and 3. The measuring system also includes ambient temperature measurement and total solar radiation intensity of south-facing vertical surface obtained by TBQ-2 pyranometer. All temperature data and radiation data are collected in real time by Agilent 34970A data collector.

3. Results and Discussions

3.1. Summer Mode. The summer tests were conducted during August 28–30, 2016, in which the outdoor vents were kept opened while the indoor vents were kept closed; the middle layer vents were shut during daytime and opened during the night, and the switching time was around 7:00 and 17:30.

3.1.1. Temperature Variation and Distribution of PCM Plates. Figure 3 shows comparison of interior side temperature difference between centers of the upper, middle, and lower PCM boards. T_{B-E} is the temperature difference between nodes B and E, and T_{D-B} is the temperature difference between nodes D and B, which are shown in Figure 2. To reduce noise in experimental data, each of the variation curve is processed based on the smooth regression analysis method as well as the following curves. As is given, during the three-day continued test, the temperature of the upper position was higher than that of the lower position most of the time. But about 0–8 o'clock each day, the temperature difference was relatively small. Besides, according to the smooth regression analysis, it shows that T_{B-E} had two evident peaks and only one peak showed in T_{D-B} contrastively. For example, in the second day, the two extrema of T_{B-E} were, respectively, 3.0°C and 3.1°C, reached at 12:00 and 19:34, respectively, meanwhile the only extremum of T_{D-B} was 4.1°C, reached at 11:28.

Figure 4 shows the temperature difference between the three nodes on the middle of PCM plate 2, shown as A, B, and C in Figure 2. T_{B-C} is the temperature difference between nodes B and C, and T_{A-B} is the temperature difference between nodes A and B. As shown during the three-day test, similar to the situation shown in Figure 3, the temperature

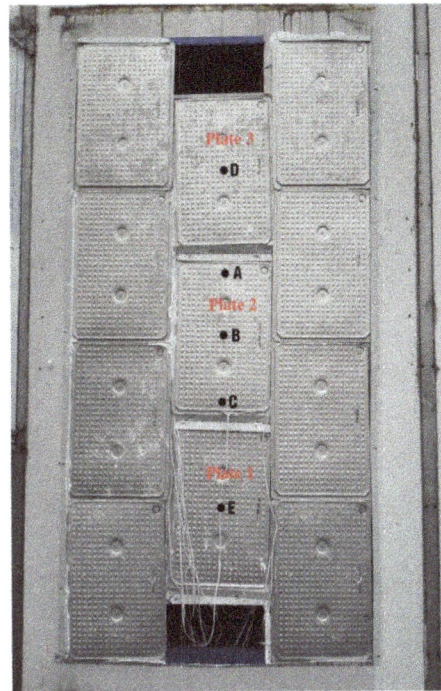

FIGURE 2: Arrangement diagram of PCM plate array and thermocouples.

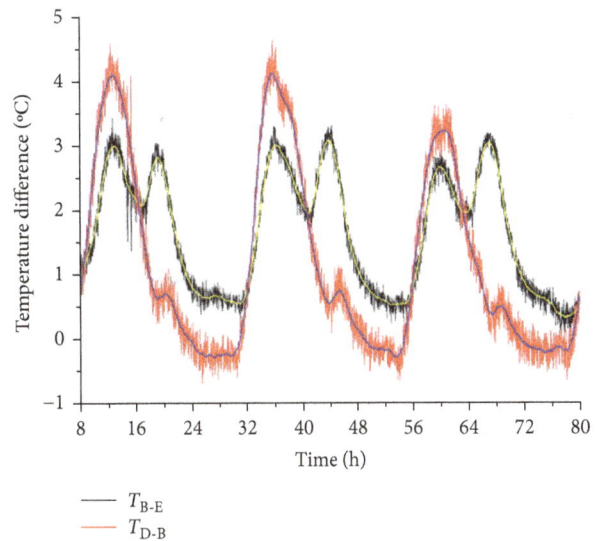

FIGURE 3: Comparison of the interior side temperature difference between the upper, middle, and lower PCM boards.

difference presented the behavior of daily fluctuation, lower position's temperature was higher than the upper position's, and temperature difference also became small during 0–8 o'clock. Otherwise, both T_{A-B} and T_{B-C} showed only one peak during daytime, and T_{B-C} became negligible at night. Also, take the second day for example, extremum of T_{A-B} was 1.4°C, reached at 11:20, meanwhile extremum of T_{B-C} was 2.3°C, reached at 14:26. It shows that the peak value of T_{B-C} had a delay phenomenon compared with that of T_{A-B}.

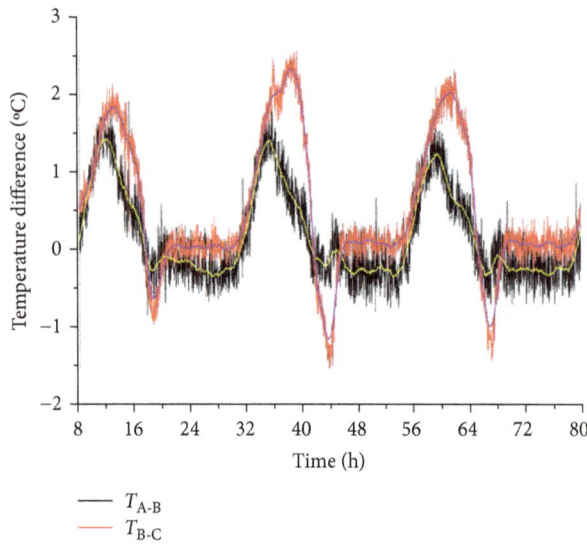

FIGURE 4: Temperature variation comparison for the nodes at the internal surface of center PCM.

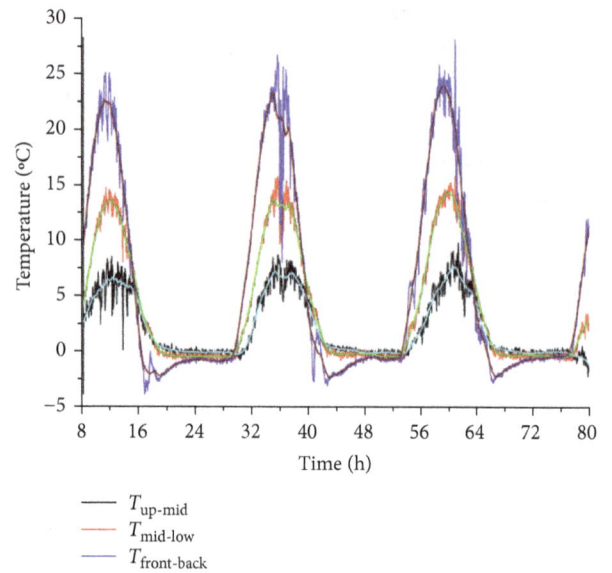

FIGURE 5: Temperature difference between the upper, middle, and lower heat-absorbing aluminum plates and back of the insulation layer.

3.1.2. Temperature Variation and Distribution of Insulated Absorbing Plate. Figure 5 represents the temperature difference between the upper, middle, and lower heat-absorbing aluminum plate and back of the insulation layer. $T_{up\text{-}mid}$ is the temperature difference between nodes "up" and "mid," $T_{mid\text{-}low}$ is the temperature difference between nodes "mid" and "low," and $T_{front\text{-}back}$ is the temperature difference between nodes "front" and "back," which are shown in Figure 1. As shown during the three-day test, $T_{up\text{-}mid}$ rose till the peak and decreased during daytime. Then at night, $T_{up\text{-}mid}$ maintained 0°C. The variation trend of $T_{mid\text{-}low}$ is similar to $T_{up\text{-}mid}$. The result demonstrates that the 3-day experiment's data was repetitive and that the temperature of the upper position on heat absorber was higher than that of the lower position during daytime. Otherwise, $T_{front\text{-}back}$ had a trough, which was different from $T_{up\text{-}mid}$ and $T_{mid\text{-}low}$. After we opened the middle layer vents, wind from the ambient came into the interior air channel and cooled the PCM, and the PCM started releasing heat. The back temperature of the insulation board would rise. For example, in the third day, the detailed data is shown in Table 1.

3.1.3. Temperature Variation and Distribution of Air Channels. Figure 6 shows variations of temperature difference between nodes of exterior air channel. There are 5 nodes at exterior air channel, which are shown in Figure 1. As shown, the temperature difference between exterior air channel was irregular, mostly because it was easily influenced by the ambient air when the outdoor vents were opened in summer. Also, take the third day for example, the detailed data is shown in Table 2.

Figure 7 shows variations of temperature difference between nodes of interior air channel. There are 5 nodes at interior air channel, which are shown in Figure 1. Different from the exterior air channel, the temperature difference of interior air channel was regular; in most time of the second day, for example, the maximum temperature difference

TABLE 1: Summarized result of temperature difference between the upper, middle, and lower of heat-absorbing aluminum plate and back of the insulation layer.

	Max	Time to reach max	Min	Time to reach min	Average
$T_{up\text{-}mid}$	7.8°C	12:36	−0.1°C	5:08	2.2°C
$T_{mid\text{-}low}$	14.5°C	12:08	−0.4°C	2:48	4.1°C
$T_{front\text{-}back}$	24.2°C	11:12	−2.2°C	18:52	6.0°C

increased with height. Besides that, the minimum and average values of $T_{5'\text{-}4'}$, $T_{4'\text{-}3'}$, $T_{3'\text{-}2'}$, and $T_{2'\text{-}1'}$ were relatively close. The data is listed in Table 3.

3.2. Winter Mode. Similarly, 2-day continuous experimental tests of winter were conducted from 9:00 on Dec. 15 to 9:00 on Dec. 17 in 2015, during which the outdoor vents were kept shut; the middle layer vents and the indoor vents were kept opened during daytime and shut during night, and the switching time was around 8:00 and 17:00.

3.2.1. Temperature Variation and Distribution of PCM Plates. Figure 8 shows variations of temperature difference between the centers of PCM plates 1, 2, and 3 in winter mode. Despite the complex variation trend, most of the time in the 2-day test, the upper position's temperature is higher than that of the lower position's. For convenience, the data of the second day was chosen for analysis. According to the smooth regression analysis, $T_{D\text{-}B}$ was approximately 0 during 0–8 o'clock. And $T_{D\text{-}B}$ reflected an upward trend and reached the extremum value of 5.4°C at 10:20. Then, a slight decrease appeared, reached the extremum value of 3.5°C at 11:27,

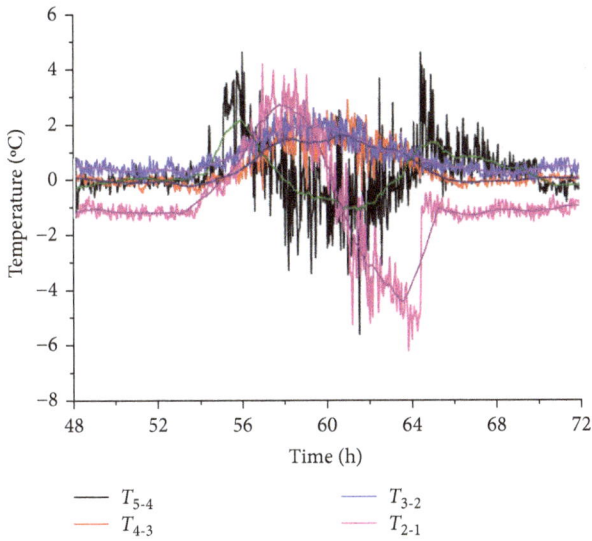

FIGURE 6: Temperature difference between the nodes of the exterior air channel.

TABLE 2: Summarized result of temperature difference between nodes of the exterior air channel.

	Max	Time to reach max	Min	Time to reach min	Average
T_{5-4}	2.2°C	7:54	−1.2°C	13:26	0.2°C
T_{4-3}	1.6°C	12:34	−0.3°C	5:24	0.4°C
T_{3-2}	2°C	11:08	0.2°C	20:08	0.8°C

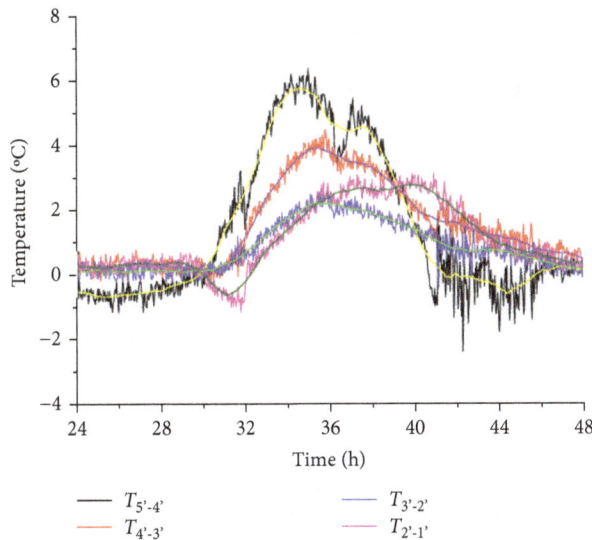

FIGURE 7: Temperature difference between the nodes of the interior air channel.

and kept rising till the extremum value reached 8.9°C at 15:25. Then, the temperature difference decreased again and reached the extremum value of −1.5°C at 18:28 and kept rising again till the extremum value reached 7.9°C at 23:04. Finally, at night, T_{D-B} maintained 0°C again similar as the

TABLE 3: Summarized result of temperature difference between nodes of the interior channel.

	Max	Time to reach max	Min	Time to reach min	Average
$T_{5'-4'}$	5.8°C	10:30	−0.7°C	1:14	1.5°C
$T_{4'-3'}$	4.0°C	11:24	0.2°C	6:08	1.6°C
$T_{3'-2'}$	2.2°C	11:17	0.1°C	5:01	0.9°C
$T_{2'-1'}$	2.8°C	15:44	−0.6°C	7:06	1.1°C

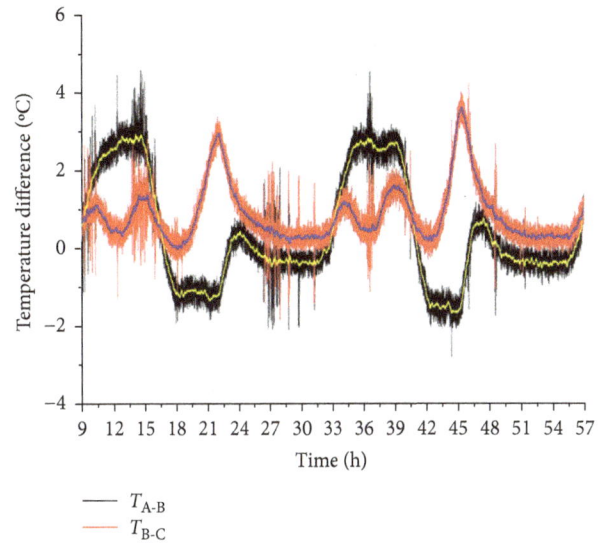

FIGURE 8: Temperature variation comparison for the nodes on the internal surface of center PCM.

first day. Its first peak occurred because till 10:20 in the second day, the top board started the phase change process. Heat absorbed from the channel changed into latent heat, so temperature difference will be diminished. The phase change time difference between panels D and B was between 10:20 and 11:27. Then, after reaching the temperature difference of 8.9°C at 15:25, T_{D-B} reached under 0°C because position B changed phase firstly that means releasing heat without temperature decrease. Then, T_{D-B} rose again because position D started the phase change process. Then, finally, temperature difference tends to be 0°C at night. T_{B-E} showed similar trends, which are only some differences between the extremum values and appear times.

Figure 9 shows temperature difference of the internal surface of the middle PCM plate 2. As shown, in the second day, during 0–8 o'clock on that day, average T_{B-C} was 0.4°C. Then, T_{B-C} rose at the peak temperature of 1.2°C at 10:00, fell off and maintained the average temperature of 0.5°C, then went up again and reached the extremum value of 1.6°C at 15:00. And it fell off again and reached extremum value of 1.6°C at 18:33 and went up again at the max temperature of 3.7°C in 21:26. Then, at night, T_{B-C} decreased to average temperature of 0.4°C again. The variation trend of T_{B-C} is similar to T_{D-B} in Figure 3. As for T_{A-B}, the average value was −0.3°C during

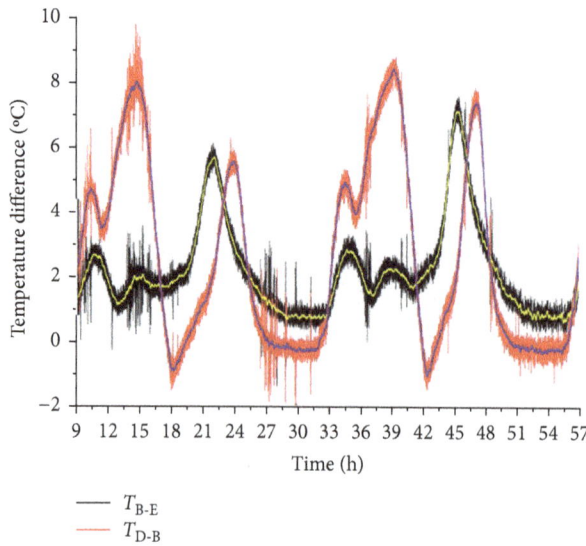

FIGURE 9: Comparison of the interior side temperature difference between the upper, middle, and lower PCM boards.

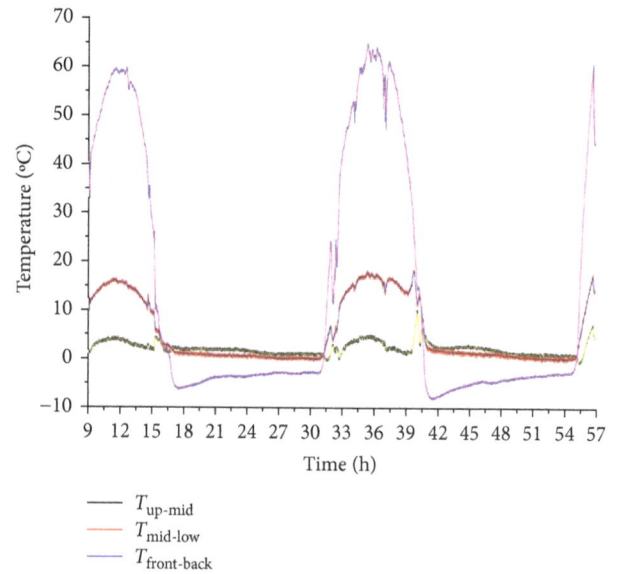

FIGURE 10: Temperature difference between the upper, middle, and lower heat-absorbing aluminum plates and back of the insulation layer.

0–8 o'clock on the second day. Then, T_{A-B} rose and reached a new balance of 2.7°C during 11–16 o'clock approximately. After that, T_{A-B} started decreasing and kept an average temperature of −1.5°C during 18–21 o'clock roughly. And it rose at the peak temperature of 0.8°C in 23:27, then finally decreased at the average temperature of 0.23°C at night. The complex variation trends of T_{A-B} and T_{B-C} both indicate the melting and freezing processes that did not occur synchronously at the different positions along vertical direction even for a single PCM plate.

3.2.2. Temperature Variation and Distribution of Insulated Absorbing Plate. Figure 10 shows temperature difference between the upper, middle, and lower heat-absorbing aluminum plates and back of the insulation layer. As is given, during the two-day test, $T_{\text{front-back}}$ was higher than that in the summer mode. On the one hand, the insulation layer prevented an amount of heat from the front. On the other hand, in the winter mode, the outdoor vents were closed, so the temperature of heat-absorbing aluminum plate was very high. The temperature of the upper position on heat-absorbing aluminum plate was higher than that of the lower position during daytime, which was similar to that of the summer mode. The particular data is listed in Table 4.

3.2.3. Temperature Variation and Distribution of Air Channels. Figure 11 shows variations of temperature difference between nodes of exterior air channel. As shown, the upper position's temperature of exterior air channel was higher than that of the lower position's. Temperature difference became high during daytime, and at night, it was approximately 0°C. Other data is listed in Table 5.

Figure 12 shows variations of temperature difference between nodes of interior air channel. As shown, $T_{5'-4'}$ was higher than others during daytime and this situation was also different from summer mode. In winter, the middle layer vents were both opened during daytime, and high

TABLE 4: Summarized result of temperature difference between the upper, middle, and lower heat-absorbing aluminum plate and back of the insulation layer.

	Max	Time to reach max	Min	Time to reach min	Average
$T_{\text{up-mid}}$	10.2°C	16:06	−0.6°C	7:36	2.2°C
$T_{\text{mid-low}}$	18.6°C	15:45	−0.7°C	6:20	5.8°C
$T_{\text{front-back}}$	65.6°C	11:22	−8.3°C	17:30	14.5°C

temperature air mixed at the top of the channel, so the temperature would be much higher at the lower position in the interior channel. The detailed data is shown in Table 6.

4. Conclusions

This paper proposes a novel solar PCM storage wall technology, that is, a dual-channel and thermal-insulation-in-the-middle type solar PCM storage wall (MSPCMW) system. By tests on a hot-box test platform, experimental tests and analyses are conducted on the system, respectively, operating in summer and winter modes. By comparison of temperature difference between the monitoring points on surfaces of PCM plates along vertical direction, temperature variation and distribution of PCM plates are studied. The following conclusions are obtained:

(1) Temperature variation and distribution of PCM plates: (a) In summer mode, during 0–8 o'clock every day, the temperature differences between centers of the upper, middle, and lower PCM plates were all relatively small. T_{B-E} had two evident peaks; meanwhile, only one peak showed in T_{D-B} during the rest of the time. Also, the temperature differences between the upper, middle, and lower positions on the middle

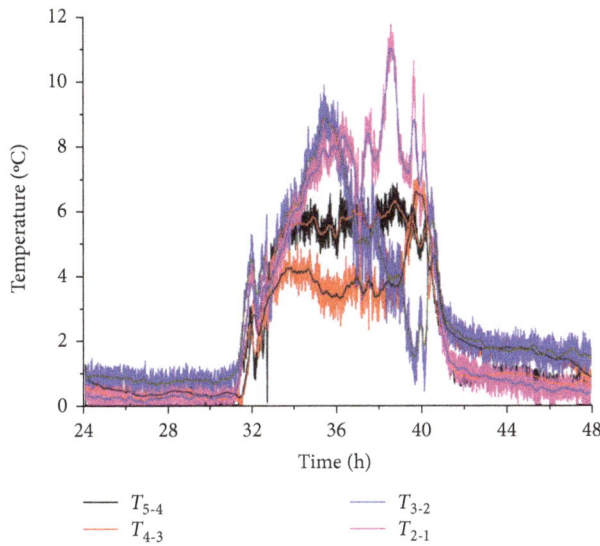

FIGURE 11: Temperature difference between the nodes of the exterior air channel.

TABLE 5: Summarized result of temperature difference between nodes of the exterior air channel.

	Max	Time to reach max	Min	Time to reach min	Average
T_{5-4}	6.4°C	14:44	0.15°C	2:45	2.4°C
T_{4-3}	6.7°C	15:46	0.22°C	7:22	2.2°C
T_{3-2}	9.0°C	11:28	0.6°C	3:20	2.9°C
T_{2-1}	11.3°C	14:33	0.1°C	2:49	2.9°C

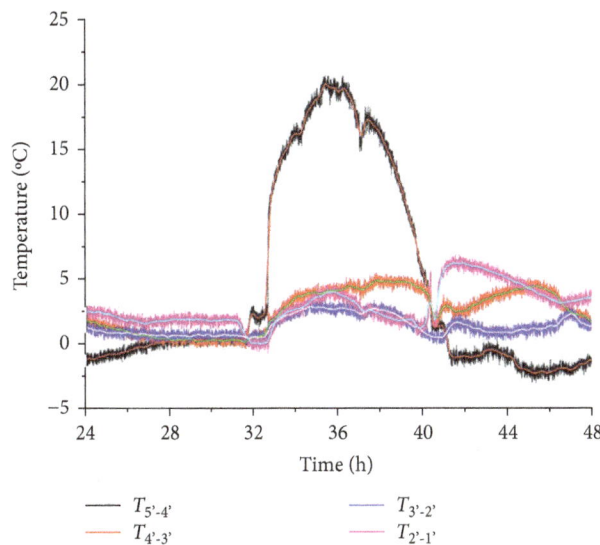

FIGURE 12: Temperature difference between the nodes of the interior air channel.

PCM plate became small during 0–8 o'clock; even the value of T_{B-C} became negative at night; and distinctively, both T_{A-B} and T_{B-C} showed only one peak

TABLE 6: Summarized result of temperature difference between nodes of the interior air channel.

	Max	Time to reach max	Min	Time to reach min	Average
$T_{5'-4'}$	20.1°C	12:14	−3.2°C	21:24	4.4°C
$T_{4'-3'}$	5.4°C	13:48	−0.5°C	7:36	2.4°C
$T_{3'-2'}$	3.3°C	11:06	−0.3°C	16:54	1.2°C
$T_{2'-1'}$	6.8°C	17:42	−0.9°C	8:36	2.8°C

during daytime. (b) In winter mode, both T_{B-E} and T_{D-B} had numerous peaks each day during the 2-day continuous experimental tests, and the peak phenomena represented there were obvious time difference of melting process between the upper and lower positions. T_{B-C} showed the same trend as T_{B-E}, and T_{A-B} showed three obvious peaks. Their differences showed that melting and freezing processes did not occur synchronously at the different positions along vertical direction even for a single PCM plate.

(2) Temperature variation and distribution of insulated absorbing plate: Both in summer and winter modes, the temperature of the upper position on heat absorber was higher than that of the lower position during daytime, while in winter mode, $T_{front-back}$ was higher than that in summer mode.

(3) Temperature variation and distribution of air channels: (a) In summer mode, the temperature difference between exterior air channel was irregular, while the temperature difference of interior air channel was regular, that is, the maximum temperature difference increased with height. (b) In winter mode, air temperature distribution in exterior and interior air channels had a similar feature with that of the upper position's air temperature which was higher than that of the lower position's. Besides, in interior air channel temperature, difference between the top position and the adjacent position was much higher than that in others during daytime and this situation was also different from summer mode.

Conflicts of Interest

The authors declare that there is no conflict of interest regarding the publication of this paper.

Acknowledgments

This study was sponsored by (1) the National Science Foundation of China (NSFC), Project no. 51408278, (2) the National Science Foundation of China (NSFC), Project no. 51366004, and (3) the Jiangxi Provincial Key Technology R&D Program, China, Project no. 20141BBE50041.

References

[1] A. Akbarzadeh, W. W. S. Charters, and D. A. Lesslie, "Thermocirculation characteristics of a Trombe wall passive test cell," *Solar Energy*, vol. 28, no. 6, pp. 461–468, 1982.

[2] G. Gan, "A parametric study of Trombe walls for passive cooling of buildings," *Energy and Buildings*, vol. 27, no. 1, pp. 37–43, 1998.

[3] G. S. Yakubu, "The reality of living in passive solar homes: a user-experience study," *Renewable Energy*, vol. 8, no. 1-4, pp. 177–181, 1996.

[4] J. Jie, Y. Hua, H. Wei, P. Gang, L. Jianping, and J. Bin, "Modeling of a novel Trombe wall with PV cells," *Building and Environment*, vol. 42, no. 3, pp. 1544–1552, 2007.

[5] J. Jie, Y. Hua, P. Gang, and L. Jianping, "Study of PV-Trombe wall installed in a fenestrated room with heat storage," *Applied Thermal Engineering*, vol. 27, no. 8-9, pp. 1507–1515, 2007.

[6] B. K. Koyunbaba, Z. Yilmaz, and K. Ulgen, "An approach for energy modeling of a building integrated photovoltaic (BIPV) Trombe wall system," *Energy and Buildings*, vol. 67, pp. 680–688, 2013.

[7] X. Hong, W. He, Z. Hu, C. Wang, and J. Ji, "Three-dimensional simulation on the thermal performance of a novel Trombe wall with venetian blind structure," *Energy and Buildings*, vol. 89, pp. 32–38, 2015.

[8] X. Xu, Y. Zhang, K. Lin, H. Di, and R. Yang, "Modeling and simulation on the thermal performance of shape-stabilized phase change material floor used in passive solar buildings," *Energy and Buildings*, vol. 37, no. 10, pp. 1084–1091, 2005.

[9] K. Peippo, P. Kauranen, and P. D. Lund, "A multicomponent PCM wall optimized for passive solar heating," *Energy and Buildings*, vol. 17, no. 4, pp. 259–270, 1991.

[10] C. Zhang, Y. Chen, L. Wu, and M. Shi, "Thermal response of brick wall filled with phase change materials (PCM) under fluctuating outdoor temperatures," *Energy and Buildings*, vol. 43, no. 12, pp. 3514–3520, 2011.

[11] Y. A. Kara and A. Kurnuç, "Performance of coupled novel triple glass unit and PCM wall," *Applied Thermal Engineering*, vol. 35, pp. 243–246, 2012.

[12] Y. Zhang, K. Lin, Y. Jiang, and G. Zhou, "Thermal storage and nonlinear heat-transfer characteristics of PCM wallboard," *Energy and Buildings*, vol. 40, no. 9, pp. 1771–1779, 2008.

[13] H. J. Alqallaf and E. M. Alawadhi, "Concrete roof with cylindrical holes containing PCM to reduce the heat gain," *Energy and Buildings*, vol. 61, pp. 73–80, 2013.

[14] A. Pasupathy, L. Athanasius, R. Velraj, and R. V. Seeniraj, "Experimental investigation and numerical simulation analysis on the thermal performance of a building roof incorporating phase change material (PCM) for thermal management," *Applied Thermal Engineering*, vol. 28, no. 5-6, pp. 556–565, 2008.

[15] N. Soares, J. J. Costa, A. R. Gaspar, and P. Santos, "Review of passive PCM latent heat thermal energy storage systems towards buildings' energy efficiency," *Energy and Buildings*, vol. 59, pp. 82–103, 2013.

[16] C. Luo, L. Xu, J. Ji, M. Liao, and D. Sun, "Experimental study of a modified solar phase change material storage wall system," *Energy*, vol. 128, pp. 224–231, 2017.

18

Phase Distribution for Subcooled Flow Boiling in an Inclined Circular Tube

Wei Bao ⓘ, JianJun Xu ⓘ, TianZhou Xie, BingDe Chen, YanPing Huang, and DianChuan Xing

CNNC Key Laboratory on Nuclear Reactor Thermal Hydraulics Technology, Nuclear Power Institute of China, Chengdu 610041, China

Correspondence should be addressed to JianJun Xu; xujjun2000@sohu.com

Academic Editor: Zhonghao Rao

An experimental investigation of phase distribution for subcooled flow boiling in an inclined circular tube (i.d. 24 mm) was conducted in this paper. The local interfacial parameters were measured by a double-sensor optical fiber probe, and the measurements were performed on three different directions in the inclined tube cross section. The experiment shows that the phase distribution under the inclined condition is different from the phase distribution under the vertical condition. The profiles skewed highly for 90° and 45° direction in the tube cross section, whereas the profile was also symmetrical at 0° direction. These results can be explained by the fact that buoyancy caused the bubbles to move toward the top of the tube cross section under inclined condition. In addition, the typical distributions were also influenced by the inclination angles.

1. Introduction

Subcooled flow boiling often appears in industrial fields, such as nuclear reactors, chemical plants, and some engineering systems. With the deepening of the research on the two-phase flow and boiling heat transfer, it has been found that distribution of the local interfacial parameter has an important influence on the flow and heat transfer characteristics of the two-phase flow, and the capability to predict the local void distribution in subcooled flow boiling is of great importance for the safety of boiling water reactor. Meanwhile, the establishment and development of the two-fluid model also require the verification with the experimental data of the local interfacial parameters. Therefore, the study on the phase distribution characteristics of the subcooled flow boiling is of great significance to the understanding of the mechanism of two-phase flow and heat transfer.

The local void fraction and interfacial area concentration (IAC) is the basic parameter determining the structure of the two-phase flow. In order to obtain a more reliable interfacial area transport equation, some experiments for the phase distribution of local parameters in two-phase flow are

indispensable. Over the past few years, a large number of the experiments regarding gas-liquid bubbly flow have been performed [1–8]. Revankar and Ishii [9] observed that the local IAC appears as a wall-peak profile in a vertical tube. In studies by Hibiki and Ishii [10], Hibiki et al. [11], and Shen et al. [12], four kinds of typical profiles of local interfacial parameters including wall peak, core peak, intermediate peak, and transition have been found in the studies.

A large number of the experiments regarding subcooled flow boiling have been performed. Some tasks for different geometrical channels have been already conducted by the previous investigators. Sekoguchi et al. [13] have used the single-sensor conductivity probe to measure both radial and axial distributions of local fraction under a subcooled boiling condition in a circular tube. Garnier et al. [14] performed the measurements of local interfacial parameters in R-12 subcooled flow boiling in a vertical channel with two-sensor optical probe; meanwhile, it is found that void fraction profiles in the experiment are concave profile, convex profile, and two-peak profile. Sun [15] reported the radial distributions of local void fraction and bubble frequency in the low-mass flux subcooled flow boiling. For vertical

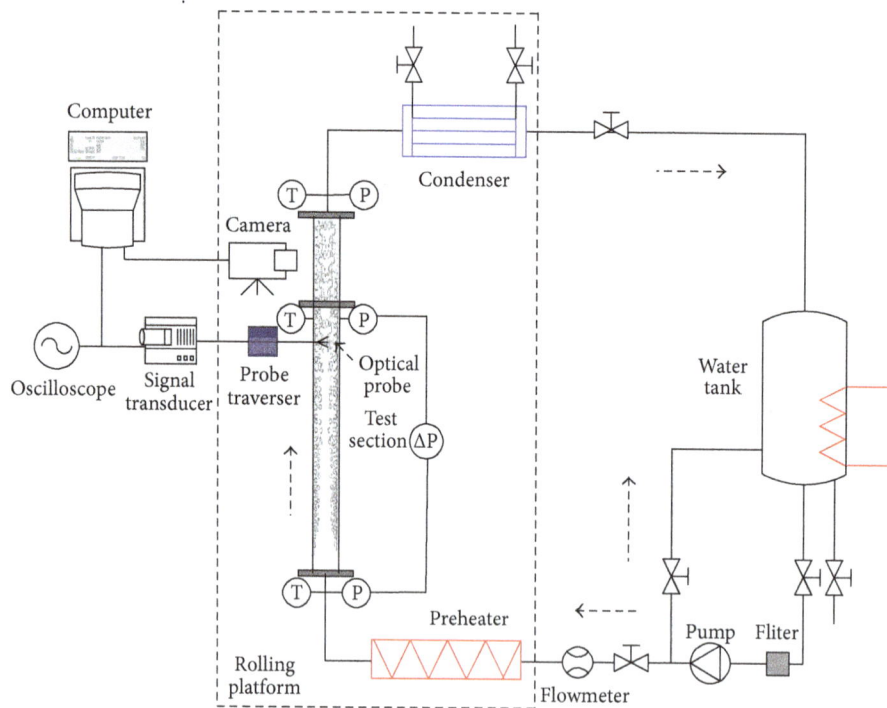

FIGURE 1: Schematic diagram of test loop.

annulus channel, Hasan et al. [16] and Roy and Velidandla [17] measured local void fraction, gas velocity, and bubble diameter in R-113 boiling flow. Recently, it has been presented in the studies of Situ et al. [18], Lee et al. [19], and Yun et al. [20] that local measurements of void fraction, bubble diameter, interfacial velocity, and liquid velocity in subcooled flow boiling were performed in annulus channel. Besides, some double-sensor conductivity probes were used for the measurements of the local interfacial parameters at three or more axial locations.

The above studies are mainly focused on the local interfacial characteristics of two-phase flow under the vertical condition. Inclined condition extensively occurred in the field of ship industry and chemical engineering. The trend of developing applications for ocean environments has attracted growing interests on two-phase flow under inclined condition. Therefore, the flow pattern, void fraction, and pressure drop of two-phase flow in inclined tubes have been studied extensively [21–25]. However, only very few literatures have focused on the local interfacial parameter distribution under inclined condition. Spindler and Hahne [26] have investigated the void fraction and bubble frequency profiles of adiabatic two-phase flow in an inclined tube with the method of optical fiber probe. Recently, Xing et al. [27] have researched the radial distribution of interfacial parameters for air-water bubbly flow in a circular tube under inclined condition with the double-sensor optical fiber probe. Only one-dimensional distribution of the local interfacial parameter in inclined bubbly flow has been experimentally studied in the previous task. Among all of the existing experiments measured along single direction of cross section in an inclined tube, the asymmetrical distributions of local parameters have been found in these experiments.

However, the measurement of multiple directions can fully reflect the three-dimensional distribution characteristics. It is regrettable that few studies aim at the phase distribution in different directions under inclined condition. Recently, Bao et al. [28] measured local interfacial parameters for subcooled flow boiling in an inclined circular tube. It is required not only by the profile for one chord of tube cross section but also by some profiles for other direction in the cross section, for it better gains the physical insight into the distribution characteristics of subcooled flow boiling under inclined condition. From this point of view, this experimental study aims to investigate the phase distribution of subcooled flow boiling in inclined circular tube under different directions.

2. Material and Methods

2.1. Experimental Loop. A schematic of the experimental loop used for this study is shown in Figure 1. The experimental system has been introduced by previous work [28]. As can be seen in Figure 1, the preheater, condenser, test section, optical probe, and probe traverser are mounted onto the rolling platform and the other apparatus are on the floor. The two parts of the test loop are connected with flexible pipe. The deionized water is stored in a tank, and the non-condensable gas in the water is removed by a heater. The water is circulated by the drive of the pump. Two regulated valves were installed separately on the bypass, and the test branch controls the flow rate through the test section. A direct electrical-heating preheater is used to regulate the liquid temperature at the inlet of the heated test section. The uniform heat flux is provided by the 80 kW DC power supply. Two-phase mixture flowing out of the test section is chilled by a condenser. The test section is a circular tube with an inner diameter of 24 mm and a heated length of 1000 mm. There are two parts in this section. One is the heated section made of stainless steel, and the other one is the visual section made of quartz glass.

The volumetric flow rate was measured with a venturi flowmeter and the accuracy of flow measurement was ±2% of the full-scale flow. Two test gauges with accuracy of ±0.1% were installed to measure the pressure at the inlet and outlet of the test section. The pressure drop across the inlet and outlet of the heated section was measured with the pressure differential pressure transmitters of range 0~0.2 MPa with accuracy of ±0.1% of full-scale pressure drop. Some N-type thermocouples were used to measure fluid temperatures at the inlet and outlet of the heated test section. A typical uncertainty associated with temperature measurement was ±1°C. The heat flux to the heated test section was obtained by measuring the current into the test section and the voltage drop across the heater. The electrical current was measured using a digital multimeter, and the voltage was measured using a multirange voltmeter. Maximum uncertainty in power measurement was ±1% of measured power. The heat loss to the ambient is estimated from the sensible heat that is gained by the fluid for single-phase heat transfer conditions.

2.2. Optical Probe System. To quantify the complicated local interfacial characteristic in subcooled flow boiling, a double-sensor optical fiber probe was applied to measure the local interfacial parameters. A signal processor generated high- or low-voltage signal corresponding to the vapor and liquid phases around the probe tip, with each pulse representing a bubble hitting the sensor tip. The double-sensor probe consists of two independent sensors, the two sensor tips space 0.7 mm along the main flow direction and upstream one is called the front sensor. The location of the optical probe was at axial position of $40D$ $(D = 24 \, \text{mm})$ distance from the entrance. The optical probe can move with the drive of a probe traverse with 0.02 mm resolution. For inclined condition, the probe was traversed in $r/R = -0.95 \sim 0.95$ to obtain the radial profiles of local parameters; r and R are the radial distance from the center and the inner radius of the heated tube, respectively.

Based on the signal from the two-sensor probe, local void fraction and bubble frequency are calculated by the signals of the front sensor. Local void fraction is equal to the ratio of all of the bubble-dwelling time measured by the front sensor to the total sampling time (T), which can be expressed by

$$\alpha = \frac{\sum (t_j - t_{j-1})}{T} = \frac{\sum \Delta t_j}{T}. \tag{1}$$

Local bubble frequency is equal to the ratio of the bubble numbers (N) passing through the front fiber tip in the measurement time to the total sampling time,

$$f_b = \frac{N}{T}. \tag{2}$$

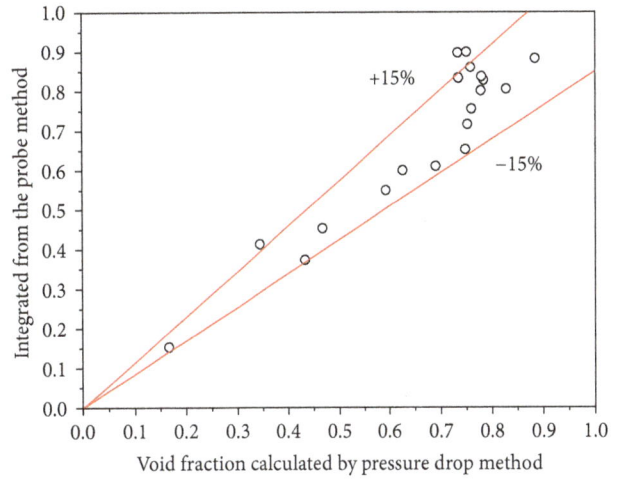

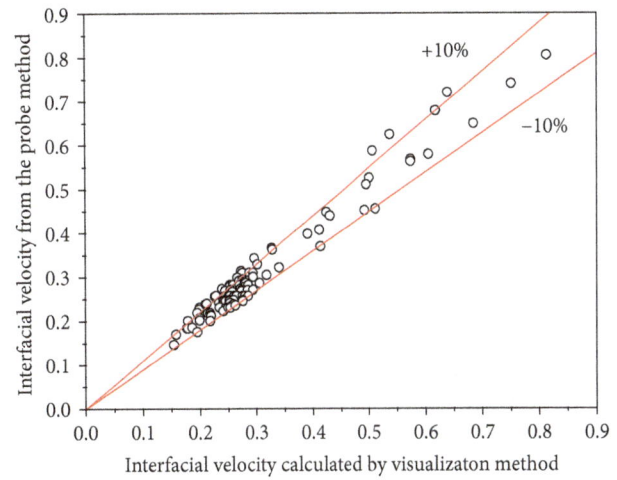

FIGURE 2: Uncertainty of the two-sensor optical probe: (a) void fraction; (b) interfacial velocity.

The interfacial velocity parallel to the flow direction can be simply calculated from the distance between the two sensor tips and the time difference, and it can be expressed as

$$v_i = \frac{\Delta s}{\Delta t_{kl}}. \tag{3}$$

Local IAC calculated by the method of Wu and Ishii [29], who considered the effect of bubble lateral motions on the IAC measured,

$$\alpha_i = \frac{2N_b}{\Delta s \Delta T} \left[2 + \left(\frac{v_b'}{v_b} \right)^{2.25} \right] \frac{\sum_j (\Delta t_j)}{N_b - N_{\text{miss}}}. \tag{4}$$

ΔT, N_b, N_{miss}, and v_b'/v_b denote the sampling time, the total numbers detected by the front sensor, the number of the missed bubbles, and the relative bubble velocity fluctuation, respectively. The missed bubbles referred to those are touched by the front sensor but not by the rear sensor, or those pass the rear sensor ahead of the front sensor due to

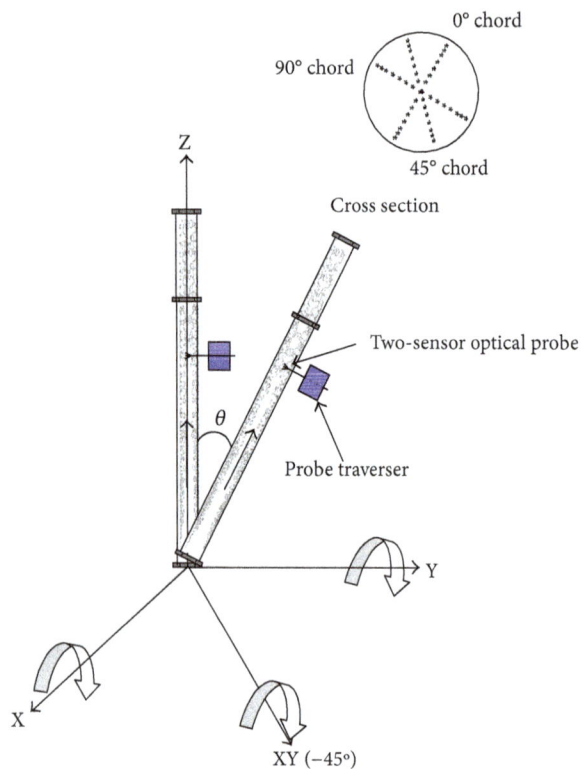

FIGURE 3: Diagram of test section under vertical and inclined condition.

bubble lateral motions. They also reported the relative bubble velocity fluctuation.

It can be seen from Figure 2 that the probe measurement results show good agreement with the void faction calculated by pressure drop method, and the maximum relative error is less than 15%. It is also found that the maximum relative error between the local interfacial velocities measured by probe and that by the visualization method is less than 10%.

2.3. Experimental Methods. The measurement method named multiangle between the direction of motion and the measuring is used in the experiment, which can obtain the local interfacial parameters in different directions of the inclined tube cross section. As shown in Figure 3, the inclination angle can be adjusted between $\theta = 0°$ (vertical) and $\theta = 30°$; in addition, the test tube can incline through different directions with the help of the moveable platform. As a result, according to the angle between the two directions of inclination axis and probe measurement, three chords (0°,45°, and 90°chord) of different directions in the tube cross section can be measured under the inclined condition. $\beta = 90°$ means that the measuring direction is perpendicular to the inclination axis direction, and $\beta = 0°$ means that the measuring direction is parallel to the inclination axis direction; $r = 0$ corresponds to the channel center, whereas $r/R > 0$ and $r/R < 0$ represent the lower and upper half part of inclined cross section of tube, respectively.

Figure 4 shows the measurement repeatability of the two-sensor probe at $r/R = 0$ for the three chords in the

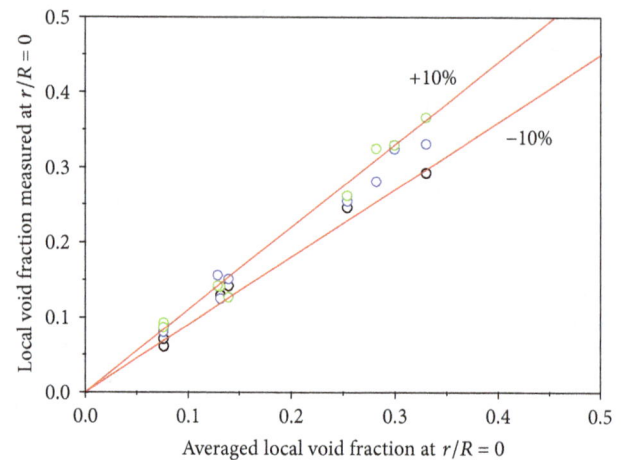

(a)

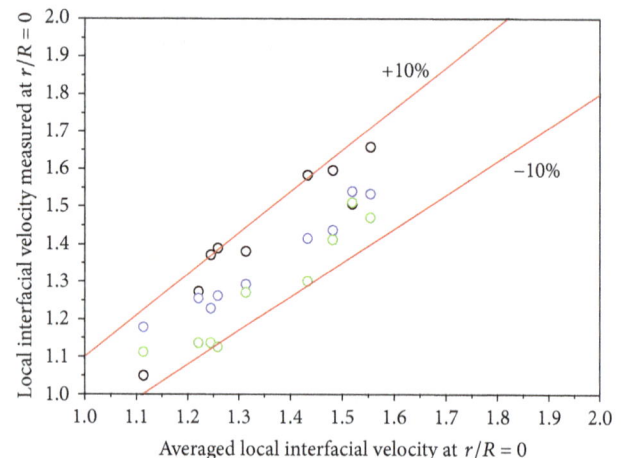

(b)

FIGURE 4: Repeatability of the two-sensor optical probe measurement at $r/R = 0$: (a) void fraction; (b) interfacial velocity.

tube cross section. It has also been found that the local void fraction and local interfacial velocity are similar regardless of the measurement chord when the tube inclined for any axis. To sum up, it clearly indicates that the multiangle measurement method is reasonable.

3. Results and Discussion

3.1. The Characteristics of Phase Distribution for Subcooled Flow Boiling under the Inclined Condition. Figure 5 clearly shows the characteristics of the phase distributions of subcooled flow boiling under inclined condition. As presented in the figures, local void distribution for inclined condition is quite different from that for vertical condition, and the phase distribution profiles are not symmetrical in inclined tube. Moreover, the different distribution profiles occurred

(a) Local void fraction

(b) Bubble Sauter diameter

(c) Interfacial area concentration

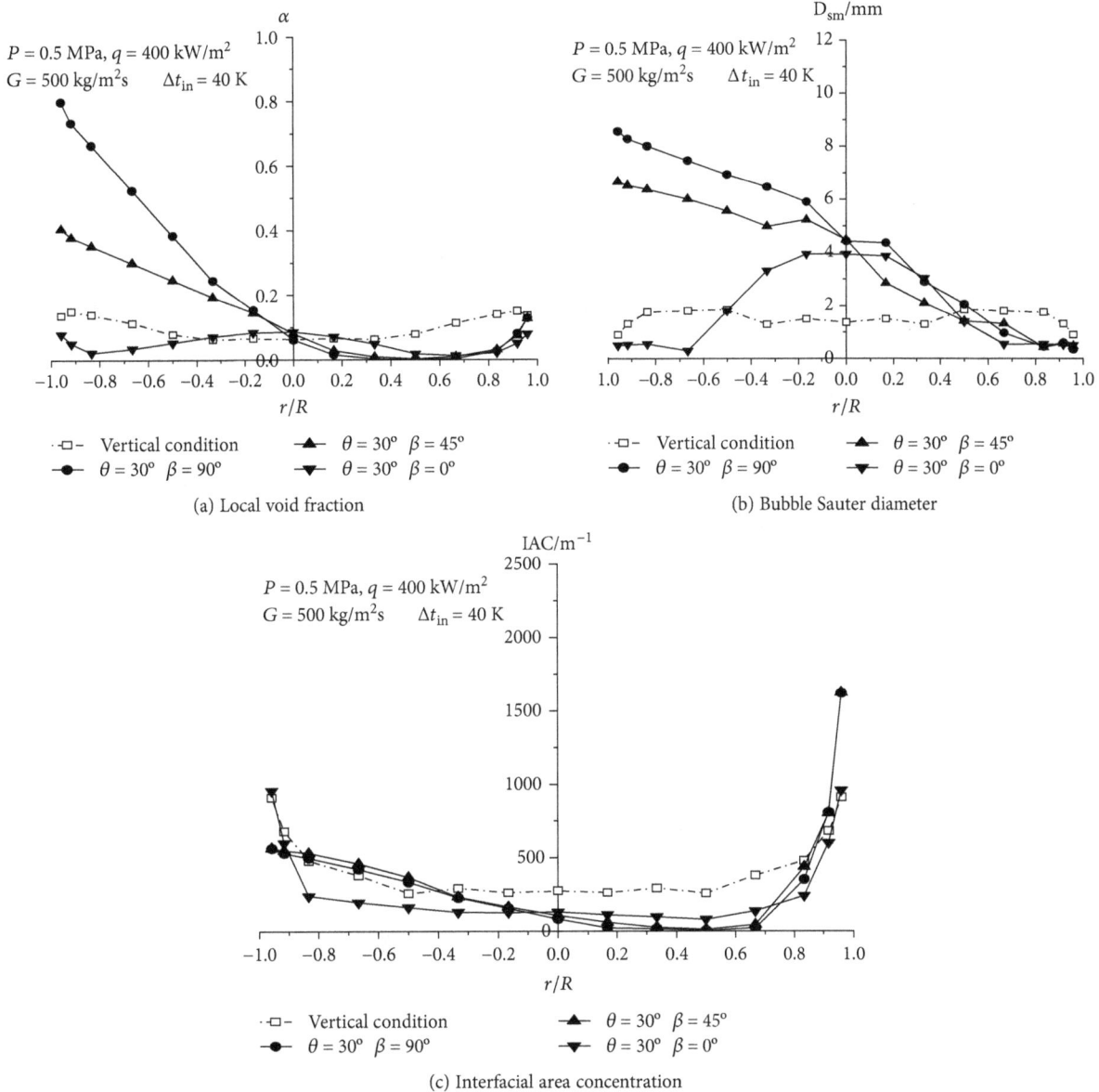

FIGURE 5: Phase distribution of subcooled flow boiling under inclined condition.

in three measured directions. That is to say, the profile of local interfacial parameters highly skewed at 90° chord and 45° chord, while the profile is still symmetrical at 0° chord. These results can be explained by the fact that buoyancy caused the bubbles to move toward the upper side of inclined tube and congregate. Owing to the lateral migration and polymerization of bubbles onto the upper wall of the channel, the peak value of the local void fraction appears in the $r/R = -1$ position of the inclined channel; in addition, the largest bubble Sauter diameter appears in this position as well. Because of the increase of the bubble size, the IAC is smaller than that under the static condition, which indicates that the heat transfer ability between the liquid and vapour phase at the top of the inclined channel becomes weaker. As the bubbles emerge from the heated wall, there are still some bubbles at the bottom of the inclined channel ($r/R = 1$). As we can see in the picture, the local void

fraction and bubble size at the bottom of the flow path are less affected by the inclination. Meanwhile, due to the migration of large bubbles to the top of the tube, the IAC can be larger than that under static condition, which indicates that the heat transfer between the liquid and vapour phase at the bottom of the inclined channel is becoming stronger.

3.2. The Influence of the Inclination Angle on the Phase Distribution for Subcooled Flow Boiling. According to the analysis above, the phase distribution on the direction of $\beta = 90°$ and $\beta = 45°$ is similar, and the phase distribution on the direction of $\beta = 0°$ is still symmetrical. As follows, the analysis of phase distribution characteristics on the directions of $\beta = 90°$ and $\beta = 0°$ has been conducted, respectively.

3.2.1. The Characteristics of the Phase Distribution on the Direction of $\beta = 90°$ in Cross Section. Figures 6–8 shows the

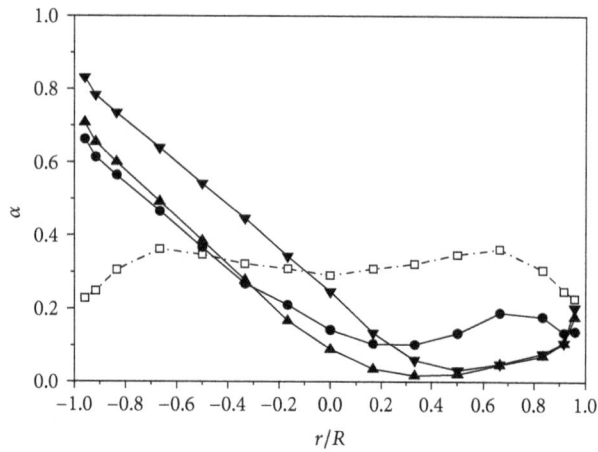

(a) $q = 400\,\text{kW/m}^2$

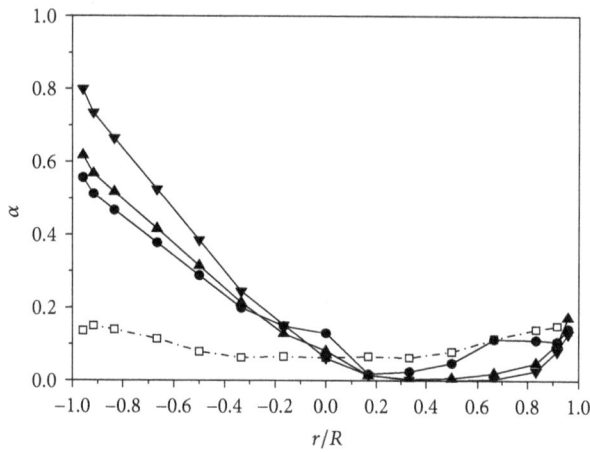

(b) $q = 450\,\text{kW/m}^2$

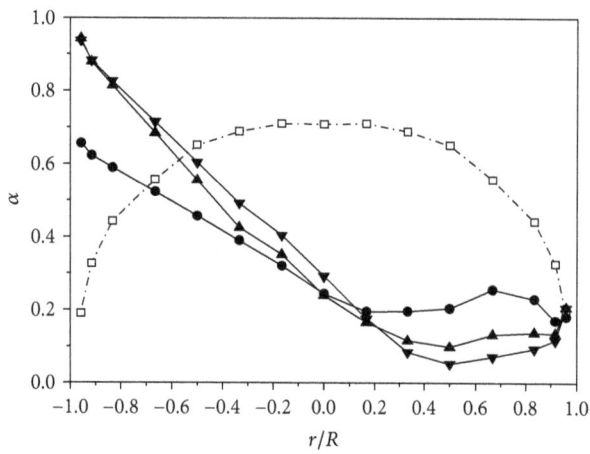

(c) $q = 500\,\text{kW/m}^2$

FIGURE 6: Inclination angle effect on the local void fraction on the direction of $\beta = 90°$ in cross section ($P = 0.5\,\text{MPa}$, $G = 500\,\text{kg/m}^2\,\text{s}$).

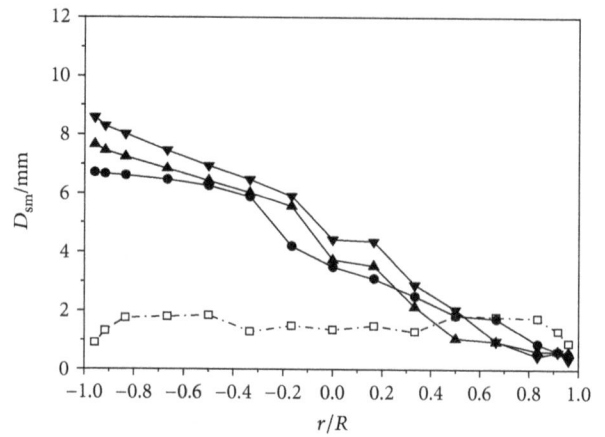

(a) $q = 400\,\text{kW/m}^2$

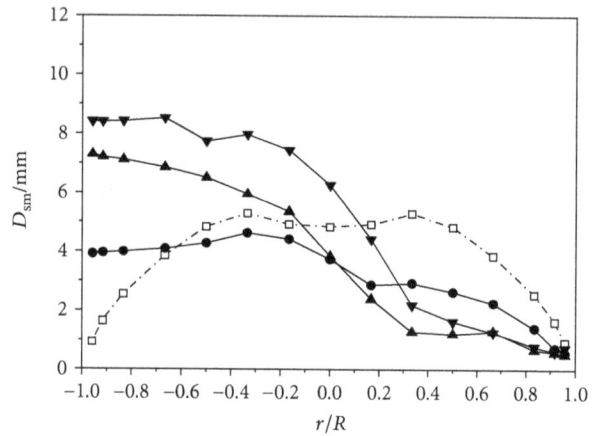

(b) $q = 450\,\text{kW/m}^2$

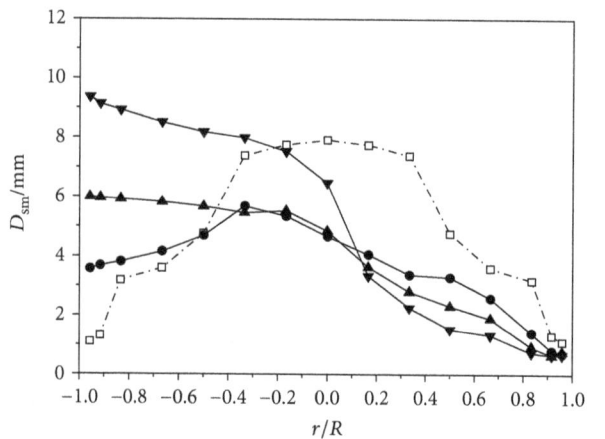

(c) $q = 500\,\text{kW/m}^2$

FIGURE 7: Inclination angle effect on the bubble Sauter diameter on the direction of $\beta = 90°$ in cross section ($P = 0.5\,\text{MPa}$, $G = 500\,\text{kg/m}^2\,\text{s}$).

(a) $q = 400\,kW/m^2$

(b) $q = 450\,kW/m^2$

(c) $q = 500\,kW/m^2$

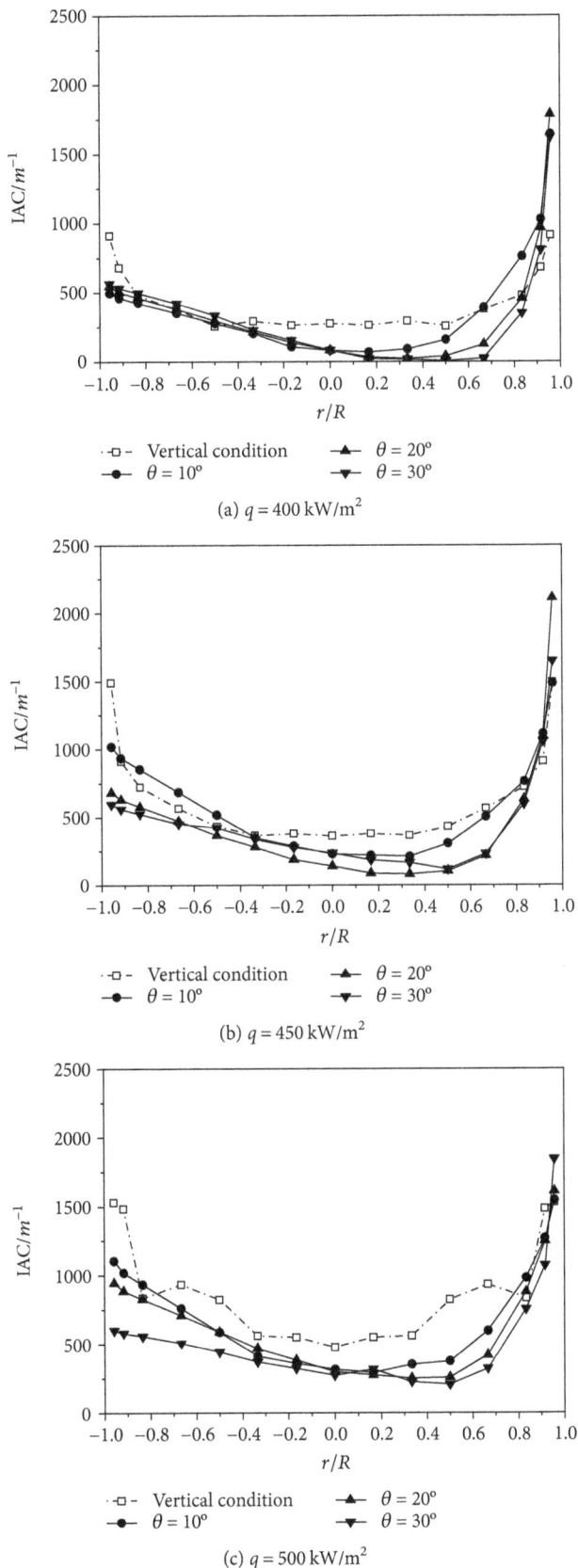

FIGURE 8: Inclination angle effect on the interfacial area concentration on the direction of $\beta = 90°$ in cross section ($P = 0.5\,MPa$, $G = 500\,kg/m^2\,s$).

influence of the inclination angle on the deviation of local void fraction, bubble size, and IAC on the direction of $\beta = 90°$. As we can see, with the increasing of the inclination angle, deviation of local void fraction, bubble size, and IAC, the peak value of local void fraction increases on the top of tube; meanwhile, the bubble Sauter diameter of also raises but IAC decreases slightly. This can be explained by the reason that the radial part of buoyancy increases with the rising of the inclination angle. Hence, it results in more bubble aggregation on the top of the flow channel and the uneven phase distribution of the subcooled flow boiling under inclined condition exacerbates.

It can be seen in Figure 6 that the peak value near the location where $r/R = -1$ increases as the inclination angle increases, while it becomes weakened near the $r/R = 0.8$ and the inclination angles will affect the typical distributions in some degree. When heat flux $q = 400\,kW/m^2$, the local void core peak profile changes into intermediate peak profile as the inclination angle increases at the lower half part of the inclined tube. Even in the case of 30° for an inclination angle, the peak near the lower side will disappear, when the heat flux is lower and the wall peak profile occurred. It is also found that local void fraction at $r/R = -1$ was independent of the inclination angle. The reason is the bubble generation controlled by the heat surface in subcooled flow boiling.

The local bubble Sauter diameter profile is illustrated in Figure 7. As clearly reported in some literature, the lift force pushes the small bubbles toward the wall in vertical upward flow, while the direction of lift force is reversed when the bubble diameter exceeds the threshold size. Different from vertical flow, the buoyancy force has a radial component normal to the tube axis. For large size bubbles, the direction of force is the same as the lift force in the lower part of the tube while it reverses to the lift force in the upper part of the tube. Therefore, large bubbles moved to the upper part of the tube and congregate. For small size bubbles, more of them moved to the upside resulting from the direction of F which reversed to the lift force in the lower part of tube.

Figure 8 shows typical profiles of the measured local IAC; there is an approximately symmetrical profile for inclined condition. Local IAC explicitly decreased near the upper side at 90° direction when heat flux was high when compared with that for vertical condition. This explains that IAC is proportional to void fraction, but inversely proportional to bubble size. And it causes the decrease of IAC in this region with the aforementioned bubbles near the upper side of the inclined tube becoming larger.

3.2.2. The Characteristics of the Phase Distribution on the Direction of $\beta = 0°$ in Cross Section.

Figure 9–11 shows the influence of the inclination angle on the local void fraction, bubble size, and IAC on the direction of $\beta = 0°$ in the flow channel. It can be clearly seen that the profile of phase distribution still maintains symmetrical. In addition, the influence of the inclination angle on the phase distribution in this direction is also of vital significance.

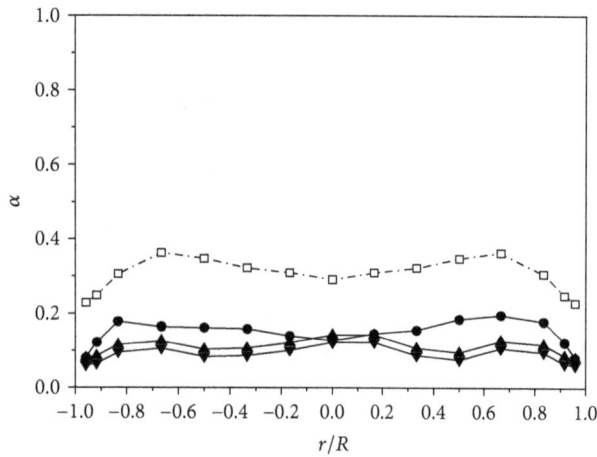

(a) $q = 400\,\text{kW/m}^2$

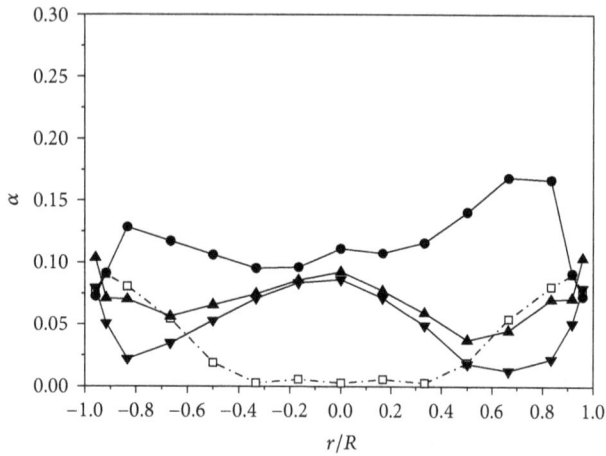

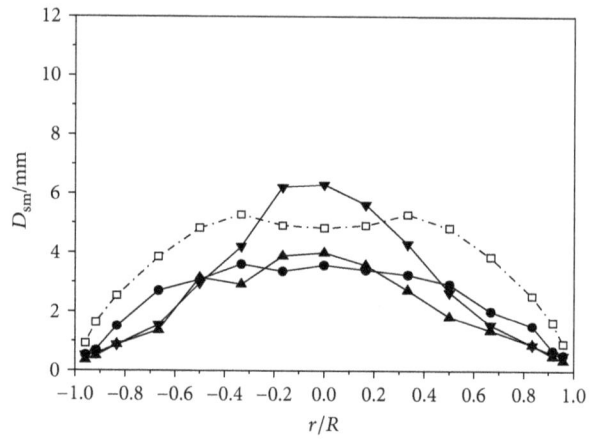

(b) $q = 450\,\text{kW/m}^2$

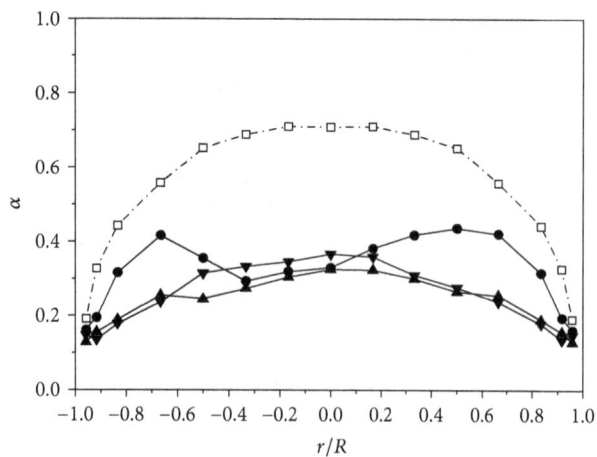

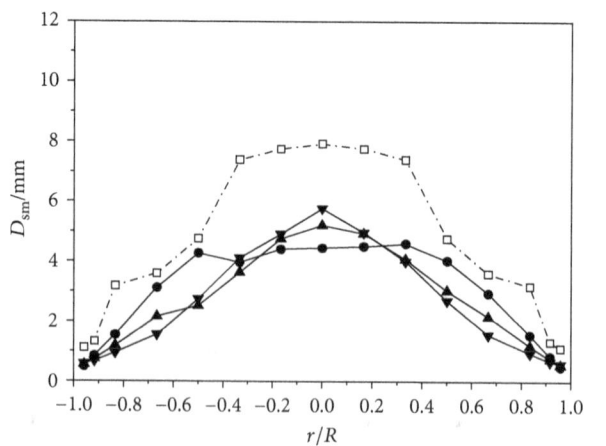

(c) $q = 500\,\text{kW/m}^2$

FIGURE 9: Inclination angle effect on the local void fraction on the direction of $\beta = 0°$ in cross section ($P = 0.5\,\text{MPa}$, $G = 500\,\text{kg/m}^2\,\text{s}$).

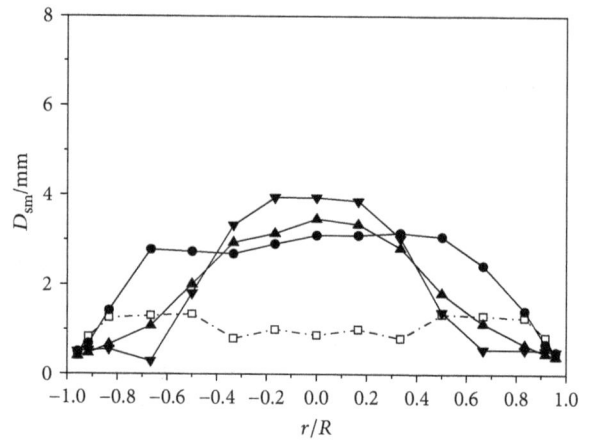

(a) $q = 400\,\text{kW/m}^2$

(b) $q = 450\,\text{kW/m}^2$

(c) $q = 500\,\text{kW/m}^2$

FIGURE 10: Inclination angle effect on the bubble Sauter diameter on the direction of $\beta = 0°$ in cross section ($P = 0.5\,\text{MPa}$, $G = 500\,\text{kg/m}^2\,\text{s}$).

(a) $q = 400\,kW/m^2$

(b) $q = 450\,kW/m^2$

(c) $q = 500\,kW/m^2$

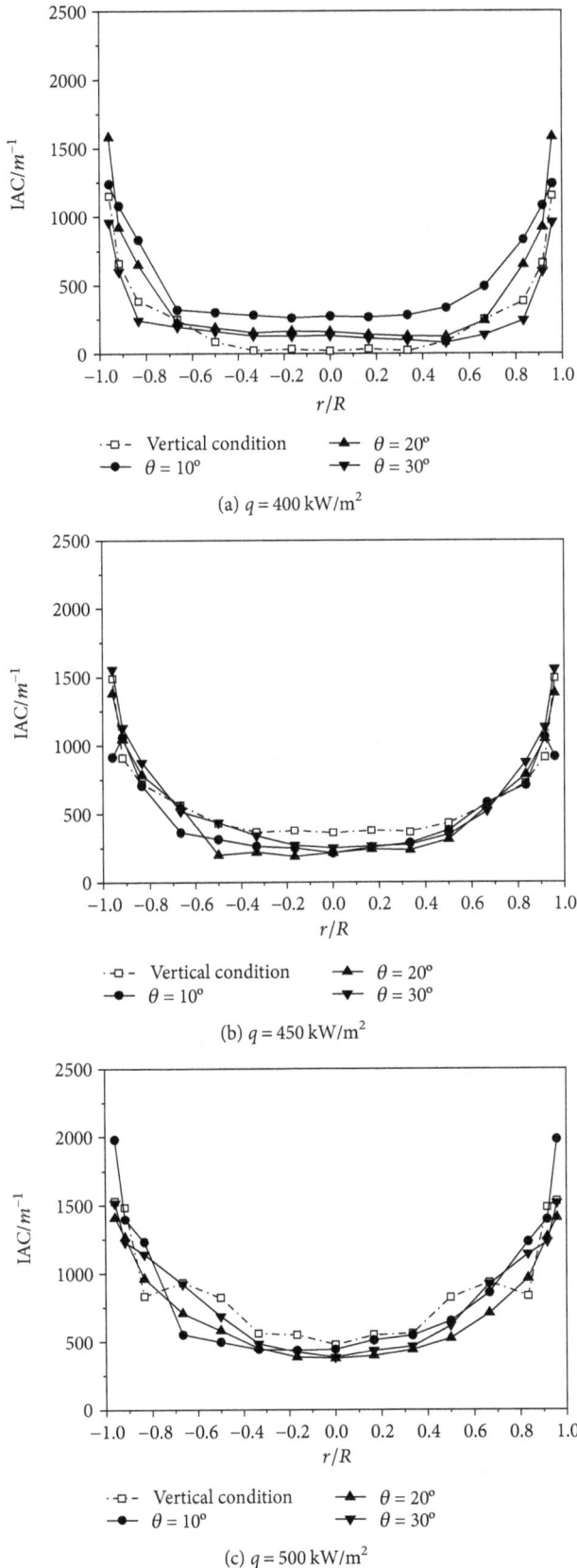

FIGURE 11: Inclination angle effect on the interfacial area concentration on the direction of $\beta = 0°$ in cross section ($P = 0.5\,MPa$, $G = 500\,kg/m^2\,s$).

Figure 9 shows the local void fraction distribution under different inclination angles. When the heat flux is $q = 400\,kW/m^2$, the bubbles rise along the radial direction under inclined condition. With the increase of the inclination angles, the near-wall peak gradually disappeared. Meanwhile, the local void fraction in the channel center is larger than that under the condition of vertical. As the heat flux increases, the local void fraction distribution on the direction of $\beta = 0°$ is more uneven. From Figure 10, it can be seen that the size of the bubble in the inclined channel is mainly the core-peak distribution. The inclination angle has little influence on the size of the bubble near the wall. With the increase of the inclination angle, the bubble size increases as well. Figure 11 shows that the inclination angle has little influence on the IAC on the direction of $\beta = 0°$ due to the reason that more large bubbles occur and gather onto the top of tube when the heat flux increases.

4. Conclusions

In this paper, local interfacial characteristics in subcooled flow boiling were investigated under vertical and inclined conditions. For vertical upward subcooled flow boilings, the local interfacial parameters presented three kinds of distribution types, including wall peak, core peak, and intermediate peak. As it is expected, the phase distribution under the inclined condition is different from the phase distribution under the vertical condition. The profiles skewed highly for 90° and 45° chord of tube cross section, whereas the profile was also symmetrical at 0° chord. These results can be explained by the fact that buoyancy caused the bubbles to move toward the top of the tube cross section under inclined condition. In addition, the typical distributions were also influenced by the inclination angles.

Nomenclature

a_i: Local interfacial area concentration (1/m)
D: Tube diameter (mm)
D_{sm}: Local bubble Sauter mean diameter (mm)
f_b: Local bubble frequency (Hz)
F_{bl}: Radial component for buoyancy force
G: Mass flux (kg/m²s)
N_b: Bubble number of sampling
N_{miss}: Missed bubble number of sampling
P: Pressure (MPa)
q: Heat flux (kW/m²s)
R: Tube radius (mm)
r: Radial position for probe measurement (mm)
v_i: Local interfacial velocity (m/s)
$\triangle T$: Samplimg time(s)/subcooling degree(°C).

Greek Symbols

α: Local void fraction
Δs: Distance between two sensor tips (m)
Δt_j: Time delay for ith interface (s)
Δt_{kl}: Time delay between the front and rear sensor (s)
ΔT_{in}: Inlet temperature.

Conflicts of Interest

The authors declare that they have no conflicts of interest.

Acknowledgments

This work is supported by the National Natural Science Foundation of China under Grant nos. 11475161 and 11505176.

References

[1] M. Higuchi and T. Saito, "Quantitative characterizations of long-period fluctuations in a large-diameter bubble column based on point-wise void fraction measurements," *Chemical Engineering Journal*, vol. 160, no. 1, pp. 284–292, 2010.

[2] S. L. Kiambi, H. K. Kiriamiti, and A. Kumar, "Characterization of two phase flows in chemical engineering reactors," *Flow Measurement and Instrumentation*, vol. 22, no. 4, pp. 265–271, 2011.

[3] T. Hibiki and M. Ishii, "Interfacial area concentration of bubbly flow systems," *Chemical Engineering Science*, vol. 57, no. 18, pp. 3967–3977, 2002.

[4] K. Sun, M. Zhang, and X. Chen, "Local measurement of gas-liquid bubbly flow with a double-sensor probe," *Chinese Journal of Chemical Engineering*, vol. 8, pp. 33–40, 2000.

[5] X. Shen and K. Mishima, "Two-phase phase distribution in a vertical large diameter pipe," *International Journal of Heat and Mass Transfer*, vol. 48, no. 1, pp. 211–225, 2005.

[6] X. Shen, Y. Saito, K. Mishima, and H. Nakamura, "A study on the characteristics of upward air-water two-phase flow in a large diameter pipe," *Experimental Thermal and Fluid Science*, vol. 31, no. 1, pp. 21–36, 2006.

[7] S. Kim, M. Ishii, Q. Wu, D. McCreary, and S. G. Beus, "Interfacial structures of confined air-water two-phase bubbly flow," *Experimental Thermal and Fluid Science*, vol. 26, no. 5, pp. 461–472, 2002.

[8] X. D. Sun, T. R. Smith, S. Kim, M. Ishii, and J. Uhle, "Interfacial area of bubbly flow in a relatively large diameter pipe," *Experimental Thermal and Fluid Science*, vol. 27, no. 1, pp. 97–109, 2002.

[9] S. T. Revankar and M. Ishii, "Local interfacial area measurement in bubbly flow," *International Journal of Heat and Mass Transfer*, vol. 35, no. 4, pp. 913–925, 1992.

[10] T. Hibiki and M. Ishii, "Experimental study on interfacial area transport in bubbly two-phase flows," *International Journal of Heat and Mass Transfer*, vol. 42, no. 16, pp. 3019–3035, 1999.

[11] T. Hibiki, M. Ishii, and Z. Xiao, "Axial interfacial area transport of vertical bubbly flows," *International Journal of Heat and Mass Transfer*, vol. 44, no. 10, pp. 1869–1888, 2001.

[12] X. Z. Shen, R. Matsui, K. Mishima, and H. Nakamura, "Distribution parameter and drift velocity for two-phase flow in a large diameter pipe," *Nuclear Engineering and Design*, vol. 240, no. 12, pp. 3991–4000, 2010.

[13] K. Sekoguchi, H. Fukui, and Y. Sato, "Flow boiling in subcooled and low quality regions heat transfer and local void fraction," in *Proceedings of Fifth International Heat Transfer Conference*, vol. 4, pp. 180–184, Tokyo, Japan, 1974.

[14] J. Garnier, E. Manon, and G. Cubizolles, "Local measurements on flow boiling of refrigerant 12 in a vertical tube," *Multiphase Science and Technology*, vol. 13, no. 1-2, p. 111, 2001.

[15] Q. Sun, *Phase Distribution for Low Mass Flux Subcooled Boiling Flow [Ph. D Thesis]*, Tsinghua University, 2004.

[16] A. Hasan, R. P. Roy, and S. P. Kalra, "Some measurements in subcooled flow boiling of refrigerant-113," *Journal of Heat Transfer*, vol. 113, no. 1, p. 216, 1991.

[17] R. P. Roy and V. Velidandla, "Local measurements in the two-phase region of turbulent subcooled boiling flow," *Journal of Heat Transfer*, vol. 116, no. 3, p. 660, 1994.

[18] R. Situ, T. Hibiki, X. D. Sun, Y. Mi, and M. Ishii, "Axial development of subcooled boiling flow in an internally heated annulus," *Experiments in Fluids*, vol. 37, no. 4, pp. 589–603, 2004.

[19] T. Lee, R. Situ, T. Hibiki, H. Park, M. Ishii, and M. Mori, "Axial developments of interfacial area and void concentration profiles in subcooled boiling flow of water," *International Journal of Heat and Mass Transfer*, vol. 52, no. 1-2, pp. 473–487, 2009.

[20] B. J. Yun, B. U. Bae, D. J. Euh, G. C. Park, and C. H. Song, "Characteristics of the local bubble parameters of a subcooled boiling flow in an annulus," *Nuclear Engineering and Design*, vol. 240, no. 9, pp. 2295–2303, 2010.

[21] J. G. Jing, M. Y. Zhang, and X. J. Chen, "A study on flow pattern transitions for gas–liquid two-phase upward flow in an inclined tube," *Journal of Xi'an Jiaotong University*, vol. 28, pp. 143–150, 1994.

[22] S. Wongwises and M. Pipathattakul, "Flow pattern, pressure drop and void fraction of two-phase gas–liquid flow in an inclined narrow annular channel," *Experimental Thermal and Fluid Science*, vol. 30, no. 4, pp. 345–354, 2006.

[23] J. Y. Xu, Y. X. Wu, Z. H. Shi, L. Y. Lao, and D. H. Li, "Studies on two-phase co-current air non-Newtonian shear-thinning fluid flows in inclined smooth pipes," *International Journal of Multiphase Flow*, vol. 33, no. 9, pp. 948–969, 2007.

[24] V. Hernandez-Perez, *Gas–Liquid Two-Phase Flow in Inclined Pipes [Ph. D Thesis]*, The University of Nottingham, 2008.

[25] A. Abdulahi, L. Abdulkareem, S. Sharaf, M. Abdulkadir, V. Hernandez Perez, and B. J. Azzopardi, "Investigating the effect of pipe inclination on two-phase gas–liquid flows using advanced instrumentation," in *The 8th Thermal Engineering Joint Conference (AJTEC2011)*, Honolubu, HI, USA, March, 2011.

[26] K. Spindler and E. Hahne, "An experimental study of the void fraction distribution in adiabatic water-air two-phase flows in an inclined tube," *International Journal of Thermal Sciences*, vol. 38, no. 4, pp. 305–314, 1999.

[27] D. Xing, C. Yan, L. Sun, J. Liu, and B. Sun, "Experimental study of interfacial parameter distributions in upward bubbly flow under vertical and inclined conditions," *Experimental Thermal and Fluid Science*, vol. 47, pp. 117–125, 2013.

[28] W. Bao, B. D. Chen, J. J. Xu, T. Z. Xie, D. C. Xing, and Y. P. Huang, "Experimental study on the local interfacial characteristic of subcooled flow boiling under inclined condition," in *23rd International Conference on Nuclear Engineering*, Chiba, Japan, 2015.

[29] Q. Wu and M. Ishii, "Sensitivity study on double-sensor conductivity probe for the measurement of interfacial area concentration in bubbly flow," *International Journal of Multiphase Flow*, vol. 25, no. 1, pp. 155–173, 1999.

19

Albendazole Degradation Possibilities by UV-Based Advanced Oxidation Processes

Davor Ljubas [ID],[1] **Mirta Čizmić** [ID],[2] **Katarina Vrbat,**[2] **Draženka Stipaničev,**[3] **Siniša Repec,**[3] **Lidija Ćurković** [ID],[4] **and Sandra Babić**[2]

[1]Department of Energy, Power Engineering and Environment, Faculty of Mechanical Engineering and Naval Architecture, University of Zagreb, Ivana Lučića 5, Zagreb, Croatia
[2]Department of Analytical Chemistry, Faculty of Chemical Engineering and Technology, University of Zagreb, Marulićev trg 20, Zagreb, Croatia
[3]Croatian Waters, Central Water Management Laboratory, Ulica grada Vukovara 220, Zagreb, Croatia
[4]Department of Materials, Faculty of Mechanical Engineering and Naval Architecture, University of Zagreb, Ivana Lučića 5, Zagreb, Croatia

Correspondence should be addressed to Davor Ljubas; davor.ljubas@fsb.hr and Mirta Čizmić; mzrncic@fkit.hr

Academic Editor: Francesco Riganti-Fulginei

Pharmaceuticals are present in an aquatic environment usually in low (ng/L) concentrations. Their continuous release can lead to unwanted effects on the nontarget organisms. The main points of their collection and release into the environment are wastewater treatment plants. The wastewater treatment plants should be upgraded by new technologies, like advanced oxidation processes (AOPs), to be able to degrade these new pollutants. In this study, the degradation of albendazole (ALB), a drug against parasitic helminths, was investigated using four UV-based AOPs: UV photolysis, UV photocatalysis (over TiO_2 film), $UV + O_3$, and $UV + H_2O_2$. The ranking of the degradation process degree of the ALB and its degradation products for studied processes is as follows: UV photolysis $<$ UV photocatalysis with $TiO_2 < UV + O_3 < UV + H_2O_2$. The fastest degradation of ALB and its degradation products was obtained by UV-C $+ H_2O_2$ process with a degradation efficiency of 99.95%, achieved in 15 minutes.

1. Introduction

Pharmaceuticals are complex molecules with different physicochemical and biological properties and functionalities. Although they are present in the aquatic environment in low (ng/L to μg/L) concentrations, they are continually being released into the environment which can lead to unwanted effects on the living organisms, especially on the nontarget organisms [1–3]. Many studies showed that the main points of collection and subsequent release of pharmaceuticals into the environment are wastewater treatment plants, suggesting that their upgrade and implementation of advanced treatment technologies are required [4, 5].

Research studies now concentrate on diverse categories of pharmaceuticals, e.g., macrolide antibiotics, hormones, endocrine-disrupting compounds, β-blockers, and anthelmintics, as well as their metabolites [6–9].

Anthelmintics are mostly used both for humans and animals, and their focused activity is the treatment of gastrointestinal parasites [10, 11]. A few different anthelmintics are commercially available today. Among them, albendazole, flubendazole, thiabendazole, and fenbendazole are usually the most commonly used ones [12].

Albendazole (ALB), as one of the most widely used benzimidazole anthelmintics, was in focus of this study, since we noticed the lack of reports on the environmental fate of anthelmintics discharged into the water environment. Degradation products and metabolites of ALB that were detected in previous studies [13–15] are albendazole sulfoxide (ALB-SX), albendazole sulfone (ALB-SF), and albendazole-

TABLE 1: Degradation conditions of ALB in experiment groups A–D.

Experiment group	Experiment label	Experiment description
A	A1	UV illumination (photolysis) with predominant wavelengths 185/254 nm (UV-C)
	A2	UV illumination (photolysis) with predominant wavelength 365 nm (UV-A)
B	B1	UV illumination in the presence of sol-gel TiO_2 film (photocatalysis) with predominant wavelengths 185/254 nm(UV-C), with continuous purging with air (O_2)
	B2	UV illumination in the presence of sol-gel TiO_2 film (photocatalysis) with predominant wavelengths 365 (UV-A) nm, with continuous purging with air (O_2)
C	C1	Ozone dosage via concentrated O_3 solution (prepared according to [24]), low dosage of O_3: 0.5 mg/L
	C2	Ozone dosage via concentrated O_3 solution (prepared according to [24]), high dosage of O_3: 1.5 mg/L
	C3	High dosage of O_3: 1.5 mg/L + UV-C radiation
	C4	High dosage of O_3: 1.5 mg/L + UV-A radiation
D	D1	H_2O_2 dosage via concentrated (30%) solution, high dosage of H_2O_2: 320 mg/L H_2O_2
	D2	H_2O_2 dosage via concentrated (30%) solution, high dosage of H_2O_2: 320 mg/L H_2O_2 + UV-C radiation
	D3	H_2O_2 dosage via diluted (1 : 4) 30% solution, low dosage of H_2O_2: 64 mg/L H_2O_2 + UV-C radiation
	D4	H_2O_2 dosage via concentrated (30%) solution, high dosage of H_2O_2: 320 mg/L H_2O_2 + UV-A radiation

2-aminosulfone (ALB-2-ASF). In addition to ALB, ALB-SX is also an important factor in the potential adverse environmental impact on organisms in the environment [15].

One of the possible solutions for the degradation and/or removal of pharmaceuticals from the wastewaters is the use of advanced oxidation processes (AOPs) as an additional treatment step. AOPs can be defined as aqueous phase oxidation methods where highly reactive species such as hydroxyl radicals are responsible for the destruction of target pollutants, e.g., pharmaceutic molecules. They can be used either alone or coupled with other physicochemical and biological processes [16–18], especially as a tertiary treatment in wastewater treatment plants [19]. Extensively investigated AOPs for the degradation of pharmaceuticals include the UV irradiation combined with H_2O_2 or O_3 as strong oxidants, the Fenton and the photo-Fenton oxidation, and the heterogeneous photocatalysis over titanium dioxide or titania (TiO_2) [20–23].

In this study, the degradation of ALB was investigated in lab-scale experiments using four UV-based processes (with 185/254 and 365 nm radiation sources): (A) UV photolysis, (B) UV photocatalysis (UV light + TiO_2 nanofilm), (C) UV + O_3 process, and (D) UV + H_2O_2 process. Screening of ALB degradation was performed using liquid chromatography with tandem mass spectrometry (HPLC-MS/MS) and with ultra-high performance liquid chromatography coupled to quadrupole time-of-flight mass spectrometry (UHPLC-QTOF-MS).

Calculation of energy use and total cost of every treatment option, as well as their comparison, will be evaluated in continuous work that will follow this study.

2. Methods and Materials

The analytical standard of ALB was obtained from Veterina Animal Health (Kalinovica, Croatia). For chromatographic analysis methanol (J. T. Baker, Deventer, Netherlands), acetonitrile (J. T. Baker, Deventer, Netherlands), and

formic acid (Merck, Darmstadt, Germany) were used. All solvents used were HPLC-grade. Ultrapure water was prepared by a Millipore Simplicity UV system (Millipore Corporation, Billerica, MA, USA) and was used for all experiments. The concentration of aqueous solution of ALB was 1 mg/L. The solutions for experiments were prepared in quantities of 1 L and kept in the dark at 4°C.

Concentrated ozone stock solutions were produced by O_2 gas through an ozone generator through ultrapure water that was cooled in an ice bath, following a procedure described in [24].

Hydrogen peroxide (H_2O_2) concentrated solution (30%) was purchased from Kemika (Zagreb, Croatia) and kept in the dark, at 4°C.

All experiments were carried out in the 0.11 L borosilicate glass cylinder reactor (with 200 mm in height and 30 mm in diameter). A scheme of the reactor set-up was published elsewhere [25]. The TiO_2 nanostructured film was deposited on an inner reactor surface by the sol-gel method and dipcoating technique, described in details elsewhere [26]. Two different UV-radiation lamps were used: model Pen-Ray 90-0019-04, with $\lambda_{max} = 365$ nm and incident photon flux $N_p = 4.295 \times 10^{-6}$ Einstein/s (UV-A lamp), and model Pen-Ray 90-0004-07 with $\lambda_{max} = 254/185$ nm (UV-C lamp) and incident photon flux $N_p = 1.033 \times 10^{-6}$ Einstein/s (UVP, Upland, CA, USA). The lamp was placed in the center of the reactor, and the UV radiation reaches the inner wall of the reactor through the solution, causing the photolytic/photocatalytic oxidation process in the reactor as described elsewhere [27]. During the experiments, samples for chromatographic analysis were taken from the reactor at particular time intervals and stored in the dark under 4°C until analysis. Only by experiments with ozone, samples were analyzed immediately (since ozone alone reacts with ALB in solution).

The experiments for ALB degradation were carried out at a temperature of 25 ± 0.2°C without adjustment of pH. Starting pH of 1 mg/L solution was in interval 5.95–6.10, using different conditions that can be seen in Table 1.

Samples from the photolytic and photocatalytic experiments were analyzed on an Agilent Series 1200 HPLC system (Santa Clara, CA, USA) connected to a triple quadrupole mass spectrometer Agilent 6410 with an ESI interface. The column used for chromatographic separation of the degradation products was Synergi Polar C18 (100 mm × 2.0 mm, particle size 2.5 μm) supplied by Phenomenex (Torrance, CA, USA). The mobile phase was MilliQ water acidified with 0.1% formic acid (A) and acetonitrile acidified with also 0.1% formic acid (B) as it was used in [27]. The gradient elution was started with 8% of B which was held for 3 min. During the next 12 min, the percentage of B was increased linearly to 95% and was held for 5 min. During 0.01 min, it was set at 0% of B and was held for 10 min for the equilibration of the column. The analyses were performed in the positive ion (PI) mode. The conditions of the ion source of the mass spectrometer were drying gas temperature 350°C, capillary voltage 4 kV, drying gas flow 11 L min^{-1}, and nebulizer pressure 35 psi. Injection volume was 5 μL. For acquisition and data processing, Agilent MassHunter software version B.01.03 was used as described elsewhere [27].

HPLC-MS and MS/MS experiments for the identification of photodegradation products in experiments with UV-H$_2$O$_2$ and UV-O$_3$ processes were performed on an LTQ-Orbitrap Velos™ coupled with the Aria TLX-1 HPLC system (Thermo Fisher Scientific Inc., USA). The sample mixture was loaded (20 μL injection volume) on an Acquity UPLC HSS T$_3$ (2.1 mm × 50 mm, 1.8 μm particle size, Waters UK) column where the chromatographic separation was achieved using an 8 min linear gradient from 5% to 95% methanol in 0.1% formic acid at the flow rate of 200 μL·min^{-1}. The sample injection, separation, and spectra acquisition were carried out automatically. The electrospray capillary voltage was set at 4 kV, the capillary temperature was at 300°C, m/z range from 100 to 1000, the instrument resolution was 100,000 at 400 m/z, and mass accuracy was within the error of ±5 ppm. Tandem mass spectrometry experiments were performed using collision-induced dissociation. Mass range was from 100 to 600 m/z, isolation width was 1 Da with a normalized collision energy of 35 V, and activation time was 30 ms. Nitrogen was used as the collision gas. The acquisition software was set up in auto MS/MS mode using three precursor ions with active exclusion on (precursor exclusion after 5 MS/MS spectra for 20 s). Data extraction and analysis were done using Thermo Xcalibur 2.2 SP1.48 (Thermo Fisher Scientific Inc., USA).

Ozone was produced from O$_2$ gas (purity 99.995%, purchased by MESSER, Croatia) with ozone generator 500 M (Fischer Technology, Germany).

Concentrations of O$_3$ and H$_2$O$_2$ in stock solutions were controlled, due to their tendency to spontaneously degrade, by a UV-VIS spectrophotometer (HEWLETT PACKARD, Model HP 8453, USA) at 254 and 240 nm, respectively, by using a 1 cm quartz cell. The dose of H$_2$O$_2$ was prepared using the Beer-Lambert law and the established value for $\varepsilon = 401/(M\cdot cm)$, as it is described in [28].

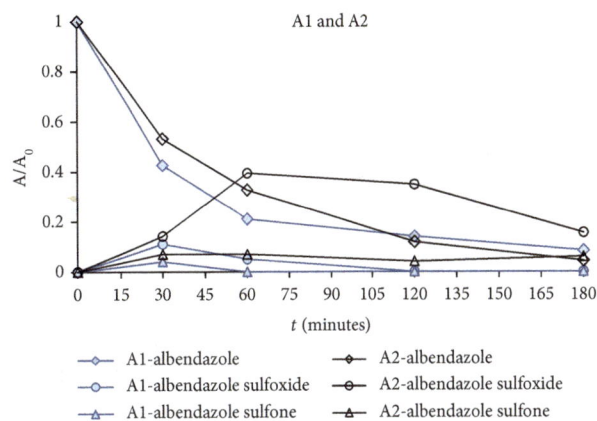

Figure 1: Photolytic degradation of ALB: A1—with UV-C radiation; A2—with UV-A radiation.

3. Results and Discussion

In following figures, the obtained results are presented as the integrated area of the chromatographic peak of specific analyte (ALB or its degradation products) at the specific time (A) divided by the integrated area of the chromatographic peak of ALB at $t = 0$ min (A_0). Three DPs were identified using high-resolution MS; they are known as ALB metabolites. All the experiments were performed in duplicate, and the final results are the average of the two replications.

Figure 1 shows the results of photolytic degradation of ALB, using only UV-C (185/254 nm radiation peaks) or UV-A (365 nm radiation peak).

Degradation with UV-C radiation was slightly more efficient than UV-A in ALB degradation, but the degradation of degradation products (albendazole sulfoxide (ALB-SX) and albendazole sulfone (ALB-SF)) is much faster: in 120 minutes, they completely disappeared with UV-C. With UV-A radiation, there is still a significant concentration of the degradation products in the solution even after 180 min of radiation exposure.

According to the studies [13, 14], ALB and its metabolites are sensitive to UV (i.e., solar) radiation and it is to expect that sunlight at the surface of a flat water body could degrade ALB by 50% per clear summer/early autumn day. However, it is not to expect, too, that every run-off from wastewater system to natural water could be exposed to the natural solar radiation at daytime, and that is the reason why wastewaters that contain ALB and its metabolites should be additionally treated.

Experiments performed "in the dark", i.e., with the lamp switched-off and with the additional aluminum foil for protection of light penetration into the reactor, confirmed that the effects of adsorption of ALB on the surface of TiO$_2$ catalyst film were negligible in the overall ALB degradation process. Figure 2 shows the photocatalytic degradation of ALB with two different lamps in the reactor.

It is obvious that the process UV-C + TiO$_2$ is much faster in ALB degradation compared to UV-A + TiO$_2$. However, degradation product, ALB-SX, remains in significant concentration after 120 minutes. In experiment B1 in $t = 10$ and 15

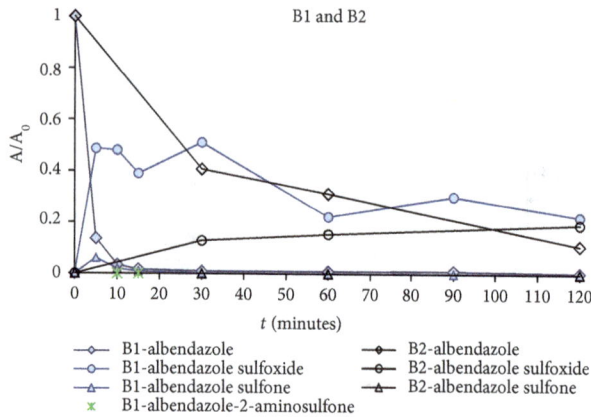

FIGURE 2: Photocatalytic degradation of ALB over TiO$_2$ film: B1—with UV-C radiation; B2—with UV-A radiation.

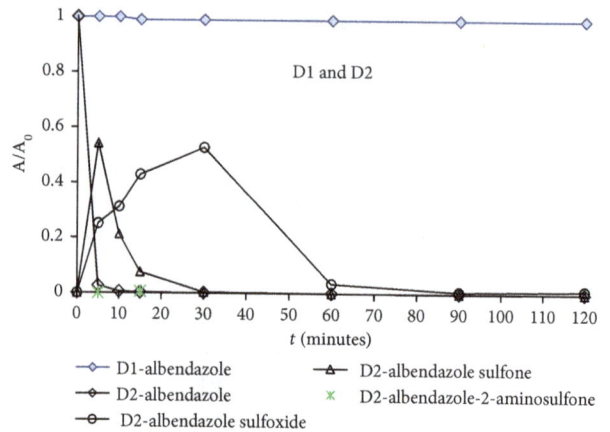

FIGURE 3: UV-based degradation of ALB with ozone: C1—ozone only, small dosage (0.5 mg/L O$_3$); C2—ozone only, high dosage (1.5 mg/L O$_3$).

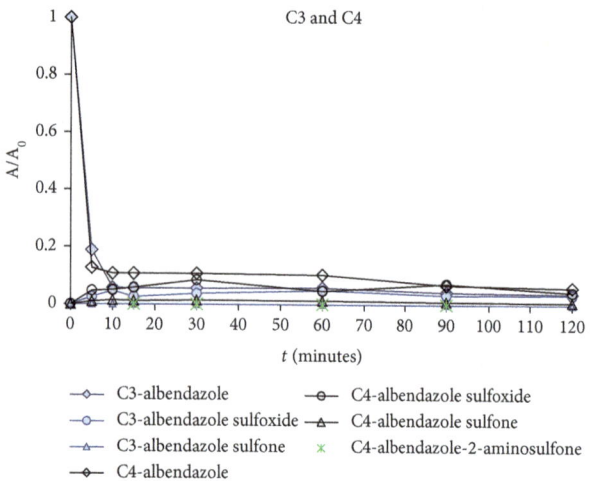

FIGURE 4: UV-based degradation of ALB with ozone: C3—ozone high dosage (1.5 mg/L O$_3$) with UV-C radiation; C4—ozone high dosage (1.5 mg/L O$_3$) with UV-A radiation.

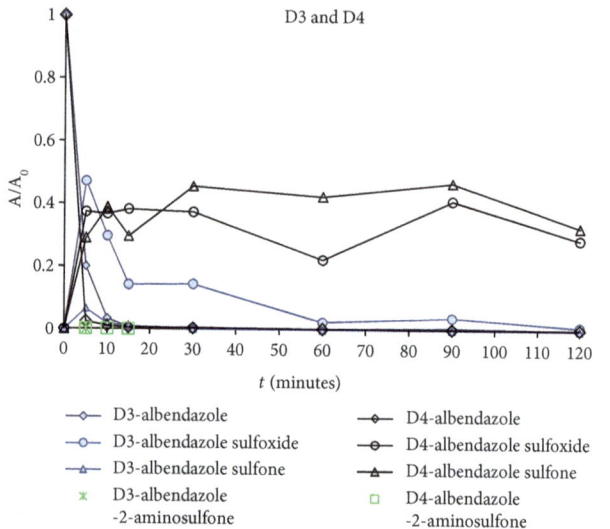

FIGURE 5: UV-based degradation of ALB with H$_2$O$_2$: D1—H$_2$O$_2$ only; D2—H$_2$O$_2$ high dosage + UV-C.

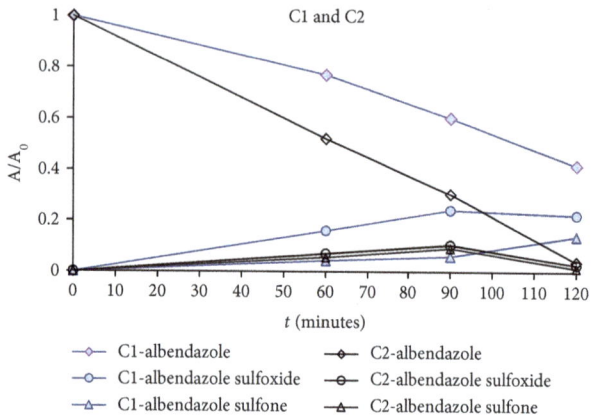

FIGURE 6: UV-based degradation of ALB with H$_2$O$_2$: D3—H$_2$O$_2$ low dosage + UV-C; D4—H$_2$O$_2$ high dosage + UV-A.

minutes, the traces of the 3rd degradation product of ALB was observed—ALB-2-ASF. Due to the fast degradation of ALB, there is a possibility for the formation of all three metabolites. The experiments B1 and B2 are faster in ALB degradation in comparison to the photolytic experiments A1 and A2 with the same radiation sources due to most probably hydroxyl radical formation over the TiO$_2$ film and its additional attack on the ALB molecule. It is especially observable when energy-higher UV-C radiation (in B1) was used.

In Figure 3, the degradation potential of ALB with ozone is shown. Ozone alone, both in high or low dosage, possesses relatively low potential for ALB and its DP degradation.

Nevertheless, when UV radiation was combined with O$_3$, again as in B group of experiments, due to hydroxyl radical formation, the degradation rate of ALB was increased (Figure 4). Rather a fast degradation of ALB, in 15 minutes, it reached around 90% of ALB removal, followed by very slow additional degradation of ALB, which implicates that the dosage of O$_3$, dosed by stock solution, was completely spent

TABLE 2: Removal efficiencies of ALB and formation quantities of ALB-SX and ALB-SF of all experiments at specific time intervals.

Experiment label	After 15 min			After 60 min			After 120 min		
	ALB, removal efficiency, %	ALB-SX, formation quantity, %	ALB-SF, formation quantity, %	ALB, removal efficiency, %	ALB-SX, formation quantity, %	ALB-SF, formation quantity, %	ALB, removal efficiency, %	ALB-SX, formation quantity, %	ALB-SF, formation quantity, %
A1	n.a.*	n.a.	n.a.	78.89	5.06	0.00	86.05	0.00	0.00
A2	n.a.	n.a.	n.a.	67.30	39.39	6.97	88.15	34.87	4.06
B1	98.37	38.95	0.88	99.08	22.14	0.02	99.50	22.20	0.02
B2	n.a.	n.a.	n.a.	68.85	15.40	0.03	89.12	19.20	0.04
C1	n.a.	n.a.	n.a.	23.01	16.00	4.20	58.29	22.15	13.86
C2	n.a.	n.a.	n.a.	48.03	7.01	5.52	96.15	2.74	1.41
C3	94.47	2.55	0.003	94.22	4.90	0.003	96.11	3.56	0.00
C4	89.31	5.79	1.22	89.72	4.72	1.37	93.99	4.44	0.9
D1	0.00	0.06	0.01	1.00	0.06	0.01	1.00	0.06	0.01
D2	99.75	42.98	7.61	99.93	3.47	0.05	99.95	1.21	0.00
D3	99.39	14.25	0.94	99.77	5.96	0.08	99.95	0.98	0.00
D4	99.50	38.21	29.60	99.77	22.30	42.40	99.82	28.76	32.72

on direct reactions with ALB or reactions to the hydroxyl radical formation, and it should be probably beneficiary to dose additional quantities of O_3 after 10 or 15 minutes.

When H_2O_2 was added to ALB solution, no reaction was observed, as can be seen in Figure 5.

In all UV-H_2O_2 processes (Figures 5 and 6), the 3rd degradation product, ALB-2-ASF, was shortly formed and after 15 minutes it disappeared, indicating that when the process of ALB degradation is fast, all three main degradation products of ALB can be observed.

Looking at all D processes, the fastest degradation of all components, ALB and its degradation products, was reached by the D2 process: in 90 minutes, practically, the water is almost free from contaminants. The D3 process is very close to D2 by its efficiency, especially when additional cost and energy requirements will be included in the evaluation of the processes.

Table 2 represents, in a short form, a comparison between all experiments, showing the ratios of the ALB and its degradation products during the treatment. It will be the base for continuing evaluation of these technologies.

4. Conclusions

The fastest degradation of ALB and its degradation products, for both UV-C radiation and UV-A radiation, was obtained by the UV-C + H_2O_2 process with removal efficiency of ALB higher than 99%, achieved in 15 minutes. However, degradation product removal requires extended time, up to 90 minutes.

In some cases, ALB was degraded more than 99% after 120 minutes but degradation products, especially ALB-SX, remained in high concentration. Such processes are characterized as not efficient enough because they do not efficiently remove all unwanted compounds that could potentially present a threat to the environment.

The ranking of the degradation process degree of the ALB and its degradation products for studied processes

is as follows: UV photolysis < UV photocatalysis with TiO_2 < UV + O_3 < UV + H_2O_2. Although the slowest degradation of ALB was obtained using UV-A processes, they have a potential for practical use: they could use natural solar radiation as a source of UV-A radiation and therewith significantly reduce the cost of the treatment step.

Disclosure

Some parts of the study were presented as a poster titled "UV-Based Advanced Oxidation Processes for Albendazole Degradation" at 4th *International Symposium on Environmental Management Towards Circular Economy* held in Zagreb, Croatia, December 7-9, 2016.

Conflicts of Interest

The authors declare that they have no conflicts of interest.

Acknowledgments

This study has been supported by the Croatian Science Foundation under the project "Fate of pharmaceuticals in the environment and during advanced wastewater treatment" (PharmaFate) (IP-09-2014-2353).

References

[1] W. C. Li, "Occurrence, sources, and fate of pharmaceuticals in aquatic environment and soil," *Environmental Pollution*, vol. 187, pp. 193–201, 2014.

[2] S. E. Jorgensen and B. Halling-Sorensen, "Drugs in the environment," *Chemosphere*, vol. 40, no. 7, pp. 691–699, 2000.

[3] F. A. Caliman and M. Gavrilescu, "Pharmaceuticals, personal care products and endocrine disrupting agents in the environment - a review," *CLEAN - Soil, Air, Water*, vol. 37, no. 4-5, pp. 277–303, 2009.

[4] M. Petrović, S. Gonzalez, and D. Barceló, "Analysis and removal of emerging contaminants in wastewater and drinking water," *TrAC Trends in Analytical Chemistry*, vol. 22, no. 10, pp. 685–696, 2003.

[5] K. Ikehata, N. Jodeiri Naghashkar, and M. Gamal El-Din, "Degradation of aqueous pharmaceuticals by ozonation and advanced oxidation processes: a review," *Ozone: Science & Engineering*, vol. 28, no. 6, pp. 353–414, 2006.

[6] S. Babić, L. Ćurković, D. Ljubas, and M. Čizmić, "TiO$_2$ assisted photocatalytic degradation of macrolide antibiotics," *Current Opinion in Green and Sustainable Chemistry*, vol. 6, pp. 34–41, 2017.

[7] J. C. Van De Steene, C. P. Stove, and W. E. Lambert, "A field study on 8 pharmaceuticals and 1 pesticide in Belgium: removal rates in waste water treatment plants and occurrence in surface water," *Science of The Total Environment*, vol. 408, no. 16, pp. 3448–3453, 2010.

[8] X. Yang, R. C. Flowers, H. S. Weinberg, and P. C. Singer, "Occurrence and removal of pharmaceuticals and personal care products (PPCPs) in an advanced wastewater reclamation plant," *Water Research*, vol. 45, no. 16, pp. 5218–5228, 2011.

[9] M. Bistan, T. Tišler, and A. Pintar, "Ru/TiO$_2$ catalyst for efficient removal of estrogens from aqueous samples by means of wet-air oxidation," *Catalysis Communications*, vol. 22, pp. 74–78, 2012.

[10] B. P. S. Capece, G. L. Virkel, and C. E. Lanusse, "Enantiomeric behaviour of albendazole and fenbendazole sulfoxides in domestic animals: pharmacological implications," *The Veterinary Journal*, vol. 181, no. 3, pp. 241–250, 2009.

[11] A. Hall and Q. Nahar, "Albendazole as a treatment for infections with *Giardia duodenalis* in children in Bangladesh," *Transactions of the Royal Society of Tropical Medicine and Hygiene*, vol. 87, no. 1, pp. 84–86, 1993.

[12] W.-J. Sim, H. Y. Kim, S. D. Choi, J. H. Kwon, and J. E. Oh, "Evaluation of pharmaceuticals and personal care products with emphasis on anthelmintics in human sanitary waste, sewage, hospital wastewater, livestock wastewater and receiving water," *Journal of Hazardous Materials*, vol. 248-249, pp. 219–227, 2013.

[13] C. A. Weerasinghe, D. O. Lewis, J. M. Mathews, A. R. Jeffcoat, P. M. Troxler, and R. Y. Wang, "Aquatic photodegradation of albendazole and its major metabolites. 1. Photolysis rate and half-life for reactions in a tube," *Journal of Agricultural and Food Chemistry*, vol. 40, no. 8, pp. 1413–1418, 1992.

[14] C. A. Weerasinghe, J. M. Mathews, R. S. Wright, and R. Y. Wang, "Aquatic photodegradation of albendazole and its major metabolites. 2. Reaction quantum yield, photolysis rate, and half-life in the environment," *Journal of Agricultural and Food Chemistry*, vol. 40, no. 8, pp. 1419–1421, 1992.

[15] L. Prchal, R. Podlipná, J. Lamka et al., "Albendazole in environment: faecal concentrations in lambs and impact on lower development stages of helminths and seed germination," *Environmental Science and Pollution Research*, vol. 23, no. 13, pp. 13015–13022, 2016.

[16] M. Klavarioti, D. Mantzavinos, and D. Kassinos, "Removal of residual pharmaceuticals from aqueous systems by advanced oxidation processes," *Environment International*, vol. 35, no. 2, pp. 402–417, 2009.

[17] V. Naddeo, D. Ricco, D. Scannapieco, and V. Belgiorno, "Degradation of antibiotics in wastewater during sonolysis, ozonation, and their simultaneous application: operating conditions effects and processes evaluation," *International Journal of Photoenergy*, vol. 2012, Article ID 624270, 7 pages, 2012.

[18] S. Norzaee, E. Bazrafshan, B. Djahed, F. Kord Mostafapour, and R. Khaksefidi, "UV activation of persulfate for removal of Penicillin G antibiotics in aqueous solution," *The Scientific World Journal*, vol. 2017, Article ID 3519487, 6 pages, 2017.

[19] A. Bernabeu, R. F. Vercher, L. Santos-Juanes et al., "Solar photocatalysis as a tertiary treatment to remove emerging pollutants from wastewater treatment plant effluents," *Catalysis Today*, vol. 161, no. 1, pp. 235–240, 2011.

[20] M. M. Huber, S. Canonica, G.-Y. Park, and U. von Gunten, "Oxidation of pharmaceuticals during ozonation and advanced oxidation processes," *Environmental Science & Technology*, vol. 37, no. 5, pp. 1016–1024, 2003.

[21] S. Babić, M. Zrnčić, D. Ljubas, L. Ćurković, and I. Škorić, "Photolytic and thin TiO$_2$ film assisted photocatalytic degradation of sulfamethazine in aqueous solution," *Environmental Science and Pollution Research*, vol. 22, no. 15, pp. 11372–11386, 2015.

[22] W. H. Glaze, J.-W. Kang, and D. H. Chapin, "The chemistry of water treatment processes involving ozone, hydrogen peroxide and ultraviolet radiation," *Ozone: Science & Engineering*, vol. 9, no. 4, pp. 335–352, 1987.

[23] C. Pablos, J. Marugán, R. van Grieken, and E. Serrano, "Emerging micropollutant oxidation during disinfection processes using UV-C, UV-C/H$_2$O$_2$, UV-A/TiO$_2$ and UV-A/TiO$_2$/H$_2$O$_2$," *Water Research*, vol. 47, no. 3, pp. 1237–1245, 2013.

[24] M. S. Elovitz and U. von Gunten, "Hydroxyl radical/ozone ratios during ozonation processes. I. The R$_{ct}$ concept," *Ozone: Science & Engineering*, vol. 21, no. 3, pp. 239–260, 1999.

[25] L. Ćurković, D. Ljubas, S. Šegota, and I. Bačić, "Photocatalytic degradation of Lissamine Green B dye by using nanostructured sol–gel TiO$_2$ films," *Journal of Alloys and Compounds*, vol. 604, pp. 309–316, 2014.

[26] S. Šegota, L. Ćurković, D. Ljubas, V. Svetličić, I. F. Houra, and N. Tomašić, "Synthesis, characterization and photocatalytic properties of sol-gel TiO$_2$ films," *Ceramics International*, vol. 37, no. 4, pp. 1153–1160, 2011.

[27] M. Čizmić, K. Vrbat, D. Ljubas, L. Ćurković, and S. Babić, "Photocatalytic degradation of macrolide antibiotic azithromycin in aqueous sample," *Proceedings of the 15th International Conference on Environmental Science and Technology (CEST 2017)*, D. F. Lekkas, Ed., , pp. 1–4, Diagramma, Athens, 2017.

[28] U. Von Gunten and Y. Oliveras, "Kinetics of the reaction between hydrogen peroxide and hypobromous acid: implication on water treatment and natural systems," *Water Research*, vol. 31, no. 4, pp. 900–906, 1997.

20

Preparation and Thermal Properties of Eutectic Hydrate Salt Phase Change Thermal Energy Storage Material

Lin Liang⑩ and Xi Chen⑩

School of Electrical and Power Engineering, China University of Mining and Technology, Xuzhou 221116, China

Correspondence should be addressed to Lin Liang; lianglin@cumt.edu.cn

Academic Editor: Hamidreza Shabgard

In this study, a new cold storage phase change material eutectic hydrate salt ($K_2HPO_4 \cdot 3H_2O$–$NaH_2PO_4 \cdot 2H_2O$–$Na_2S_2O_3 \cdot 5H_2O$) was prepared, modified, and tested. The modification was performed by adding a nucleating agent and thickener. The physical properties such as viscosity, surface tension, cold storage characteristics, supercooling, and the stability during freeze-thaw cycles were studied. Results show that the use of nucleating agents, such as sodium tetraborate, sodium fluoride, and nanoparticles, are effective. The solidification temperature and latent heat of these materials which was added with 0, 3, and 5 wt% thickeners were −11.9, −10.6, and −14.8°C and 127.2, 118.6, 82.56 J/g, respectively. Adding a nucleating agent can effectively improve the nucleation rate and nucleation stability. Furthermore, increasing viscosity has a positive impact on the solidification rate, supercooling, and the stability during freeze-thaw cycles.

1. Introduction

The development of cities is accompanied by huge energy consumption; people have gradually realized that the energy storage technology has very good ability to improve this situation, which can effectively improve the energy utilization ratio and reduce losses. Phase change material (PCM) is an effective latent heat thermal storage material. It has been widely used as a thermal functional material in thermal and cold energy storage fields, like solar energy storage [1], industrial waste heat utilization [2], building heating and air conditioning [3], thermal management of mobile devices [4], and so on. All of these are aimed at realizing the conversion of energy beyond the restriction of time and space. Thus, the study of PCM is of great significance for energy storage [5]. PCM can be mainly divided into two categories of organic and inorganic. However, organic PCMs such as paraffin wax [6] and some organic compounds [7] are poor in low thermal conductivity and have flammability, which greatly limits the application of PCMs [8]. Therefore, the study of inorganic PCM has attracted much attention of researchers.

Salt hydrates [8] as an inorganic PCM have an advantage in higher heat storage capacity, suitable cold storage temperature [9], nonflammable, and so on. Thus, it would be more suitable and safer than organic PCM in energy storage application especially cryogenic storage. The traditional ice storage and water storage cannot reach the temperature of low-temperature cold storage. The temperature requirement of low cold storage is between −20 and −30°C, and the high-temperature cold storage is between 0 and 4°C. However, the ice storage and water storage systems can only reach 0°C, which cannot meet requirements of low-temperature application. Adding inorganic salts in the water can ensure that the amount of phase change latent heat almost unchanged and reduce the phase change temperature of cool storage material at the same time. Compound salts can not only further reduce the melting point of solidification but also optimize and modify the overall physical properties of certain materials. Many researchers have studied the compounded eutectic hydrate salt to achieve more suitable phase change temperature and better cold storage performance. Li et al. [10] have studied the preparation, characterization, and modification of a new phase change material $CaCl_2 \cdot 6H_2O$–$MgCl_2 \cdot 6H_2O$ eutectic hydrate salt. As a result, its phase change temperature and latent heat are 21.4°C and 102.3 J/g, respectively. Liu et al. [11] studied the energy storage characteristics of a novel binary hydrated salt by means of SEM, XRD, and DSC techniques. It was concluded that the phase transition temperature was 27.3°C, the degree of supercooling was 3.67°C, and the enthalpy of phase transition was 220.2 J/g. Efimova et al. [12] proposed a ternary hydrated salt mixture suitable for use in air conditioning systems and carried out the thermal analysis. It was concluded that the material has a melting temperature of 18–21°C and an enthalpy of fusion of 110 kJ/kg. However, different from organic PCM [13], inorganic PCM such as salt hydrates has a significant supercooling [14, 15] and relatively poor stability, which greatly limit the life of phase change material and energy storage performance. In general, methods that can be used to effectively enhance the stability of the material are additions of the thickener [16], nucleating agent [17, 18], microencapsulation [19] or nanocomposite modification [20], and so on. Tyagi [21] et al. studied the supercooling behavior of $CaCl_2 \cdot 6H_2O$ and the effect of pH value, which indicates that the supercooling of PCM can be removed or decreased by adjusting the pH value. Bilen et al. [22] studied the modification of $CaCl_2 \cdot 6H_2O$ PCM system and selected different nucleating agents. Pilar et al. [23] used $SrCO_3$ and $Sr(OH)_2$ as nucleating agents for $MgCl_2 \cdot 6H_2O$; the results show that the addition of 1 wt% $SrCO_3$ and 0.5 wt% $Sr(OH)_2$ almost fully suppress the supercooling and improve the performance of this PCM system. For phase separation, Wang et al. [24] studied the thermal stability of a novel eutectic ternary system by placing the salt mixture in an argon atmosphere with a constant heating rate. In order to determine the accurate upper limit of the working temperature of the ternary salt, Sharma et al. [25] used a differential scanning calorimeter to carry out 1500 times melting and freezing cycles to study the changes in thermal properties of thermal energy storage materials. At present, the inorganic salt hydrate phase change cold storage materials have a good application prospect in refrigerator energy-saving field.

It can be seen above that most of the research on hydrated salt cold storage has focused on the development of energy storage materials for air conditioning. There is a lack of research on the energy storage materials with phase change temperature below 0°C. Otherwise, it is a meaningful attempt to study the improvement of supercooling and cycle stability of ternary eutectic system from the point of fluid viscosity and different nucleating agents added. In this paper, firstly, a $K_2HPO_4 \cdot 3H_2O$–$NaH_2PO_4 \cdot 2H_2O$–$Na_2S_2O_3 \cdot 5H_2O$ ternary salt system was prepared, modified, and synthesized. Then the contrast experimental method, step cooling method, and DSC technology were proceeded to investigate the thermal storage property and solidification behavior of eutectic hydrate when adding different amounts of nucleating agent and thickener.

2. Materials and Experimental Methods

2.1. Preparation of Hydrated Salt-Based PCMs.
Preparation of hydrated salt phase change material (HSPCM) is the major key step to ensure the performance of PCM in cold storage

application. The proper mixing and stabilization are required in order to achieve stable HSPCM. In the present study, deionized water (DI water) as the base PCM; sodium dihydrogen phosphate dihydrate, dipotassium hydrogen phosphate trihydrate, and sodium thiosulfate pentahydrate as hydrated salts; nanoactivated carbon, sodium tetraborate, and sodium fluoride as nucleating agents; polyethylene glycol 400 as a dispersant; and sodium alginate as a thickener were used to prepare the HSPCM. The used materials are listed in Table 1. After a large number of experimental analysis, the hydrated salts with the same mass ratio (6 wt%) were mixed with DI water by a magnetic stirrer (MVP22–1) for 10 min to make the mixed solution system uniform. Before using the nanoparticle as a nucleating agent, the solution needs to be adjusted to weak alkaline by 10 wt% NaOH, followed by being mixed with nanoparticles and dispersant through magnetic stirring for 10 min. Afterwards, this mixture was ultrasonicated for 60 min at a frequency of 40 kHz and another 20 min stirring to make the nanoparticles dispersed as evenly as possible. The preparation process of nanocomposite hydrated salts is shown in Figure 1. The nanocomposite fluid can be considered accessible when it is observed that the formulated nanofluid undergoes no visual sedimentation over time.

2.2. Experimental Setup.
Table 2 shows the uncertainty to measure results caused by experimental equipment. Figure 2 shows the schematic of the experimental setup to conduct the studies during the solidification of the HSPCM. The cooling chamber of 0.108 m³ has the cooling ability to freeze 5 kg reagent from 25°C to −18°C in 24 hours. In the first experiment, the temperatures were measured by the armored thermocouple (T-Type) floating at the same location in the containers filled with reagents. The signal of temperature was monitored continuously for every 10 s, and the experiment was continued until the PCM completely solidified. When measuring the percentage of solidification, three samples of different viscosity were put in a square container with a small thickness. On the upper surface, the square container was divided into a grid scale, and then the three samples were put into the cooling bath during the first trial. The solidified mass was calculated by observing the grid value change of the solid area from the initial state of the liquid PCMs. The thermal properties of eutectic mixtures were obtained by using a differential scanning calorimeter (DSC, NETZSCH 200F3). The surface tension was tested by a surface tension meter (BZY-1). The step cooling analysis method was employed to examine the values of supercooling. In addition, the thermal stability was measured by the freeze-thaw cycles test, which was performed in three containers with a volume of 150 ml. The hydrated salt PCM mixture was cycled repeatedly through 20, 30, and 50 times within a freezer. Temperature data acquisition system was used to study the supercooling change after several freeze-thaw processes. The weighing of drug used a high precision electronic balance with an accuracy of ±0.002 g. All the armored thermocouples are connected to a data logger of Agilent 34970A to store the continuous data generated during the experiments.

TABLE 1: Experimental materials.

Experimental materials	Purity	Application
Sodium dihydrogen phosphate dihydrate ($NaH_2PO_4·2H_2O$,)	AR	Hydrated salt
Sodium alginate	CP	Thickener
Sodium thiosulfate pentahydrate ($Na_2S_2O_3·5H_2O$)	AR	Hydrated salt
Dipotassium hydrogen phosphate trihydrate phosphate ($K_2HPO_4·3H_2O$)	AR	Hydrated salt
Nanoactivated carbon (100 nm, heat treatment)	AR	Nucleating agent
Sodium tetraborate	AR	Nucleating agent
Polyethylene glycol 400	AR	Dispersant
Sodium fluoride	AR	Nucleating agent

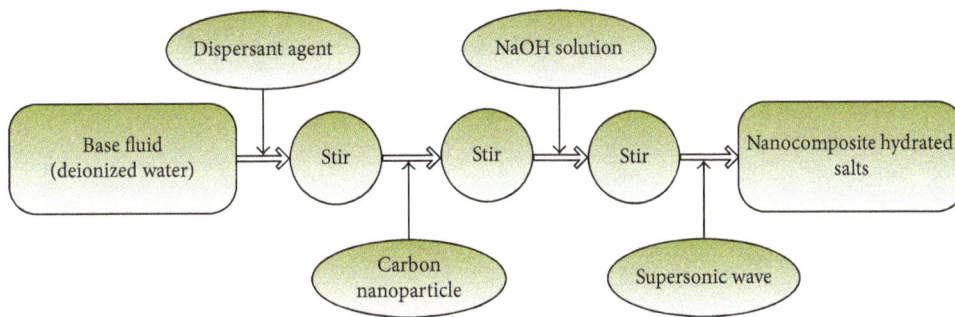

FIGURE 1: Nanocomposite PCM preparation flow chart.

TABLE 2: Result uncertainty analysis of experimental equipment.

Measured quantities	Deviation
Latent heat	±6 kJ/kg
Mass	±0.002 g
Temperature data logger	±0.06%
Volume (10 ml, 20 ml)	±0.06 ml
Volume (100 ml)	±0.02 ml
Thermocouple	±1.0%
Surface tension	±0.01 mN/m
Viscosity	±2% mps

3. Results and Discussion

3.1. The Thermal Performance of Salt Hydrate PCM. In order to assess the effect of thickener concentration on onset/end temperature and phase change heat absorption capacity during the phase change process, the thermal analysis was conducted in different cases of 0, 3, and 5 wt% sodium alginates, while the latent heat of melting process and onset/end temperature was measured with a differential scanning calorimeter (DSC), using measuring temperature range from −25 to 5°C and constant heating rates of 2 K·min⁻¹. In the process of measurement, the temperature of the material rises constantly, and the heat flux of the sample was measured continuously. The heat flux is proportional to the instantaneous specific heat of the material. Melting phase change enthalpy was calculated through the area of the

endothermic peak in the DSC picture. In addition, the specific heat, heating rate (dH_m/dt), phase change enthalpy, and specific heat satisfy certain relationships as follows:

$$\frac{dH_m}{dt} = m \cdot C_p \cdot \frac{dT}{dt},$$
$$C_p = \frac{H_p}{\Delta t},$$

(1)

where H_m, H_p, C_p, m, T, and Δt represent the heat flux, melting phase change enthalpy, specific heat, sample mass, and the temperature difference, respectively, between the beginning and end of the melting process and time interval.

Figure 3 shows the DSC heat absorption curves of three eutectic hydrate salt nitrates with 0, 3, and 5 wt% thickeners separately. It is obviously seen that the onset and end temperature of the absorption peak become higher with the increase of the viscosity when the mass fraction is 3%. After that, with more thickener added, there is a substantial reduction with the durative increase of the viscosity. Through comparing the three samples, it was easily seen that several microparticles from the industrial sodium alginate scattered in the solution that may have the function of improving the solidification point in the case of low viscosity. However, when the fluid viscosity is increasing, the solidification point decreased adversely; the impact of microparticles gradually disappeared. Furthermore, the height of the absorption peak and the phase change enthalpy reached the maximum without the thickener and gradually decreased after adding the thickener. This is because the fluid viscosity affects the

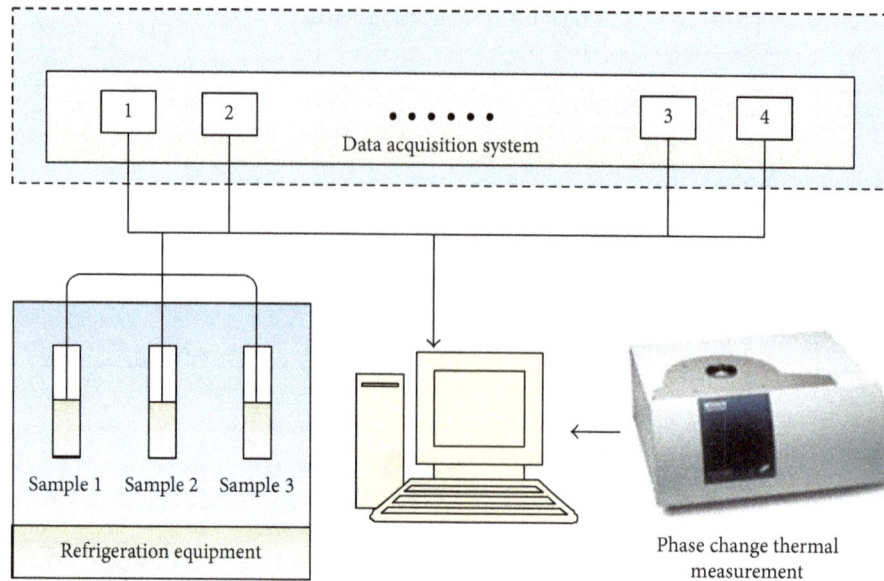

FIGURE 2: Schematic diagram of the experimental system.

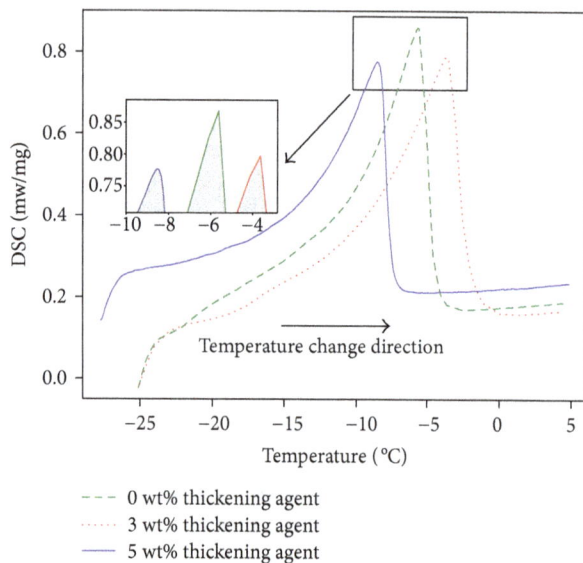

- - - 0 wt% thickening agent
...... 3 wt% thickening agent
— 5 wt% thickening agent

FIGURE 3: DSC curves of eutectic hydrate salt with three different amount of thickening agent.

TABLE 3: DSC measurements.

Data items	Thickener (wt%)		
	0	3	5
Area (J/g)	82.56	118.6	127.2
Peak sample temperature (°C)	−8.3	−2.4	−5.0
Onset temperature (°C)	−14.8	−10.6	−11.9
End temperature (°C)	−7.5	−2.3	−4.6

TABLE 4: Melting latent heat of water and three samples.

Material	Melting latent heat (kJ/kg)
Ice	335
Hydrated salt with 0 wt% thickener added	127.2
Hydrated salt with 3 wt% thickener added	118.6
Hydrated salt with 5 wt% thickener added	82.56

convective heat transfer performance within the fluid. When the viscosity of the fluid increases, the effect of the microconvection of the fluid and the particles which can enhance heat conduction is gradually suppressed.

From Table 3, it is clearly observed that the onset temperature of the three samples is −11.9, −10.6, and −14.8°C while the end temperature is −4.6, −2.3, and −7.5°C separately. The peak temperature of the three sample is −5.0, −2.4, and −8.3°C in turn. By calculating the DSC phase change area which is divided by the phase change DSC curve and the onset/end baseline, it is concluded that the phase change enthalpies of the three samples are 127.2, 118.6, and 82.56 J/g, respectively. Compared with similar studies [9], this material has an edge in storage ability.

Table 4 shows the melting latent heat change of the three samples with different additions of the thickener. It is not difficult to find that the melting latent heat of cold storage materials was reduced by 62%, 64.6%, and 75.3% as compared with ice. The increase of viscosity will reduce the heat absorption amount, thereby affecting the cold storage ability. Therefore, the thickener needs to be controlled in a suitable amount.

3.2. The Nucleation and Supercooling Characteristics of Hydrated Salt PCM. Supercooling is a process closely related to the crystallization process. Before the temperature returns to the original freezing point, the liquid solidifies below its normal freezing point and continues to decrease until complete solidification. When the degree of supercooling increased, the degree of the deviation from the equilibrium state has risen, and the critical dimension of the ice core and the formation energy also decrease dramatically, which ultimately increases the probability of forming the nucleus. Calculating the absolute degree of supercooling following relation was used to measure the thermal storage performance of materials:

$$T_{sc} = T_m - T_s, \qquad (2)$$

where T_{sc}, T_m, and T_s represent the supercooling degree, melting point, and solidification temperatures, respectively.

The supercooling degree was measured by temperature change curves during the cooling process by putting a thermocouple at the three sample's center separately. During the solidification and crystallization process, the general liquid crystal is divided into the stable region (noncrystalline), substable region, and unstable region as shown below in Figure 4. Crystallization phenomenon only occurred in the unstable area. The viscosity and temperature change of the inorganic hydrated salt are both important factors which affect the nucleation.

Figure 5 shows four hydrated salts of different viscosities without adding the nucleating agent. The three-step cooling curve of ternary hydrate salt solution with different viscosities can be obtained to observe the supercooling. Comparing the four curves, the recovery process after the reduction of the temperature curve gradually decreases during the solidification process with the increase of the viscosity, which can consider that the supercooling is gradually reduced. This is because the increase in liquid viscosity will affect the diffusion of the fluid, reducing the driving force of crystallization, and thus, the supercooling is suppressed. Therefore, increasing the fluid viscosity will not only reduce the latent heat of phase transformation but also improve the stability of the phase transition and reduce the degree of supercooling. When the inorganic PCM is applied to the practical application, the viscosity should be appropriately increased to find a middle value of the best performance.

3.3. The Physical Characteristics of Hydrated Salt PCM. In this experiment, the characteristics of the fluid itself were significantly changed with a constant increase in viscosity. From Table 5, we can find that ternary hydrate salt itself is more viscous than water whose viscosity is 2.04, 6.3, and 8.77 mpa·s, respectively. The viscosity of three solutions with 1.5, 2.5, and 3 wt% nucleating agent added was changed by adding a thickener. Results show that the salt solution with higher viscosity has a relatively lower degree of supercooling.

Figure 6 shows that the degree of supercooling is almost eliminated with the thickener and nucleating agent added simultaneously. It was seen that both the nucleating agent and the thickener have a dampening effect on supercooling. However, the amount of thickener should not be too much; otherwise, the heat capacity will drop dramatically; this is not economical. Therefore, rationally, selecting the amount of nucleating agent and thickener is improving the cold storage performance of inorganic solution better.

Figure 7 shows the mass fraction solidified at various time intervals during the solidification process with 0, 3, and 5 wt% thickeners added, respectively. Solidification time is

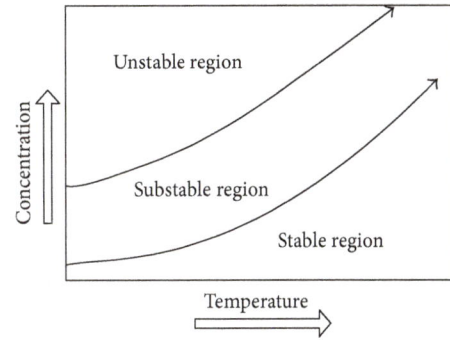

FIGURE 4: Supersaturation and super solubility curves of the solution.

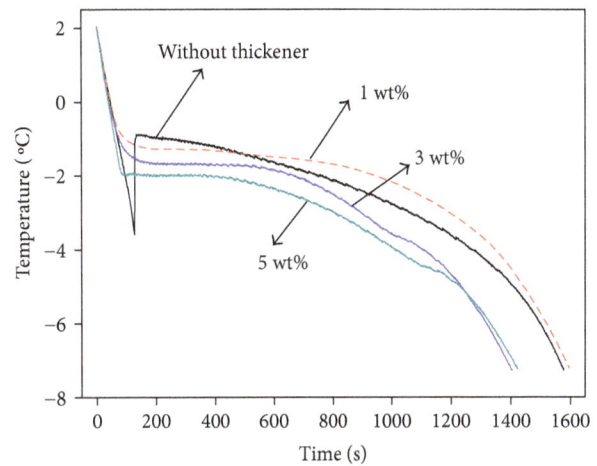

FIGURE 5: Step cooling curves of adding 1, 3, and 5 wt% thickeners in the solution without the nucleating agent.

TABLE 5: Viscosity of water and hydrated salts.

Material	Viscosity (mpa·s)
Ice	1
Hydrated salt with 0 wt% thickener added	2.04
Hydrated salt with 3 wt% thickener added	6.3
Hydrated salt with 5 wt% thickener added	8.77

also an important parameter which was used to measure the response time of a cold storage system. Under the same phase change enthalpy and time, storage medium which reached a higher percentage of solidification earlier has the quality to start next phase change circulation quickly, so as to improve the energy storage efficiency. It has been noted from Figure 7 that the solidification started about 40 minutes later from the start of the experiment and the hydrated salts with 5 wt% thickener appear crystal firstly, followed by the sample with 3 wt% thickener added. The above two samples began to appear crystal almost at the same time, only the sample without thickener crystallized slower which started at about 50–55 minutes. The crystallization rate of the three samples was also different in the solidification process. From

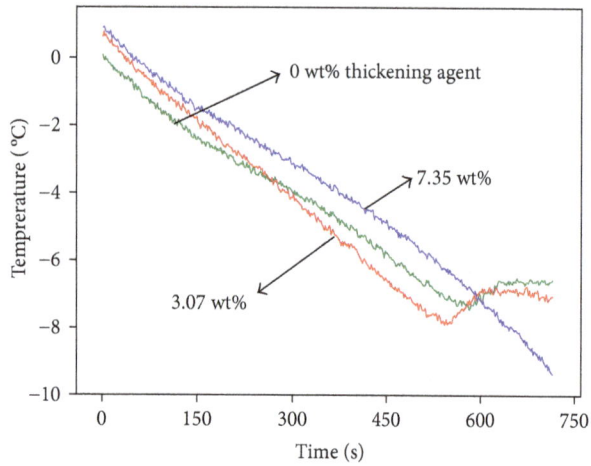

FIGURE 6: Step cooling curves of adding 0, 3.12, and 7.46 wt% thickeners and 2.5 wt% nucleating agent in the solution.

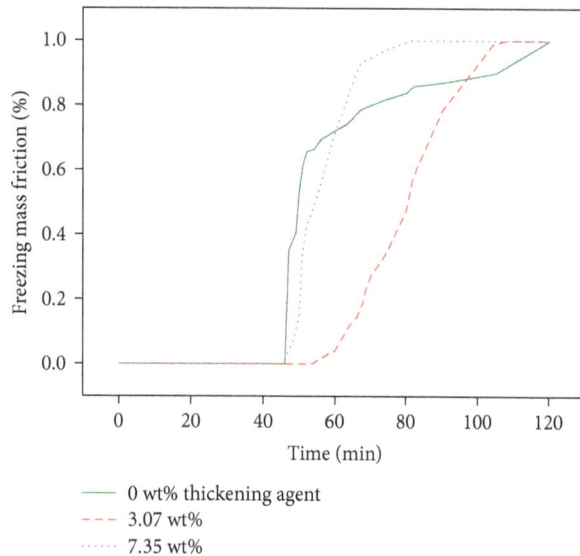

FIGURE 7: Solidification mass friction curve of three samples with different nucleating agents added.

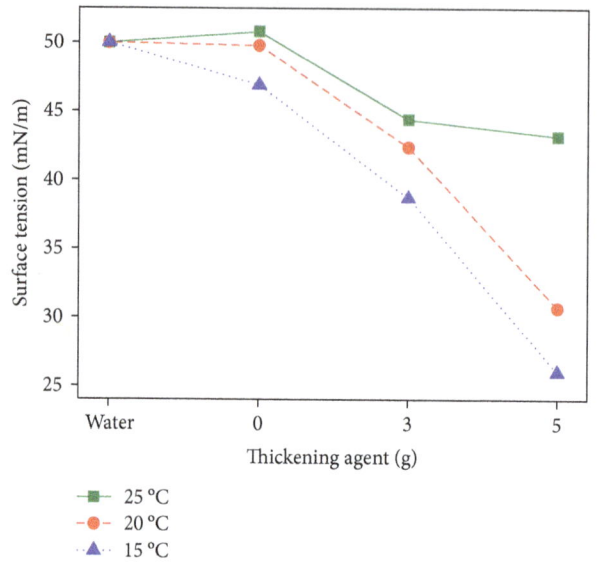

FIGURE 8: Surface tension of three samples within three different temperatures.

The viscosity of the hydrated salt affects not only the internal thermophysical properties of the fluid but also the properties of the fluid surface. Figure 8 shows the variation of surface tension change of hydrated salts and single water. It was found that the surface tension of hydrated salt is lower than that of the water, and the surface tension decreases as the viscosity increases gradually. This is due to the fluid with larger viscosity which has higher surface shrinkage resistance. In addition, the effect of temperature on the surface tension is very obvious; the surface tension decreases with the increase of temperature. This is because the increase of liquid temperature leads to the increase of molecular internal energy and the decrease of interaction force between molecules. Owing to the surface tension which originates from the attraction between molecules on the surface of the liquid, the decrease of attractive force leads to the decrease of surface tension.

3.4. The Influence of Different Nucleating Agents to Hydrated Salt PCM. In the above test, it was found that the supercooling phenomenon is a serious defect of the inorganic phase change material. Through many studies of the nucleation mechanism, it is proved that the supercooling also improved effectively by adding a nucleating agent. According to the crystal nucleation theory, for homogeneous nucleation process, the formation of the crystal needs to be larger than the critical size of the particles, but for heterogeneous nucleation, the surface affinity should also be considered [26]. Nevertheless, the effective crystallization depends on the value of the surface free energy. The aim of adding nucleation agent in the solution is to increase the surface affinity or reduce the critical dimension so that the surface free energy is reduced. From Figure 9(c), it was discovered that the degree of supercooling ranges from about 3°C to 1.5°C as the mass fraction of $Na_2B_4O_7 \cdot 10H_2O$ increases from 1.5 wt% to 3 wt%. Therefore, sodium tetraborate is an ideal agent for

45 to 50 minutes, the sample with 3 wt% and 5 wt% thickeners added had the similar crystallization rate. However, the crystallization rate of the sample without adding a thickener was obviously slowed down while the rate of the other two samples kept stable all the time. Finally, the time of the complete solidification of these three samples was about 120, 100, and 75 minutes, respectively. It can be concluded that the increase of the viscosity of the cold storage fluid can effectively shorten the time of the complete solidification of the material. This is because the solidified PCM offers increasing thermal resistance between the solidified PCM and the surrounding heat transfer fluid as the freezing front moves away from the surface. From the previous conclusion, it is drawn that viscosity has the same effect on the storage capacity of phase change materials, so the viscosity selection should ensure both the storage capacity and the solidification rate at a right value.

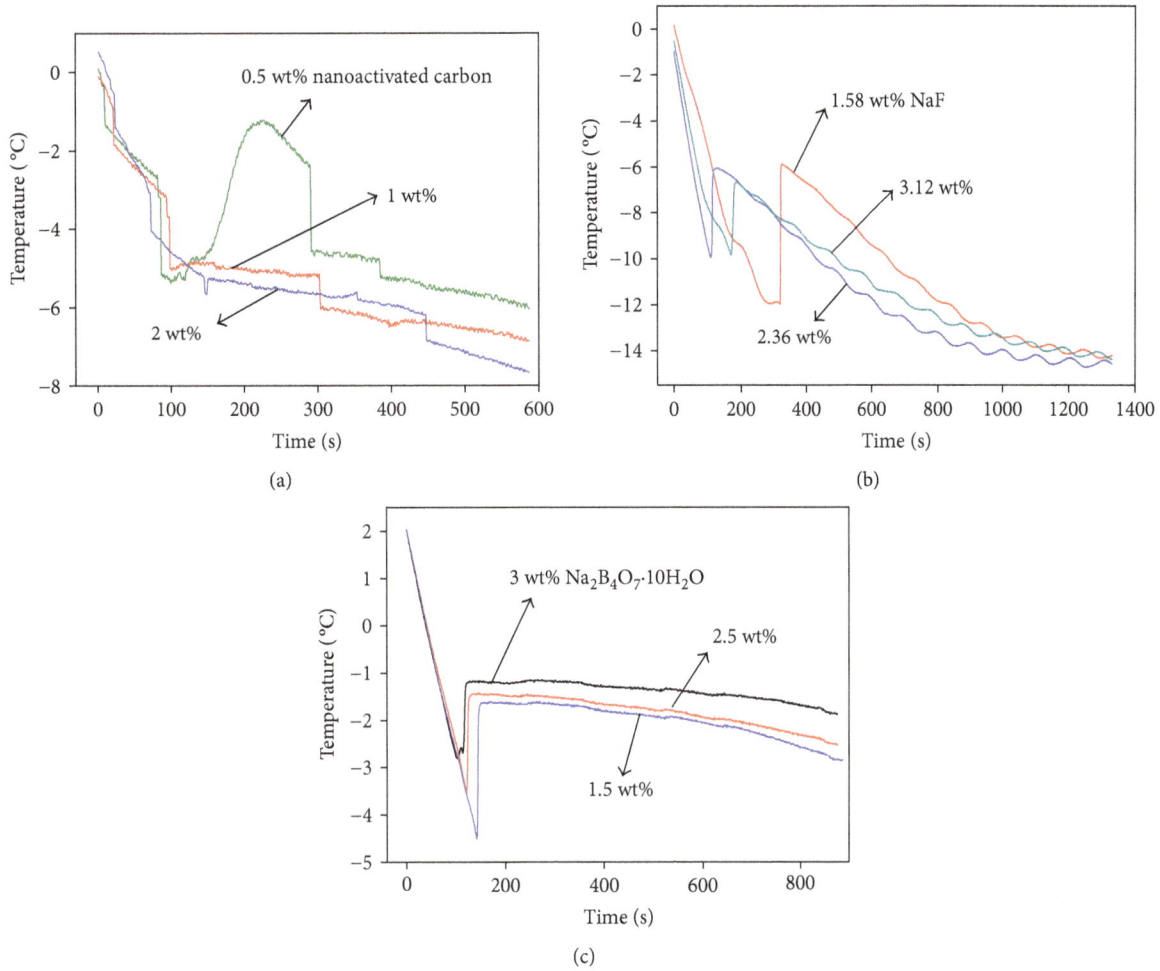

FIGURE 9: (a), (b), and (c) represent step cooling curves of adding three different kinds of nucleating agent (nanoactivated carbon, NaF, and $Na_2B_4O_7 \cdot 10H_2O$) in ternary eutectic hydrate salts without a thickener.

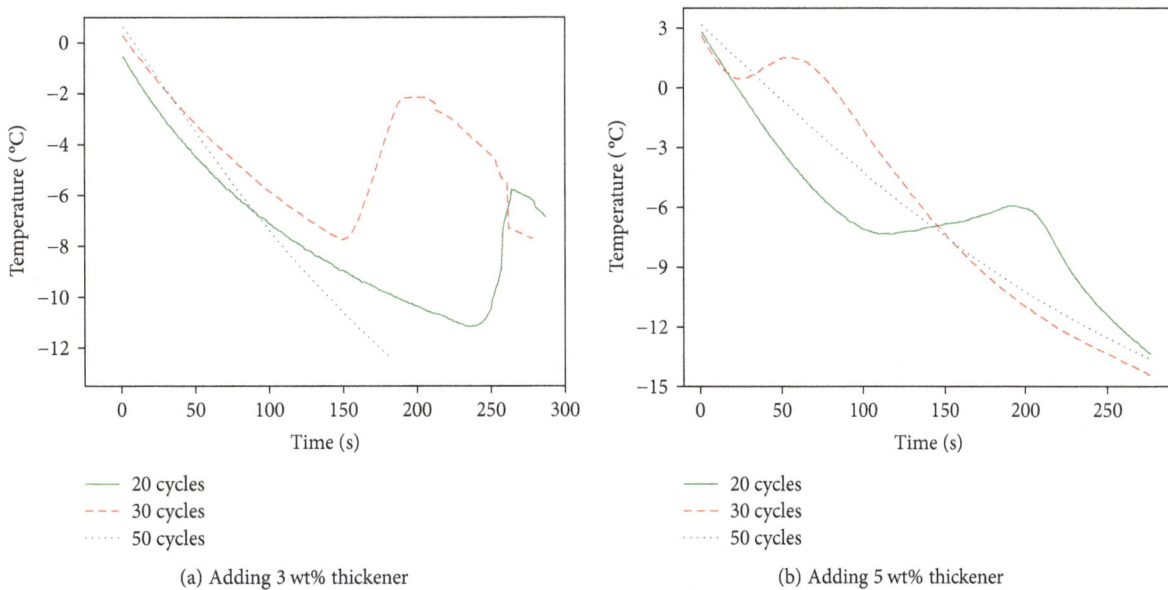

(a) Adding 3 wt% thickener

(b) Adding 5 wt% thickener

FIGURE 10: (a) and (b) represent the change of step cooling temperature curve after 20, 30, and 50 freeze-thaw cycles.

the inhibition of the supercooling. In order to further study the influence of nucleating agent on inorganic phase change materials, two other nucleating agents were chosen and three different proportioning were selected to add to the HSPCMs. In Figure 9(a), 0.5, 1, and 2 wt% nanoactivated carbon particles with a particle size of 100 nm were separately added into the ternary eutectic hydrate salts. It was seen that the addition of 0.5 wt% particles cannot inhibit the super-cooling obviously. With the increase of the content of nano-particles, the degree of supercooling is almost zero. Through the analysis of Figure 9(b), the amount of nucleating agent NaF added is 1.5, 2.5, and 3 wt%; when the 3 wt% NaF is added, the degree of supercooling is almost half of that of 1.5 wt% and the total supercooling degree is reduced to about 3°C, so the effect is remarkable. Then, increasing the content of the same kind of nucleating agent, the effect almost remains unchanged, compared to the result curve of the three nucleating agents. It is found that an excessive dose of the nucleating agent does little influence to the inhibition of supercooling and these three different nucleating agents all reduced the supercooling significantly. Adding inorganic salt, hydrated salt and nanoparticles all contribute to the nonhomogeneous nucleation.

3.5. Thermal Cycling Test of Hydrated Salt PCM. When using a cold storage material, life span is an important factor to measure the stability. In this experiment, adding a thickener in inorganic hydrated salt is an effective method to enhance the working life of materials and prevent phase separation phenomenon. The salt solution which was added 3 wt% and 5 wt% thickeners still maintained good stability, and there was no obvious phase separation phenomenon. However, a thin layer of crystal at the bottom after about 10 times of freeze-thaw cycles in the salt solution without a thickener appeared. As the freeze-thaw cycles continue, crystal thick-ness was increased, which severely affected the cold storage characteristics of the salt solution. After stirring or heating, the sample changed back to its initial state and no rapid deg-radation or irreversibility was observed. Therefore, to avoid the shortcoming of the phase separation mentioned above, the mixture can be stabilized by adding a thickener.

Figure 10 shows that the HSPCMs showed better stability as the number of freeze-thaw cycles increased. The cooling curve changed from an undulating curve into a smooth decline gradually after 50 or more cycles. The performance of the phase change material was more stable, and the supercooling phenomenon will also be reduced gradually. PCM is easier to crystallize than never to solidify due to the residual crystal which is not completely melted during the last process which will reduce the required energy of the next crystallization process during freeze-thaw cycles.

4. Conclusions

This experimental study was conducted on the preparation, modification, and characterization of a new ternary eutectic of inorganic salt hydrate phase change materials with 0, 3, and 5 wt% thickeners added into salt hydrate solution, and the effect of viscosity and nucleating agents on the thermophysical properties and cold storage ability of PCM was measured by DSC and temperature acquisition system. The results of this study can be summarized as follows:

(1) The increase of the viscosity will reduce the phase change enthalpy, which reduces the storage capacity. Meanwhile, the phase stability, supercooling, and phase separation will be significantly improved.

(2) NaF, $Na_2B_4O_7 \cdot 10H_2O$, and nanoactivated carbon were all proved to be effective nucleating agents. With the increasing content of a nucleating agent, the degree of cooling can be obviously reduced or even eliminated.

(3) During the stability test of the inorganic material, the supercooling of inorganic materials will become smaller after 20, 30, and 50 freeze-thaw cycles.

Nomenclature

T_{sc}:	The degree of supercooling
H_m:	Heat flux
H_p:	Melting phase change enthalpy
T_m:	The melting point temperature
T_s:	The freezing point temperature
AR:	Analytically pure reagent
CP:	Chemically pure reagent
DI:	Deionized water
PCM:	Phase change materials
HSPCM:	Hydrated salt-based PCMs
STA:	Simultaneous thermal analyzer
XRD:	X-ray diffraction
DSC:	Differential scanning calorimeter.

Conflicts of Interest

The authors declare that they have no conflicts of interest.

Acknowledgments

This work is financially supported by the Fundamental Research Funds for the Central Universities (Grant no. 2014QNB05).

References

[1] W. Su, J. Darkwa, and G. Kokogiannakis, "Development of microencapsulated phase change material for solar thermal energy storage," *Applied Thermal Engineering*, vol. 112, pp. 1205–1212, 2017.

[2] Z. Huang, Z. Luo, X. Gao, X. Fang, Y. Fang, and Z. Zhang, "Investigations on the thermal stability, long-term reliability of LiNO3/KCl–expanded graphite composite as industrial waste heat storage material and its corrosion properties with metals," *Applied Energy*, vol. 188, pp. 521–528, 2017.

[3] X. Q. Zhai, X. L. Wang, T. Wang, and R. Z. Wang, "A review on phase change cold storage in air-conditioning system: materials and applications," *Renewable and Sustainable Energy Reviews*, vol. 22, pp. 108–120, 2013.

[4] C.-V. Hémery, F. Pra, J.-F. Robin, and P. Marty, "Experimental performances of a battery thermal management system using a phase change material," *Journal of Power Sources*, vol. 270, pp. 349–358, 2014.

[5] J. P. da Cunha and P. Eames, "Thermal energy storage for low and medium temperature applications using phase change materials – a review," *Applied Energy*, vol. 177, pp. 227–238, 2016.

[6] R. Yang, H. Xu, and Y. Zhang, "Preparation, physical property and thermal physical property of phase change microcapsule slurry and phase change emulsion," *Solar Energy Materials and Solar Cells*, vol. 80, no. 4, pp. 405–416, 2003.

[7] S. Behzadi and M. M. Farid, "Long term thermal stability of organic PCMs," *Applied Energy*, vol. 122, pp. 11–16, 2014.

[8] F. Souayfane, F. Fardoun, and P.-H. Biwole, "Phase change materials (PCM) for cooling applications in buildings: a review," *Energy and Buildings*, vol. 129, pp. 396–431, 2016.

[9] N. Pfleger, M. Braun, M. Eck, and T. Bauer, "Assessment of novel inorganic storage medium with low melting point," *Energy Procedia*, vol. 69, pp. 988–996, 2015.

[10] G. Li, B. B. Zhang, X. Li, Y. Zhou, Q. G. Sun, and Q. Yun, "The preparation, characterization and modification of a new phase change material: $CaCl_2 \cdot 6H_2O$–$MgCl_2 \cdot 6H_2O$ eutectic hydrate salt," *Solar Energy Materials and Solar Cells*, vol. 126, pp. 51–55, 2014.

[11] Y. Liu and Y. Yang, "Preparation and thermal properties of $Na_2CO_3 \cdot 10H_2O$–$Na_2HPO_4 \cdot 12H_2O$ eutectic hydrate salt as a novel phase change material for energy storage," *Applied Thermal Engineering*, vol. 112, pp. 606–609, 2017.

[12] A. Efimova, S. Pinnau, M. Mischke, C. Breitkopf, M. Ruck, and P. Schmidt, "Development of salt hydrate eutectics as latent heat storage for air conditioning and cooling," *Thermochimica Acta*, vol. 575, pp. 276–278, 2014.

[13] M. Li, H. Kao, Z. Wu, and J. Tan, "Study on preparation and thermal property of binary fatty acid and the binary fatty acids/diatomite composite phase change materials," *Applied Energy*, vol. 88, no. 5, pp. 1606–1612, 2011.

[14] A. Safari, R. Saidur, F. A. Sulaiman, Y. Xu, and J. Dong, "A review on supercooling of phase change materials in thermal energy storage systems," *Renewable and Sustainable Energy Reviews*, vol. 70, pp. 905–919, 2017.

[15] R. Al-Shannaq, J. Kurdi, S. Al-Muhtaseb, M. Dickinson, and M. Farid, "Supercooling elimination of phase change materials (PCMs) microcapsules," *Energy*, vol. 87, pp. 654–662, 2015.

[16] X. Li, Y. Zhou, H. G. Nian et al., "Preparation and thermal energy storage studies of $CH_3COONa \cdot 3H_2O$–KCl composites salt system with enhanced phase change performance," *Applied Thermal Engineering*, vol. 102, pp. 708–715, 2016.

[17] P. Chandrasekaran, M. Cheralathan, V. Kumaresan, and R. Velraj, "Solidification behavior of water based nanofluid phase change material with a nucleating agent for cool thermal storage system," *International Journal of Refrigeration*, vol. 41, pp. 157–163, 2014.

[18] R. Androsch, A. Monami, and J. Kucera, "Effect of an alpha-phase nucleating agent on the crystallization kinetics of a propylene/ethylene random copolymer at largely different supercooling," *Journal of Crystal Growth*, vol. 408, pp. 91–96, 2014.

[19] F. Cao and B. Yang, "Supercooling suppression of microencapsulated phase change materials by optimizing shell composition and structure," *Applied Energy*, vol. 113, pp. 1512–1518, 2014.

[20] Y. D. Liu, J. Q. Wang, C. J. Su, S. C. Geng, Y. K. Gao, and Q. G. Peng, "Nucleation rate and supercooling degree of water-based graphene oxide nanofluids," *Applied Thermal Engineering*, vol. 115, pp. 1226–1236, 2017.

[21] V. V. Tyagi, S. C. Kaushik, A. K. Pandey, and S. K. Tyagi, "Experimental study of the supercooling and *p*H behavior of a typical phase change material for thermal energy storage," *Indian Journal of Pure & Applied Physics*, vol. 49, pp. 117–125, 2011.

[22] K. Bilen, F. Takgil, and K. Kaygusuz, "Thermal energy storage behavior of $CaCl_2 \cdot 6H_2O$ during melting and solidification," *Energy Sources, Part A: Recovery, Utilization, and Environmental Effects*, vol. 30, no. 9, pp. 775–787, 2008.

[23] R. Pilar, L. Svoboda, P. Honcova, and L. Oravova, "Study of magnesium chloride hexahydrate as heat storage material," *Thermochimica Acta*, vol. 546, pp. 81–86, 2012.

[24] T. Wang, D. Mantha, and R. G. Reddy, "Novel high thermal stability LiF–Na_2CO_3–K_2CO_3 eutectic ternary system for thermal energy storage applications," *Solar Energy Materials and Solar Cells*, vol. 140, pp. 366–375, 2015.

[25] R. K. Sharma, P. Ganesan, V. V. Tyagi, and T. M. I. Mahlia, "Accelerated thermal cycle and chemical stability testing of polyethylene glycol (PEG) 6000 for solar thermal energy storage," *Solar Energy Materials and Solar Cells*, vol. 147, pp. 235–239, 2016.

[26] A. Mariaux and M. Rappaz, "Influence of anisotropy on heterogeneous nucleation," *Acta Materialia*, vol. 59, no. 3, pp. 927–933, 2011.

21

Exergy Analysis of Two Kinds of Solar-Driven Cogeneration Systems in Lhasa, Tibet, China

Haofei Zhang(ID)**, Bo Lei**(ID)**, Tao Yu, and Zhida Zhao**(ID)

School of Mechanical Engineering, Southwest Jiaotong University, Chengdu 610031, China

Correspondence should be addressed to Bo Lei; lbswjtu@163.com

Academic Editor: Giulia Grancini

In this study, an exergy analysis of two kinds of solar-driven cogeneration systems consisting of solar collectors and an organic Rankine cycle (ORC) is presented for series mode and parallel mode. Three kinds of solar collectors are considered: flat-plate collectors (FPC), evacuated tube collectors (ETC), and parabolic trough collectors (PTC). This study mainly compares the exergy output of the two kinds of solar cogeneration systems under different temperatures of the return heating water and different inlet temperatures of the solar collectors. This study shows that, from the perspective of W_{net} or E , the parallel mode is superior to the series mode. From the perspective of E_z, the parallel mode is superior to the series mode when the solar collector is FPC; however, the series mode is superior to the parallel mode when the solar collector is PTC. When the solar collector is ETC, the result depends on the temperature of the return heating water. When the temperature of the return heating water is low (below 46°C), the series mode is better, and when the temperature of the return heating water is high (above 46°C), the parallel mode is better.

1. Introduction

A solar water heating system is a solar energy application that has drawn great attention among researchers in this field. In Lhasa, Tibet, China, known as "the city of sunlight," a number of buildings use solar water heating systems for heating in winter. However, the sunlight is very strong, and the cycle water is sometimes heated to boiling, as shown in Figure 1. Wang et al. [1] analyzed the problems of current solar heating systems in Lhasa and noted that sunlight exposure was the most common cause.

As there is no need for a cold supply in Lhasa in the summer, a good solution might be to transform the strong solar radiation into electrical power through solar collectors and thermal power plants during the nonheating season and parts of the heating season. This not only improves the reliability of the solar water heating system but also reduces the local consumption of fossil fuels and their impact on the ecological environment. However, the outlet temperature of solar collectors for heating systems is low for thermal power generation based on the Rankine cycle. The organic

Rankine cycle (ORC) enables efficient power generation units from low-grade heat sources by replacing water with organic working fluids, such as refrigerants or hydrocarbons. Thus, solar collectors and the ORC can form solar thermal power generation systems in the nonheating season and solar-driven cogeneration systems in the heating season. This paper mainly discusses the solar-driven cogeneration system consisting of solar collectors and the ORC.

A few studies have examined solar collectors integrated with an ORC for electrical power production. Wang et al. [2, 3] proposed and tested a low-temperature solar ORC system utilizing R245fa as the working fluid. The overall power generation efficiency was 4.2% when the solar collectors were evacuated tube collectors (ETC), while it was 3.2% when using flat plate collectors (FPC). Al-Sulaiman et al. [4–6] studied a novel trigeneration system using parabolic trough collectors (PTC) and an ORC. The maximum electrical efficiency for the solar mode was 15%, the solar and storage mode was 7%, and the storage mode was 6.5%; the maximum electrical-exergy efficiency was 7%, the solar and storage mode was 3.5%, and the storage mode was 3%. He et al. [7]

FIGURE 1: Overheating of solar collector systems in Lhasa.

built a model for a typical thermal power generation system with PTC and an ORC within the transient energy simulation package TRNSYS and found that pentane had the best performance among three organic working fluids: R113, R123, and pentane. Pei et al. [8, 9] reported that the overall electrical efficiency was approximately 8.6% when a solar irradiation of $750 \, \text{W/m}^2$ was assumed for a low-temperature solar thermal electric generation system based on compound parabolic collectors (CPC) and an ORC. In a different study, Li et al. [10] noted that the optimal evaporation temperature and corresponding annual power output for Lhasa are 116°C and $163.42 \, \text{kWh/m}^2$, respectively, for a low-temperature solar thermal electric generation system mainly consisting of CPC and an ORC with R123.

It can be observed from the literature that the main form of the solar-driven cogeneration system is in series mode, as shown in Figure 2. The heating source for the heating system is the heat of condensation of the ORC system. However, the efficiency of collectors decreases with increasing inlet temperature of the fluid, and the efficiency of the ORC system decreases with increasing condensing temperature. Therefore, the series mode might not be the optimal form.

This paper proposes a new form of the solar-driven cogeneration system, the parallel mode, as shown in Figure 3. The solar water heating system and the solar ORC power generation system are connected in parallel. A portion of the solar collectors operate at low temperatures for heating, and the rest operate at high temperatures for power production. The solenoid valve VM is used to regulate the area of the collectors for heating or power generation. In this way, the efficiency of the collectors in the solar water heating system can be higher. In addition, the condensing temperature of the solar ORC power generation system is lower, which provides a higher power generation efficiency.

However, in parallel mode, a large amount of condensing heat is discharged into the environment. Therefore, it is not known whether the parallel mode has better thermodynamic performance than the series mode.

This paper mainly analyzes and compares the thermodynamic properties of the two kinds of solar-driven cogeneration systems: series mode and parallel mode.

2. System Descriptions

Figures 2 and 3 show the examined systems. Figure 2 shows the series mode, while Figure 3 shows the parallel mode. The examined systems are located in Lhasa, Tibet, China. Each mode is separated into 3 main parts: the solar collector field, the ORC system, and the heating system. All these parts are operated simultaneously to convert solar energy into electricity and heat.

The solar collector field consists of a number of solar collectors (FPC, ETC, or PTC). The working fluid is pressurized water to remain in the liquid phase in all cases, operating with a pressure of 5 bar. The water remains in the liquid phase even when its temperature reaches up to 150°C under this pressure [11]. To obtain an efficient ORC, the organic fluid should be carefully selected. A number of researches [12–16] have studied the selection of the ORC system. One of the typical recommended organic fluid types used to operate the ORC is R245fa. Hence, it is selected as the working fluid of the ORC.

The power generation of the two modes is similar, except that the condensing temperature of the ORC system is different. There are three main operating modes, which are controlled by the solenoid valves V1~V6:

(1) Solar mode: V2, V5, pump-1, on; V1, V3, V4, V6, pump-2, off

(2) Solar and storage mode: V1, V2, V4, V5, pump-1, on; V3, V6, pump-2, off

(3) Storage mode: V1, V2, V4, V5, pump-1, off; V3, V6, pump-2, on

The heating system in the two modes is different; the heating source of the series mode is the heat of condensation of the ORC system, and the source for the parallel mode is the solar collector system.

There are three main operating modes of the heating system in the series mode, which are controlled by the solenoid valves V7~V10:

(1) Heating mode: V7, V8, V9, V10, off; pump-3, on

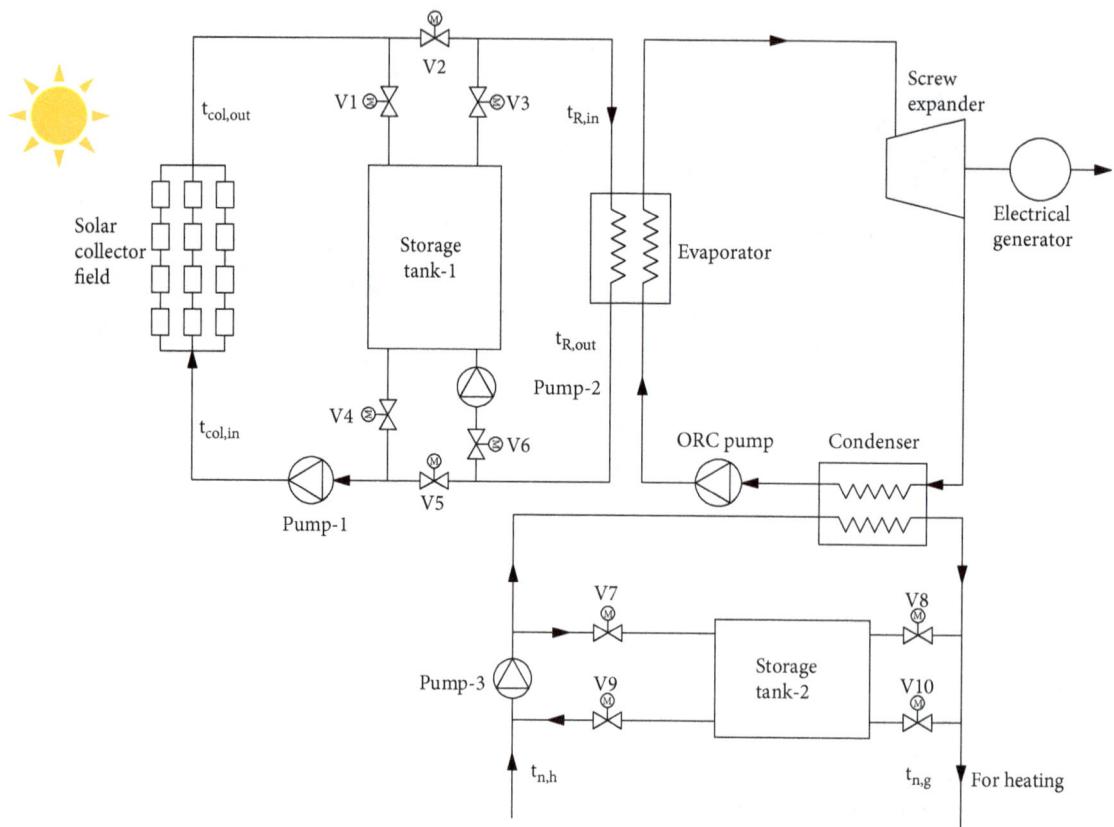

FIGURE 2: The series mode of the solar-driven cogeneration system.

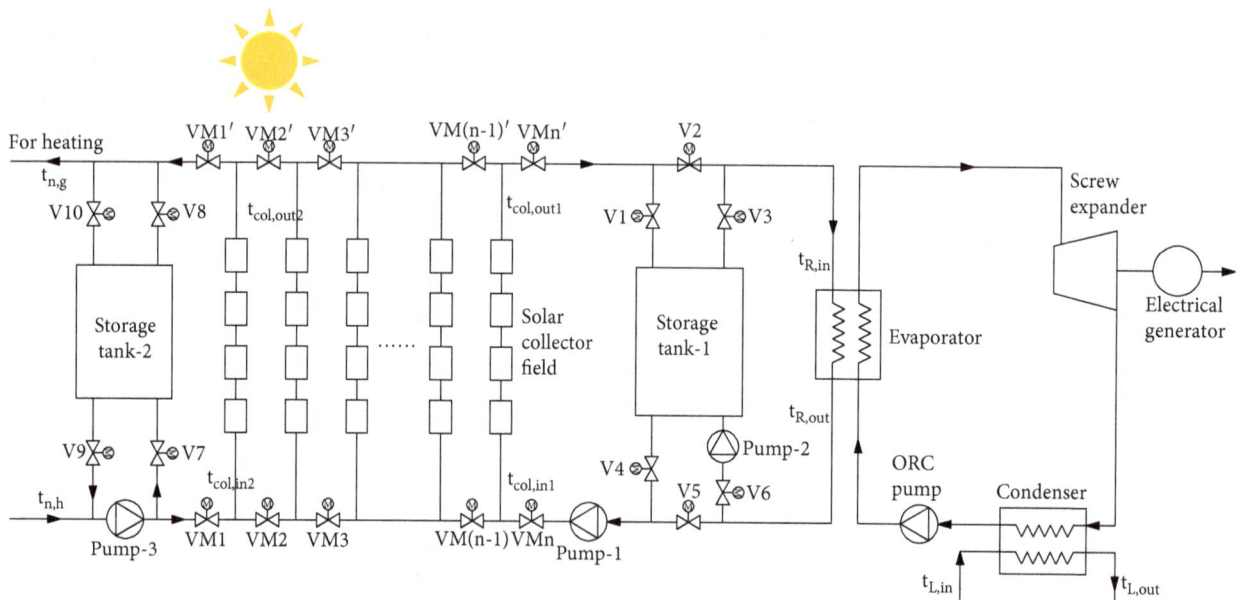

FIGURE 3: The parallel mode of the solar-driven cogeneration system.

(2) Heating and storage mode: V9, V10, pump-3, on; V7, V8, off

(3) Storage mode: V9, V10, off; V7, V8, pump-3, on

There are three main operating modes of the heating system in the parallel mode, which are controlled by the solenoid valves V7~V10 and VM~VM′:

(1) Heating mode: V7, V8, V9, V10, off; pump-3, VM1~VM(n) and VM1′~VM(n)′, on

(2) Heating and storage mode: V7, V8, off; V9, V10, pump-3, VM1~VM(n) and VM1′~VM(n)′, on

(3) Storage mode: V9, V10, VM1 and VM1′, off; V7, V8, pump-3, on

3. Mathematical Modeling

3.1. Solar Collector. Solar energy is the energy source of the analyzed system. The solar energy potential of the solar field can be calculated from the collector aperture and the effective radiation. Equation (1) shows the available solar energy.

$$Q_{sol} = A_{col} \cdot G_{eff}. \tag{1}$$

The effective radiation is different for every collector. FPC and ETC use both beam and diffuse radiation, while PTC only uses beam radiation. The effective radiation of every collector is presented as follows:

$$G_{eff} = \begin{cases} G_{til} & \text{for FPC and ETC,} \\ G_{PTC} & \text{for PTC.} \end{cases} \tag{2}$$

The radiation in the tilted surface is given from (3) [17]. The symbol G is used for the solar irradiance. The subscripts b, n, dif, and tot are as follows: b and n refer to the beam radiation on a plane normal to the direction of propagation; dif refers to the diffuse radiation; and tot refers to the total radiation on a horizontal surface. The symbol θ represents the angle of incidence, and β represents the slope.

$$G_{til} = G_{b,n} \cdot \cos(\theta) + G_{dif} \cdot \left(\frac{1 + \cos(\beta)}{2}\right) + G_{tot} \cdot \rho \cdot \left(\frac{1 - \cos(\beta)}{2}\right). \tag{3}$$

The radiation that PTC exploit is given as follows:

$$G_{PTC} = G_{b,n} \cdot \cos(\theta). \tag{4}$$

For a tilted surface, θ is given as follows:

$$\begin{aligned} \cos(\theta) = {} & \sin(\delta)\sin(\phi)\cos(\beta) \\ & - \sin(\delta)\cos(\phi)\sin(\beta)\cos(\gamma) \\ & + \cos(\delta)\cos(\phi)\cos(\beta)\cos(\omega) \\ & + \cos(\delta)\sin(\phi)\sin(\beta)\cos(\gamma)\cos(\omega) \\ & + \cos(\delta)\sin(\beta)\sin(\gamma)\sin(\omega). \end{aligned} \tag{5}$$

For a surface that can rotate around an axis in the east-west direction, θ is given as follows:

$$\cos(\theta) = \sqrt{1 - \cos^2(\delta)\sin^2(\omega)}, \tag{6}$$

where δ, ϕ, γ, and ω represent declination, latitude, azimuth angle, and solar hour angle, respectively. The declination δ can be found from the equation by Cooper [18]. The symbol n refers to the day of the year.

$$\delta = 23.45° \sin\left(360° \times \frac{284 + n}{365}\right). \tag{7}$$

The solar hour angle ω can be calculated from the following:

$$\begin{aligned} \omega &= 0.25°(\text{AST} - 720), \\ \text{AST} &= \text{LST} + \text{ET} - 4(\text{SL} - \text{LL}), \\ \text{ET} &= 9.87 \sin(2B) - 7.53 \cos(B) - 1.5 \sin(B), \\ B &= 360° \times \frac{(n - 81)}{364}, \end{aligned} \tag{8}$$

where AST and LST refer to solar time and local standard time, respectively. The symbols SL and LL refer to the standard meridian for the local time zone and the longitude of the location, respectively.

The thermal efficiency of the collector is the comparison of the useful energy that the working fluid absorbs to the solar energy delivered to the collector. Equation (9) presents this efficiency:

$$\eta_{col} = \frac{Q_{use}}{Q_{sol}} = \frac{c_{p,col} \cdot m_{col} \cdot (t_{col,out} - t_{col,in})}{A_{col} \cdot G_{eff}}. \tag{9}$$

For each collector type, a typical efficiency curve from the literature is selected [19–21], and (10) presents their efficiencies:

$$\eta_{col} = \begin{cases} 0.79 - 6.67 \cdot \left(\dfrac{t_{col,in} - t_a}{G_{tit}}\right) & \text{for FPC,} \\[3mm] 0.76 - 2.19 \cdot \left(\dfrac{t_{col,in} - t_a}{G_{tit}}\right) & \text{for ETC,} \\[3mm] 0.762 - 0.2125 \cdot \left(\dfrac{t_{col,in} - t_a}{G_{PTC}}\right) - 0.001672 \cdot G_{PTC} \cdot \left(\dfrac{t_{col,in} - t_a}{G_{PTC}}\right)^2 & \text{for PTC.} \end{cases} \tag{10}$$

3.2. ORC System. The thermal process of the ORC system is shown in Figure 4.

Processes 5–1 in the evaporator are given by (11). The pinch point temperature of the evaporator can be described by (12).

$$Q_{\text{evap}} = m_{\text{org}} \cdot (h_1 - h_5) = c_{p,\text{col}} \cdot m_{\text{col}} \cdot (t_{R,\text{in}} - t_{R,\text{out}}), \quad (11)$$

$$\Delta t_{\text{evap}} = t_8 - t_6. \quad (12)$$

Processes 1–2 in the screw expander are given by

$$W_{\text{SE}} = m_{\text{org}} \cdot (h_1 - h_2) \cdot \eta_{\text{SE}}. \quad (13)$$

Processes 2–4 in the condenser are given by

$$Q_{\text{cond}} = m_{\text{org}} \cdot (h_2 - h_4) = c_{p,\text{wat}} \cdot m_{\text{wat}} \cdot (t_{L,\text{out}} - t_{L,\text{in}}). \quad (14)$$

The pinch point temperature of the condenser can be described by

$$\Delta t_{\text{cond}} = t_3 - t_9. \quad (15)$$

An evaporative condenser exchanges heat by both heat and mass transfer on the outside surface of the condenser tubes. A large part of the heat exchanged by the condenser comes from the evaporating water, so an evaporative condenser is mainly a wet bulb sensitive device [22]. The condensing temperature can be calculated by (16), where T_{wb} is the web bulb temperature of the outside air.

$$t_{\text{cond}} = t_{\text{wb}} + (t_9 - t_{L,\text{in}}) + \Delta t_{\text{cond}}. \quad (16)$$

According to the literature [23], the power consumption of the evaporative condenser is given by (17), where W_{pu} is the power consumption of the evaporative condenser per unit cooling load.

$$W_{\text{cond}} = Q_{\text{cond}} \cdot W_{\text{pu}}. \quad (17)$$

Processes 4–5 in the ORC pump are given by

$$W_{\text{OP}} = \frac{m_{\text{org}} \cdot (h_5 - h_4)}{\eta_{\text{OP}}}. \quad (18)$$

The net power output of the ORC system is

$$W_{\text{ORC}} = W_{\text{SE}} - W_{\text{OP}} - W_{\text{cond}}. \quad (19)$$

The thermal efficiency of the ORC system can be calculated as

$$\eta_{\text{ORC}} = \frac{W_{\text{ORC}}}{Q_{\text{evap}}} = \frac{W_{\text{SE}} - W_{\text{OP}} - W_{\text{cond}}}{c_{p,\text{col}} \cdot m_{\text{col}} \cdot (t_{R,\text{in}} - t_{R,\text{out}})}. \quad (20)$$

3.3. Exergy Analysis. For the solar-driven cogeneration system, the energy level of the two kinds of output, heat and electricity, is different. Therefore, this paper uses the exergy

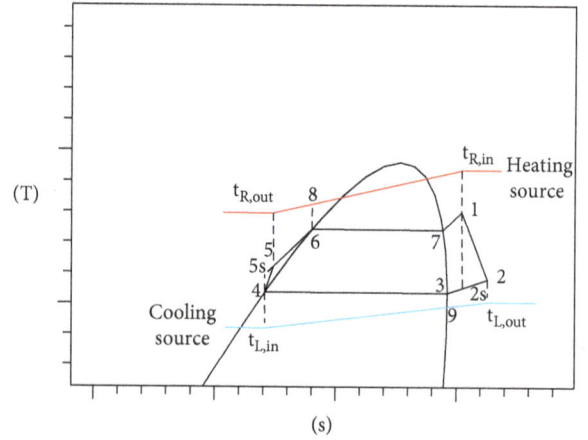

FIGURE 4: T-s diagram of an ORC.

TABLE 1: Other parameters of the analysis system.

Parameter	Value	Parameter	Value
ϕ	29.5°	Δt_{cond}	3°C
LL	91.1°	W_{pu}	11.12 W/kW
SL	120°	η_{SE}	0.85 [13]
n	172	H_1	15 m
γ	0	H_2	15 m
β	40°	H_3	30 m
ρ	0.2	r	9.8 kN/m³
Δt_{evap}	3°C	$\eta_{p(i)}$	0.6

analysis method to compare the two kinds of solar-driven cogeneration systems.

The exergy of the output electricity is

$$W_{\text{net}} = W_{\text{ORC}} - W_{p1} - W_{p2} - W_{p3}, \quad (21)$$

where W_{p1}, W_{p2}, and W_{p3} represent the power consumption of the pumps (pump-1, pump-2, and pump-3, respectively, in Figures 2 and 3).

$$W_{p(i)} = \frac{r Q_{(i)} H_{(i)}}{\eta_{p(i)}}, \quad (22)$$

where $Q_{p(i)}$, $H_{p(i)}$, and $\eta_{p(i)}$ represent the flow rate, hydraulic head, and efficiency of the pumps, respectively, and r represents the volume weight of the working fluid.

The exergy of the output heat is

$$E_n = c_n \cdot m_n \cdot \left[(t_{n,g} - t_{n,h}) - t_a \cdot \ln \left(\frac{t_{n,g}}{t_{n,h}} \right) \right], \quad (23)$$

where c_n represents the specific heat of the heating water, in kJ/kg·K; m_n represents the mass flow rate of the heating water, in kg/s; and $t_{n,g}$ and $t_{n,h}$ represent the temperature of the supply and return heating water, in K.

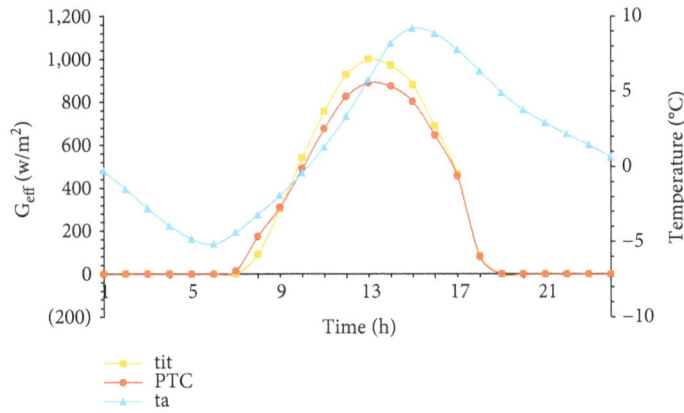

FIGURE 5: Average meteorological data for the heating season.

The total exergy is

$$E_z = W_{net} + E_n. \tag{24}$$

3.4. Simulation Parameters. For the meteorological data in the simulation, the values of $G_{b,n}$, G_{dif}, G_{tot}, and ambient temperature use TMY data that are taken from Meteonorm software. The other parameters (the latitude of the place, the slope angle of the collectors, and the ground reflectance) are shown in Table 1. FPC and ETC are sloped, while PTC are horizontal with the axis in the east-west direction and a tracking system for movement in the north-south direction.

As there is no document to specify the design day, this paper takes the average meteorological data of the heating season for analysis. Figure 5 shows the average meteorological data.

The optimization parameter of this system is the inlet temperature of the pressurized water in the collectors. This parameter determines the collector efficiency and influences the outlet temperature of the collectors. The outlet and inlet temperatures of the collectors are the inlet and outlet temperatures of the evaporator of the ORC system, respectively, which determines the efficiency of the ORC system. A high inlet temperature for the collectors decreases the collector thermal efficiency but increases the efficiency of the ORC system. Thus, an optimization of this temperature is needed to design an optimum system. Four different simulations, one for each collector type, are presented in this study. It is essential to mention that the solar potential is different from case to case because each collector type utilizes the diffuse radiation in a different way, which has been presented in (2). The analysis is performed under steady-state conditions using MATLAB and REFPROP.

4. Results and Discussions

All the simulation results are presented and analyzed to determine which kind of solar-driven cogeneration system has better thermodynamic properties.

4.1. Collector Performance. The collector efficiency depends on the operating conditions, among which the effective radiation and the water inlet temperature are two crucial parameters. Figure 5 shows the effective irradiance of every collector. Apparently, the effective radiation of the collectors changes with time. Before 7:00 and after 19:00, there was no effective irradiance. The effective irradiance increases in the morning from 7:00 until it reaches a maximum value and then decreases until 19:00. The maximum effective irradiance of every collector occurs at 13:00. The effective irradiance of PTC is higher than that of FPC and ETC from 7:00 to 9:00 but lower from 9:00 to 19:00.

Figure 6 shows the efficiency of each collector under different inlet temperature conditions. It is obvious that the efficiency decreases with increasing inlet temperature. PTC are the most efficient technology with ETC being the second. FPC follow with a lower efficiency. For higher inlet temperature levels, the performance of FPC is very low because the heat losses are very large. PTC and ETC use evacuated tubes, so the heat losses remain low for the entire examined temperature region.

Figure 7 shows the collected heat of collectors with a 1 m² collector area at different inlet temperatures on the analyzed day. It can be seen from the figure that the collected heat decreases with increasing inlet temperature, and FPC decreases the most while PTC decreases the least. Therefore, for the solar-driven cogeneration system with FPC, the parallel mode can significantly improve the collected heat. However, for ETC or PTC, the improvement is not as significant.

4.2. ORC Performance. For the ORC system, the temperatures of the heat source and the cold source are two crucial parameters for the thermal efficiency. Figure 8 depicts the influence of inlet temperature on the thermal efficiency of the ORC system in parallel mode and series mode.

A higher temperature of the heat source leads to a higher efficiency. A higher temperature of the heat source corresponds to a higher inlet temperature of the solar collectors. Apparently, the thermal efficiency increases with the inlet temperature up to 145°C because the critical temperature of R245fa is 154°C.

Figure 8 shows that the efficiency of the ORC system in the parallel mode is significantly higher than that in the series mode. As the heating temperature increased, the efficiency of

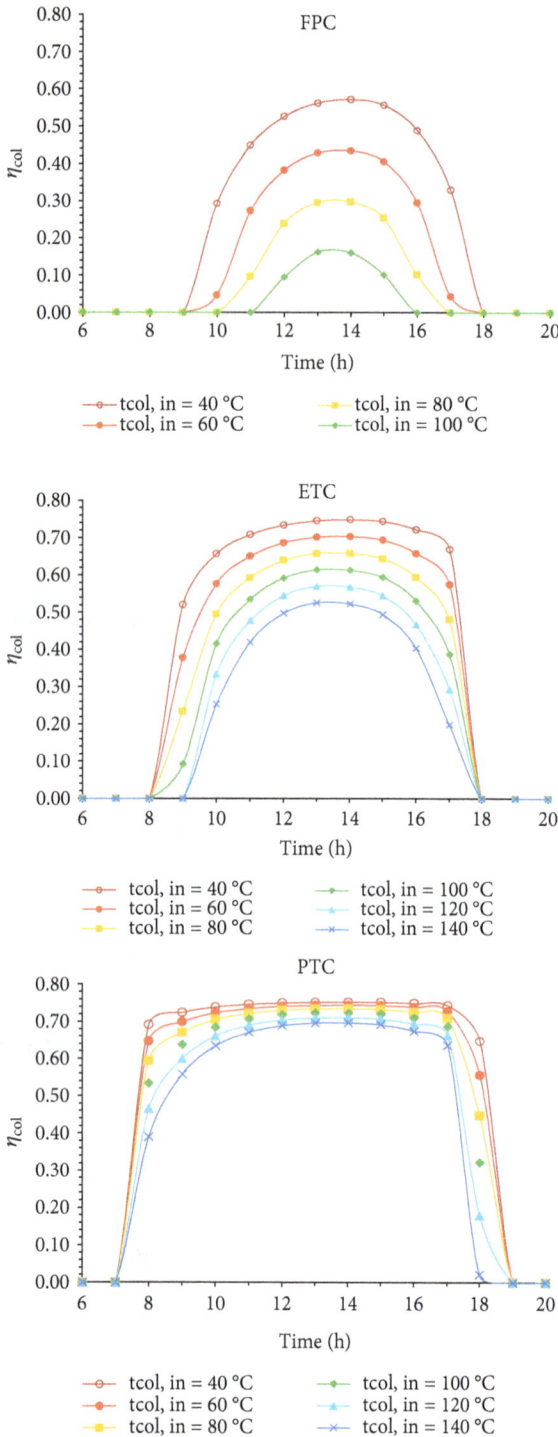

FIGURE 6: Efficiency of each collector for various inlet temperatures.

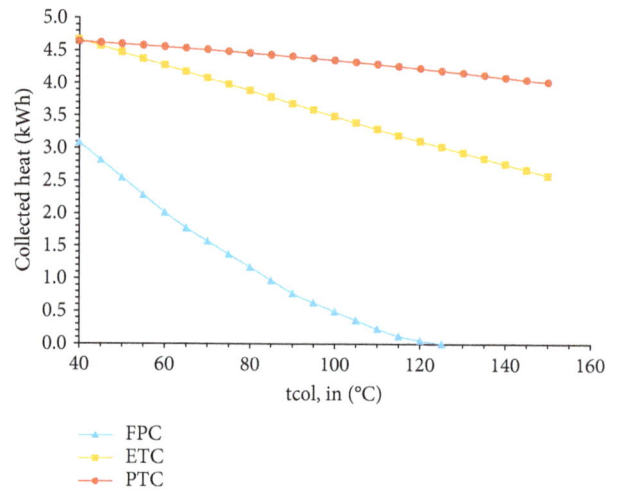

FIGURE 7: Collected heat by a 1 m^2 collector area on the analyzed day.

the return heating water are two crucial parameters for the thermal performance. Figure 9 depicts the influence of inlet temperature on the exergy output of the series mode of the solar-driven cogeneration system when the temperature of the return heating water is 40°C.

Figure 9 shows that for the three types of solar collectors, the output exergy could be ranked from high to low: $E_z > E_n > W_{net}$.

For FPC, with the increase of the inlet temperature of the collectors, W_{net} increases until it reaches a maximum value and then decreases. E_n and E_z decrease with the increase of the inlet temperature of the collectors. The optimum inlet temperature for FPC is 65°C, and the maximum E_z is 0.265 kWh.

For ETC and PTC, with the increase of the inlet temperature of the collectors, W_{net} and E_z increase until they reach a maximum value and then decrease. E_n decreases with the increase of the inlet temperature of the collectors. The optimum inlet temperature for ETC is 85°C, and the maximum E_z is 0.657 kWh. The optimum inlet temperature for PTC is 135°C, and the maximum E_z is 0.850 kWh.

4.4. Parallel Mode of the Solar-Driven Cogeneration System. For the parallel mode of the solar-driven cogeneration system, a portion of the solar collectors operate at low temperatures for heating, and the rest operate at high temperatures for power production. The inlet temperature of the collectors for heating is the temperature of the return heating water, which is 40°C. The inlet temperature of the collectors for power production is the optimum temperature, which is 60°C for FPC, 100°C for ETC, and 140°C for PTC.

For the parallel mode of the solar-driven cogeneration system, the proportion of the solar collector area that is used for heating and power production is a crucial parameter for the thermal performance.

Figure 10 shows that for the three types of solar collectors, E_n and E_z increase while W_{net} decreases with an increasing proportion of solar collector area for heating. That is, the exergy output of the solar heating system is higher than that

the ORC system in the series mode gradually decreased. The reason is that the condensing temperature of the ORC system in the parallel mode is the wet bulb temperature of the air, which is much lower than that in the series mode.

4.3. Series Mode of the Solar-Driven Cogeneration System. For the series mode of the solar-driven cogeneration system, the inlet temperature of the collectors and the temperature of

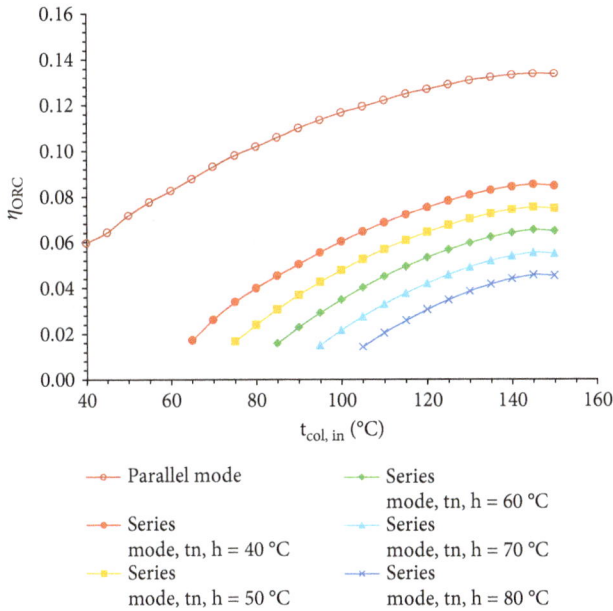

FIGURE 8: Efficiency of the ORC system.

of the solar power production system when the solar collector area is the same. E_z (E_n) reaches the maximum when all the collectors are used for heating, and W_{net} reaches the maximum when all the collectors are used for power production.

4.5. Comparison of the Two Kinds of Solar-Driven Cogeneration Systems.

Figure 11 shows the maximum exergy output of the two modes of the solar-driven cogeneration system with a 1 m² collector area on the analyzed day at different temperatures of the return heating water. The value beside each point in Figure 11 indicates the optimum inlet temperature of the collectors for the maximum exergy output in series mode. The optimum proportion of solar collector area for heating is 1 for E_n (E_z) and 0 for W_{net} in parallel mode.

In the series mode, for the three types of solar collectors, the maximum W_{net} decreases with the increase of the temperature of the return heating water. The reason is that the condensing temperature increases with increasing temperature of the return heating water, which decreases the power generation efficiency. For FPC, the maximum E_z decreases with increasing temperature of the return heating water. When $t_{n,h} = 40°C$, the maximum E_z is 0.265 kWh; when $t_{n,h} = 60°C$, the maximum E_z is 0.194 kWh; and when $t_{n,h} = 80°C$, the maximum E_z is 0.088 kWh. For ETC and PTC, the maximum E_z increases with increasing temperature of the return heating water. For ETC, when $t_{n,h} = 40°C$, the maximum E_z is 0.657 kWh; when $t_{n,h} = 60°C$, the maximum E_z is 0.754 kWh; and when $t_{n,h} = 80°C$, the maximum E_z is 0.823 kWh. For PTC, when $t_{n,h} = 40°C$, the maximum E_z is 0.850 kWh; when $t_{n,h} = 60°C$, the maximum E_z is 0.976 kWh; and when $t_{n,h} = 80°C$, the maximum E_z is 1.088 kWh.

In the parallel mode, for the three types of solar collectors, the maximum W_{net} remains the same with the increase

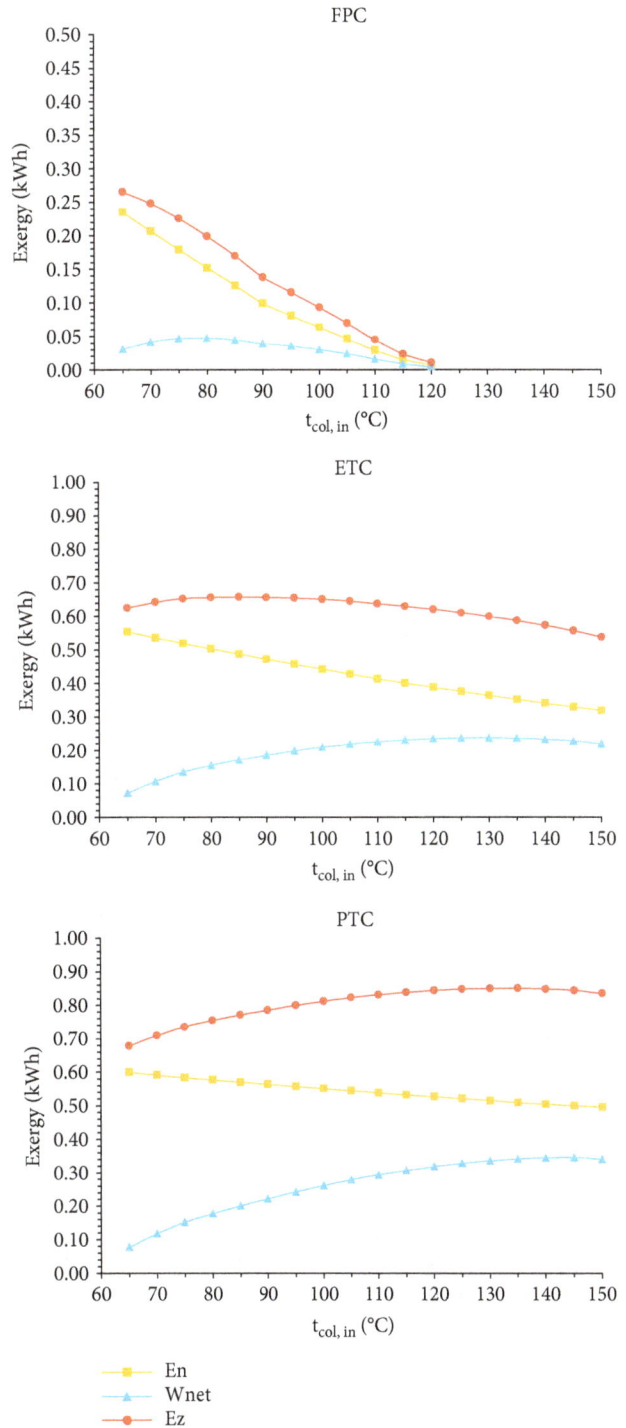

FIGURE 9: Output exergy for a 1 m² collector area on the analyzed day ($t_{n,h} = 40°C$).

of the temperature of the return heating water because all the solar collectors are used for power production and none for heating. For FPC, the maximum E_z decreases with increasing temperature of the return heating water. When $t_{n,h} = 40°C$, the maximum E_z is 0.415 kWh; when $t_{n,h} = 60°C$, the maximum E_z is 0.376 kWh; and when $t_{n,h} = 80°C$, the maximum E_z is 0.272 kWh. For ETC and PTC, the maximum E_z

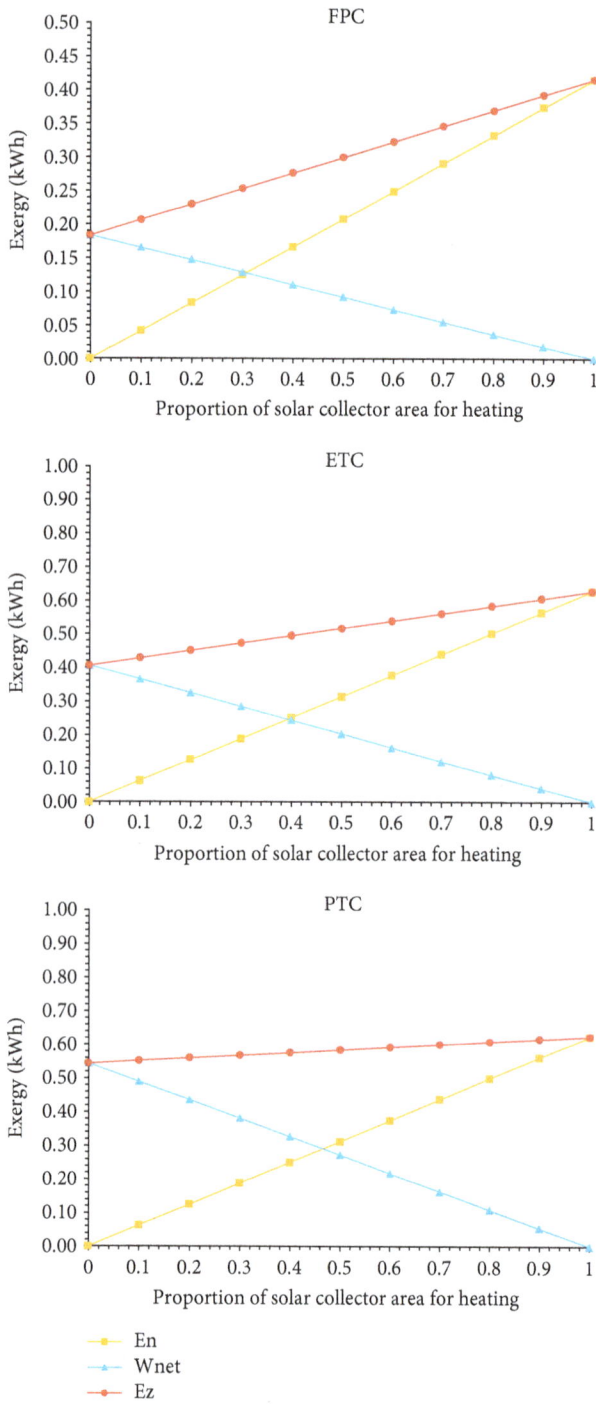

FIGURE 10: Output exergy for a 1 m² collector area on the analyzed day ($t_{n,h} = 40°C$).

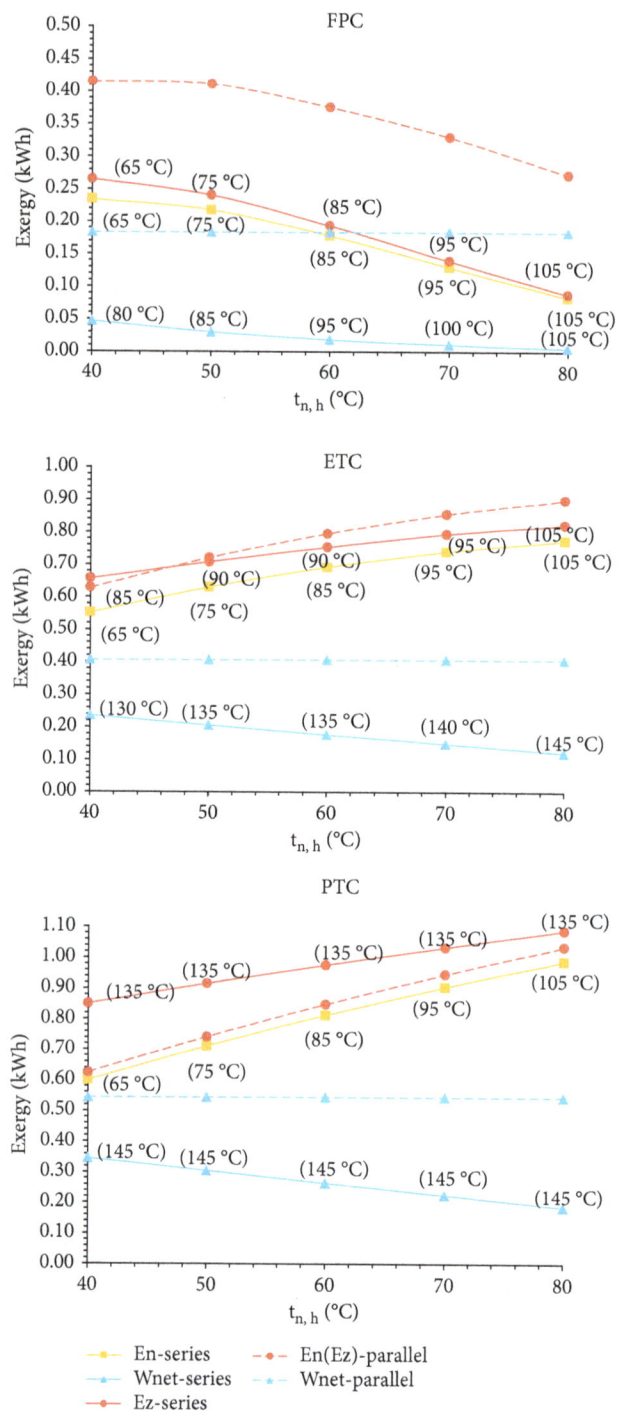

FIGURE 11: Maximum output exergy of two kinds of solar-driven cogeneration systems with a 1 m² collector area on the analyzed day.

increases with increasing temperature of the return heating water. For ETC, when $t_{n,h} = 40°C$, the maximum E_z is 0.628 kWh; when $t_{n,h} = 60°C$, the maximum E_z is 0.795 kWh; and when $t_{n,h} = 80°C$, the maximum E_z is 0.900 kWh. For PTC, when $t_{n,h} = 40°C$, the maximum E_z is 0.624 kWh; when $t_{n,h} = 60°C$, the maximum E_z is 0.848 kWh; and when $t_{n,h} = 80°C$, the maximum E_z is 1.035 kWh.

Make $\Delta E_n = (E_n - \text{series}) - (E_n - \text{parallel})$, $\Delta W_{net} = (W_{net} - \text{series}) - (W_{net} - \text{parallel})$, and $\Delta E_z = (E_z - \text{series}) - (E_z - \text{parallel})$, and then ΔE_n, ΔW_{net}, and ΔE_z of the three collector types are shown in Figure 12. From Figure 12, it can be seen that from the perspective of W_{net} or E_n, the parallel mode is superior to the series mode. From the perspective of E_z, the parallel mode is superior to the series mode when the solar collector is FPC; however, the series mode

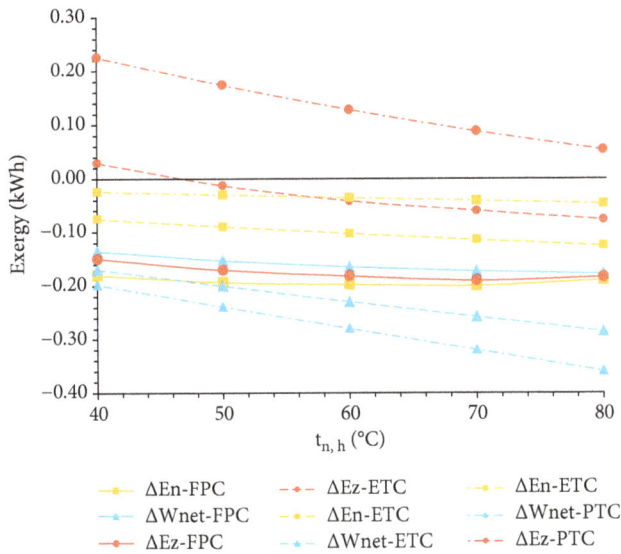

FIGURE 12: ΔE_n, ΔW_{net}, and ΔE_z of the three collector types.

FPC; however, the series mode is superior to the parallel mode when the solar collector is PTC. When the solar collector is ETC, the result depends on the temperature of the return heating water. When the temperature of the return heating water is low (below 46°C), the series mode is better, and when the temperature of the return heating water is high (above 46°C), the parallel mode is better

Nomenclature

A: Area (m²)
AST: Solar time (minutes)
c: Specific heat (kJ/kg·K)
E: Exergy (kwh)
G: Solar irradiance (W/m²)
h: Enthalpy (kJ/kg)
LL: Longitude (°)
LST: Local standard time (minutes)
m: Mass flow rate (kg/s)
Q: Heat rate (W)
s: Entropy (kJ/kg)
SL: Standard meridian for the local time zone (°)
t: Temperature (°C)
W: Power (kWh)

Greek Letters

β: Collector slope (°)
γ: Azimuth angle (°)
δ: Declination angle (°)
η: Efficiency
θ: Angle of incidence (°)
ρ: Ground reflectance
ω: Solar hour angle (°)
ϕ: Latitude (°)

Subscripts and Superscripts

a: Ambient
b,n: Beam radiation on a plane normal to the direction of propagation
col: Collector
cond: Condenser
dif: Diffuse radiation
eff: Effective
evap: Evaporator
in: Inlet
L: Cooling source
OP: Organic pump
org: Organic fluid
out: Outlet
R: Heating source
SE: Screw expander
sol: Solar energy
til: Tilted surface
tot: Total radiation on a horizontal surface
use: Useful
wat: Water
wb: Wet bulb.

is superior to the parallel mode when the solar collector is PTC. When the solar collector is ETC, the result depends on the temperature of the return heating water. When the temperature of the return heating water is low (below 46°C), the series mode is better, and when the temperature of the return heating water is high (above 46°C), the parallel mode is better.

5. Conclusions

The performance of two kinds of solar-driven cogeneration systems consisting of solar collectors and an ORC is compared and analyzed for series mode and parallel mode. Three kinds of solar collectors are considered: FPC, ETC, and PTC. The exergy outputs of the two kinds of solar cogeneration systems under different temperatures of the return heating water and different inlet temperatures of the solar collectors are calculated and compared. The main results of this study are summarized below:

(1) In the series mode, with the increase of the inlet temperature of the collectors, the total exergy output decreases for FPC and increases until it reaches a maximum value and then decreases for ETC and PTC

(2) In the parallel mode, the total exergy output increases with the increase of the proportion of solar collector area for heating. The exergy output of the solar heating system is higher than that of the solar power production system when the solar collector areas are the same

(3) From the perspective of W_{net} or E_n, the parallel mode is superior to the series mode

(4) From the perspective of E_z, the parallel mode is superior to the series mode when the solar collector is

Conflicts of Interest

The authors declare that there is no conflict of interest regarding the publication of this paper.

Acknowledgments

This work was supported by the Sichuan Province Youth Science and Technology Innovation Team of Building Environment and Energy Efficiency (no. 2015TD0015).

References

[1] L. Wang, L. Yuan, C. M. Zhu, Z. R. Li, and N. Y. Yu, "Necessity of the whole process commissioning for active solar heating systems," *Heating, Ventilating and Air Conditioning*, vol. 42, pp. 53–56, 2012.

[2] X. D. Wang, L. Zhao, J. L. Wang, W. Z. Zhang, X. Z. Zhao, and W. Wu, "Performance evaluation of a low-temperature solar Rankine cycle system utilizing R245fa," *Solar Energy*, vol. 84, no. 3, pp. 353–364, 2010.

[3] X. D. Wang, L. Zhao, and J. L. Wang, "Experimental investigation on the low-temperature solar Rankine cycle system using R245fa," *Energy Conversion and Management*, vol. 52, no. 2, pp. 946–952, 2011.

[4] F. A. Al-Sulaiman, I. Dincer, and F. Hamdullahpur, "Exergy modeling of a new solar driven trigeneration system," *Solar Energy*, vol. 85, no. 9, pp. 2228–2243, 2011.

[5] F. A. Al-Sulaiman, F. Hamdullahpur, and I. Dincer, "Performance assessment of a novel system using parabolic trough solar collectors for combined cooling, heating, and power production," *Renewable Energy*, vol. 48, pp. 161–172, 2012.

[6] F. A. Al-Sulaiman, "Exergy analysis of parabolic trough solar collectors integrated with combined steam and organic Rankine cycles," *Energy Conversion and Management*, vol. 77, pp. 441–449, 2014.

[7] Y. L. He, D. H. Mei, W. Q. Tao, W. W. Yang, and H. L. Liu, "Simulation of the parabolic trough solar energy generation system with organic Rankine cycle," *Applied Energy*, vol. 97, pp. 630–641, 2012.

[8] G. Pei, J. Li, and J. Ji, "Analysis of low temperature solar thermal electric generation using regenerative organic Rankine cycle," *Applied Thermal Engineering*, vol. 30, no. 8-9, pp. 998–1004, 2010.

[9] P. Gang, L. Jing, and J. Jie, "Design and analysis of a novel low-temperature solar thermal electric system with two-stage collectors and heat storage units," *Renewable Energy*, vol. 36, no. 9, pp. 2324–2333, 2011.

[10] L. Jing, P. Gang, and J. Jie, "Optimization of low temperature solar thermal electric generation with organic Rankine cycle in different areas," *Applied Energy*, vol. 87, no. 11, pp. 3355–3365, 2010.

[11] E. Bellos, C. Tzivanidis, and K. A. Antonopoulos, "Exergetic, energetic and financial evaluation of a solar driven absorption cooling system with various collector types," *Applied Thermal Engineering*, vol. 102, pp. 749–759, 2016.

[12] B. T. Liu, K. H. Chien, and C. C. Wang, "Effect of working fluids on organic Rankine cycle for waste heat recovery," *Energy*, vol. 29, no. 8, pp. 1207–1217, 2004.

[13] B. Saleh, G. Koglbauer, M. Wendland, and J. Fischer, "Working fluids for low-temperature organic Rankine cycles," *Energy*, vol. 32, no. 7, pp. 1210–1221, 2007.

[14] P. J. Mago, L. M. Chamra, and C. Somayaji, "Performance analysis of different working fluids for use in organic Rankine cycles," *Proceedings of the Institution of Mechanical Engineers, Part A: Journal of Power and Energy*, vol. 221, no. 3, pp. 255–263, 2007.

[15] P. J. Mago, K. K. Srinivasan, L. M. Chamra, and C. Somayaji, "An examination of exergy destruction in organic Rankine cycles," *International Journal of Energy Research*, vol. 32, no. 10, pp. 926–938, 2008.

[16] H. Hajabdollahi, A. Ganjehkaviri, and M. N. Mohd Jaafar, "Thermo-economic optimization of RSORC (regenerative solar organic Rankine cycle) considering hourly analysis," *Energy*, vol. 87, pp. 369–380, 2015.

[17] S. Kalogirou, *Solar Energy Engineering*, Academic Press, Boston, 2009.

[18] J. A. Duffie and W. A. Beckman, Eds., *Solar Engineering of Thermal Processes,*, Wiley, Hoboken, NJ, USA, third edition, 2006.

[19] S. Kalogirou, "The potential of solar industrial process heat applications," *Applied Energy*, vol. 76, no. 4, pp. 337–361, 2003.

[20] S. A. Kalogirou, "Solar thermal collectors and applications," *Progress in Energy and Combustion Science*, vol. 30, no. 3, pp. 231–295, 2004.

[21] V. Dudley, *SANDIA Report Test Results for Industrial Solar Technology Parabolic Trough Solar Collector [SAND94-1117]*, Sandia National Laboratory, Albuquerque, USA, 1995.

[22] K. Manske, *Performance Optimization of Industrial Refrigeration Systems [M.S. Thesis]*, United States: Mechanical Engineering, Solar Energy Laboratory, University of Wisconsin Madison, 1999.

[23] Y. M. Zhuang, "The energy consumption and economical analysis for evaporative condenser compared to shell-tube water-cooling condensers," *Refrigeration*, vol. 20, pp. 48–51, 2001.

Permissions

The contributors of this book come from diverse backgrounds, making this book a truly international effort. This book will bring forth new frontiers with its revolutionizing research information and detailed analysis of the nascent developments around the world.

We would like to thank all the contributing authors for lending their expertise to make the book truly unique. They have played a crucial role in the development of this book. Without their invaluable contributions this book wouldn't have been possible. They have made vital efforts to compile up to date information on the varied aspects of this subject to make this book a valuable addition to the collection of many professionals and students.

This book was conceptualized with the vision of imparting up-to-date information and advanced data in this field. To ensure the same, a matchless editorial board was set up. Every individual on the board went through rigorous rounds of assessment to prove their worth. After which they invested a large part of their time researching and compiling the most relevant data for our readers.

The editorial board has been involved in producing this book since its inception. They have spent rigorous hours researching and exploring the diverse topics which have resulted in the successful publishing of this book. They have passed on their knowledge of decades through this book. To expedite this challenging task, the publisher supported the team at every step. A small team of assistant editors was also appointed to further simplify the editing procedure and attain best results for the readers.

Apart from the editorial board, the designing team has also invested a significant amount of their time in understanding the subject and creating the most relevant covers. They scrutinized every image to scout for the most suitable representation of the subject and create an appropriate cover for the book.

The publishing team has been an ardent support to the editorial, designing and production team. Their endless efforts to recruit the best for this project, has resulted in the accomplishment of this book. They are a veteran in the field of academics and their pool of knowledge is as vast as their experience in printing. Their expertise and guidance has proved useful at every step. Their uncompromising quality standards have made this book an exceptional effort. Their encouragement from time to time has been an inspiration for everyone.

The publisher and the editorial board hope that this book will prove to be a valuable piece of knowledge for researchers, students, practitioners and scholars across the globe.

List of Contributors

Alsanossi M. Aboghrara
Mechanical Department, Faculty of Engineering, UPM, 43400 Serdang, Selangor, Malaysia
Physics Department, Faculty of Science Traghen, University of Sebha, Sebha, Libya

M. A. Alghoul
Center of Research Excellence in Renewable Energy (CoRERE), King Fahd University of Petroleum and Minerals (KFUPM), Dhahran 31261, Saudi Arabia
Energy and Building Research Center, Kuwait Institute for Scientific Research, Safat, 13109 Kuwait City, Kuwait

B. T. H. T. Baharudin and A. A. Hairuddin
Mechanical Department, Faculty of Engineering, UPM, 43400 Serdang, Selangor, Malaysia

A. M. Elbreki, A. A. Ammar and K. Sopian
Solar Energy Research Institute, Universiti Kebangsaan Malaysia, 43600 Bangi, Selangor, Malaysia

Aymen Jemaa, Ons Zarrad and Mohamed Nejib Mansour
Unit of Industrial Systems Study and Renewable Energy (ESIER), National Engineering School, University of Monastir, Monastir, Tunisia

Mohamed Ali Hajjaji
Laboratory of Electronic and Microelectronic, University of Monastir, Monastir, Tunisia
Higher Institute of Applied Sciences and Technology of Kairouan, University of Kairouan, Kairouan, Tunisia

Juan José Martínez
Mechatronics Engineering Department, Technological Institute of Celaya, Av. Tecnológico y G. Cubas, s/n, 38010 Celaya, GTO, Mexico

José Alfredo Padilla-Medina, Juan Prado and Alejandro I. Barranco
Electronics Engineering Department, Technological Institute of Celaya, Av. Tecnológico y G. Cubas, s/n, 38010 Celaya, GTO, Mexico

Sergio Cano-Andrade
Department of Mechanical Engineering, Universidad de Guanajuato, 36885 Salamanca, GTO, Mexico

Agustín Sancen
Department of Engineering Sciences, Technological Institute of Celaya, Av. Tecnológico y G. Cubas, s/n, 38010 Celaya, GTO, Mexico

Syed Furqan Rafique
Department of Electrical Engineering, National University of Science and Technology, Islamabad, Pakistan
Department of Electrical Engineering, North China Electric Power University, Beijing, China
Goldwind Technology, Beijing, China

Zhang Jianhua and Jing Guo
Department of Electrical Engineering, North China Electric Power University, Beijing, China

Rizwan Rafique
Department of Electrical Engineering, Norwegian University of Science and Technology, Trondheim, Norway

Irfan Jamil
College of Energy & Electrical Engineering, Hohai University, Nanjing, China

Tao Zhang
School of Energy and Mechanical Engineering, Shanghai University of Electric Power, Shanghai 200090, China

Khaled Bataineh
Department of Mechanical Engineering, Jordan University of Science and Technology, Irbid, Jordan

Rubeena Kousar, Muzaffar Ali, Shabaz Khan Lodhi and Muhammad Rameez ud Din
University of Engineering and Technology, Taxila, Punjab, Pakistan

Muhammad Amar and Khuram Pervez Amber
Mirpur University of Science and Technology (MUST), Mirpur, 10250 AJK, Pakistan

Ghulam Qadar Chaudhary and Allah Ditta
University of Engineering and Technology, Taxila, Punjab, Pakistan
Mirpur University of Science and Technology (MUST), Mirpur, 10250 AJK, Pakistan

Xiaoyan Li
Beijing University of Technology, Beijing 100124, China

Hao Zhuang, Xixi Huang and Likai Jiang
CECEP Solar Energy Technology (Zhenjiang) Co., Ltd., Zhenjiang 212132, China

Shaoliang Wang
Beijing Jiaotong University, Beijing 100044, China

Xianfang Gou
Beijing University of Technology, Beijing 100124, China
CECEP Solar Energy Technology (Zhenjiang) Co., Ltd., Zhenjiang 212132, China

Gebretinsae Yeabyo Nigussie, Gebrekidan Mebrahtu Tesfamariam, Berhanu Menasbo Tegegne, Yemane Araya Weldemichel and Desta Gebremedhin Gebrehiwot
Department of Chemistry, College of Natural and Computational Sciences, Mekelle University, Mekelle, Ethiopia

Tesfakiros Woldu Gebreab
Department of Physics, College of Natural and Computational Sciences, Mekelle University, Mekelle, Ethiopia

Gebru Equar Gebremichel
Department of Biology, College of Natural and Computational Sciences, Mekelle University, Mekelle, Ethiopia

Chika Terada, Takahiro Imamura, Seiko Tatehara, Reiko Tokuyama-Toda, Shigeo Yamachika, Nagataka Toyoda and Kazuhito Satomura
Department of Oral Medicine and Stomatology, School of Dental Medicine, Tsurumi University, 2-1-3 Tsurumi, Tsurumi-ku, Yokohama 230-8501, Japan

Tomoko Ohshima and Nobuko Maeda
Department of Oral Microbiology, School of Dental Medicine, Tsurumi University, 2-1-3 Tsurumi, Tsurumi-ku, Yokohama 230-8501, Japan

Xiaoxiao Tong
Horticulture & Landscape College, Hunan Agricultural University, Changsha 410128, China
School of Architecture & Design, China University of Mining and Technology, Xuzhou 221116, China

Xingyao Xiong
Institute of Vegetables and Flowers, Chinese Academy of Agricultural Sciences, Beijing 100081, China

Jing Qin, Jie Ji, Wenzhu Huang, Mawufemo Modjinou and Guiqiang Li
Department of Thermal Science and Energy Engineering, University of Science and Technology of China, No. 96 Jinzhai Road, Hefei, Anhui, China

Hong Qin
Guangdong University of Technology, Guangzhou, China

Alejandra Sánchez-Sánchez, Ivonne Linares-Hernández, Verónica Martínez-Miranda and Reyna María Guadalupe Fonseca-Montes de Oca
Centro Interamericano de Recursos del Agua, Universidad Autónoma del Estado de México, Carretera Toluca-Atlacomulco, Km 14.5 Unidad San Cayetano, 50200 Toluca, MEX, Mexico

Moisés Tejocote-Pérez
Centro de Investigación en Ciencias Biológicas Aplicadas, Universidad Autónoma del Estado de México, Carretera Toluca-Atlacomulco, Km 14.5 Unidad San Cayetano, 50200 Toluca, MEX, Mexico

Rosa María Fuentes-Rivas
Facultad de Geografía, Universidad Autónoma del Estado de México, Cerro de Coatepec s/n, Ciudad Universitaria, 50110 Toluca, MEX, Mexico

Marco A. García-Morales, Julio César González Juárez and Sonia Martínez-Gallegos
Instituto Nacional de México, Instituto Tecnológico de Toluca, Av. Tecnológico s/n, Col. Agrícola Buenavista, 52149 Toluca, MEX, Mexico

Gabriela Roa-Morales, Eduardo Martin del Campo López, Carlos Barrera-Díaz and Verónica Martínez Miranda
Facultad de Química, Paseo Colón s/n, Residencial Colón, Universidad Autónoma del Estado de Mexico (UAEMéx), 50120 Toluca de Lerdo, MEX, Mexico

Ever Peralta
Universidad del Mar, Campus Puerto Angel, Ciudad Universitaria s/n, 70902 Puerto Angel, OAX, Mexico

Teresa Torres Blancas
Instituto de Química, Carretera Toluca-Atlacomulco Km 14.5, Universidad Nacional Autónoma de México and Centro Conjunto de Investigación en Química Sustentable UAEM-UNAM, 50200 Toluca, MEX, Mexico

Mehmet Ali Özçelik
Technical Science, Electric and Energy Department, Gaziantep University, Şehitkamil, 27310 Gaziantep, Turkey

Chee Mun Chong, Stuart Wenham, Jingjia Ji, Ly Mai, Sisi Wang and Brett Hallam
School of Photovoltaic and Renewable Energy Engineering, University of New South Wales, Sydney, NSW, Australia

Hua Li
LONGi Lerri Solar, Xi'an, China

Chenglong Luo, Mengyin Liao and Dan Sun
Institute of Energy Research, Jiangxi Academy of Sciences, Nanchang 330096, China

Lijie Xu and Jie Ji
Department of Thermal Science and Energy Engineering, University of Science and Technology of China, Hefei 230027, China

Wei Bao, JianJun Xu, TianZhou Xie, BingDe Chen, YanPing Huang and DianChuan Xing
CNNC Key Laboratory on Nuclear Reactor Thermal Hydraulics Technology, Nuclear Power Institute of China, Chengdu 610041, China

Davor Ljubas
Department of Energy, Power Engineering and Environment, Faculty of Mechanical Engineering and Naval Architecture, University of Zagreb, Ivana Lučića 5, Zagreb, Croatia

Mirta Čizmić, Katarina Vrbat and Sandra Babić
Department of Analytical Chemistry, Faculty of Chemical Engineering and Technology, University of Zagreb, Marulićev trg 20, Zagreb, Croatia

Draženka Stipaničev and Siniša Repec
Croatian Waters, Central Water Management Laboratory, Ulica grada Vukovara 220, Zagreb, Croatia

Lidija Ćurković
Department of Materials, Faculty of Mechanical Engineering and Naval Architecture, University of Zagreb, Ivana Lučića 5, Zagreb, Croatia

Lin Liang and Xi Chen
School of Electrical and Power Engineering, China University of Mining and Technology, Xuzhou 221116, China

Haofei Zhang, Bo Lei, Tao Yu and Zhida Zhao
School of Mechanical Engineering, Southwest Jiaotong University, Chengdu 610031, China

Index